The Maillard Reaction

Recent Advances in Food and Biomedical Sciences

ANNALS OF THE NEW YORK ACADEMY OF SCIENCES
Volume 1126

The Maillard Reaction
Recent Advances in Food and Biomedical Sciences

Edited by
ERWIN SCHLEICHER, VERONIKA SOMOZA, AND PETER SCHIEBERLE

Published by Blackwell Publishing on behalf of the New York Academy of Sciences
Boston, Massachusetts
2008

Library of Congress Cataloging-in-Publication Data

The Maillard Reaction: Recent Advances in Food and Biomedical Sciences/editors, Erwin Schleicher, Veronika Somoza, and Peter Schieberle.
    p.; cm. – (Annals of the New York Academy of Sciences, ISSN 0077-8923)
    Includes bibliographical references.
    ISBN-13: 978-1-57331-719-1 (paper: alk. paper)
    ISBN-10: 1-57331-719-5 (paper: alk. paper)
    1. Maillard reaction–Congresses. I. Schleicher, Erwin. II. Somoza, Veronika. III. Schieberle, Peter. IV. New York Academy of Sciences. V. Series.
    [DNLM: 1. Maillard Reaction–Congresses. W1 AN626YL v.1126 2007 / QZ 40 M2195 2007]

RB171.M338 2007
612'.0157–dc22

2007050340

The *Annals of the New York Academy of Sciences* (ISSN: 0077-8923 [print]; ISSN: 1749-6632 [online]) is published 28 times a year on behalf of the New York Academy of Sciences by Blackwell Publishing with offices at (US) 350 Main St., Malden, MA 02148-5020, (UK) 9600 Garsington Road, Oxford, OX4 2ZG, and (Asia) 165 Cremorne St., Richmond VIC 3121, Australia. Blackwell Publishing was acquired by John Wiley & Sons in February 2007. Blackwell's program has been merged with Wiley's global Scientific, Technical, and Medical business to form Wiley-Blackwell.

MAILING: *Annals* is mailed Standard Rate. Mailing to rest of world by IMEX (International Mail Express). Canadian mail is sent by Canadian publications mail agreement number 40573520. POSTMASTER: Send all address changes to *Annals of the New York Academy of Sciences*, Blackwell Publishing Inc., Journals Subscription Department, 350 Main St., Malden, MA 02148-5020.

Disclaimer: The Publisher, the New York Academy of Sciences and Editors cannot be held responsible for errors or any consequences arising from the use of information contained in this publication; the views and opinions expressed do not necessarily reflect those of the Publisher, the New York Academy of Sciences and Editors.

Copyright and Photocopying: © 2008 New York Academy of Sciences. All rights reserved. No part of this publication may be reproduced, stored or transmitted in any form or by any means without the prior permission in writing from the copyright holder. Authorization to photocopy items for internal and personal use is granted by the copyright holder for libraries and other users registered with their local Reproduction Rights Organization (RRO), e.g. Copyright Clearance Center (CCC), 222 Rosewood Drive, Danvers, MA 01923, USA (www.copyright.com), provided the appropriate fee is paid directly to the RRO. This consent does not extend to other kinds of copying such as copying for general distribution, for advertising or promotional purposes, for creating new collective works or for resale. Special requests should be addressed to: journalsrights@oxon.blackwellpublishing.com

Blackwell Publishing is now part of Wiley-Blackwell.

Information for subscribers: For ordering information, claims, and any inquiry concerning your subscription please contact your nearest office:

UK: Tel: +44 (0)1865 778315; Fax: +44 (0) 1865 471775
USA: Tel: +1 781 388 8599 or 1 800 835 6770 (toll free in the USA & Canada); Fax: +1 781 388 8232 or Fax: +44 (0) 1865 471775
Asia: Tel: +65 6511 8000; Fax: +44 (0)1865 471775,
Email: customerservices@blackwellpublishing.com

Subscription prices for 2008 are: Premium Institutional: US$4265 (The Americas), £2370 (Rest of World). Customers in the UK should add VAT at 7%; customers in the EU should also add VAT at 7%, or provide a VAT registration number or evidence of entitlement to exemption. Customers in Canada should add 5% GST or provide evidence of entitlement to exemption. The Premium institutional price also includes online access to the current and all online back files to January 1, 1997, where available. For other pricing options, including access information and terms and conditions, please visit www.blackwellpublishing.com/nyas.

Delivery Terms and Legal Title: Prices include delivery of print publications to the recipient's address. Delivery terms are Delivered Duty Unpaid (DDU); the recipient is responsible for paying any import duty or taxes. Legal title passes to the customer on despatch by our distributors.

Membership information: Members may order copies of *Annals* volumes directly from the Academy by visiting www.nyas.org/annals, emailing membership@nyas.org, faxing +1 212 298 3650, or calling 1 800 843 6927 (toll free in the USA), or +1 212 298 8640. For more information on becoming a member of the New York Academy of Sciences, please visit www.nyas.org/membership. Claims and inquiries on member orders should be directed to the Academy at email: membership@nyas.org or Tel: 1 800 843 6927 (toll free in the USA) or +1 212 298 8640.

Printed in the USA. Printed on acid-free paper.

Annals is available to subscribers online at Blackwell Synergy and the New York Academy of Sciences Web site. Visit www.blackwell-synergy.com or www.annalsnyas.org to search the articles and register for table of contents e-mail alerts.

The paper used in this publication meets the minimum requirements of the National Standard for Information Sciences Permanence of Paper for Printed Library Materials, ANSI Z39.48 1984.

ISSN: 0077-8923 (print); 1749-6632 (online)
ISBN-10: 1-57331-719-5 (paper: alk. paper); ISBN-13: 978-1-57331-719-1 (paper: alk. paper)

A catalogue record for this title is available from the British Library.

ANNALS OF THE NEW YORK ACADEMY OF SCIENCES

Volume 1126
April 2008

The Maillard Reaction

Recent Advances in Food and Biomedical Sciences

Editors
ERWIN SCHLEICHER, VERONIKA SOMOZA, AND PETER SCHIEBERLE

This volume is the result of the **9th International Symposium on the Maillard Reaction**, held September 1–5, 2007 in Munich, Germany.

CONTENTS

Preface. *By* Erwin Schleicher, Veronika Somoza, and Peter Schieberle xiii

Part I. Keynote Papers

The Sense of Smell: Reception of Flavors. *By* H. Breer 1

Receptor for Advanced Glycation End Products: Fundamental Roles in the Inflammatory Response: Winding the Way to the Pathogenesis of Endothelial Dysfunction and Atherosclerosis. *By* Ravichandran Ramasamy, Shi Fang Yan, Kevan Herold, Raphael Clynes, and Ann Marie Schmidt 7

Central Nervous System Regulation of Energy Metabolism: Ghrelin versus Leptin. *By* Ruben Nogueiras, Matthias H. Tschöp, and Jeffrey M. Zigman .. 14

Determination of N^ε-(Carboxymethyl)lysine in Foods and Related Systems. *By* Jennifer M. Ames 20

Part II. Plenary Lectures

Food Anoxia and the Formation of Either Flavor or Toxic Compounds by Amino Acid Degradation Initiated by Oxidized Lipids. *By* Francisco J. Hidalgo and Rosario Zamora 25

Post-Schiff Base Chemistry of the Maillard Reaction: Mechanism of Imine Isomerization. *By* Fong Lam Chu and Varoujan A. Yaylayan 30

Usefulness of Antibodies for Evaluating the Biological Significance of AGEs. *By* Ryoji Nagai, Yukio Fujiwara, Katsumi Mera, Keita Motomura, Yasunori Iwao, Keiichiro Tsurushima, Mime Nagai, Kazuhiro Takeo, Makiko Yoshitomi, Masaki Otagiri, and Tsuyoshi Ikeda 38

Receptor for Advanced Glycation End Product Expression in Experimental Diabetic Retinopathy. *By* Yumei Wang, Franziska vom Hagen, Frederick Pfister, Angelika Bierhaus, Yuxi Feng, Reinhold Gans, and Hans-Peter Hammes 42

Advanced Glycation End Product Homeostasis: Exogenous Oxidants and Innate Defenses. *By* Helen Vlassara, Jaime Uribarri, Weijing Cai, and Gary Striker ... 46

Formation Mechanisms of Melanoidins and Fluorescent Pyridinium Compounds as Advanced Glycation End Products. *By* Fumitaka Hayase, Teruyuki Usui, Yoriyuki Ono, Yoshinobu Shirahashi, Tomomi Machida, Takashi Ito, Nozomu Nishitani, Kazuhito Shimohira, and Hirohito Watanabe 53

Advanced Glycation as a Basis for Understanding Retinal Aging and Noninvasive Risk Prediction. *By* Anna M. Pawlak, Josephine V. Glenn, James R. Beattie, John J. McGarvey, and Alan W. Stitt 59

The Aroma Side of the Maillard Reaction. *By* Christoph Cerny 66

Methylglyoxal: Its Presence in Beverages and Potential Scavengers. *By* Di Tan, Yu Wang, Chih-Yu Lo, Shengmin Sang, and Chi-Tang Ho 72

The RAGE Pathway: Activation and Perpetuation in the Pathogenesis of Diabetic Neuropathy. *By* Ivan K. Lukic, Per M. Humpert, Peter P. Nawroth, and Angelika Bierhaus 76

The Role of the Amadori Product in the Complications of Diabetes. *By* Vincent M. Monnier, David R. Sell, Zhenyu Dai, Ina Nemet, Francois Collard, and Jianye Zhang 81

Mitigation Strategies to Reduce Acrylamide Formation in Fried Potato Products. *By* Francisco Morales, Edoardo Capuano, and Vincenzo Fogliano 89

Therapeutic Interruption of Advanced Glycation in Diabetic Nephropathy: Do All Roads Lead to Rome? *By* Karly C. Sourris, Josephine M. Forbes, and Mark E. Cooper 101

The Other Side of the Maillard Reaction. *By* Ram H. Nagaraj, Ashis Biswas, Antonia Miller, Tomoko Oya-Ito, and Manjunatha Bhat 107

Dicarbonyls Stimulate Cellular Protection Systems in Primary Rat Hepatocytes and Show Anti-inflammatory Properties. *By* Timo M. Buetler, Hélia Latado, Alexandra Baumeyer, and Thierry Delatour 113

N-terminal Glycation of Proteins and Peptides in Foods and *in Vivo*: Evaluation of *N*-(2-Furoylmethyl)Valine in Acid Hydrolyzates of Human Hemoglobin. *By* Ilka Penndorf, Changhao Li, Uwe Schwarzenbolz, and Thomas Henle 118

The Dicarbonyl Proteome: Proteins Susceptible to Dicarbonyl Glycation at Functional Sites in Health, Aging, and Disease. *By* Naila Rabbani and Paul J. Thornalley 124

Peroxyl Radicals Are Essential Reagents in the Oxidation Steps of the Maillard Reaction Leading to Generation of Advanced Glycation End Products. *By* Gerhard Spiteller 128

Application of Mass Spectrometry for the Detection of Glycation and Oxidation Products in Milk Proteins. *By* Jasmin Meltretter and Monika Pischetsrieder 134

Inhibition of Advanced Glycation End Products: An Implicit Goal in Clinical Medicine for the Treatment of Diabetic Nephropathy? *By* Toshio Miyata and Yuko Izuhara 141

Inflammation and the Redox-sensitive AGE–RAGE Pathway as a Therapeutic Target in Alzheimer's Disease. *By* Annette Maczurek, Kirubakaran Shanmugam, and Gerald Münch 147

Part III. Oral Presentations

Some Natural Compounds Enhance N^ε-(Carboxymethyl)lysine Formation. *By* Yukio Fujiwara, Naoko Kiyota, Keita Motomura, Katsumi Mera, Motohiro Takeya, Tsuyoshi Ikeda, and Ryoji Nagai 152

Immunological Detection of N^ω-(Carboxymethyl)arginine by a Specific Antibody. *By* Katsumi Mera, Yukio Fujiwara, Masaki Otagiri, Noriyuki Sakata, and Ryoji Nagai 155

Isolation and Partial Characterization of Four Fluorophores Formed by Nonenzymatic Browning of Methylglyoxal and Glutamine-derived Ammonia. *By* Celine Niquet, Serge Pilard, David Mathiron, and Frederic J. Tessier 158

Receptor for Advanced Glycation End Product Polymorphisms and Type 2 Diabetes: The CODAM Study. *By* Katrien H.J. Gaens, Carla J.H. van der Kallen, Marleen M.J. van Greevenbroek, Edith J. Feskens, Coen D.A. Stehouwer, and Casper G. Schalkwijk 162

Pentosidine Effects on Human Osteoblasts *in Vitro*. *By* Roberta Sanguineti, Daniela Storace, Fiammetta Monacelli, Alberto Federici, and Patrizio Odetti 166

Strategy for the Study of the Health Impact of Dietary Maillard Products in Clinical Studies: The Example of the ICARE Clinical Study on Healthy Adults. *By* Philippe Pouillart, Hélène Mauprivez, Lamia Ait-Ameur, Amélie Cayzeele, Jean-Michel Lecerf, Frédéric J. Tessier, and I. Birlouez-Aragon 173

Plasma Concentration and Urinary Excretion of N^ε-(Carboxymethyl)lysine in Breast Milk– and Formula-fed Infants. *By* Katarína Šebeková, Giselle Saavedra, Cornelia Zumpe, Veronika Somoza, Kristína Klenovicsová, and Ines Birlouez-Aragon 177

Maillard Reaction Products in the *Escherichia coli*–derived Therapeutic Protein Interferon Alfacon-1. *By* Roumyana Mironova, Angelina Sredovska, Ivan Ivanov, and Toshimitsu Niwa 181

Acrolein Induces Inflammatory Response Underlying Endothelial Dysfunction: A Risk Factor for Atherosclerosis. *By* Yong Seek Park and Naoyuki Taniguchi 185

Oxidative Stress and Advanced Glycation in Diabetic Nephropathy. *By* Melinda T. Coughlan, Amy L. Mibus, and Josephine M. Forbes 190

Vitamin C–mediated Maillard Reaction in the Lens Probed in a Transgenic-mouse Model. *By* Xingjun Fan and Vincent M. Monnier 194

N^ε-(Carboxymethyl)lysine during the Early Development of Hypertension. *By* Marcus Baumann, Coen Stehouwer, Jean Scheijen, Uwe Heemann, Harry Struijker Boudier, and Casper Schalkwijk 201

Aging, Diabetes, and Renal Failure Catalyze the Oxidation of Lysyl Residues to 2-Aminoadipic Acid in Human Skin Collagen: Evidence for Metal-catalyzed Oxidation Mediated by α-Dicarbonyls. *By* David R. Sell, Christopher M. Strauch, Wei Shen, and Vincent M. Monnier 205

α-Dicarbonyl Compounds—Key Intermediates for the Formation of Carbohydrate-based Melanoidins. *By* Lothar W. Kroh, Thorsten Fiedler, and Janine Wagner 210

Approaches to Wine Aroma: C_1 Transfer during the Reaction between Diacetyl and Cysteine. *By* John Almy and Gilles de Revel 216

Antioxidant Activity and Chemical Properties of Crude and Fractionated Maillard Reaction Products Derived from Four Sugar–Amino Acid Maillard Reaction Model Systems. *By* Xiu-Min Chen and David D. Kitts 220

Advanced Glycation Endproducts in Chronic Heart Failure. *By* Andries J. Smit, Jasper W. L. Hartog, Adriaan A. Voors, and Dirk J. van Veldhuisen 225

Methylglyoxal and Methylglyoxal-arginine Adducts Do Not Directly Inhibit Endothelial Nitric Oxide Synthase. *By* Olaf Brouwers, Tom Teerlink, Jan van Bezu, Rob Barto, Coen D.A. Stehouwer, and Casper G. Schalkwijk 231

Kinetic Study of the Reaction of Glycolaldehyde with Two Glycation Target Models. *By* Miquel Adrover, Bartolomé Vilanova, Francisco Muñoz, and Josefa Donoso 235

Origin and Yields of Acetic Acid in Pentose-based Maillard Reaction Systems. *By* Tomas Davidek, Elisabeth Gouézec, Stéphanie Devaud, and Imre Blank ... 241

The Peptide-catalyzed Maillard Reaction: Characterization of ^{13}C Reductones. *By* Leif Alexander Garbe, Alexander Würtz, Christian T. Piechotta, and Roland Tressl 244

Model Studies on Protein Glycation: Influence of Cysteine on the Reactivity of Arginine and Lysine Residues toward Glyoxal. *By* Uwe Schwarzenbolz, Susann Mende, and Thomas Henle 248

Analysis of Amadori Peptides Enriched by Boronic Acid Affinity Chromatography. *By* Andrej Frolov and Ralf Hoffmann 253

Induction of Heat Shock Proteins and the Proteasome System by Casein-N^ε-(Carboxymethyl)lysine and N^ε-(Carboxymethyl)lysine in Caco-2 Cells. *By* Karoline Schmid, Martin Haslbeck, Johannes Buchner, and Veronika Somoza 257

Reversal of Hyperglycemia-induced Angiogenesis Deficit of Human Endothelial Cells by Overexpression of Glyoxalase 1 *in Vitro*. *By* Usman Ahmed, Darin Dobler, Sarah J. Larkin, Naila Rabbani, and Paul J. Thornalley 262

Pathophysiological Role of the Glyoxalase System in Renal Hypoxic Injury. *By* Takanori Kumagai, Masaomi Nangaku, and Reiko Inagi 265

A419C (E111A) Polymorphism of the Glyoxalase I Gene and Vascular Complications in Chronic Hemodialysis Patients. *By* Marta Kalousová, Alexandra Germanová, Marie Jáchymová, Oto Mestek, Vladimír Tesař, and Tomáš Zima 268

Succination of Proteins by Fumarate: Mechanism of Inactivation of Glyceraldehyde-3-Phosphate Dehydrogenase in Diabetes. *By* Matthew Blatnik, Suzanne R. Thorpe, and John W. Baynes 272

Dietary Advanced Glycation Endproducts and Oxidative Stress: *In Vivo* Effects on Endothelial Function and Adipokines. *By* Alin Stirban, Monica Negrean, Christian Götting, Jaime Uribarri, Thomas Gawlowski, Bernd Stratmann, Knut Kleesiek, Theodor Koschinsky, Helen Vlassara, and Diethelm Tschoepe 276

Preparation of Nucleotide Advanced Glycation Endproducts—Imidazopurinone Adducts Formed by Glycation of Deoxyguanosine with Glyoxal and Methylglyoxal. *By* Thomas Fleming, Naila Rabbani, and Paul J. Thornalley 280

Maillard Products as Biomarkers in Cancer. *By* Beatrice E. Bachmeier, Andreas G. Nerlich, Helmut Rohrbach, Erwin D. Schleicher, and Ulrich Friess 283

Glycation of Plasma Lipoprotein Lipid Membrane and Screening for Lipid Glycation Inhibitor. *By* Kiyotaka Nakagawa, Daigo Ibusuki, Shinji Yamashita, and Teruo Miyazawa 288

Analysis of Amadori-glycated Phosphatidylethanolamine in the Plasma of Healthy Subjects and Diabetic Patients by Liquid Chromatography–Tandem Mass Spectrometry. *By* Teruo Miyazawa, Daigo Ibusuki, Shinji Yamashita, and Kiyotaka Nakagawa 291

Nonenzymatically Glycated Lipoprotein ApoA-I in Plasma of Diabetic and Nephropathic Patients. *By* Annunziata Lapolla, Maura Brioschi, Cristina Banfi, Elena Tremoli, Chiara Cosma, Luciana Bonfante, Simone Cristoni, Roberta Seraglia, and Pietro Traldi 295

Evaluating the Extent of Protein Damage in Dairy Products: Simultaneous Determination of Early and Advanced Glycation-induced Lysine Modifications. *By* Jörg Hegele, Véronique Parisod, Janique Richoz, Anke Förster, Sarah Maurer, René Krause, Thomas Henle, Timo Bütler, and Thierry Delatour 300

A Novel Yellow Pigment, Furpipate, Derived from Lysine and Furfural. *By* Masatsune Murata, Hana Totsuka, and Hiroshi Ono 307

Time-dependent Component-specific Regulation of Gastric Acid Secretion-related Proteins by Roasted Coffee Constituents. *By* M. Rubach, R. Lang, T. Hofmann, and V. Somoza 310

Maillard Reaction versus Other Nonenzymatic Modifications in Neurodegenerative Processes. *By* Reinald Pamplona, Ekaterina Ilieva, Victoria Ayala, Maria Josep Bellmunt, Daniel Cacabelos, Esther Dalfo, Isidre Ferrer, and Manuel Portero-Otin 315

Suppression of Renal α-Dicarbonyl Compounds Generated following Ureteral Obstruction by Kidney-Specific α-Dicarbonyl/L-Xylulose Reductase. *By* Hiroko Odani, Jun Asami, Aiko Ishii, Kayoko Oide, Takako Sudo, Atsushi Nakamura, Noriyuki Miyata, Noboru Otsuka, Kenji Maeda, and Junichi Nakagawa 320

Comparison of Pharmacokinetics between Highly and Mildly Modified AGE Proteins in Mice. *By* Ryoji Nagai, Katsumi Mera, Yukio Fujiwara, Mime Nagai, and Masaki Otagiri 325

Modification of Vimentin: A General Mechanism of Nonenzymatic Glycation in Human Skin. *By* Thomas Kueper, Tilman Grune, Gesa-Meike Muhr, Holger Lenz, Klaus-Peter Wittern, Horst Wenck, Franz Stäb, and Thomas Blatt 328

Erratum for Ann. N.Y. Acad. Sci. 1072: 386–388. Inflammatory Bowel Disease: Genetics, Barrier Function, Immunologic Mechanisms, and Microbial Pathways. 333

Erratum for Ann. N.Y. Acad. Sci. 1088: A1–A10. Neuroendocrine and Immune Crosstalk. 334

Erratum for Ann. N.Y. Acad. Sci. 1108: 505–514. Autoimmunity, Part D: Autoimmune Disease, Annus Mirabilis. 335

Errata for Ann. N.Y. Acad. Sci. 1111: 442–454. Coccidioidomycosis: Sixth International Symposium. 336

Index of Contributors 337

The New York Academy of Sciences believes it has a responsibility to provide an open forum for discussion of scientific questions. The positions taken by the participants in the reported conferences are their own and not necessarily those of the Academy. The Academy has no intent to influence legislation by providing such forums.

Preface

The 9th International Symposium on the Maillard Reaction, held in Munich in September of 2007, offered a wealth of stimulating talks and discussions. More than 200 delegates from academia and industry and from more than 20 nations actively contributed to the success of the symposium. Following earlier trends, the interdisciplinary nature of the meeting was impressively emphasized by the wide range of subjects addressed. These encompassed food science, food technology and processing, and toxicology of processed food, as well as taste, flavor, and satiety-sensing mechanisms. In food science, the majority of contributions circled around the elucidation of the molecular mechanism of the Maillard reaction, the identification of adverse pathways and their mitigation. In the biomedical field, the interest focused on the development of diabetic complications, particularly affecting the eye, nerves, kidneys, and also the macrovascular system, and on the pathogenesis of neurodegenerative diseases. Moreover, several contributions provided evidence for the involvement of the Maillard reaction products/advanced glycation end products (AGEs) and their receptors in the immune system, in cancer, and, unexpectedly, in anxiety. The improvements in new analytical techniques, particularly chromatography coupled with mass spectrometry, greatly enhance the sensitivity and specificity of analysis, leading to the characterization of new Maillard reaction products in food and biological samples and, thus, potentially leading to the identification of new biomarkers for associated diseases.

Because no other international conferences encompass the scope of the Maillard symposium and because the area is spreading very rapidly, it has been decided that future meetings will take place biannually. The next location will be Australia for the meeting in 2009, and the 2011 symposium will be held in France in honor of the 100th anniversary of Louis Camille Maillard's first publication describing the "browning reaction" that now carries his name.

We also wish to express our sincere appreciation to the local organizing committee members, as well as to the international advisory board and all other contributors. We especially thank our sponsors for their generous support in helping to make this symposium successful (list is in alphabetical order): Deutsche Forschungsanstalt für Lebensmittelchemie, DiagnOptics, Firmenich, Gambro, the German Research Foundation (DFG), the German Society of Clinical Chemistry and Laboratory Medicine (DGKL), the International Maillard Reaction Society, L'Oreal, Nationales Aktionsforum Diabetes mellitus (NAFDM), the National Institute of Diabetes and Digestive and Kidney Diseases, Nestlé, NeoMPS, Procter & Gamble, Roche, the Technical University of Munich, and the University Clinic of Tübingen.

ERWIN SCHLEICHER
University of Tübingen
Tübingen, Germany

VERONIKA SOMOZA
German Research Center for Food Chemistry
Garching, Germany

PETER SCHIEBERLE
Technical University of Munich
Garching, Germany

The Sense of Smell
Reception of Flavors

H. BREER

University of Hohenheim, Institute of Physiology, Stuttgart, Germany

The sensory and hedonic evaluation of most food-related flavors is mainly dependent on olfactory perception. The sense of smell is able to recognize and discriminate myriads of airborne molecules with great accuracy and sensitivity. The primary processes of odor perception are mediated by the chemosensory olfactory neurons in the nasal epithelium, which upon interaction with appropriate odorants elicit a chemo–electrical transduction process converting the chemical signal into electrical impulses. The encoded information is conveyed onto distinct glomeruli, inducing topographic activity patterns in the olfactory bulb. The emerging chemotopic maps are decoded in the olfactory cortex, leading to the perception of distinct flavors.

Key words: olfaction; olfactory neurons; odorant receptor; signal transduction; axonal wiring; odor coding; information processing

Introduction

The flavors for most of our food and beverages are generated during processing of raw materials, such as roasting, cooking, and baking, when chemical reactions not only produce the characteristic brown and golden colors but also produce a large number of odorous compounds. Especially the nonenzymatic browning reaction—the Maillard reaction—leads to the formation of multiple volatile compounds, thus creating the characteristic flavor of foods and beverages.[1] Maillard reaction products are formed when, under high temperature conditions, reducing sugar molecules react via their carbonyl group with the NH_2 group of amino acids. Since the amino acid types determine the flavor of the resulting compounds, the diversity of proteins and sugars in foods accounts for literally hundreds of odorous compounds, creating the characteristic flavor of roasted coffee, grilled meat, toasted bread, fried onions—to mention a few. The Maillard products determine the profoundly different flavors of, for example, meat cooked in boiling water compared to meat cooked in a fryer at a temperature above 120°C.[2] A precise correlation between the composition of Maillard products in processed foods and the human perception is still elusive. However, it can be assumed that most of the compounds are sensed by the nose. In fact, it has been estimated that in approximately 80–90% of what is perceived as food, "taste" actually is due to the sense of smell.

Process of Olfaction

Our nose has an enormous capacity to recognize and discriminate myriads of small volatile compounds of many chemical classes and structural diversity, ranging from short-chain aliphatic compounds to complex aromates with multiple side chains and functional groups. In fact, even novel compounds, which were designed and synthesized by chemists, are immediately recognized by the olfactory system and perceived as distinct odor. Moreover, the nose is also a very sensitive chemodetector; it recognizes odorous compounds at concentrations as low as a few parts per trillion.[3] An understanding of the mechanisms underlying the remarkable sensory capacity of the nose, the principle for encoding the complex sensory information, as well as how the brain reconstructs these stimuli into a "smell map" of the world is a major objective for research in olfaction and has greatly advanced over the past decades.[4,5] The perception of an olfactory stimulus is accomplished by two main processes: the primary signal transduction events in the nasal neuroepithelium and the processing of sensory information in the olfactory bulb and higher brain centers. The process of olfactory perception begins when inhaled odorous compounds dissolve in the mucus that covers and protects the epithelium. Odorants are recognized by millions of chemosensory cells residing in the nasal neuroep-

Address for correspondence: Heinz Breer, University of Hohenheim, Institute of Physiology, Garbenstrasse 30, D-70599 Stuttgart, Germany. Voice: +49-(0) 711-459-23566; fax: +49-(0) 711-459-23720.
breer@uni-hohenheim.de

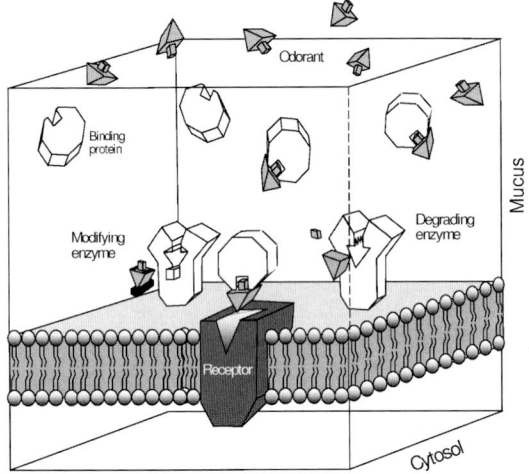

FIGURE 1. Processes in the mucus layer of the olfactory epithelium influence the entry, exit, or residence time of odorous molecules in the receptor environment. These ancillary processes are integral components of the chemical-sensing systems; they include the interaction with soluble-binding proteins, which may act as possible shuttles of the volatile, lipophilic, odorous molecules through the aqueous mucus layer, as well as the inactivation of odorants by degrading and/or biotransformation enzymes, thus clearing the system between consecutive sniffs. (From Breer.[31] Reproduced with permission.)

ithelium within the posterior cavity of the nose. The bipolar olfactory neurons send an axon to the olfactory bulb and extend a dendritic process to the nasal lumen. The tip of the apical dendrite carries 5–20 cilia that are embedded in the protecting nasal mucus and exposed to the external environment; they are the sites of primary olfactory events.

Perireceptor Processes

The airborne primarily lipophilic odorants must traverse the aqueous milieu of the mucus layer covering the nasal epithelium before contacting the olfactory cilia. The entry, exit, and residence time of lipophilic odorants in the receptor environment are considered an important part of the chemical-sensing process, although the mechanisms for these "perireceptor events" are still poorly understood (FIG. 1). The discovery of abundant, small, globular proteins in the mucus fluid surrounding the sensory dendrite and cilia, which are produced by the glands of the nasal cavity, has led to the concept that these odorant-binding proteins, which are members of the lipocaline family, may accommodate hydrophobic molecules in an aqueous environment and enhance their access to the receptor

sites.[6] For a continuous monitoring of the chemical environment, a rapid inactivation and clearance of odorous molecules is necessary to maintain the capability of the olfactory system to receive iterative incoming signals with every breathing airstream. This seems to be accomplished by the action of biotransformation enzymes.[7] The reaction of phase I enzymes (e.g., cytochrome P-450 mono-oxygenases), which introduce chemical changes, such as hydroxylation, is followed by phase II enzymes, such as UDP-glucuronosyl transferase or glutathione-S-transferase catalyzing the conjugation of glucuronic acid or of glutathione to phase I-modified odorants. Odorous molecules modified by this sequential biotransformation are no longer lipid soluble and are incapable of receptor stimulation.

Chemo–electrical Transduction

Upon interaction of odorants with the chemosensory cilia of appropriate olfactory cells, the processes of chemo–electrical signal transduction are elicited, leading to an inwardly depolarizing current that is converted to a distinct frequency of action potentials which are conveyed, via the axon, to the olfactory bulb and deciphered by higher brain centers.[8] Thus, the strength and duration of odorant stimuli are encoded into patterns of neuronal signals. In most of the ciliated olfactory sensory neurons, signal transduction seems to be mediated via the G_{olf}/adenylyl cyclase III (ACIII)/cyclic adenosine monophosphate (cAMP)-pathway (FIG. 2). The interaction of suitable odorants with distinct receptor proteins in the ciliary membrane activates trimeric G-proteins, which stimulate the specific ACIII, thus efficiently generating cAMP. This leads to a very rapid and transient increase in cAMP concentration with a time course of a few hundred milliseconds. Starting from a presumed micromolar concentration of cAMP, the observed five- to tenfold increase brings the cAMP level well above the kd value determined for the cAMP-gated channels in the ciliary membrane. Thus, the elevated cAMP level triggers membrane depolarization via opening of cyclic-nucleotide-gated ion channels preferentially permeable for Ca^{2+} ions, which causes an elevation of intraciliary Ca^{2+} concentration.[9] In olfactory cilia, elevation of Ca^{2+} concentration activates ion channels that permeate chloride ions. Interestingly, the calcium-activated chloride conductance further depolarizes the cell. This untypical reaction is based on the unusually high intracellular Cl^- concentration in olfactory neurons, leading to an efflux of Cl^- ions through Ca^{2+}-activated-chloride channels.[10]

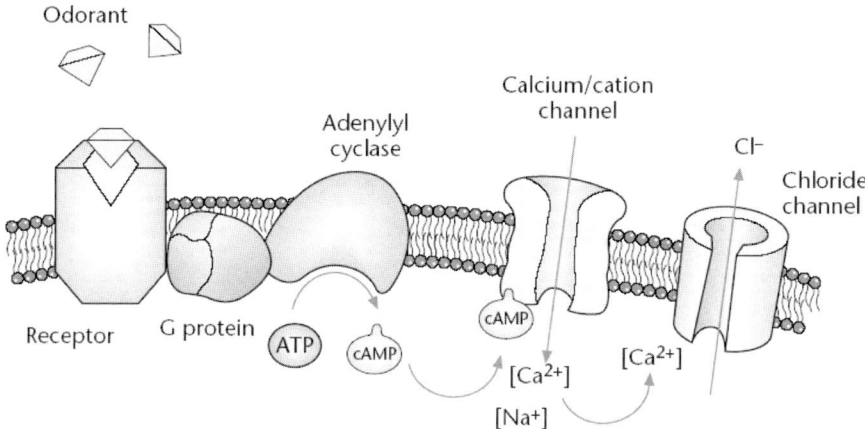

FIGURE 2. Schematic representation of the chemo–electrical transduction pathway in olfactory sensory neurons. Upon binding of appropriate odorous ligands, odorant receptors in the ciliary membrane act through specific G-proteins (G_{olf}) to stimulate adenylyl cyclase (type III) generating cAMP. The resulting elevated second-messenger levels elicit the activation of cation channels, allowing the influx of sodium and especially calcium ions. Calcium ions, in turn, activate Ca^{2+}-dependent chloride channels. Because of the characteristic equilibrium potential for chloride in olfactory neurons, the induced Ca^{2+} current is depolarizing, thus resulting in a significant amplification of the primary odor-induced current. (From Breer.[31] Reproduced with permission.)

The Ca^{2+}-mediated linkage from cyclic-nucleotide-gated channels to chloride channels has been considered a mechanism that contributes to the high-gain and low-noise amplification of olfactory neuron responses. The transduction cascade is reset by terminating the formation of cAMP via feedback reactions including receptor phosphorylation, GTPase activity of $G\alpha$ subunits, Ca^{2+} block of CNG channels, and by hydrolysis of cAMP in $5'$-AMP catalyzed by phosphodiesterase. The influx of calcium via the cAMP-gated channels is also important for the process of olfactory adaptation, i.e., decrease of a stimulus perception upon stimulation over longer periods, which appears to be a result of a decaying responsiveness of olfactory sensory neurons. Calcium ions entering the olfactory cilia interact with calmodulin, and the newly formed calcium–calmodulin complex closes the ion channel even in the presence of cAMP.[11] This negative-feedback loop turns off the electrical response of an olfactory neuron even in the presence of appropriate odorants. This mechanism is supplemented by kinase reactions, leading to phosphorylation of transduction elements and thereby inducing longer lasting desensitization effects.

Olfactory Receptors

How thousands of different odorants which vary widely in size and structure are readily detected and discriminated has been a long-standing puzzle. The accuracy of odor discrimination depends on the specificity with which odorants activate olfactory neurons via distinctive receptor proteins residing in the ciliary membrane. Thus, olfactory receptor proteins can be considered as molecular entities at the interface between the environment and the nervous system. Accordingly, understanding the nature, diversity, and specificity of receptors for odorants has always been considered key for understanding the molecular basis of olfaction. The discovery of a large family of genes, which encode heptahelical transmembrane proteins and are expressed in the olfactory epithelium,[12] was the ground-breaking work for a wealth of studies indicating that odorant receptors (ORs) are members of the G-protein-coupled receptor superfamily; they are encoded by a multigene family comprising as many as a thousand distinct genes that are organized in clusters at many different loci spread over all but a few chromosomes in the mouse genome. In the human genome, about 900 OR genes have been identified, but two-thirds of these turned out to be nonfunctional or "pseudogenes," which have lost their function during evolution; a total of 347 putative, functional OR genes in the human genome was determined.[13] In spite of the reduced number of OR subtypes, the human olfactory system retained the ability to recognize a broad spectrum of chemicals; however, its discriminatory capacity is probably reduced compared to mice. The high proportion of pseudogenes[14] and the unusually high rate of single nucleotide polymorphisms in human receptor genes[15] indicate a variable repertoire of functional OR genes in the human population. Many specific anosmia (i.e., the inability to smell particular odors) could be a result of hereditary defects of OR genes.

Response Specificity of an Odorant Receptor

Physiological recordings have demonstrated that individual olfactory sensory neurons typically respond to a variety of different odorants and that each cell shows a unique order of agonist potency, indicating that olfactory neurons are highly diverse and broadly tuned.[16,17] Based on the notion that each olfactory sensory cell expresses only one OR subtype, it seems likely that a relatively nonspecific ligand spectrum is a characteristic feature of ORs. The assessment of the responsiveness of a distinct receptor type requires the expression in heterologous cells. The problems encountered in this approach were circumvented by an *in vivo* expression system transfected by means of recombinant adenovirus and assessed by electrophysiological recordings[18] or by *in vitro* systems using engineered OR chimeric receptors.[19] The general consensus from these studies is that a distinct OR type is activated by multiple odorants and that the range of dissimilar ligands for a distinct OR subtype resembles that of individual olfactory sensory neurons.[20] Not only can a distinct receptor type recognize multiple odorants but, at the same time, a single odorant is capable of activating multiple receptor types. Thus, all data point to the concept that the nose uses a combinatorial coding scheme to discriminate the vast number of different smells. Analogous to the visual system, which uses three receptor types (three opsin subtypes of the three-cone populations) to make sense of all perceivable colors, the olfactory system computes information from combinations involving any of hundreds of OR types. The numerous possible combinations explain the capacity of the system to encode an unlimited number of odors. Instead of dedicating an individual odor receptor to a specific odor, the olfactory system uses an "alphabet" of receptors to create a specific odor response[21]; in this view, a distinct receptor type participates in encoding very different odors, much the same way as a distinct letter participates in forming very different words. The principle of combinatorial coding implies that odorants of nearly identical structure are recognized by different but overlapping sets of receptors, thus explaining why even a slight change in the structure of an odorant can cause a dramatic shift in its perceived odor; for example, when the hydroxyl group of octanol is replaced by a carboxyl group to make octanoic acid, its odor changes from orange to rancid. Combinatorial coding may also explain the phenomenon that the perceived quality of an odorant can differ with a change in its concentration; for example, indole is perceived as floral at low concentration but smells putrid at higher doses.

Spatial Expression Patterns of Odorant Receptors

Despite the large number of receptor genes, a given sensory neuron seems to express only one type of receptor derived from a single allele. This notion implies that the supposedly about 20 million olfactory neurons of a mouse can be subdivided into about a thousand subpopulations. Within the olfactory epithelium, subpopulations (i.e., groups of 10,000–20,000 olfactory sensory neurons expressing the same receptor type) are confined to one of several broad expression zones. Within a zone, the neurons expressing a distinct receptor type are broadly distributed and surrounded by cells expressing different receptor types.[22,23] The functional implications of this zonal segregation are still elusive, but it is maintained in the olfactory bulb with each zone of the epithelium projecting to a distinct region of the olfactory bulb. In contrast to the zonal distribution of most subpopulations, olfactory sensory neurons expressing receptor subtypes of the OR37 subfamily are segregated in a small area on the tip of central turbinates spatially organized in clusters.[24] The functional relevance of this unique distribution pattern is unknown, but it is maintained in a distinct projection area in the bulb.[25]

Projection Patterns of Olfactory Neurons

To inform the brain, each olfactory neuron projects a single unbranched axon into the olfactory bulb where it terminates in rounded regions of neuropil, termed *glomeruli*, which are considered anatomical and functional units for olfactory processing (FIG. 3). This spherical neuropil structure of about 100 μm in diameter and surrounded by glia cell processes consists of the arborizing preterminal fibers and terminal boutons of the olfactory axons synapsing onto the distal dendritic tufts of the mitral cells as well as the periglomerular neurons. Thousands of olfactory neurons project their axon onto one glomerulus, thus achieving a high degree of convergence. In mice, the 20 million olfactory neurons converge onto some 2000 glomeruli giving an average convergence of 10,000:1. Recent studies have demonstrated that all neurons expressing the same receptor type converge their axons onto the same glomerulus, usually two glomeruli which are located on the lateral and medial hemisphere of the bulb, respectively.[26] Thus, olfactory neurons that express the same receptor type are widely scattered in a distinct zone of the nasal neuroepithelium, but their axons converge at two specific sites in the olfactory bulb (FIG. 3). Exploring the projection of two neuron populations that express highly related receptor genes, it was found that neuron populations with very similar receptor types nevertheless project to distinct glomeruli; however, the glomeruli are located in the immediate vicinity.[25] The wiring processes and the interplay between receptor-specific axon populations in finding a common target area in the bulb and projecting onto receptor-specific glomeruli are main topics of contemporary research. Recent evidence points to an immediate role of the receptor proteins themselves. Indeed, receptor proteins are found not only in the dendrites but also in the axonal fibers[27] where they could take part in the cell–cell interactions underlying axonal guidance and target recognition, essential processes for the formation of precise receptor-specific connections between sensory neurons and the appropriate neuronal circuitry in the bulb. In this view, the array of glomeruli can be considered as a map reflecting the receptor types expressed in the epithelium; consequently, a distinct odor elicits a characteristic pattern of glomerular activation,[28] representing an odor map. It is only in the olfactory bulb where the complex neuronal processing of olfactory information begins, primarily mediated by inhibitory and excitatory microcircuits. It is performed at two stages, one at the glomerular level and one involving the granule cells.[29] Besides the intraglomerular processing, there is also an interaction among adjacent glomeruli mediated by periglomerular neurons as well as modulatory influences from higher brain centers. Nevertheless, the precise identification of an odor requires further processing in cortical regions of the brain. The output of the olfactory bulb is conveyed by the mitral cell axons, which course through the olfactory tracts. They project directly to distinct regions of the cerebral cortex, the main area being the piriform cortex. From here the information is relayed to the thalamus and finally to the neocortex, where presumably the conscious perception of odors takes place, and to limbic areas, where emotional and behavioral responses are elicited.[30]

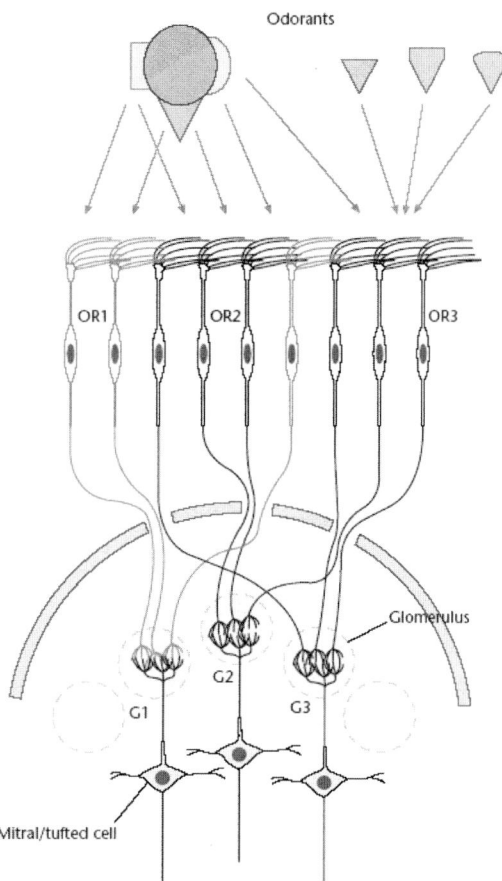

FIGURE 3. Topographic projection of olfactory sensory neurons expressing distinct receptor types (OR1–OR3) to distinct glomeruli (G1–G3) in the olfactory bulb. The chemo-specific responsiveness of olfactory sensory cells as well as the targeting of their axons in the olfactory bulb is determined by the receptor type they express. Structurally different odorants that share a defined feature (odotope) are supposed to interact with the same receptor-specific cells, thus leading to an activation of the same glomerulus. Complex odorous compounds comprising multiple odotopes may interact with various receptor types eliciting the activation of several glomeruli. In this way, each odorant would be represented spatially in the bulb by a unique ensemble of active glomeruli, representing the combinatorial code of an odor. (From Breer.[31] Reproduced with permission.)

Conclusion

The enormous sensory capacity of our nose to recognize and discriminate thousands of extraneous volatile compounds has long been considered as a scientific mystery. However, due to modern biological approaches and a meticulous exploration at the molecular, cellular, and network level, the field of olfaction has witnessed much progress in deciphering several of the principles and mechanisms underlying the sense of smell. The basic units of odor information appear to be the epitopes of odorant molecules interacting with specific sites of distinct receptors. Activation of receptors is linked via second-messenger cascades, amplifying the olfactory signal into the electrical response of the sensory neuron. Thus, the various epitopic sites of an odor molecule are mapped into activation of distinct subpopulations of olfactory neurons and, subsequently, into activation of characteristic combinations of glomeruli. Such activity maps in the olfactory bulb are ultimately decoded in the olfactory cortex, thus allowing the perception of distinct scents.

Acknowledgment

The work from this laboratory was supported by the Deutsche Forschungsgemeinschaft.

Conflict of Interest

The author declares no conflicts of interest.

References

1. VAN BOEKEL, M.A. 2006. Formation of flavour compounds in the Maillard reaction. Biotechnol. Adv. **24:** 230–233.
2. FARMER, L.J. 1994. The role of nutrients in meat flavour formation. Proc. Nutr. Soc. **53:** 327–333.
3. SHEPHERD, G.M. 2004. The human sense of smell: are we better than we think? PLoS. Biol. **2:** 572–575.
4. FIRESTEIN, S. 2001. How the olfactory system makes sense of scents. Nature **413:** 211–218.
5. BUCK, L.B. 2005. Unraveling the sense of smell (Nobel lecture) Angew. Chem. Int. Ed Engl. **44:** 6128–6140.
6. LOBEL, D., M. JACOB, M. VOLKNER & H. BREER. 2002. Odorants of different chemical classes interact with distinct odorant binding protein subtypes. Chem, Senses **27:** 39–44.
7. LAZARD, D., K. ZUPKO, Y. PORIA, et al. 1991. Odorant signal termination by olfactory UDP glucuronosyl transferase. Nature **349:** 790–793.
8. MENINI, A., L. LAGOSTENA & A. BOCCACCIO. 2004. Olfaction: from odorant molecules to the olfactory cortex. News Physiol. Sci. **19:** 101–104.
9. FRINGS, S. 2001. Chemoelectrical signal transduction in olfactory sensory neurons of air-breathing vertebrates. Cell Mol. Life Sci. **58:** 510–519.
10. KLEENE, S.J. 1993. Origin of the chloride current in olfactory transduction. Neuron **11:** 123–132.
11. KURAHASHI, T. & A. MENINI. 1997. Mechanism of odorant adaptation in the olfactory receptor cell. Nature **385:** 725–729.
12. BUCK, L. & R. AXEL. 1991. A novel multigene family may encode odorant receptors: a molecular basis for odor recognition. Cell **65:** 175–187.
13. ZOZULYA, S., F. ECHEVERRI & T. NGUYEN. 2001. The human olfactory receptor repertoire. Genome Biol. **2:** research0018.1-research0018.12.
14. MENASHE, I., R. ALONI & D. LANCET. 2006. A probabilistic classifier for olfactory receptor pseudogenes. BMC. Bioinformatics **7:** 393.
15. GILAD, Y., D. SEGRE, K. SKORECKI, et al. 2000. Dichotomy of single-nucleotide polymorphism haplotypes in olfactory receptor genes and pseudogenes. Nat. Genet. **26:** 221–224.
16. MALNIC, B., J. HIRONO, T. SATO & L.B. BUCK. 1999. Combinatorial receptor codes for odors. Cell **96:** 713–723.
17. SICARD, G. & A. HOLLEY. 1984. Receptor cell responses to odorants: similarities and differences among odorants. Brain Res. **292:** 283–296.
18. ZHAO, H., L. IVIC, J.M. OTAKI, et al. 1998. Functional expression of a mammalian odorant receptor. Science **279:** 237–242.
19. KRAUTWURST, D., K.W. YAU & R.R. REED. 1998. Identification of ligands for olfactory receptors by functional expression of a receptor library. Cell **95:** 917–926.
20. ARANEDA, R.C., A.D. KINI & S. FIRESTEIN. 2000. The molecular receptive range of an odorant receptor. Nat. Neurosci **3:** 1248–1255.
21. BREER, H. 2003. Olfactory receptors: molecular basis for recognition and discrimination of odors. Anal. Bioanal. Chem. **377:** 427–433.
22. RESSLER, K.J., S.L. SULLIVAN & L.B. BUCK. 1993. A zonal organization of odorant receptor gene expression in the olfactory epithelium. Cell **73:** 597–609.
23. STROTMANN, J., I. WANNER, T. HELFRICH, et al. 1994. Rostro-caudal patterning of receptor-expressing olfactory neurons in the rat nasal cavity. Cell Tissue Res. **278:** 11–20.
24. STROTMANN, J., I. WANNER, J. KRIEGER, et al. 1992. Expression of odorant receptors in spatially restricted subsets of chemosensory neurones. NeuroReport **3:** 1053–1056.
25. STROTMANN, J., S. CONZELMANN, A. BECK, et al. 2000. Local permutations in the glomerular array of the mouse olfactory bulb. J. Neurosci. **20:** 6927–6938.
26. MOMBAERTS, P., F. WANG, C. DULAC, et al. 1996. Visualizing an olfactory sensory map. Cell **87:** 675–686.
27. STROTMANN, J., O. LEVAI, J. FLEISCHER, et al. 2004. Olfactory receptor proteins in axonal processes of chemosensory neurons. J Neurosci. **24:** 7754–7761.
28. LUO, M. & L.C. KATZ. 2001. Response correlation maps of neurons in the mammalian olfactory bulb. Neuron **32:** 1165–1179.
29. CHEN, W.R. & G.M. SHEPHERD. 2005. The olfactory glomerulus: a cortical module with specific functions. J Neurocytol. **34:** 353–360.
30. SHEPHERD, G.M. 2005. Outline of a theory of olfactory processing and its relevance to humans. Chem. Senses **30**(Suppl 1): i3–i5.
31. BREER, H. 2001. Olfaction. In Encyclopedia of Life Sciences. Nature Publishing Group/www.els.net. John Wiley & Sons Limited.

Receptor for Advanced Glycation End Products

Fundamental Roles in the Inflammatory Response: Winding the Way to the Pathogenesis of Endothelial Dysfunction and Atherosclerosis

RAVICHANDRAN RAMASAMY,[a] SHI FANG YAN,[a] KEVAN HEROLD,[b] RAPHAEL CLYNES,[c] AND ANN MARIE SCHMIDT[a]

Departments of [a]Surgery and [c]Medicine, Columbia University Medical Center, New York, New York, USA

[b]Department of Medicine and Immunobiology, Yale University School of Medicine, New Haven, Connecticut, USA

The multiligand receptor for advanced glycation end products (RAGE) of the immunoglobulin superfamily is expressed on multiple cell types implicated in the immune–inflammatory response and in atherosclerosis. Multiple studies have elucidated that ligand–RAGE interaction on cells, such as monocytes, macrophages, and endothelial cells, mediates cellular migration and upregulation of proinflammatory and prothrombotic molecules. In addition, recent studies reveal definitive rules for RAGE in effective T lymphocyte priming *in vivo*. RAGE ligand AGEs may be formed in diverse settings; although AGEs are especially generated in hyperglycemia, their production in settings characterized by oxidative stress and inflammation suggests that these species, in part via RAGE, may contribute to the pathogenesis of atherosclerosis. In murine models of atherosclerosis, vascular inflammation is a key factor and one which is augmented, in parallel with even further increases in RAGE ligands, in diabetic macrovessels. The findings that antagonism and genetic disruption of RAGE in atherosclerosis-susceptible mice strikingly reduces vascular inflammation and atherosclerotic lesion area and complexity link RAGE intimately to these processes and suggest that RAGE is a logical target for therapeutic intervention in aberrant inflammatory mechanisms and in atherosclerosis.

Key words: receptor for advanced glycation end products; inflammation; atherosclerosis; adaptive immunity; T cell priming

RAGE: A Multiligand Receptor

Our evolving understanding of the biology of the receptor for advanced glycation end products (RAGE) has been facilitated by the demonstration that RAGE is a multiligand member of the immunoglobulin superfamily.[1] RAGE was first described as a receptor for the products of nonenzymatic glycation and oxidation of proteins, the advanced glycation end products or AGEs.[2] Although initially considered as modified species that are formed to accelerated degrees in hyperglycemia, it has been shown that AGE formation may be stimulated even in normoglycemia. For example, stresses, such as renal failure, oxidative stress, inflammation, and aging, may provoke AGE generation.[1]

In addition to AGEs, natural ligands of the receptor, including certain members of the S100/calgranulin family[3,4] and amphoterin (or high mobility group box-1 [HMGB1]),[5,6] also interact with RAGE. Interestingly, HMGB1 was originally described as an intracellular (nuclear) protein. Its release by stimulated cells, however, suggested new roles for this molecule in inflammation and cell stress.[7] Interaction of HMGB1 with RAGE increases expression of proinflammatory molecules. *In vivo* studies supported the critical role of HMGB1 and RAGE in inflammation. For example, blockade of either ligand–RAGE interaction or HMGB1 (in the latter case, using antibodies to HMGB1) suppressed inflammation in a rodent model of inflammatory arthritis.[8,9] Indeed, intra-articular

Address for correspondence: Dr. Ann Marie Schmidt, Department of Surgery, Columbia University Medical Center, 630 West 168th Street, P&S 17-401, New York, NY 10032. Voice: +1-212 305 6406; fax: +1-212 305 5337.

ams11@columbia.edu

injection of HMGB1 was sufficient to induce arthritis in rodents.[10]

Furthermore, RAGE has been shown to be a receptor for Mac-1, thus establishing additional mechanisms by which ligation of RAGE may mediate inflammatory cell migration and activation in susceptible foci.[11] Taken together, these lessons, learned from the multiligand nature of RAGE, suggest it plays key roles in the inflammatory response.

RAGE Is a Signal Transduction Receptor

A central mechanism by which ligand–RAGE interaction mediates cell stress and upregulates inflammatory pathways is via activation of signal transduction pathways. Multiple signaling pathways have been shown to be activated by RAGE. For example, ligand–RAGE interaction stimulates activation of mitogen-activated protein kinases (MAP kinases) such as p44/42, p38, and JNK MAPK.[12,13] Further, activation of Jak/STAT pathways, rho and rac GTPases, and p21^{ras} have been linked to RAGE.[14–16] Definitive studies both *in vitro* and *in vivo* indicate that the impact of RAGE ligands is due to activation of signal transduction. When the cytoplasmic domain of RAGE is deleted, resulting in a construct in which ligands may bind at the extracellular domain, the loss of the intracellular domain evokes a dominant negative (DN) effect such that when ligands bind RAGE, they are unable to modulate gene expression in the absence of interaction with intracellular signaling partners.

The variety of signaling pathways triggered by RAGE ligands likely reflects both the specific cell type and the time course and duration of activation. RAGE is expressed by multiple cell types at low levels in homeostasis across a varied array of cell types.[17] In most cases, expression of RAGE is upregulated in disease states, such as in diabetic aorta, kidney, and retina.[18–20] Interestingly, in lung tissue even at base line, high levels of RAGE expression have been observed. The precise roles for RAGE in lung homeostasis and disease have yet to be elucidated.

RAGE and Inflammation: First Insights in Animal Models

The discovery of S100/calgranulins and HMGB1 as RAGE ligands shed first light on potential roles for RAGE in diabetes-independent inflammatory responses. We and others used soluble RAGE (sRAGE), the extracellular ligand-binding decoy of RAGE, and F(ab′)$_2$ fragments prepared from anti-RAGE IgG or anti-S100A12 IgG to test if blockade of the ligand–RAGE interaction suppressed delayed type hypersensitivity reactions in mice sensitized and challenged with methylated bovine serum albumin. Compared to appropriate vehicles and controls, blockade of ligand–RAGE suppressed the challenge phase of footpad edema. Infiltration of inflammatory cells, granuloma formation, and edema responses were significantly suppressed in the presence of such blockade.[3] Consistent with marked suppression of proinflammatory mechanisms, nuclear extracts retrieved from the treated mouse food pads revealed strikingly diminished activation of nuclear factor kappa B (NF-κB).[3] In other experiments, Interleukin (IL)-10-deficient mice displayed significantly reduced gut inflammation and activation of NF-κB when treated with sRAGE compared to treatment with a vehicle.[3]

These studies suggested for the first time that RAGE played important roles in the immune response. In this review, representative experiments underscoring such roles for RAGE in immunity are discussed.

Roles for RAGE in the Pathogenesis of Type 1 Diabetes

T lymphocytes play essential roles in the pathogenesis of autoimmune diabetes. In addition to macrophages, RAGE is expressed in CD4$^+$ and CD8$^+$ T lymphocytes and B lymphocytes.[21] Roles for RAGE in inflammation linked to the pathogenesis of type 1 diabetes were established by administration of sRAGE to nonobese diabetic (NOD)/scid mice subjected to adoptive transfer of diabetogenic spleen cells (from NOD mice). Upon transfer of these splenocytes, increased expression of both RAGE and proinflammatory S100A12 in the islets was observed.

When the mice receiving diabetogenic splenocytes were treated with sRAGE, a significantly increased time to hyperglycemia was noted. In addition to delays in development of diabetes, downregulation of IL-1β and tumor necrosis factor (TNF)-α in the islet tissue was observed compared to vehicle treatment. Further, IL-10 and TGF-β transcripts were increased in the islets in the presence of sRAGE compared to vehicle treatment.[21]

These experiments established for the first time that RAGE might play roles in the pathogenesis of type 1 diabetes and linked this receptor to tissue-damaging inflammation.

Roles for RAGE in the Alloimmune Response

Based on these findings in NOD mice, we sought to extend these concepts and probe if RAGE contributed to alloimmune responses in organ transplantation. We employed an established model of heterotopic allogeneic heart transplantation in which fully mismatched grafts (donor, H2^q) were transplanted into C57BL/6 recipients (H2^b). sRAGE was administered to the animals and graft survival was assessed. The mean graft survival time in vehicle phosphate-buffered saline solution-treated mice was 7.3 ± 0.7 days. However, mice treated with 100 μg/day sRAGE displayed significantly increased graft survival at 11.7 ± 1.7 days. At a higher dose of sRAGE (200 μg/day), even greater graft survival time was observed (19.5 ± 2.8 days).[22]

Extensive analysis of the allograft tissue samples revealed that inflammatory cell infiltration (especially that of T lymphocytes), expression of RAGE and its ligands, edema, and necrosis in the sRAGE-treated hearts were greatly reduced. To directly probe roles for RAGE in alloimmune responses relevant to this setting, purified T cells and MHC class II$^+$ antigen-presenting cells (APCs) were retrieved from MHC-mismatched mice. Incubation with sRAGE resulted in a dose-dependent decrease in lymphocyte proliferation versus IgG control-treated cultures. In addition, cells in the mouse allogeneic mixed lymphocyte culture were incubated with blocking antibodies to RAGE. Compared to nonimmune IgG, incubation with monoclonal anti-RAGE IgG resulted in a significant dose-dependent decrease in lymphocyte proliferation.[22] Similar results using sRAGE and antibodies to RAGE were observed in human mixed-lymphocyte reaction responses *in vitro*.[22]

Roles for RAGE in Experimental Autoimmune Encephalomyelitis: Implications for Roles for RAGE in T Lymphocyte Migration

One key feature of RAGE-dependent roles in the inflammatory response implicates this receptor in T lymphocyte migration and infiltration into vulnerable foci. These concepts were tested in a murine model of experimental autoimmune encephalomyelitis (EAE). Administration of sRAGE suppressed EAE induced by myelin basic protein (MBP) or when EAE occurred spontaneously in T cell receptor (TCR)-transgenic mice devoid of endogenous TCR-alpha and TCR-beta chains. These studies revealed that in sRAGE-treated animals, significantly decreased infiltration of the spinal cord by immune and inflammatory cells was evident. As inflammation and spinal cord injury are intimately linked to T lymphocyte responses, we directly addressed the role of CD4 T lymphocyte RAGE signal transduction in EAE and inflammatory damage to the spinal cord.

Transgenic mice were generated with targeted overexpression of DN RAGE in CD4$^+$ T cells. As indicated above, signal transduction initiated by ligand–RAGE interaction is required for changes in gene expression to occur in a RAGE-dependent manner. Thus, wild-type littermate mice were compared to hemizygous CD4-DN RAGE-transgenic mice. Compared to littermates, transgenic CD4 DN RAGE mice were resistant to MBP-induced EAE.[23] Migration of T lymphocyte effector cells into spinal cord was strikingly suppressed when CD4 T lymphocyte RAGE signaling was suppressed.

RAGE Is Required for Effective T Lymphocyte Priming *In Vivo*: Definitive Studies *In Vitro* and *In Vivo*

Taken together, these representative studies in animal models of inflammation and autoimmunity strongly suggested that T lymphocyte RAGE responses were integral to the host response and led us to test if RAGE played direct roles in T lymphocyte priming. To accomplish this, we employed I-A^b-restricted ovalbumin (OVA)-specific OT-II CD4 TCR-transgenic cells. RAGE was expressed in primary OT-II CD4 cells.[24]

As RAGE is also expressed in dendritic cells (DC) and macrophages,[25] it was necessary to dissect the distinct contributions of RAGE on T lymphocytes and host non-T lymphocytes in adoptive transfer systems. Labeled RAGE-expressing and null OT-II CD4$^+$ CD45.1$^+$ cells were transferred into syngeneic C57 BL.6 CD45.2$^+$ RAGE-expressing or null hosts and the mice were subsequently immunized intraperitoneally with complete Freund's adjuvant (CFA)/OVA. Although homing of naive T lymphocytes to secondary lymphoid organs was comparable in all unimmunized mice, significant roles for RAGE in priming were evident. Deletion of RAGE on T lymphocytes dramatically reduced T lymphocyte proliferative responses upon OVA exposure in wild-type recipients. RAGE expression on host cells also contributed to T lymphocyte priming as transferred RAGE-expressing OT-II cells divided to a greater degree in RAGE-expressing mice than in RAGE-null recipients.[24] The effects were additive as deletion of RAGE on both host cells and transferred OT-II cells resulted in the lowest proliferative response. *In vitro* experiments extended these findings, as RAGE deficient T lymphocytes showed strikingly impaired proliferative responses to nominal and alloantigen. Cytokine production of interferon (IFN)-γ and IL-2 was significantly reduced in the absence of RAGE.[24]

Taken together, these data provided supportive evidence that RAGE contributes to early T lymphocyte expansion during priming *in vivo*. Further, these studies also suggest important roles for additional cellular contributions (beyond T lymphocytes) in adaptive immune responses.

In this context and in contrast to other reports,[25] our studies did not reveal significant roles for RAGE in APC/DC responses, as RAGE-deficient DC did not reveal functional impairment in antigen presentation, maturation, or migratory capacities.[24] In addition to T lymphocytes, macrophages, and DC, RAGE is expressed on B lymphocytes. Studies are in progress to discern if expression of RAGE in B lymphocytes contributes to adaptive immunity. Lastly, it is essential to consider that endothelial cells (EC) RAGE responses are critically linked to the integrated cellular responses evoked in priming of T lymphocytes.

Endothelial Dysfunction: Integral to the Inflammatory Response and the Pathogenesis of Endothelial Dysfunction and Atherosclerosis

Testing the Role of RAGE in Atherosclerosis: Effects of sRAGE

Studies in endothelial dysfunction and atherosclerosis were originally approached from the context of diabetes and acceleration of macrovascular disease that typifies this disorder. In our first studies, mice deficient in apolipoprotein E were used, as these animals displayed spontaneous hypercholesterolemia and increases in atherosclerotic lesion area and complexity versus wild-type mice. Importantly, these animals develop atherosclerosis on normal rodent chow. To address the impact of diabetes on atherosclerosis, relative insulin-deficient (type 1) diabetes in these mice was induced with streptozotocin (stz). In hyperglycemic animals, significantly increased atherosclerosis and vascular inflammation were observed compared to nondiabetic apo E-null mice of the same age.[18,26] To block ligand–RAGE interaction in these animals, sRAGE was administered once daily by intraperitoneal administration to apoE-null mice rendered diabetic with stz. Soluble RAGE was begun in the diabetic mice immediately at the time of documentation of hyperglycemia and was continued for 6 weeks. Compared to vehicle-treated mice, sRAGE-treated animals displayed a dose-dependent decrease in atherosclerosis area at the aortic root in parallel with decreased features of lesion complexity.[18] RAGE actions were directed at amplification mechanisms linked to the acceleration of vascular inflammation as RAGE blockade did not affect levels of lipids or glucose, thereby suggesting that RAGE acted downstream of these key risk factors.[18]

As these studies illustrated the effects of sRAGE in prevention of early progression, the role of RAGE in diabetic apo E-null mice with established atherosclerosis was then addressed. sRAGE was administered to diabetic mice commencing only after 6 weeks of sustained hyperglycemia; sRAGE or vehicles were continued for an additional 6 weeks. Control diabetic animals received murine serum albumin. These studies revealed that sRAGE-treated mice displayed significant stabilization of lesion area at the aortic root compared to vehicle-treated diabetic mice.[27] Of particular importance, vascular inflammation and oxidant stress were markedly attenuated in the aortas of sRAGE-treated animals as reflected by decreased cox-2 and nitrotyrosine epitopes, JE-MCP-1, tissue factor antigens; matrix metalloproteinase (MMP)9 antigen activity; and phosphorylated p38 MAP kinase.[27] As in the case of animals treated with sRAGE commencing immediately at the onset of hyperglycemia, in these animals, there were no differences in glucose, insulin, or lipid number or profile.[27]

Administration of stz induces a relative insulin-deficient diabetes. Thus, an essential step was to test the premise that RAGE modulated atherosclerosis in murine models of type 2 diabetes. To accomplish this, apoE-null mice were bred into the db/db background. In apo E-null db/db mice, atherosclerosis was accelerated compared to littermate apo E-null mice without diabetes.[28] Administration of sRAGE from age 8 to 11 weeks resulted in a highly significant decrease in atherosclerotic lesion area in parallel with decreased vascular expression of proinflammatory RAGE ligand S100/calgranulins and vascular cell adhesion molecule-1 (VCAM-1) and MMPs.[28]

Of note, in animal models and in human cardiovascular disease, there is increased accumulation of RAGE ligands in atherosclerotic plaques even in the absence of diabetes. These considerations are not surprising as vascular inflammation and oxidative stress drive generation of AGEs as well as increased expression of proinflammatory S100/calgranulins and HMGB1. Thus, when sRAGE was administered to nondiabetic apoE-null mice (C57BL/6 background) and to apoE-null mice in the db background, sRAGE-treated nondiabetic mice displayed significantly decreased atherosclerosis and vascular inflammation compared to vehicle-treated control mice.[27,28]

Therefore, although diabetes represents a state of highly exaggerated RAGE ligand generation, even in nondiabetic atherosclerotic disease, oxidative stress evoked by hyperlipidemia and inflammation evoke

AGE formation and, thus, clearly contribute integrally to the pathogenesis of atherosclerosis.

Testing the Role of RAGE in Atherosclerosis: Effects of Genetic Modulation of RAGE

Studies in RAGE-modified Mice

Two distinct genetic strategies were employed to dissect the role of the ligand–RAGE interaction in apoE-null mice. First, homozygous RAGE-null mice were bred into the apoE-null background. Second, to probe the role of endothelial RAGE in atherosclerosis, a transgenic mouse was prepared to express human DN RAGE specifically in EC by the pre-proendothelin-1 (PPET) promoter.[29]

At age 14 weeks, compared to apoE-null mice, apoE-null mice in the RAGE-null background and in the transgenic PPET DN RAGE background displayed significantly less atherosclerotic lesion area and complexity; there was no differences in plasma cholesterol or triglyceride among the three groups of mice.[29]

Fundamental protection from endothelial dysfunction in apoE-null mice was observed by modulation of RAGE. Exposure of aortic rings to increasing doses of acetylcholine revealed that endothelium-dependent relaxation was significantly improved in rings retrieved from apoE-null/RAGE-null and apoE null/Tg PPET DN RAGE mice compared to apoE-null mice aortic rings.[29]

Deletion of RAGE or introduction of DN RAGE in EC in apoE-null mice resulted in significantly lower levels of mediators of vascular inflammation in aorta tissue as observed by decreased levels of VCAM-1, MCP-1, MMP-2 protein and activity, IL-10, and CD40 compared to apoE-null aorta. In plasma, levels of sVCAM-1 were significantly lower in apoE-null/RAGE-null and Tg PPET DN RAGE/apoE-null mice versus apoE-null mice.

Studies in RAGE-modified EC

To specifically dissect the signal transduction mechanisms linking RAGE to endothelial dysfunction, EC were isolated and purified from the aortas of wild-type C57BL/6 and RAGE- null and Tg PPET DN RAGE mice.[29]

Wild-type, RAGE-null, and DN RAGE EC were stimulated with the prototypic RAGE ligand S100b. Increased VCAM-1 antigen was observed in a manner dependent on RAGE and RAGE signaling. Key roles for RAGE in activation of JNK MAP kinase were illustrated by significant reduction in S100b-induced phosphorylation of this MAP kinase in RAGE-null or Tg PPET DN RAGE EC versus stimulated wild-type cells. Pretreatment of the EC with the JNK MAP kinase inhibitor, SP600125, resulted in highly significant reduction of S100b-mediated VCAM-1 upregulation. Introduction of small interfering (si)RNA to knockdown JNK expression blunted the effect of S100b on upregulation of VCAM-1 as well. In contrast, introduction of scrambled siRNA or the ERK MAP kinase inhibitor, PD98059, had no effect on S100b stimulation of mouse primary EC.

These concepts were extended to human aortic EC. Lentiviral gene transduction was employed to introduce full-length RAGE (to further increase human RAGE expression) or DN RAGE into primary cultures of human aortic EC. S100b induced a significant increase in monolayer permeability of full-length RAGE-expressing EC compared to unstimulated, full-length, RAGE-expressing cells.[29] In DN RAGE-expressing human EC, S100b failed to increase monolayer permeability.[29] In addition, siRNA was employed to suppress RAGE expression in human aortic EC. RAGE siRNA suppressed S100b-stimulated upregulation of VCAM-1 compared to cells treated with scramble siRNA and S100b. The impact of S100b on upregulation of VCAM-1 was dependent on JNK MAP kinase signaling, as pretreatment with SP600125 significantly reduced VCAM-1 antigen, as did introduction of siRNA to knockdown JNK expression.[29]

Conclusions and Future Directions

Studies from atherosclerosis to the complications of diabetes to experiments in purified inflammatory and EC strongly implicate RAGE in amplification of inflammatory mechanisms linked to cellular injury in each of these settings. Most importantly, experiments in atherosclerosis demonstrate that RAGE regulates expression of VCAM-1 and MCP-1 in the atherosclerosis-vulnerable aorta. Thus, the link between activated inflammatory cells and primed endothelium suggest that RAGE is a key interface mediating vascular permeability, infiltration, and activation of inflammatory cells into the vessel wall, and then subsequently injury phenotypes. In aorta-derived vascular rings, deletion of RAGE or blockade of RAGE signaling in EC significantly improved the response to acetylcholine in apoE-null vascular tissue. In atherosclerosis, experiments using pharmacological antagonism of ligand–RAGE, and genetic modulation of RAGE support the premise that RAGE critically impacts vascular inflammation.

The next steps in dissecting the role of RAGE in vascular inflammation and atherosclerosis include intense focus on the role of RAGE and RAGE signaling in T and B lymphocytes, macrophages, and DC in vascular injury, both in the absence and presence of hyperglycemia. To dissect the underlying mechanisms, we must take the next step forward by dissecting, in intricate detail, the role of RAGE and RAGE signaling in each of these cell types alone and via their interaction with EC in the pathogenesis of vascular inflammation and atherosclerosis.

Acknowledgments

The authors gratefully acknowledge the Juvenile Diabetes Research Foundation and the United States Public Health Service for support of this work.

Conflict of Interest

The authors declare no conflicts of interest.

References

1. RAMASAMY, R., S.J. VANNUCCI, S.S. YAN, et al. 2005. Advanced glycation end products and RAGE: a common thread in aging, diabetes, neurodegeneration, and inflammation. Glycobiology **15:** 16R–28R.
2. SCHMIDT, A.M., M. VIANNA, M. GERLACH, et al. 1992. Isolation and characterization of binding proteins for advanced glycosylation endproducts from lung tissue which are present on the endothelial cell surface. J. Biol. Chem. **267:** 14987–14997.
3. HOFMANN, M.A., S. DRURY, C. FU, et al. 1999. RAGE mediates a novel proinflammatory axis: a central cell surface receptor for S100/calgranulin polypeptides. Cell **97:** 889–901.
4. FOELL, D., H. WITTKOWSKI & J. ROTH. 2007. Mechanisms of disease: a 'DAMP' view of inflammatory arthritis. Nat. Clin. Pract. Rheumatol. **3:** 382–390.
5. HORI, O., J. BRETT, T. SLATTERY, et al. 1995. The receptor for advanced glycation endproducts (RAGE) is a cellular binding site for amphoterin: mediation of neurite outgrowth and co-expression of RAGE and amphoterin in the developing nervous system. J. Biol. Chem. **270:** 25752–25761.
6. TAGUCHI, A., D.C. BLOOD, G. DEL TORO, et al. 2000. Blockade of amphoterin/RAGE signalling suppresses tumor growth and metastases. Nature **405:** 354–360.
7. WANG, H., O. BLOOM, M. ZHANG, et al. 1999. HMG-1 as a late mediator of endotoxin lethality in mice. Science **285:** 248–251.
8. KOKKOLA, R., J. LI, E. SUNDBERG, et al. 2003. Successful treatment of collagen-induced arthritis in mice and rats by targeting extracellular high mobility group box chromosomal protein 1 activity. Arthritis Rheum. **48:** 2052–2058.
9. HOFMANN, M.A., S. DRURY, B.I. HUDSON, et al. 2002. RAGE and arthritis: the G82S polymorphism amplifies the inflammatory response. Genes Immun. **3:** 123–135.
10. PULLERITS, R., I.M. JONSSON, M. VERDRENGH, et al. 2003. High mobility group box chromosomal protein 1, a DNA binding cytokine, induces arthritis. Arthritis Rheum. **48:** 1693–1700.
11. CHAVAKIS, T., A. BIERHAUS, N. AL-FAKHRI, et al. 2003. The Pattern Recognition Receptor (RAGE) is a counterreceptor for leukocyte integrins: a novel pathway for inflammatory cell recruitment. J. Exp. Med. **198:** 1507–1515.
12. MARSCHE, G., M. SEMLITSCH, A. HAMMER, et al. 2007. Hypochlorite-modified albumin colocalizes with RAGE in the artery wall and promotes MCP-1 expression via the RAGE-ERK 1/2 MAP-kinase pathway. FASEB J. **21:** 1145–1152.
13. ZENG, S., N. FEIRT, M. GOLDSTEIN, et al. 2004. Blockade of receptor for advanced glycation endproducts (RAGE) attenuates ischemia and reperfusion injury in the liver in mice. Hepatology **39:** 422–432.
14. SAKAGUCHI T., S.F. YAN, S.D. YAN, et al. 2003. Arterial restenosis: central role of RAGE-dependent neointimal expansion. J. Clin. Invest. **111:** 959–972.
15. BIANCHI, R., C. ADAMI, I. GIAMANCO, et al. 2007. S100b binding to RAGE in microglia stimulates cox-2 expression. J. Leukoc. Biol. **81:** 108–118.
16. LANDER, H.L., J.M. TAURAS, J.S. OGISTE, et al. 1997. Activation of the Receptor for Advanced Glycation Endproducts triggers a MAP Kinase pathway regulated by oxidant stress. J. Biol. Chem. **272:** 17810–17814.
17. BRETT, J., A.M. SCHMIDT, Y.S. ZOU, et al. 1993. Tissue distribution of the receptor for advanced glycation endproducts (RAGE): expression in smooth muscle, cardiac myocytes, and neural tissue in addition to the vasculature. Am. J. Pathol. **143:** 1699–1712.
18. PARK, L., K.G. RAMAN, K.J. LEE, et al. 1998. Suppression of accelerated diabetic atherosclerosis by soluble Receptor for AGE (sRAGE). Nat. Med. **4:** 1025–1031.
19. WENDT, T.M., N. TANJI, J. GUO, et al. 2003. RAGE drives the development of glomerulosclerosis and implicates podocyte activation in the pathogenesis of diabetic nephropathy. Am. J. Pathol. **162:** 1123–1137.
20. BARILE, G.R., S.I. PACHYDAKI, S.R. TARI, et al. 2005. The RAGE axis in early diabetic retinopathy. Invest. Ophthalmol. Vis. Sci. **46:** 2916–2924.
21. CHEN, Y., S.S. YAN, J. COLGAN, et al. 2004. Blockade of late stages of autoimmune diabetes by inhibition of the receptor for advanced glycation end products. J. Immunol. **173:** 1399–1405.
22. MOSER, B., M.J. SZABOLCS, H.J. ANKERSMIT, et al. 2007. Blockade of RAGE suppresses alloimmune reactions in vitro and delays allograft rejection in heart transplantation. Am. J. Transplant. **7:** 293–302.
23. YAN, S.S.D., Z.Y. WU, H.P. ZHANG, et al. 2003. Suppression of experimental autoimmune encephalomyelitis by selective blockade of encephalitogenic T-cell infiltration of the central nervous system. Nat. Med. **9:** 287–293.
24. MOSER, B., D. D. DESAI, M. P. DOWNEY, et al. 2007. Receptor for advanced glycation endproducts expression on T cells

contributes to antigen-specific cellular expansion in vivo. J. Immunol. **179:** 8051–8058.
25. DUMITRIU, I.E., P. BARUAH, B. VALENTINIS, *et al.* 2005. Release of high mobility group box 1 by dendritic cells controls T cell activation via the receptor for advanced glycation end products. J. Immunol. **174:** 7506–7515.
26. KISLINGER T, N. TANJI, T. WENDT, *et al.* 2001. RAGE mediates inflammation and enhanced expression of tissue factor in the vasculature of diabetic apolipoprotein E null mice. Arterio. Thromb. Vasc. Biol. **21:** 905–910.
27. BUCCIARELLI, L.G., T. WENDT, W. QU, *et al.* 2002. RAGE blockade stabilizes established atherosclerosis in diabetic apolipoprotein E null mice. Circulation **106:** 2827–2835.
28. WENDT, T., E. HARJA, L. BUCCIARELLI, *et al.* 2006. RAGE modulates vascular inflammation and atherosclerosis in a murine model of type 2 diabetes. Atherosclerosis **185:** 70–77.
29. HARJA, E., D. X. BU, B.I. HUDSON, *et al.* 2008. Vascular and inflammatory stresses mediate atherosclerosis via RAGE and its ligands in apoE null mice. J. Clin. Invest. **118:** 183–194.

Central Nervous System Regulation of Energy Metabolism

Ghrelin versus Leptin

RUBEN NOGUEIRAS,[a] MATTHIAS H. TSCHÖP,[a,b] AND JEFFREY M. ZIGMAN[c]

[a]*Obesity Research Center, Department of Psychiatry, University of Cincinnati, Cincinnati, Ohio, USA*

[b]*Department of Pharmacology, German Institute of Human Nutrition Potsdam-Rehbruecke, Nuthetal, Germany*

[c]*Division of Hypothalamic Research and Division of Endocrinology & Metabolism, Department of Internal Medicine, University of Texas Southwestern Medical Center, Dallas, Texas, USA*

In this brief review, we introduce some major themes in the regulation of energy, lipid, and glucose metabolism by the central nervous system (CNS). Rather than comprehensively discussing the field, we instead will discuss some of the key findings made regarding the interaction of the hormones ghrelin and leptin with the CNS.

Key words: ghrelin; leptin; central nervous system; energy balance equation; energy partitioning; melanocortin

Ghrelin

Ghrelin is a peptide hormone first described in 1999 as the endogenous ligand of the growth hormone secretagogue receptor (GHSR; ghrelin receptor).[1] Endocrine cells lining the stomach and proximal small intestine serve as the source of most of the ghrelin that circulates in the blood. Although ghrelin was initially identified because of its potent GH secretory actions, a number of other potential functions for ghrelin have since been elucidated, including roles in gastrointestinal motility, gastric acid secretion, and various cardiovascular, immunological, reproductive, and behavioral processes.[2] Ghrelin also seems to be an important regulator of glucose homeostasis. For instance, infusions of ghrelin increase blood glucose, reduce glucose tolerance, and restrict insulin secretion via mechanisms that involve, at least in part, direct action on the pancreatic islets of Langerhans.[3] On the other hand, deletion of ghrelin augments glucose-induced insulin secretion and increases peripheral insulin sensitivity.[4]

Ghrelin Regulates Energy Balance

Perhaps most attention on ghrelin action has been placed on its role in regulating body weight. From a thermodynamic perspective, the regulation of body weight can be described as a linear equation balancing both food intake and energy expenditure (energy balance) to derive the amount of fat stored. Under normal circumstances, a balance in energy intake and energy expenditure results in body weight maintenance. Weight loss would occur if there is a relative increase in energy expenditure and/or a relative decrease in food intake. On the other hand, weight gain would be expected if there is a relative increase in food intake and/or a relative decrease in energy expenditure. Very shortly after its discovery, it was reported that intracerebroventricular injections of ghrelin strongly stimulated feeding in rats.[5] Furthermore, ghrelin was shown to decrease oxygen consumption.[6] As one would predict for an orexigenic agent that also decreases energy expenditure, both peripherally and centrally administered ghrelin produce a positive energy balance and lead to increased body weight gain; this seems to be primarily a result of an increase in fat mass.[5,7]

We now know that ghrelin's effects on body weight likely include many actions in addition to its direct effects on food intake and decreased energy

Address for correspondence: Jeffrey M. Zigman, UT Southwestern Medical Center, 5323 Harry Hines Boulevard, MC 9077, Dallas, TX 75390-9077. Voice: +1-214-648-6422; fax: +1-214-648-5612.

jeffrey.zigman@utsouthwestern.edu

expenditure. These include effects on energy partitioning, which indirectly influences the other factors in the energy balance equation. As an example of energy partitioning, fat may be directed to either oxidative processes in skeletal muscle or brown adipose tissue (BAT) or to triglyceride storage in white adipose tissue (WAT). Recent reports have shown that central ghrelin administration increases the respiratory quotient, thus indicating decreased use of lipids for the generation of energy.[8] Furthermore, ghrelin increases the expression of fat storage enzymes in WAT and decreases the expression of thermogenesis-related uncoupling proteins in BAT.[8] Importantly, these latter effects were shown to occur after intracerebroventricular administration of ghrelin (and therefore are likely centrally mediated via activation of the sympathetic nervous system) and occurred independently from ghrelin-induced hyperphagia.[8] Finally, ghrelin has also been shown to shift food preference toward diets high in fat.[9]

Role of the Endogenous Ghrelin Signaling System in Body Weight Homeostasis

Although the previously described experiments demonstrate that ghrelin has the ability to affect many behaviors and processes connected to energy metabolism, they primarily rely on exogenous administration of ghrelin and therefore do not address the importance or requirement of ghrelin signaling for those processes. Some of the initial clues as to the role of the endogenous ghrelin signaling system come from studies in which healthy subjects who were provided with meals on a fixed schedule were shown to have plasma ghrelin levels that increased nearly twofold immediately before each meal and then fell within 1 h after eating.[10] The clear preprandial rise and postprandial fall in plasma ghrelin levels support the hypothesis that ghrelin plays a physiological role in meal initiation in humans.

This question has also been examined by the use of various animal models in which either ghrelin or GHSR has been genetically deleted. Using GHSR-KO mice, it could be convincingly shown that GHSR is the relevant receptor for the stimulatory effects of ghrelin on food intake and GH secretion, since in contrast to wild-type mice, acute treatment of GHSR-KO mice with ghrelin had no effect on GH release or food intake.[11,12] Although the first few loss-of-function studies were not supportive of a significant role for endogenous ghrelin in the metabolic adaptation to nutrient availability,[13,14] subsequent studies have demonstrated that intact ghrelin signaling systems do appear to be required for normal body weight homeostasis and, in particular, the full development of diet-induced obesity. For instance, when fed a high-fat diet, both female and male GHSR-null mice ate less food, stored less of their consumed calories, preferentially used fat as an energy substrate, and accumulated less body weight and adiposity than control mice; similar effects on body weight and adiposity were also observed in female, but not male, GHSR-null mice fed standard chow.[12] Male ghrelin-KO mice were also protected from the rapid weight gain induced by early exposure to a high-fat diet; this reduced weight gain was associated with decreased adiposity and increased energy expenditure and locomotor activity.[14] The main findings of decreased adiposity and body weight, decreased food intake, and/or increased energy expenditure observed in the most recent ghrelin and GHSR deletion studies have also been observed in other models of the ghrelin signaling blockade, including RNA-mediated and vaccine-mediated neutralization of acylated ghrelin and use of GHSR antagonists.[15–17]

Interestingly, increased signaling through GHSR has not been shown to be the cause for most forms of human obesity. Most obese humans tend to have ghrelin levels that are lower than those in lean individuals.[18] In fact, there seems to be a negative relationship between ghrelin levels and body mass index. This inverse relationship remains intact even during diet-induced weight loss, which increases the circulating level of ghrelin in both healthy individuals[19] and obese subjects.[20] One interpretation of these findings is that ghrelin secretion is reduced in obesity states to reduce orexigenic stimulation. Further support for this model is derived from studies in patients with anorexia nervosa who have elevated plasma ghrelin concentrations that return to a normal range after partial weight gain.[21]

Ghrelin and Food-seeking Behaviors

One exception to the negative correlation between body weight and circulating ghrelin level occurs in Prader-Willi Syndrome (PWS). This disorder, which is characterized by a constellation of certain facial features, cognitive dysfunction, hypogonadism, and severe obesity plus hyperphagia, is also associated with marked elevations of ghrelin.[22] The hyperphagia of PWS is extreme and is fueled by a nearly constant state of hunger that is evidenced by maladaptive behaviors, including pica, binge eating, hoarding, and foraging for food.[23] These extreme food-seeking behaviors are reminiscent of the types of motivated behaviors observed in individuals with addictions to drugs of abuse and suggest the possibility that the high ghrelin levels may be engaging neurocircuits similar to those activated by drugs of abuse. One such circuit involves a mesolimbic dopaminergic pathway that starts at the ventral tegmental area (VTA) and projects to the nucleus accumbens (NAc). Studies mainly from the drug addiction field have identified the NAc and its dopaminergic inputs from the VTA as playing a critical role in reward-seeking behavior.[24] Virtually all drugs of abuse increase dopaminergic transmission in the NAc, and this is thought to contribute to the acute rewarding effects of the drugs.[24]

Several pieces of evidence suggest that one mechanism by which increases in ghrelin lead to increased food intake is via direct engagement of midbrain dopaminergic pathways. For instance, GHSR is highly expressed in tyrosine hydroxylase neurons within the VTA and substantia nigra.[25] Microinjection of ghrelin into the VTA increases food intake, while injection of GHSR antagonists into the VTA blocks the acute stimulation of food intake induced by ghrelin.[26,27] Ghrelin administration also induces dopamine overflow in the NAc.[27,28] Taken together, all these data suggest that the mesolimbic reward circuitry is a target of ghrelin in the physiological regulation of feeding.

The Hypothalamic Arcuate Nucleus: A Key Site for Ghrelin Action

The VTA–NAc is not the end of the story. In fact, various histological analyses of the distribution of GHSR suggest that there are several other potential sites of ghrelin action, which include several possible sites for ghrelin's actions on food intake and body weight. Included among these is the arcuate nucleus (Arc) located within the hypothalamus. Within the Arc, two distinct ghrelin-responsive cell groups exist.[29,30] The first is orexigenic and coexpresses the peptides neuropeptide Y (NPY) and agouti-related gene product (AgRP). The second distinct population of neurons is identified by the coexpression of proopiomelanocortin (POMC) and cocaine- and amphetamine-regulated transcript (CART) and is often referred to as an anorexigenic population. GHSRs are expressed predominantly on the NPY/AgRP neurons and in much fewer numbers on the POMC/CART neurons.[29,30] These two Arc subpopulations interact with one another and with similar downstream target neurons to effect changes that ultimately help to regulate body weight. Investigations by a number of investigators have demonstrated that ghrelin directly binds to its receptor on the NPY/AgRP neurons to activate them. This has been shown not only by electrophysiology but also by demonstration of ghrelin induction of intracellular calcium influxes and induction of c-fos and NPY + AgRP gene transcription. Activation of the NPY/AgRP neurons has at least three relevant effects: 1) the potent orexigen NPY is released and this results in both increased food intake and decreased energy expenditure; 2) it causes the release of AgRP, which serves as both a competitive antagonist and an inverse agonist to the melanocortin 4 receptor (MC4R) and thus inhibits downstream melanocortin pathways that otherwise decrease food intake and increase energy expenditure; and 3) there is an increased release of the inhibitory neurotransmitter GABA onto neighboring POMC/CART neurons, which in turn causes a decreased release of alpha-melanocyte-stimulating hormone (MSH) (the endogenous ligand for the MC4R) and thus a further reduction in melanocortinergic tone.

Leptin

The same Arc circuitry described above as being important for ghrelin action is engaged by other neurotransmitters and hormones known to affect body weight. These include leptin, which is a hormone whose discovery initiated an explosion in our current knowledge of the central nervous system (CNS) circuits responsible for energy, glucose, and lipid metabolism. The story of the discovery of leptin is an unusual one and likely can be traced back at least to a set of exquisite parabiosis experiments reported by Coleman in 1973.[31] These experiments involved linking the circulatory systems of different combinations of wild-type mice, ob/ob mice (which are obese, hyperphagic, and have increased adiposity), and db/db mice (which are diabetic with obesity, high insulin, and hyperglycemia). Upon parabiosis between a db/db mouse and a wild-type mouse, the wild-type mouse lost weight and eventually died of starvation, while the db/db mouse remained unaffected. Upon parabiosis between an ob/ob mouse and a wild-type mouse, the obese ob/ob mouse manifested decreased food intake (as well as decreased insulin levels and blood sugar), while the wild-type mouse was unaffected. These results were interpreted as evidence that the phenotype of the ob/ob mouse resulted from the failure to produce a satiety factor, whereas the phenotype of the db/db mouse resulted from insensitivity to otherwise elevated levels of the satiety factor. This interpretation was confirmed when an ob/ob mouse and a db/db mouse were parabiosed; while the db/db mouse was unaffected, the ob/ob mouse became hypophagic and died of starvation. The genes causing the extreme phenotypes of the ob/ob mouse and the db/db mouse were cloned in 1994[32] and 1995,[33] respectively. The ob gene product was named leptin and was found to be an anorexigenic hormone produced by fat. The ob/ob mouse resulted from an ob gene mutation causing an absence of functional leptin. The db gene product was found to be the receptor for leptin. The high circulating levels of leptin in the db/db mouse were the effect of compensation for a lack of functional leptin receptor and thus an extreme case of leptin resistance.

The Hypothalamic Arcuate Nucleus: Integration of Leptin and Ghrelin Pathways and Involvement in Glucose Metabolism

As mentioned above, leptin has direct effects on the same hypothalamic Arc neurons engaged by ghrelin. However, as one might predict by the end result of leptin action, leptin's effects on the Arc circuitry are counter to those of ghrelin's. Leptin receptors are found on both POMC/CART neurons and NPY/AgRP neurons. Leptin directly activates the POMC/CART neurons causing release of alpha-MSH, which in turn engages the melanocortin circuitry that decreases food intake and increases energy expenditure. Leptin also directly inhibits NPY/AgRP neurons, which leads to decreased NPY release as well as decreased AgRP and GABA release, the latter causing an increase in melanocortinergic tone.[29] Thus, leptin impacts the energy balance equation by decreasing intake and increasing energy expenditure, which in turn causes a decreased body weight.

The presence of diabetes in the previously described obese mouse models also suggests a role for leptin (and in particular leptin action in the Arc) in glucose homeostasis. In fact, the leptin-engaged melanocortin system plays an important role in the regulation of glucose handling by the body. This has now been shown by a number of studies. For instance, central infusion of MC4R agonists not only decreases intra-abdominal fat but also increases insulin sensitivity—as evidenced by increased peripheral glucose uptake and decreased hepatic glucose production.[34] As another example, restoration of leptin signaling in the Arc of mice that were otherwise deficient in leptin was found to markedly improve hyperinsulinemia and normalize blood glucose levels, thus demonstrating that leptin signaling in the Arc is sufficient for mediating leptin's effects on glucose homeostasis.[35]

Relevance of Leptin-engaged Circuitry to Obesity and Diabetes in Humans

As one might predict from the phenotypes of the ob/ob and db/db mice, interference with the functioning of leptin-engaged circuitry at a number of different places may also lead to obesity and problems with glucose metabolism in humans. Although not common, leptin deficiency does occur in humans and results in hyperphagia, severe obesity, and alterations in immune function and delayed puberty, all of which improve with leptin administration.[36,37] The human equivalent of the db/db mouse occurs much more commonly. In fact, the prevalence of pathogenic leptin receptor mutations in a cohort of 300 subjects with severe early onset obesity was found to be 3%.[38,39] Even more common is a resistance, or rather impaired responsiveness, to the effects of leptin, which is observed in most obese humans. Human MC4R mutation carriers also exist and have severe obesity, increased lean mass, increased linear growth, hyperphagia, and severe hyperinsulinemia. Mutations in the MC4R appear to be the commonest monogenic cause of obesity thus far described in humans.[40]

Conclusion

We have come a long way from the seminal parabiosis experiments of Coleman[31] in our knowledge of the regulation of energy, lipid, and glucose metabolism by the CNS. These early mouse experiments as well as more recent ones with genetic and pharmacologic manipulation of key components of signaling pathways shared by the hormones ghrelin and leptin have revealed crucial information that could be translated into new and effective treatment strategies for obesity and diabetes in the future.

Conflict of Interest

The authors declare no conflicts of interest.

References

1. KOJIMA, M., H. HOSODA, Y. DATE, et al. 1999. Ghrelin is a growth-hormone-releasing acylated peptide from stomach. Nature **402:** 656–660.
2. VAN DER LELY, A.J., M. TSCHOP, M.L. HEIMAN & E. GHIGO. 2004. Biological, physiological, pathophysiological, and

pharmacological aspects of ghrelin. Endocr. Rev. **25:** 426–457.
3. DEZAKI, K., H. HOSODA, M. KAKEI, *et al.* 2004. Endogenous ghrelin in pancreatic islets restricts insulin release by attenuating Ca2+ signaling in beta-cells: implication in the glycemic control in rodents. Diabetes **53:** 3142–3151.
4. SUN, Y., M. ASNICAR, P.K. SAHA, *et al.* 2006. Ablation of ghrelin improves the diabetic but not obese phenotype of ob/ob mice. Cell Metab. **3:** 379–386.
5. NAKAZATO, M., N. MURAKAMI, Y. DATE, *et al.* 2001. A role for ghrelin in the central regulation of feeding. Nature **409:** 194–198.
6. ASAKAWA, A., A. INUI, T. KAGA, *et al.* 2001. Ghrelin is an appetite-stimulatory signal from stomach with structural resemblance to motilin. Gastroenterology **120:** 337–345.
7. TSCHOP, M., D.L. SMILEY & M.L. HEIMAN. 2000. Ghrelin induces adiposity in rodents. Nature **407:** 908–913.
8. THEANDER-CARRILLO, C., P. WIEDMER, P. CETTOUR-ROSE, *et al.* 2006. Ghrelin action in the brain controls adipocyte metabolism. J. Clin. Invest. **116:** 1983–1993.
9. SHIMBARA, T., M.S. MONDAL, T. KAWAGOE, *et al.* 2004. Central administration of ghrelin preferentially enhances fat ingestion. Neurosci. Lett. **369:** 75–79.
10. CUMMINGS, D.E., J.Q. PURNELL, R.S. FRAYO, *et al.* 2001. A preprandial rise in plasma ghrelin levels suggests a role in meal initiation in humans. Diabetes **50:** 1714–1719.
11. SUN, Y., P. WANG, H. ZHENG & R.G. SMITH. 2004. Ghrelin stimulation of growth hormone release and appetite is mediated through the growth hormone secretagogue receptor. Proc. Natl. Acad. Sci. USA **101:** 4679–4684.
12. ZIGMAN, J.M., Y. NAKANO, R. COPPARI, *et al.* 2005. Mice lacking ghrelin receptors resist the development of diet-induced obesity. J. Clin. Invest. **115:** 3564–3572.
13. SUN, Y., S. AHMED & R.G. SMITH. 2003. Deletion of ghrelin impairs neither growth nor appetite. Mol Cell Biol. **23:** 7973–7981.
14. WORTLEY, K.E., J.P. DEL RINCON, J.D. MURRAY, *et al.* 2005. Absence of ghrelin protects against early-onset obesity. J. Clin. Invest. **115:** 3573–3578.
15. SHEARMAN, L.P., S.P. WANG, S. HELMLING, *et al.* 2006. Ghrelin neutralization by a ribonucleic acid-SPM ameliorates obesity in diet-induced obese mice. Endocrinology **147:** 1517–1526.
16. ZORRILLA, E.P., S. IWASAKI, J.A. MOSS, *et al.* 2006. Vaccination against weight gain. Proc. Natl. Acad. Sci. USA **103:** 13226–13231.
17. ESLER, W.P., J. RUDOLPH, T.H. CLAUS, *et al.* 2007. Small-molecule ghrelin receptor antagonists improve glucose tolerance, suppress appetite, and promote weight loss. Endocrinology **148:** 5175–5185.
18. TSCHOP, M., C. WEYER, P.A. TATARANNI, *et al.* 2001. Circulating ghrelin levels are decreased in human obesity. Diabetes **50:** 707–709.
19. RAVUSSIN, E., M. TSCHOP, S. MORALES, *et al.* 2001. Plasma ghrelin concentration and energy balance: overfeeding and negative energy balance studies in twins. J. Clin. Endocrinol. Metab. **86:** 4547–4551.
20. HANSEN, T.K., R. DALL, H. HOSODA, *et al.* 2002. Weight loss increases circulating levels of ghrelin in human obesity. Clin. Endocrinol. (Oxf.) **56:** 203–206.
21. OTTO, B., U. CUNTZ, E. FRUEHAUF, *et al.* 2001. Weight gain decreases elevated plasma ghrelin concentrations of patients with anorexia nervosa. Eur. J. Endocrinol. **145:** 669–673.
22. CUMMINGS, D.E., K. CLEMENT, J.Q. PURNELL, *et al.* 2002. Elevated plasma ghrelin levels in Prader Willi syndrome. Nat. Med. **8:** 643–644.
23. GOLDSTONE, A.P. 2004. Prader-Willi syndrome: advances in genetics, pathophysiology and treatment. Trends Endocrinol. Metab. **15:** 12–20.
24. HYMAN, S.E., R.C. MALENKA & E.J. NESTLER. 2006. Neural mechanisms of addiction: the role of reward-related learning and memory. Annu. Rev. Neurosci. **29:** 565–598.
25. ZIGMAN, J.M., J.E. JONES, C.E. LEE, *et al.* 2006. Expression of ghrelin receptor mRNA in the rat and the mouse brain. J. Comp. Neurol. **494:** 528–548.
26. NALEID, A.M., M.K. GRACE, D.E. CUMMINGS & A.S. LEVINE. 2005. Ghrelin induces feeding in the mesolimbic reward pathway between the ventral tegmental area and the nucleus accumbens. Peptides **26:** 2274–2279.
27. ABIZAID, A., Z.W. LIU, Z.B. ANDREWS, *et al.* 2006. Ghrelin modulates the activity and synaptic input organization of midbrain dopamine neurons while promoting appetite. J. Clin. Invest. **116:** 3229–3239.
28. JERLHAG, E., E. EGECIOGLU, S.L. DICKSON, *et al.* 2006. Ghrelin stimulates locomotor activity and accumbal dopamine-overflow via central cholinergic systems in mice: implications for its involvement in brain reward. Addict. Biol. **11:** 45–54.
29. ZIGMAN, J.M. & J.K. ELMQUIST. 2003. Minireview: from anorexia to obesity–the yin and yang of body weight control. Endocrinology **144:** 3749–3756.
30. MORTON, G.J., D.E. CUMMINGS, D.G. BASKIN, *et al.* 2006. Central nervous system control of food intake and body weight. Nature **443:** 289–295.
31. COLEMAN, D.L. 1973. Effects of parabiosis of obese with diabetes and normal mice. Diabetologia **9:** 294–298.
32. ZHANG, Y., R. PROENCA, M. MAFFEI, *et al.* 1994. Positional cloning of the mouse obese gene and its human homologue. Nature **372:** 425–432.
33. TARTAGLIA, L.A., M. DEMBSKI, X. WENG, *et al.* 1995. Identification and expression cloning of a leptin receptor, OB-R. Cell **83:** 1263–1271.
34. OBICI, S., Z. FENG, J. TAN, *et al.* 2001. Central melanocortin receptors regulate insulin action. J. Clin. Invest. **108:** 1079–1085.
35. COPPARI, R., M. ICHINOSE, C.E. LEE, *et al.* 2005. The hypothalamic arcuate nucleus: a key site for mediating leptin's effects on glucose homeostasis and locomotor activity. Cell Metab. **1:** 63–72.
36. MONTAGUE, C.T., I.S. FAROOQI, J.P. WHITEHEAD, *et al.* 1997. Congenital leptin deficiency is associated with severe early-onset obesity in humans. Nature **387:** 903–908.
37. FAROOQI, I.S., S.A. JEBB, G. LANGMACK, *et al.* 1999. Effects of recombinant leptin therapy in a child with congenital leptin deficiency. N. Engl. J. Med. **341:** 879–884.
38. FAROOQI, I.S., E. BULLMORE, J. KEOGH, *et al.* 2007. Leptin regulates striatal regions and human eating behavior. Science **317:** 1355.

39. CLEMENT, K., C. VAISSE, N. LAHLOU, et al. 1998. A mutation in the human leptin receptor gene causes obesity and pituitary dysfunction. Nature **392:** 398–401.

40. FAROOQI, I.S., J.M. KEOGH, G.S. YEO, et al. 2003. Clinical spectrum of obesity and mutations in the melanocortin 4 receptor gene. N. Engl. J. Med. **348:** 1085–1095.

Determination of N$^\varepsilon$-(Carboxymethyl)lysine in Foods and Related Systems

JENNIFER M. AMES

School of Biological Sciences, Queen's University Belfast, Belfast, Northern Ireland

The sensitive and specific determination of advanced glycation end products (AGEs) is of considerable interest because these compounds have been associated with pro-oxidative and proinflammatory effects *in vivo*. AGEs form when carbonyl compounds, such as glucose and its oxidation products, glyoxal and methylglyoxal, react with the ε-amino group of lysine and the guanidino group of arginine to give structures including N$^\varepsilon$-(carboxymethyl)lysine (CML), N$^\varepsilon$-(carboxyethyl)lysine, and hydroimidazolones. CML is frequently used as a marker for AGEs in general. It exists in both the free or peptide-bound forms. Analysis of CML involves its extraction from the food (including protein hydrolysis to release any peptide-bound adduct) and determination by immunochemical or instrumental means. Various factors must be considered at each step of the analysis. Extraction, hydrolysis, and sample clean-up are all less straight forward for food samples, compared to plasma and tissue. The immunochemical and instrumental methods all have their advantages and disadvantages, and no perfect method exists. Currently, different procedures are being used in different laboratories, and there is an urgent need to compare, improve, and validate methods.

Key words: N$^\varepsilon$-(carboxymethyl)lysine; analysis; advanced glycation end products; dietary N$^\varepsilon$-(carboxymethyl)lysine

Formation of Dietary Ages

When sugars or their oxidation products, e.g., glyoxal (GO), methylglyoxal (MGO), and 3-deoxyglucosone (3-DG), react with protein, various adducts are formed on the side chains of amino acid residues, especially the ε-amino group of lysine and the guanidino group of arginine. These adducts are called advanced glycation end products (AGEs), and the structures that have been identified in food have been reviewed.[1,2] Some of the better known AGEs are N$^\varepsilon$-(carboxymethyl)lysine (CML), N$^\varepsilon$-(carboxyethyl)lysine (CEL), and the hydroimidazolones (HIs).[3] FIGURE 1 summarizes the chemistry of formation of these AGEs from glucose. When glucose itself condenses with the ε-amino group of lysine, the Amadori rearrangement product (ARP) fructoselysine (FL) is produced as an unstable intermediate. It subsequently undergoes oxidation to form CML. GO can react directly with lysine giving another route to CML. In an analogous reaction, MGO forms CEL. GO, MGO, and 3-DG also react with the guanidino group of arginine to give a series of HIs.

The formation of AGEs in food and other biological systems is of great interest because they have been associated with pro-oxidative, proinflammatory, and other effects, especially in diabetes and chronic renal insufficiency.[4] CML has been reported to be a relatively abundant AGE in the biomedical field, and it has been extensively studied since its identification more than 20 years ago as an oxidation product of FL.[5] For these reasons, it is frequently selected as the compound of choice in studies when a single AGE is used as a marker of total AGEs in a system.

CML was the first AGE to be identified in food (milk and milk products)[6] and is formed by a number of pathways in food systems (FIG. 2). In addition to its formation from the ARP or GO formed from sugar, lipid and ascorbic acid are precursors of CML. GO formed during lipid peroxidation can yield CML,[7] while ascorbic acid may undergo oxidation to yield erythrose, which may itself take part in Schiff base and ARP formation and oxidation leading to CML.[8]

Significance of Dietary Ages

The impact of dietary AGEs on human health is currently a hotly debated topic that is addressed by other

---

Address for correspondence: Professor J.M. Ames, School of Biological Sciences, Queen's University Belfast, David Keir Building, Stranmillis Road, Belfast BT9 5AG Northern Ireland. Voice: +44 (0)28 9097 6539; fax: +44 (0)28 9097 6513.

j.m.ames@qub.ac.uk

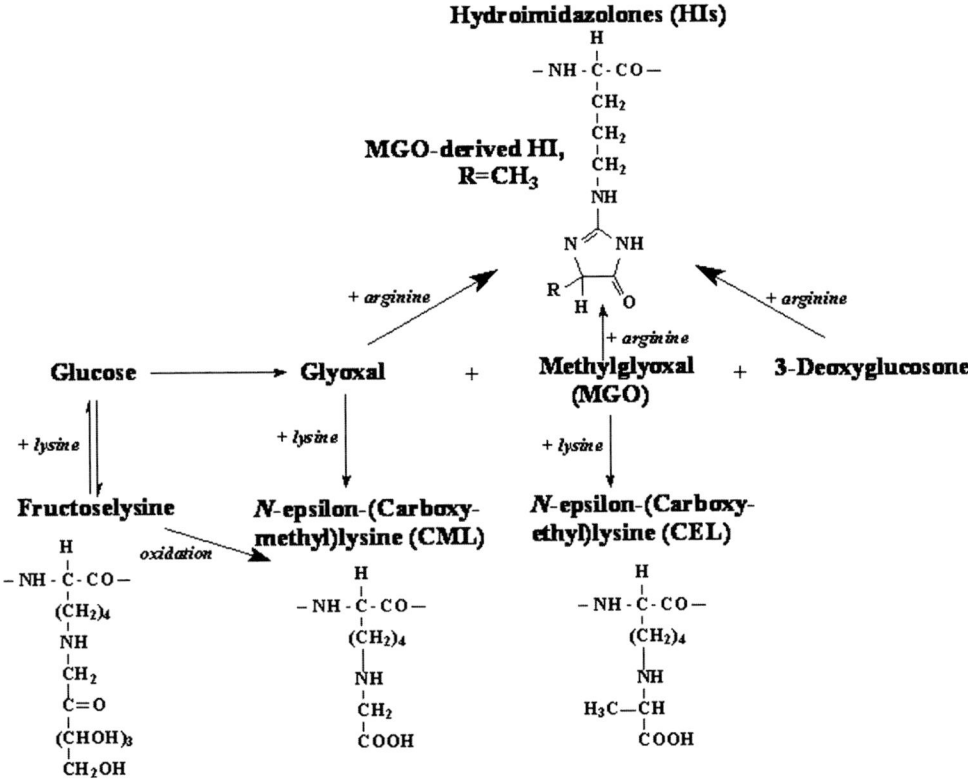

FIGURE 1. Summary of the formation of N^ε-(carboxymethyl)lysine (CML), N^ε-(carboxyethyl)lysine (CEL), and hydroimidazolones (HIs) from glucose and lysine or arginine.

papers in this volume and was discussed at a European COST Action 927 workshop in 2006.[4,9] Controversy exists because although a heat-treated dietary pattern can lead to increased levels of plasma reactive oxygen species and proinflammatory markers, the reasons for these effects are less obvious. Also, heat processing can result in beneficial effects (e.g., generation of antioxidant activity is attributed to reductones and melanoidins that are Maillard reaction products,[10,11] although their significance *in vivo* remains to be established). Adverse effects may be due to increased levels of absorbed dietary AGEs, but other factors may be responsible. For example, increased levels of dietary oxidized lipid, decreased levels of dietary antioxidants or other micronutrients, or a combination of these are consequences of heat processing of food. Nevertheless, even if dietary AGEs, *per se*, do not exert effects that may impact on human health, the other thermally induced dietary modifications may lead to different levels of endogenously formed AGEs. For example, dietary lipid may react with physiological protein, or lowered levels of endogenous antioxidants, such as glutathione, may promote the formation of reactive oxygen species, which in turn can increase oxidation of glucose and lipid, yielding carbonyls that are precursors of AGEs.

Analysis of N^ε-(carboxymethyl)lysine in Foods

Extraction of CML

For most foods, some procedure capable of extracting the CML will be required to eliminate matrix interferences, e.g., from lipid, before its determination in the extract. Less sample clean-up, yielding less 'clean' extracts, may be adequate for subsequent determination using immunochemical methods (e.g., enzyme-linked immunosorbent assay [ELISA]), while rigorous clean-up, including passage through a solid-phase extraction column, will be necessary for analysis by instrumental procedures (gas chromatography [GC], HPLC). Thus, samples, such as skimmed milk, may require no clean-up prior to analysis by ELISA.

CML exists in food in two forms: free and bound to lysine residues within peptides and protein. Free CML may be extracted from liquid foods, such as fruit juices and meat gravies, by filtration through paper

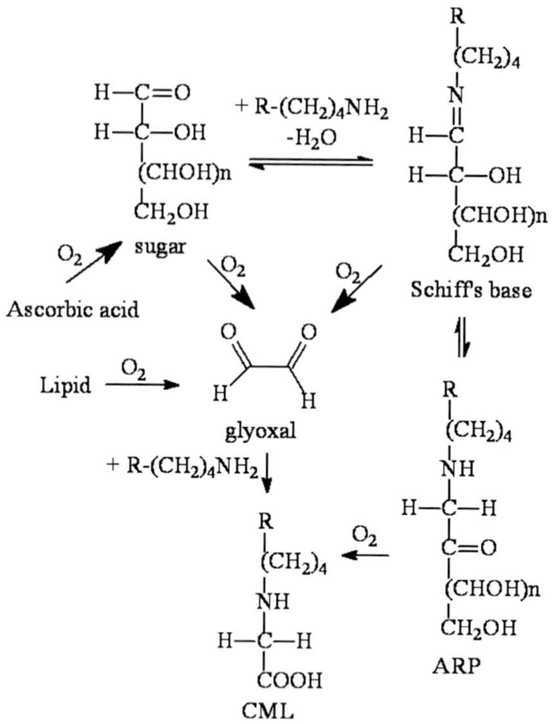

FIGURE 2. Pathways to CML in food systems.

to remove suspended particles, followed by ultrafiltration to remove any macromolecular material. When solid foods have to be analyzed, any free CML will be removed by extraction with acidified water.

When the lysine residues in protein have been modified to CML, the first step in the preparation of a CML extract is to isolate the protein. Various methods are available, including ultrafiltration, precipitation of the protein in 20% trichloroacetic acid (TCA), and the Folch extraction. Ultrafiltration without further cleanup is only useful when the lipid content of the liquid food is very low and there are no other high-molecular mass components, but it has been successfully applied to defatted milk.[3] When an aqueous solution or suspension of the food can be produced and the food is not lipid-rich, precipitation in 20% TCA works well. For foods with a high lipid content, such as cheese with a lipid content of >20%, a good choice is the Folch extraction involving extraction with a mixture of chloroform and methanol.[12]

Bound CML is released from protein by hydrolysis in 6 mol L^{-1} HCl or enzymatic digestion. Acid hydrolysis is the easier, less expensive, and more reliable procedure. However, one reported disadvantage is that the ARP is oxidized to CML during hydrolysis using 6 mol L^{-1} HCl.[6,13] Nevertheless, Charisson et al.[14] have recently reported that CML did not form from the ARP during acid hydrolysis when a range of foods was analyzed. Even so, most researchers prepare a suspension of the food or the protein isolated from it and incubate this with sodium borohydride, prior to acid hydrolysis. This stablizes the ARPs by reducing them to their corresponding alcohol forms.

Enzymatic digestion is worthy of consideration when acid-labile AGEs, e.g., HIs and pyrraline, need to be determined in addition to those that are acid-stable (e.g., CML, CEL, and pentosidine). Enzymatic digestion using a cocktail of enzymes has been used successfully in the biomedical field, notably by Thornalley and co-workers.[15] However, while it is likely to be of value for foods in which little protein cross-linking has occurred, for more heavily processed foods (e.g., breadcrust and well-done beefsteak), the protein is unlikely to be completely digested by proteolytic enzymes. When acid-labile AGEs need to be determined, an alternative approach may be to measure levels of acid degradation products of the AGEs.

Determination of N$^\varepsilon$-(carboxymethyl)lysine

Both immunochemical and instrumental methods are available for the determination of CML in food, and the published studies are summarized in TABLE 1. Reports have focused on dairy foods, including infant formulas. The determination of AGEs in infant formulas is of particular interest since these items form the sole or principal source of nutrition. The immunochemical and instrumental methods each have advantages and disadvantages. In general, the immunochemical methods may be less specific and more subject to matrix interference but are more rapid and less expensive. The instrumental methods generally provide more reliable data but require highly trained personnel. The challenge is to provide methods that are sensitive, specific, rapid, reliable, and inexpensive. No current method for CML fulfills all of these criteria.

Liquid chromatography (LC), combined with either fluorescence detection or mass spectrometry (MS), has been used by several groups to detect CML (see TABLE 1). The very first published methods used LC with fluorescence detection and they are still in use more than 20 years later.[20] The conversion to fluorescent derivatives is required when analysis is by fluorescence detection, while volatile derivatives are needed when a GC procedure is used. The use of LC combined with MS increases the sensitivity of the determination while avoiding the need to derivatize adducts. The operation of the mass spectrometer in multiple reaction monitoring (MRM) mode is largely responsible for the improved sensitivity and also selectivity of the

TABLE 1. Summary of methods available for the determination of CML in foods

Method	Key points	Foods	Reference
ELISA	Monoclonal anti-CML antibody	Wide range	16
ELISA	Monoclonal anti-CML antibody (Alteon[a])	Infant formulas and milks	17
ELISA	Monoclonal anti-CML antibody) (MicroCoat Biotechnologies[b])	Infant formulas and breast milk	18
HPLC	Fluorescence detection of OPA derivative	Milks and milk products, cereal foods, Asparagus soup	6, 19
HPLC	Fluorescence detection of OPA derivative	Whole meals	20
GC	Heptafluorobutyryl isobutyl ester	Infant formulas	13
GC-MS	SIM of trifluoroacetyl methyl ester	Milks, infant formulas, cookies, beef, salmon	14
LC-MS/MS	MRM	Milks, colas	3
LC-MS/MS	MRM	Milks, infant formulas	21

Abbreviations: CML, N^ε-(carboxymethyl)lysine; ELISA, enzyme-linked immunosorbent assay; GC, gas chromatography; HPLC, high-performance liquid chromatography; LC, liquid chromatography; MRM, multiple reaction monitoring; MS, mass spectrometry; OPA, o-phthaldialdehyde; SIM, single ion monitoring.
[a] Ramsey, NJ.
[b] Bernried, Germany.

method. MRM experiments are conducted using a triple quadrupole mass spectrometer.[22] In the first mass filter, the parent ion of the adduct is selected for fragmentation. In the second mass filter, a single prominent fragment ion is selected for analysis. This results in most of the ions being eliminated before the collision cell (giving less background and enhanced sensitivity). Theoretically, only the AGE of interest should possess the same parent ion and daughter ion at the relevant retention time (giving increased selectivity). The reliability of the method is improved further by spiking samples with deuterated standards.

A relatively new innovation is ultrapressure liquid chromatography (UPLC).[23] UPLC packings have a particle size of 1.7 μm, compared to typically 5 μm for conventional LC columns. UPLC columns can withstand pressures of approximately 85 MPa and operate using a flow rate of typically 0.2 mL min^{-1}. These features lead to improved sensitivity, selectivity, and speed of analysis compared to conventional LC. UPLC shows considerable promise for the determination of CML and other AGEs.

Conclusion

A range of methods exists for the determination of CML in foods. They all have their advantages and disadvantages, and improvements are steadily being introduced. Currently, there is a need to compare data obtained for the same samples using the different available methods so that methods may be improved and validated.

Acknowledgments

I thank John Baynes and Suzanne Thorpe (University of South Carolina, USA), Thierry Delatour and Timo Buetler (Nestle Research Centre, Switzerland), Monika Pischetsrieder (University of Erlangan, Germany), and Shima Assar, Permal Deo, and Maria Lima (Queen's University Belfast) for many helpful discussions.

Conflict of Interest

The author declares no conflicts of interest.

References

1. HENLE, T. 2003. Ages in food – do they play a role in uremia? Kidney Int. **63**(s84): 145–147.
2. HENLE, T. 2005. Protein-bound advanced glycation endproducts (AGEs) and bioactive amino acid derivatives in foods. Amino Acids **29**: 313–322.
3. AHMED et al. 2005. Assay of advanced glycation endproducts in selected beverages and food by liquid chromatography with tandem mass spectrometric Detection. Mol. Nutr. Food Res. **49**: 691–699.
4. SEBEKOVA, K. & V. SOMOZA. 2007. Dietary advanced glycation endproducts (AGEs) and their health effects – PRO. Mol. Nutr. Food Res. **51**: 1079–1084.
5. AHMED, M.U. et al. 1986. Identification of N^ε-carboxymethyllysine as a degradation product of fructoselysine in glycated protein. J. Biol. Chem. **261**: 4889–4994.
6. HARTKOPF, J. et al. 1994. Determination of N^ε-carboxymethyllysine by a reversed-phase high-performance liquid chromatography method. J. Chrom. A **672**: 242–246.

7. Fu, M.X. et al. 1996. The advanced glycation endproduct, N$^\varepsilon$-(carboxymethyl)lysine, is a product of both lipid peroxidation and glycoxidation reactions. J. Biol. Chem. **271:** 9982–9986.
8. Dunn, J.A. et al. 1990. Reaction of ascorbate with lysine and protein under autoxidizing conditions: Formation of N$^\varepsilon$-(carboxymethyl)lysine by reaction between lysine and products of autoxidation of ascorbate. Biochem. **29:** 10964–10970.
9. Ames, J.M. 2007. Evidence against dietary advanced glycation endproducts being a risk to human health. Mol. Nutr. Food Res. **51:** 1085–1090.
10. Lindenmeir, M. et al. 2002. Structural and functional characterization of pronyl-lysine, a novel protein modification in bread crust melanoidins showing *in vitro* antioxidative and phase I/II enzyme modulating activity. J. Agric. Food Chem. **50:** 6997–7006.
11. Somoza, V. et al. 2005. Influence of feeding malt, bread crust and a pronylated protein on the activity of chemopreventative enzymes and antioxidative defense parameters *in vivo*. J. Agric. Food Chem. **53:** 8176–8182.
12. Folch, J. et al. 1957. A simple method for the isolation and purification of total lipids from animal tissues. J. Biol. Chem. **226:** 497–509.
13. Bueser, W. et al. 1987. Identification and determination of N-ε-carboxymethyllysine by gas-liquid chromatography. J. Chrom. **387:** 515–519.
14. Charisson, A. et al. 2007. Evaluation of a gas chromatography/mass spectrometry method for the quantification of carboxymethyllysine in food samples. J. Chrom. A **1140:** 189–194.
15. Ahmed, N. et al. 2002. Assay of advanced glycation endproducts (AGEs): surveying AGEs by chromatographic assay with derivatisation by aminoquinolyl-N-hydroxysuccimidyl-carbamate and application to N$^\varepsilon$-carboxymethyl-lysine- and N$^\varepsilon$-(1-carboxyethyl)lysine-modified albumin. Biochem. J. **364:** 1–14.
16. Goldberg, T. et al. 2004. Advanced glycoxidation end products in commonly consumed foods. J. Am. Diet. Assoc. **104:** 1287–1291.
17. Birlouez-Aragon, I. et al. 2004. Assessment of protein glycation markers in infant formulas. Food Chem. **87:** 253–259.
18. Dittrich, R. et al. 2006. Concentrations of N$^\varepsilon$-carboxymethyllysine in human breast milk, infant formulas, and urine of infants. J. Agric. Food Chem. **54:** 6924–6928.
19. Drusch, S. et al. 1999. Determination of N$^\varepsilon$-carboxymethyllysine in milk products by a modified reversed-phase HPLC method. Food Chem. **65:** 547–553.
20. Delgado-Andrade, C. et al. 2007. Maillard reaction indicators in diets usually consumed by adolescent population. Mol. Nutr. Food Res. **51:** 341–351.
21. Fenaille, F. et al. 2006. Modifications of milk constituents during processing: a preliminary benchmarking study. Int. Dairy J. **16:** 728–739.
22. Kinter, M. & N.E. Sherman. 2000. Protein Sequencing and Identification Using Tandem Mass Spectrometry. Wiley-Interscience. New York, NY.
23. http://www.waters.com [last accessed 6 October 2007].

Food Anoxia and the Formation of Either Flavor or Toxic Compounds by Amino Acid Degradation Initiated by Oxidized Lipids

FRANCISCO J. HIDALGO AND ROSARIO ZAMORA

Instituto de la Grasa, Consejo Superior de Investigaciones Científicas, Seville, Spain

Amino acid degradation plays an important role in the formation of both flavors and toxic compounds during food processing. These reactions are produced, to a significant extent, as a consequence of the Maillard reaction among amino acids and carbohydrates. However, recent studies have shown that lipids also take part in these reactions. This article reviews the current knowledge of the contribution of lipids to both flavor and toxic-compound formation by amino acid degradation, describing the formation of Strecker aldehydes and the conversion of amino acids into their vinylogous derivatives as a consequence of lipid–amino acid reactions. Current data suggest that amino acids can be converted into either Strecker aldehydes or vinylogous derivatives by many lipid derivatives, which exhibit diverse reactivities for both reactions. Reaction conditions, including the presence of oxygen, also play a major role in the Strecker aldehyde/vinylogous derivative ratio obtained. Nevertheless, the high number of lipid derivatives involved, the different alternative pathways, and the existence of both positive and negative synergisms between lipids and carbohydrates make it hard to predict the effect of reaction conditions in the Strecker aldehyde/vinylogous derivative ratio obtained in complex food systems.

Key words: acrylamide; carbonyl–amine reactions; lipid oxidation; nonenzymatic browning; lipid oxidation products; Strecker degradation; vinylogous derivatives; amino acids

Introduction

Amino acid degradation is an important pathway for the formation of both flavor and toxic compounds in the Maillard reaction.[1] Strecker degradation of amino acids is one of the most important reactions leading to final aroma compounds in foods because it not only produces Strecker aldehydes but also is the origin of pyrazines, among other heterocyclic derivatives.[2] In addition, the Maillard reaction is related to the formation of toxic derivatives. Among them, the conversion of asparagine into its vinylogous derivative acrylamide has received considerable attention in recent years.[3] Nevertheless, different studies have shown that oxidized lipids react with amino compounds analogously to carbohydrates, and identical (or similar) products are frequently produced by the same (or similar) carbonyl–amine reactions.[4] Therefore, oxidized lipids might also be contributing to amino acid degradation in food systems. This article reviews the current knowledge of the contribution of lipids to both flavor and toxic-compound formation by amino acid degradation.

Strecker Degradation of Amino Acids Initiated by Lipids

Tertiary lipid oxidation products (hydroxyalkenals, epoxyalkenals, and unsaturated epoxy keto fatty esters, among others) are able to degrade amino acids to the corresponding Strecker aldehydes analogously to the α-dicarbonyl compounds produced in the Maillard reaction.[5] This reaction is believed to be produced by formation of an imine followed by an electronic rearrangement to produce either the Strecker aldehyde or the α-keto acid, depending on the decarboxylation of the amino acid (FIG. 1).[6]

Secondary lipid oxidation products (alkadienals and ketodienes, among others) are also able to produce this reaction, but these compounds need further oxidation to degrade the amino acid (FIG. 1).[7] Once oxidized, the tertiary lipid oxidation products degrade the amino acids as described above.

Address for correspondence: Francisco J. Hidalgo, Instituto de la Grasa, Avenida Padre García Tejero 4, 41012-Seville, Spain. Voice: +34 954 611 550; fax: +34 954 616 790.
fhidalgo@ig.csic.es

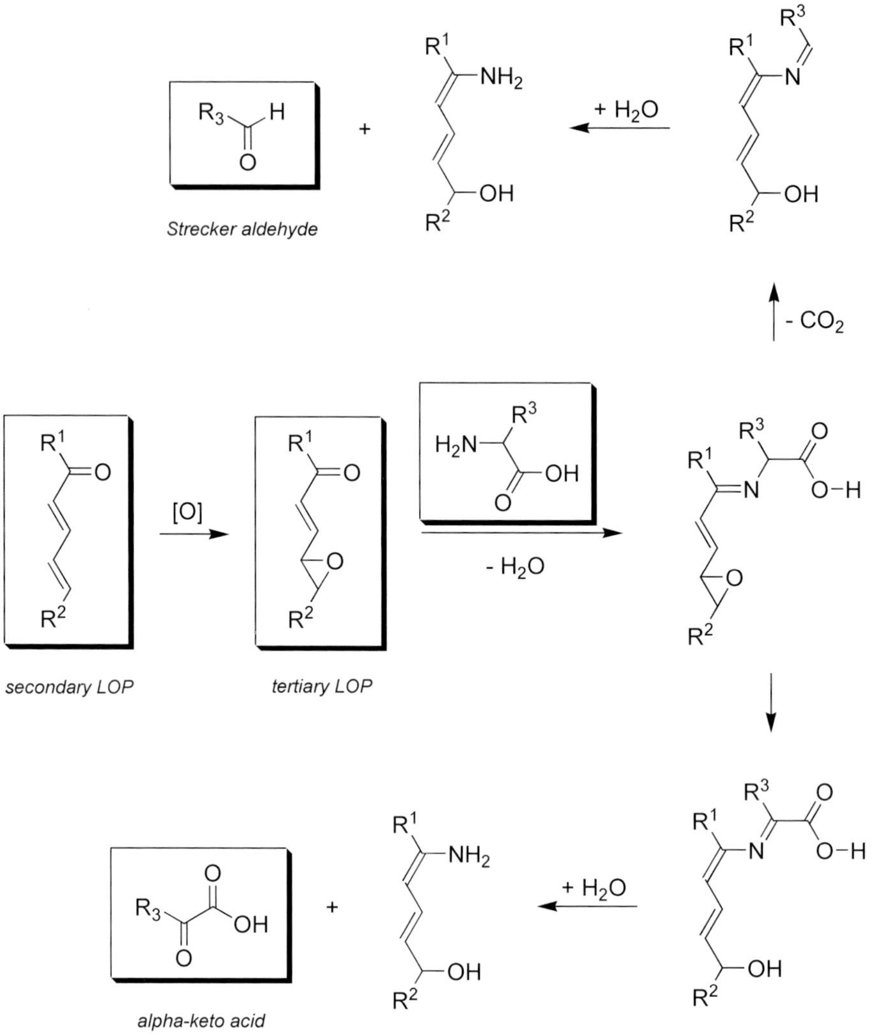

FIGURE 1. Oxidation of secondary lipid oxidation products (LOP) to tertiary LOP, and the later reaction of tertiary LOP with amino acids to produce Strecker aldehydes and α-keto acids.

Primary lipid oxidation products are also able to convert amino acids into their corresponding Strecker aldehydes, and the same has been observed for unsaturated lipids in both model systems and food products.[8]

Degradation of Amino Acids to Their Vinylogous Derivatives Initiated by Lipids

In addition to the formation of Strecker aldehydes, lipid oxidation products are able to degrade amino acids to their corresponding vinylogous derivatives. The most reactive lipid oxidation product for this reaction is the decadienal,[9] but other primary, secondary, and tertiary lipid oxidation products also produce the reaction to different extents.[10] The reaction between decadienal and asparagine is believed to be produced by formation of an imine followed by a β-elimination reaction (FIG. 2).

The reactivity found for different lipid oxidation products for the conversion of asparagine into acrylamide is[10]:

$$DD \gg ED > MeLCO > MeLOOH$$
$$\gg OA \sim MeLEPCO \gg MeLEPOH$$
$$> MeLOH = MeL = 0$$

where DD is 2,4-decadienal, ED is 4,5-epoxy-2-decenal, MeLCO is methyl 13-oxooctadeca-9,11-dienoate, MeLOOH is methyl 13-hydroperoxy-

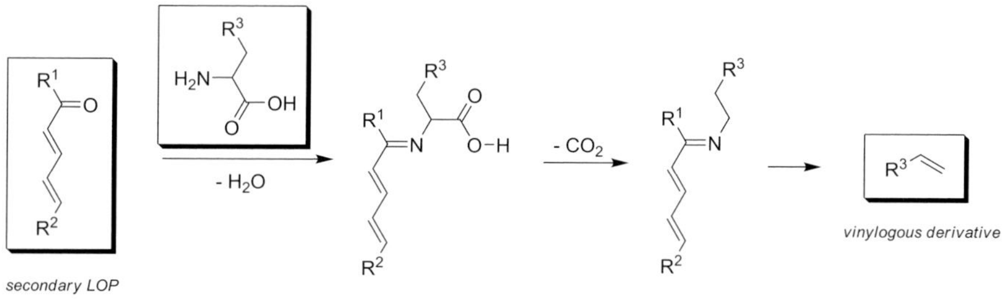

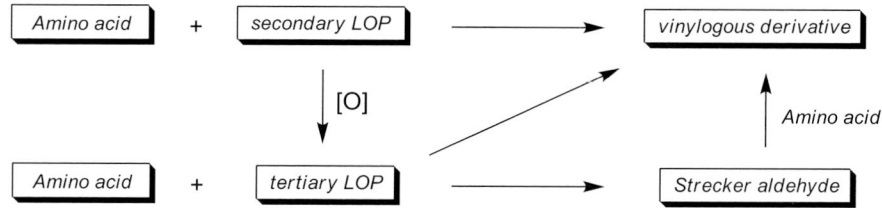

FIGURE 2. Reaction of secondary lipid oxidation products (LOP) with amino acids to produce vinylogous derivatives.

FIGURE 3. Conversion of amino acids into Strecker aldehydes or vinylogous derivatives by reaction with secondary and tertiary lipid oxidation products (LOP) and the role of oxygen in these reactions.

octadeca-9,11-dienoate, OA is 2-octenal, MeLEPCO is methyl 9,10-epoxy-13-oxo-11-octadecenoate, MeLEPOH is methyl 9,10-epoxy-13-hydroxy-11-octadecenoate, MeLOH is methyl 13-hydroxy-octadeca-9,11-dienoate, and MeL is methyl linoleate.

In addition to lipid oxidation products, other aldehydes have also been shown to convert amino acids into their vinylogous derivatives. In particular, the reactivity of certain Strecker aldehydes (e.g., phenylacetaldehyde) for this reaction (Zamora *et al.*, unpublished results) can also play a role in the Strecker aldehyde/vinylogous derivative ratio produced.

The Role of Food Anoxia in the Formation of Either Strecker Aldehydes or Vinylogous Derivatives of Amino Acids

The presence of oxygen plays an important role in the conversion of amino acids into either Strecker aldehydes or vinylogous derivatives.[9] In the absence of oxygen, secondary lipid oxidation products favor the conversion of amino acids into their vinylogous derivatives. However, in the presence of oxygen, secondary lipid oxidation products are oxidized, and the formed tertiary lipid oxidation products degrade the amino acids to the corresponding Strecker aldehydes (FIG. 3). The existence of alternative pathways for converting amino acids into their vinylogous derivatives has also been shown. Thus, as described above, tertiary lipid oxidation products and Strecker aldehydes are able to convert amino acids into their vinylogous derivatives. Both reactions are produced in the absence of oxygen to a higher extent than under aerobic conditions.

As a general rule, the presence of oxygen always decreases the amount of vinylogous derivative produced by a lipid oxidation product having either a keto or a hydroperoxy function. On the other hand, the presence of oxygen increases the amount of vinylogous derivatives of amino acids produced by unoxidized lipids or oxidized lipids with no keto or hydroperoxy functions. This is likely because all these lipid derivatives need an oxidation to produce a lipid oxidation product adequate for degrading the amino acid.

The Role of Other Food Constituents

The presence of other food constituents also influences the Strecker aldehyde/vinylogous derivative ratio. Thus, carbohydrates and lipids exhibit a negative synergism for the Strecker degradation of amino acid, and the amount of Strecker aldehydes produced is lower than the addition of the Strecker aldehydes produced by both carbohydrates and lipids when tested independently.[8] This negative synergism is lower when using unoxidized lipids than secondary lipid oxidation products. Nevertheless, the highest negative synergism is observed for tertiary lipid oxidation products.

Carbohydrates and lipids also exhibit a synergism for the conversion of amino acids into their vinylogous derivatives. However, this synergism could be positive or negative depending on the lipid compound. Thus, the synergism exhibited by unoxidized lipids is positive, and the synergism exhibited by alkadienals is negative.[10] This behavior might be related to the formation of free radicals in the reaction between the carbohydrate and the amino acid,[11] which would induce lipid oxidation. The oxidation of unoxidized lipids would convert these compounds into more reactive derivatives for this reaction. On the other hand, any further oxidation of alkadienals would decrease the reactivity of these lipid oxidation products for this reaction.

The presence of oxygen also influences the synergism of carbohydrates and lipids in the formation of vinylogous derivatives of the amino acids. In the presence of oxygen, the synergism between carbohydrates and unoxidized lipids is more positive, and the synergism between alkadienals and carbohydrates is less negative. In the first case, the presence of oxygen should increase the oxidation of the lipid and, therefore, facilitate the degradation of the amino acid. The increase of the synergism between alkadienals and carbohydrates for the conversion of amino acids into their vinylogous derivatives in the presence of oxygen might be related to the negative synergism observed for the same mixtures in the conversion of amino acids into their Strecker derivatives (see above).

This information suggests that the scheme in FIGURE 3 is much more complex than indicated because not only can other lipid oxidation products produce both degradations but also the interactions between lipids and other food constituents will influence the Strecker aldehyde/vinylogous derivative ratio.

Conclusions

The above review suggests that amino acids can be converted into either Strecker aldehydes or vinylogous derivatives. In addition, reaction conditions, including the presence of oxygen, play a major role in the Strecker aldehyde/vinylogous derivative ratio obtained. However, numerous lipid oxidation products are able to produce these reactions, and different alternative pathways are competing among them, making results difficult to predict. For example, although the presence of oxygen will decrease the amount of vinylogous derivative produced by most oxidized lipids when tested in binary mixtures with the amino acid, in complex food systems the results may be different. The increase in synergism between lipids and carbohydrates for the conversion of amino acids into their vinylogous derivatives and the negative synergism observed for the conversion of amino acids into their Strecker aldehydes in the presence of oxygen, may result in a decrease of the Strecker aldehyde/vinylogous derivative ratio when the amount of oxygen is increased. This is new evidence confirming that lipid oxidation and the Maillard reaction are so interrelated that they should be considered simultaneously to understand the products formed in complex food systems.[12]

Acknowledgments

This study was supported in part by the European Union (FEDER funds), the Plan Nacional de I + D of the Ministerio de Educación y Ciencia of Spain (Project AGL2006-01092), and the Junta de Andalucía (Project AGR-135). We are indebted to José L. Navarro for technical assistance.

Conflict of Interest

The authors declare no conflicts of interest.

References

1. BAYNES, J.W. et al., Eds. 2005. The Maillard Reaction. Chemistry at the Interfase of Nutrition, Aging, and Disease. The New York Academy of Sciences. New York, NY.
2. REINECCIUS, G. 2006. Flavor Chemistry and Technology, 2nd ed. Taylor & Francis. Boca Raton, FL.
3. FRIEDMAN, M. & D. MOTTRAM, Eds. 2005. Chemistry and Safety of Acrylamide in Food. Springer. New York, NY.
4. ZAMORA, R. & F.J. HIDALGO. 2005. Coordinate contribution of lipid oxidation and Maillard reaction to the nonenzymatic food browning. Crit. Rev. Food Sci. Nutr. **45:** 49–59.
5. HIDALGO, F.J. & R. ZAMORA. 2004. Strecker-type degradation produced by the lipid oxidation products 4,5-epoxy-2-alkenals. J. Agric. Food Chem. **52:** 7126–7131.
6. ZAMORA, R. et al. 2006. Chemical conversion of α-amino acids into α-keto acids by 4,5-epoxy-2-decenal. J. Agric. Food Chem. **54:** 6101–6105.
7. ZAMORA, R., E. GALLARDO & F.J. HIDALGO. 2007. Strecker degradation of phenylalanine initiated by 2,4-decadienal or methyl 13-oxooctadeca-9,11-dienoate in model systems. J. Agric. Food Chem. **55:** 1308–1314.
8. HIDALGO, F.J., E. GALLARDO & R. ZAMORA. 2007. Comparative Strecker degradation of amino acids produced by either carbohydrates or oxidized lipids. Presented at the 9th International Maillard Reaction Symposium. Munich, Germany, September 2.
9. ZAMORA, R. & F.J. HIDALGO. 2007. Conversion of phenylalanine into styrene by 2,4-decadienal in model systems. J. Agric. Food Chem. **55:** 4902–4906.

10. HIDALGO, F.J. & R. ZAMORA. 2007. Contribution of oxidized lipids to the formation of acrylamide in thermally processed foods. Presented at the 234th ACS National Meeting. Boston, MA, August 22.
11. YIM, H.S. *et al.* 1995. Free-radicals generated during the glycation reaction of amino acids by methylglyoxal—A model study of protein-cross-linked free radicals. J. Biol. Chem. **270:** 28228–28233.
12. HIDALGO, F.J. & R. ZAMORA. 2005. Interplay between the Maillard reaction and lipid peroxidation in biochemical systems. Ann. N.Y. Acad. Sci. **1043:** 319–326.

Post-Schiff Base Chemistry of the Maillard Reaction

Mechanism of Imine Isomerization

FONG LAM CHU AND VAROUJAN A. YAYLAYAN

Department of Food Science and Agricultural Chemistry, McGill University, Quebec, Canada

Schiff bases play a critical role, not only in initiating the Maillard reaction, but also in its propagation. Little attention has been paid so far to the ability of these imines to undergo isomerization and thus contribute to the diversity of Maillard reaction products. In this study, imine isomerization through 5-oxazolidinone formation was explored in a phenylalanine/glyceraldehyde model system, and spectroscopic evidence was provided for its formation by taking advantage of the strong carbonyl absorption band centered at 1784 cm^{-1}. The importance of 5-oxazolidinone formation lies in its ability to decarboxylate to azomethine ylide and subsequently form two isomeric imines, each capable of producing distinct Maillard products. Evidence for the formation of such ylides was also provided through their ability to undergo 1,3-dipolar cycloaddition with dipolarophiles.

Key words: azomethine ylide; 2-azaallyl anion; 5-oxazolidinone; transamination; imine isomerization mechanisms; Schiff base; phenylalanine; phenylpyruvic aicd; glyceraldehyde; FTIR

Introduction

The initial step in the interaction between reducing sugars and amino acids is the formation of the so-called Schiff base intermediate (see FIG. 1). Schiff bases, or imines, play a critical role, not only in initiating the Maillard reaction, but also in its propagation in the later stages of the reaction because of their chemical conversion into various reactive moieties able to interact and form new imines that are capable of further transformations. Little attention has been paid so far to the ability of these imines to isomerize and thus contribute to the diversity of Maillard reaction products. The possibility that the initial imine formed between a sugar and an amino acid will be converted into its isomeric imines during the Maillard reaction was first proposed by Høltermand[1] and has been referred to as transamination reaction. Similar isomerization of the imine, formed between α-keto acids and amino acids, was observed by Herbst and Engel[2] and its mechanism characterized by Cram and Guthrie[3] as a base-catalyzed methylene–azomethine rearrangement. Davidek et al.[4] also suggested the conversion of amino acids through an imine isomerization mechanism into 2-keto acids in the presence of glyoxal during the Maillard reaction. The first systematic study to confirm the isomerization of imines formed in alanine/glycolaldehyde and pyruvic acid/aminoethanol model systems was carried out by Yaylayan and Wnorowski[5] using ^{13}C-labeled precursors. More detailed studies were carried out using Fourier-transform infrared spectroscopy (FTIR) and gas chromatography–tandem mass spectrometry (GC–MS/MS) to monitor the formation of the imine in the pyruvic acid/aminopropanediol model system and its isomerization and subsequent rearrangement into the Amadori product.[6] Recently, Yaylayan et al.[7,8] have suggested a mechanism of decarboxylation of amino acids under dry conditions, in which the Schiff base undergoes intramolecular cyclization by the action of the carboxylate anion to form a 5-oxazolidinone intermediate capable of losing CO_2 and forming a relatively stable azomethine ylide (FIG. 2 pathway b). Interestingly, this ylide can undergo a 1,2-prototropic shift and generate two isomeric imines, providing a second mechanistic pathway of imine isomerization in addition to transamination. In this study, imine isomerization mechanisms are explored in the phenylalanine/glyceraldehyde model system, and spectroscopic evidence is provided for the formation of the 5-oxazolidinone intermediate and subsequent generation of azomethine ylide.

Address for correspondence: V. A. Yaylayan, McGill University, Department of Food Science and Agricultural Chemistry, 21,111 Lakeshore, Ste. Anne de Bellevue, Quebec, Canada, H9X 3V9. Voice: 514-398-7918; fax: 514-398-7977.

varoujan.yaylayan@mcgill.ca

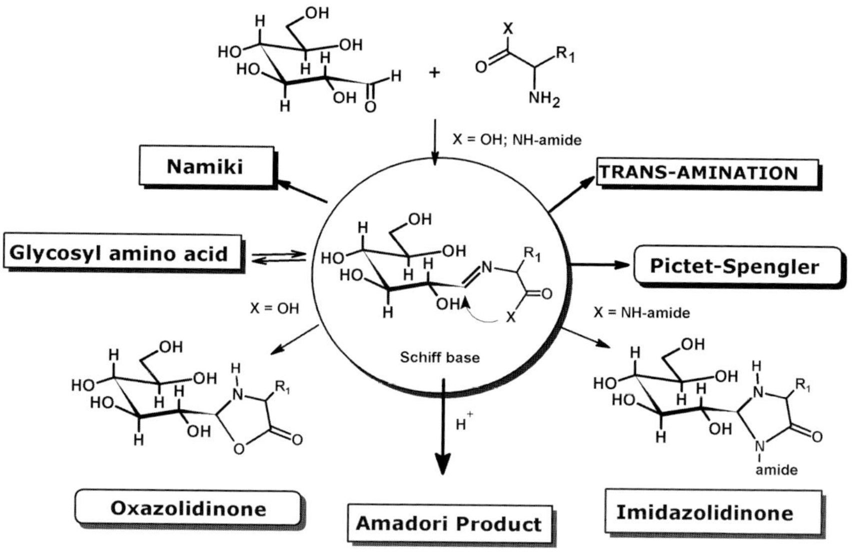

FIGURE 1. Chemical transformations of Schiff bases.

Materials and Methods

All reagents and chemicals were purchased from Aldrich Chemical Company (Milwaukee, WI) and used without further purification. The labeled [^{13}C-1] and [^{15}N]phenylalanine were purchased from Cambridge Isotope Laboratories (Andover, MA).

Pyrolysis GC/MS Analysis

Pyrolysis GC/MS (Py–GC/MS) analyses were conducted using a Varian CP-3800 GC coupled with a Varian Saturn 2000 ion trap mass spectrometry detector (Varian, Walnut Creek, CA). The pyrolysis unit included a CDS 1500 valved interface and a CDS Pyroprobe 2000 unit (CDS Analytical, Oxford, PA), was installed onto the GC injection port. The GC is equipped with a sample preconcentration trap (SPT) filled with Tenax GR. About one milligram of sample mixture was packed inside a quartz tube (0.3 mm thickness), which was plugged with quartz wool, inserted inside the coil probe, and pyrolyzed for 20 s at a temperature range from 175°C to 250°C. The volatiles after pyrolysis were concentrated on the SPT at 50°C for 4 min and subsequently desorbed at 100°C to the GC column for separation. A DB-5MS capillary column (J&W Scientific, Folsom, CA; 50 m × 0.2 mm i.d; coating thickness, 0.33 μm) was used under the following conditions: a pressure pulse of 70 psi was set for the first 4 min and later maintained with a constant flow of 1.5 mL/min for the rest of the run, regulated by an electronic flow controller. The GC oven temperature was set at −5°C for 5 min, using CO_2 as the cryogenic cooling source. The temperature was increased to 50°C at a rate of 50°C/min and then to 180°C at a rate of 5°C/min. The oven temperature was further increased to 280°C at a rate of 20°C/min and was kept at that temperature for 9.5 min. MS data were collected using the electron impact ionization mode under the following conditions: MS transfer line temperature, 250°C; MS manifold temperature, 50°C; MS ion trap temperature, 175°C; ionization voltage, 70 eV; electron multiplier voltage, 1700 V; scan range, 20–650 m/z. Compound identification was performed by using AMDIS (v2.62) and the National Institute of Standards and Technology (NIST) Standard Reference Database (v05).

Generation and FTIR Monitoring of Oxazolidinone Intermediate

An equimolar mixture (10 mg) of sugar and amino acid was heated in toluene (methanol and/or p-toluene-sulfonic acid could be added to help dissolve insoluble models) for 10 min or until most reactants dissolved at 115°C in an open vial (2 mL). The solution was filtered immediately and the sample was applied to the ATR crystal and scanned after evaporation of the solvent. Infrared spectra were recorded on a Nicolet single bounce ATR. A total of 64 scans at 4 cm^{-1} resolution were co-added. Processing of the FTIR data was performed using GRAMS/32 AI (ThermoGalactic, Waltham, MA).

Browning Measurement by UV-VIS

An equimolar solution of reactants (0.03 M each) in dimethyl sulfoxide was stirred in the presence and

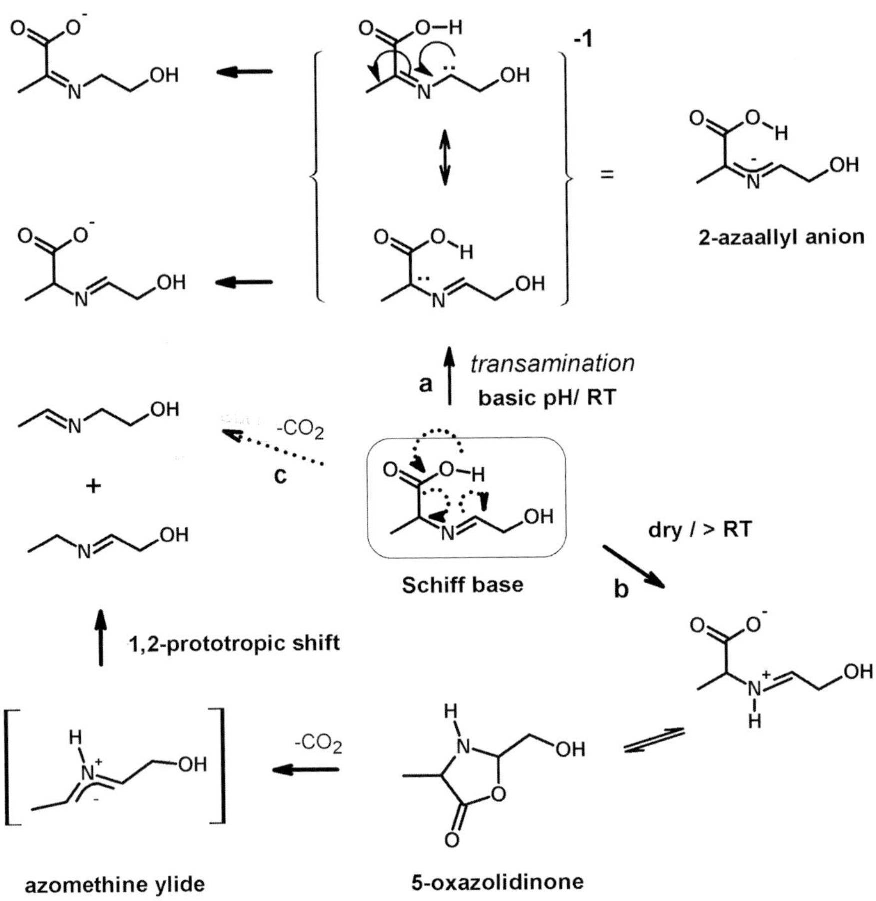

FIGURE 2. Proposed mechanisms of imine isomerization in the Maillard reaction: (a) base-catalyzed transamination, (b) oxazolidinone pathway, and (c) decarboxylative transamination. RT = room temperature.

absence of dimethyl fumarate (0.03 M) at 80°C for 30 min. The solution was cooled and browning was measured by scanning between 400 and 600 nm using the Evolution 300 scanning spectrophotometer from Thermo Electron Corporation (Madison, WI).

Results and Discussion

Imines formed between amino acids and reducing sugars are generally referred to as Schiff bases (FIG. 1). The fate of this initial Schiff base is mainly dependent on the moisture content of the system, the pH, and the temperature. Under intermediate moisture conditions and pH between 5 and 7, Schiff bases are converted into Amadori products and the Maillard reaction cascade is initiated. Under lower-moisture conditions, Schiff bases tend to prevail and subsequently undergo several transformations (FIG. 1) depending on the pH. Under basic pH and at room temperature, transamination[6] can generate isomeric imines (FIG. 2). Such isomeric imines can also be formed under pyrolytic conditions.[5,6] However, under slightly acidic or neutral pH, Schiff bases can undergo intramolecular cyclization and produce decarboxylated isomeric imines through 5-oxazolidinone formation, as shown in FIGURE 2. Other reactions of Schiff bases are summarized in FIGURE 1.

Isomerization of Imines through Transamination

Transamination reactions usually involve a base-catalyzed tautomerization of the imine and formation of a delocalized 2-azaallyl anion[3] (FIG. 2). This process has been recognized as a prototype of the biochemical transamination between amino acids and pyridoxal.[9] Recently, these intermediates have been generated under mild conditions at room temperature through deprotonation of imines using potassium *tert*-butoxide.[10]

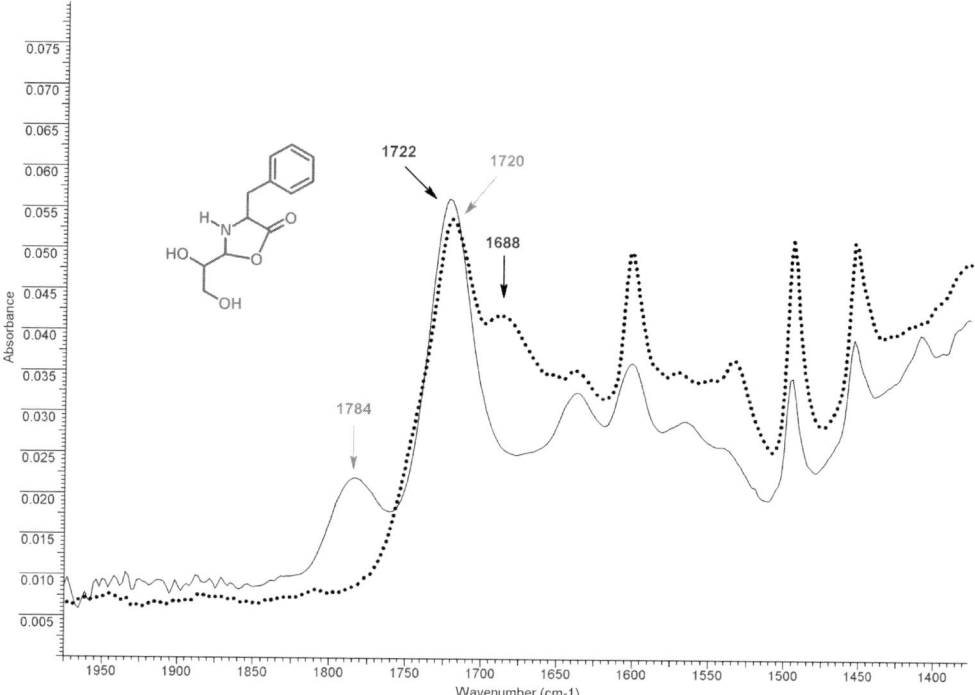

FIGURE 3. FTIR spectrum of phenylalanine/glyceraldehyde (solid line) and [^{13}C-]phenylalanine/glyceraldehyde (dotted line) acquired after heating for 10 min in toluene at 110°C.

A concerted mechanism was proposed[3] for this isomerization, in which the base removes a proton from one α carbon synchronously with the donation of a proton to the other carbon by its conjugate acid. In the specific case of pyruvic acid and ethanolamine, the initial imine formation and its subsequent isomerization through a 1,3-prototropic shift occurs very rapidly at room temperature, and the reaction goes to completion within 12 min as indicated by FTIR analysis.[6] After hydrolysis, the two isomeric imines in this case, can generate alanine and glycolaldehyde, indicating the transfer of an amino group between a donor amino compound and an acceptor keto acid (transamination).

Isomerization of Imines through 5-oxazolidinone and Azomethine Ylide Formation

The transamination pathway of imine isomerization described above applies to Schiff bases formed between any amino group and any carbonyl moiety. However, the second mechanism that can lead to the formation of isomeric imines is specific to Schiff bases formed between amino acids and any carbonyl-containing compounds and is accompanied by loss of a CO_2 from the amino acid. FIGURE 2 depicts the formation of azomethine ylide[11] after the loss of CO_2 from the 5-oxazolidinone intermediate. In dry systems, the Schiff bases of reducing sugars and amino acids are prone to undergo intramolecular cyclization to form either 5-oxazolidinone or glycosylamines, in contrast to high-moisture systems, where they tend to undergo Amadori rearrangement, the more stable isomer. However, the formation of 5-oxazolidinone and the subsequent generation of azomethine ylides have so far been verified only in model systems consisting of amino acids and simple aldehydes.[11,12] Indirect evidence for the involvement of Schiff bases in assisting the decarboxylation process was obtained earlier[8] when [^{13}C-4]-aspartic acid was pyrolyzed alone and in the presence of glucose. Analysis of the data showed exclusive decarboxylation of aspartic acid form C-1 when pyrolyzed in the presence of glucose, indicating a preference for the formation of 5-oxazolidinone as an intermediate for the decarboxylation step. However, to provide direct evidence for the formation of 5-oxazolidinone, the amino acid/sugar reactions were analyzed by FTIR to monitor the formation of a peak in the range between 1780 and 1810 cm^{-1}, where 5-oxazolidinones are known to exhibit a strong absorption band.[12] Spectroscopic studies using the glyceraldehyde/amino acid model systems in toluene heated to 110°C clearly indicated the formation of an intense

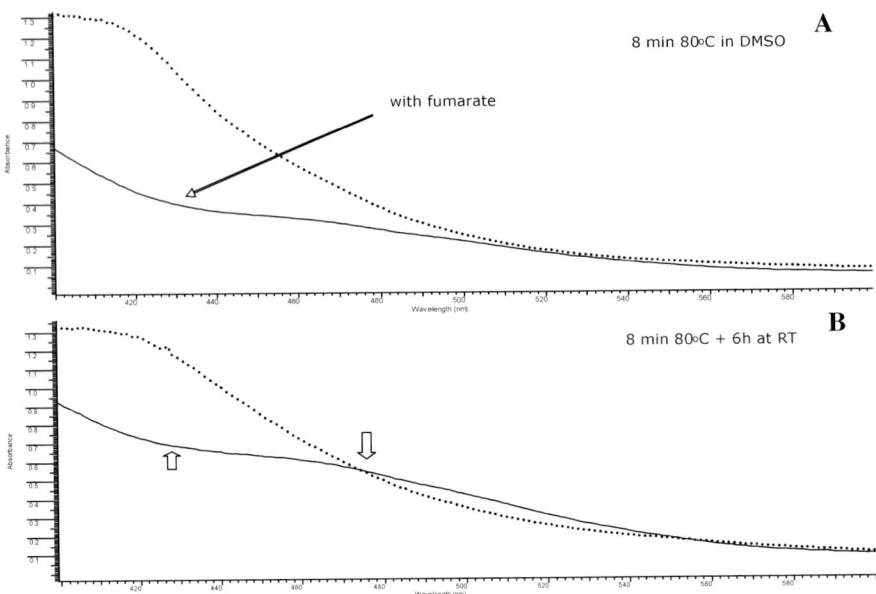

FIGURE 4. UV-VIS spectrum of the phenylalanine/glyceraldehyde solution **(A)** acquired after heating for 8 min in dimethylsulfoxide (DMSO) at 80°C and **(B)** after additional storage at room temperature for 6 h.

TABLE 1. Effect of the addition of carbonyl compounds on the decarboxylation efficiency[a] of phenylalanine at 250°C

Model system	Relative efficiency[b] (phenethylamine formation)
Phenylalanine	1
Phenylalanine + pyruvic acid	19
Phenylalanine + glyceraldehyde	99
Phenylalanine + 2,3-butanedione	115
Phenylalanine + phenylacetaldehyde	303
Phenylalanine + phenylpyruvic acid	300

[a]Estimated from the chromatographic peak area of phenethylamine produced per mole of phenylalanine.
[b]Based on three replicates with coefficient of variation < 15%.

TABLE 2. Effect of temperature on the decarboxylation efficiency[a] of phenylalanine

Temperature (°C)	Relative efficiency[b] (phenethylamine formation)
175	0
200	1
250	191

[a]Estimated from the chromatographic peak area of phenethylamine produced per mole of phenylalanine
[b]Based on three replicates with coefficient of variation <15%.

peak in the range of 1780–1810 cm^{-1}, depending on the amino acid. FIGURE 3 shows the carbonyl absorption peak centered at 1784 cm^{-1} for the phenylalanine/glyceraldehyde model system. The identity of the peak was verified by observing the expected 40 cm^{-1} shift when [^{13}C-1]-labeled phenylalanine was used. Furthermore, evidence for the formation of azomethine ylides was also provided using their specific ability to undergo 1,3-dipolar cycloadditions with dipolarophiles.[11] The addition of dipolarophiles, such as dimethyl fumarate, to the heated phenylalanine/glyceraldehyde model system led to a significant drop in the intensity of the Maillard browning (FIG. 4), indicating the importance of the azomethine ylide as a browning precursor in the Maillard reaction. In addition to direct spectroscopic evidence for oxazolidinone formation, decarboxylation efficiencies can provide indirect evidence for the involvement of Schiff bases in the decarboxylation process. In fact, the efficiency of the decarboxylation of phenylalanine, as measured by the amount of phenethylamine produced per mole of the precursor during Py–GC/MS analysis, increased significantly in the presence of carbonyl-containing compounds (TABLE 1). In addition, phenylalanine can undergo decarboxylation in the absence of carbonyls. The efficiency of this independent decarboxylation process is a function of temperature, as shown in TABLE 2. The addition of carbonyl compounds not only increases

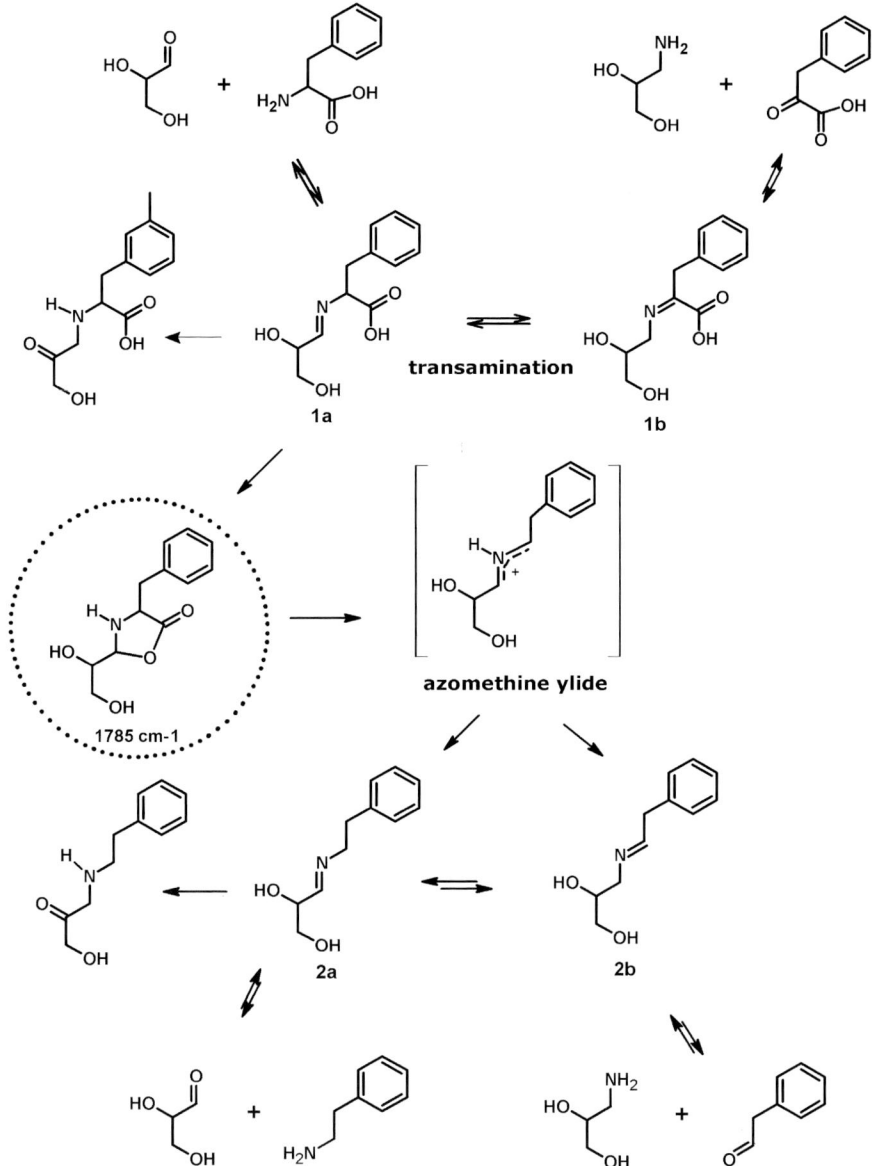

FIGURE 5. Isomerizations of the Schiff base formed between phenylalanine and glyceraldehyde through transamination and 5-oxazolidinone formation.

the efficiency of decarboxylation, but also allows decarboxylation to occur at lower temperatures (results not shown).

Evidence of Imine Isomerizations in the Phenylalanine/Glyceraldehyde Model System

To investigate the occurrence of imine isomerization in Maillard model systems, the phenylalanine/glyceraldehyde model was chosen because of the availability of the hydrolysis products of the four possible imines (1a, 1b, 2a, and 2b in FIG. 5) that could be formed in this model system. When the initial imine (1a) undergoes isomerization through transamination, the resulting imine (1b) can be obtained through the interaction of readily available 1-aminopropanediol and phenylpyruvic acid. Alternatively, if the imine 1a undergoes isomerization through the oxazolidinone pathway, the resulting imines, 2a and 2b, could also be obtained through the interaction of readily available

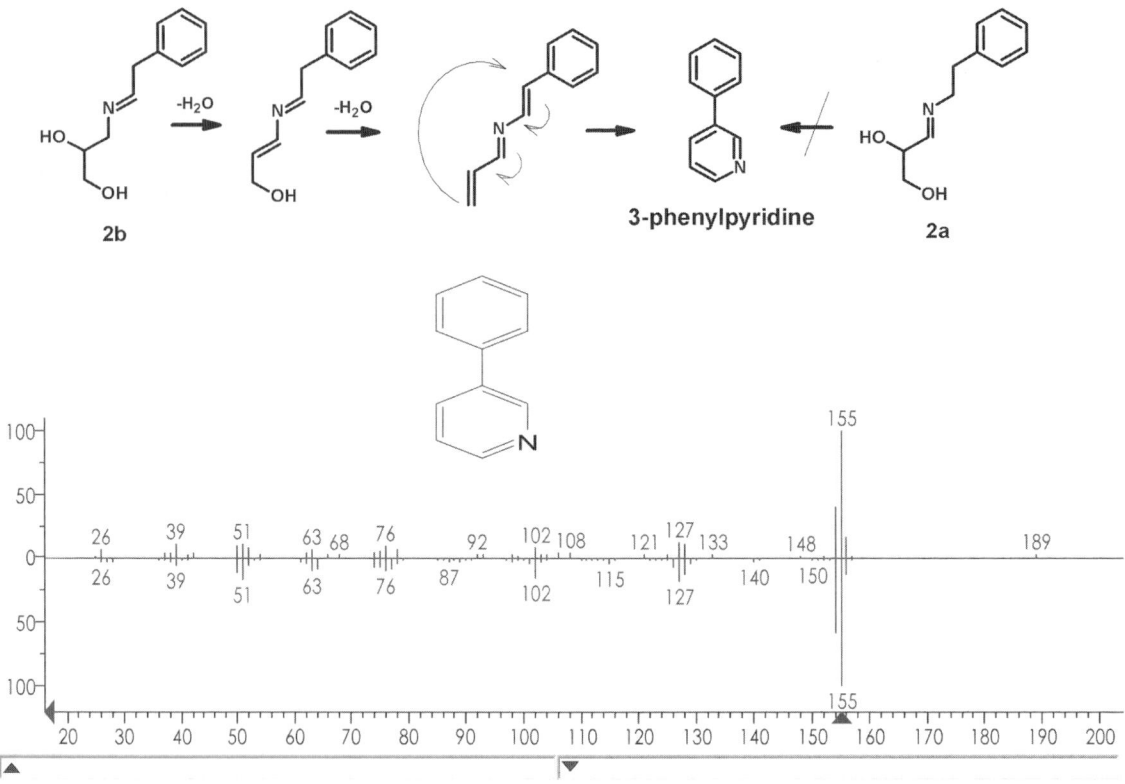

FIGURE 6. Proposed mechanism of the formation of 3-phenylpyridine and its mass spectrum compared with the authentic NIST library spectrum in head-to-tail fashion.

precursors such as glyceraldehyde, phenethylamine, 1-aminopropanediol, and phenylacetaldehyde. The formation of 5-oxazolidinone in this particular system has already been confirmed, as described above, using the absorption band centered at 1785 cm^{-1}. The four model systems generating imines 1a, 1b, 2a, and 2b shown in FIGURE 5 were analyzed using Py–GC/MS, and the formation of selected products was used to indicate the occurrence of specific isomerization. For example, detection of phenylacetaldehyde and phenethylamine in the pyrolysis products of the glyceraldehyde/phenylalanine model system may indicate the occurrence of isomerization between 2a and 2b, as shown in FIGURE 5. However, phenylacetaldehyde can also be formed through the Strecker reaction, and phenethylamine can also be formed through independent decarboxylation of the amino acid without reactants necessarily passing through the 5-oxazolidinone intermediate. To eliminate this possibility, the model systems were also studied at 175°C, a temperature at which independent decarboxylation was not observed (see TABLE 2). In addition, to identify a unique product indicating formation of imine 2b, the precursors of the two imines, 2a and 2b, were studied, and the products formed in both model systems were compared. The analysis of the data indicated a significant formation of 3-phenylpyridine in the model system of 1-aminopropanediol/phenethylaldehyde and its complete absence in the glyceraldehyde/phenethylamine system. FIGURE 6 shows a possible mechanism of formation of 3-phenylpyridine, starting from 2b through two dehydration steps and an electrocyclic ring closure step. This mechanism is consistent with the inability of the 2a to form this compound. The identity of this peak was further verified by observing an identical retention time (25.05 min) with the commercially obtained 3-phenylpyridine; furthermore, the incorporation of one nitrogen atom in the structure when [^{15}N]-phenylalanine was used is consistent with the proposed structure. Similarly, detection of 3-phenylpyridine and phenethylamine in the pyrolysis products of the 1-aminopropanediol/phenylpyruvaldehyde model system (FIG. 5) can also indicate the occurrence of isomerization, not only between 1b and 1a, but also between 2a and 2b. Pyrolysis of the phenylalanine/glyceraldehyde and 1-aminopropanediol/phenylpyruvic acid model systems indicated the formation of both indicator compounds, phenethylamine and 3-phenylpyridine, confirming the formation of 1b, 1a, 2a, and 2b. In addition, these results indicate that transamination of 2a to 2b is difficult to achieve in imines lacking a carboxylic acid moiety in their structures that is able to catalyze prototropic shifts,[5,6] as in the case of isomerization of 1b to 1a, further supporting the intermediacy of azomethine ylide in the two model systems. Similar imine isomerizations were also observed by Granvogl et al.[13] in phenylalanine dicarbonyl mixtures. They proposed a decarboxylative transamination (pathway c in FIGURE 2) mechanism similar to the suggestion of Bishop et al.[9] in explaining the transamination reaction between amino acids and α-keto acids.

Acknowledgments

The authors acknowledge funding for this research by the Natural Sciences and Engineering Research Council of Canada.

Conflict of Interest

The authors declare no conflicts of interest.

References

1. HØLTERMAND, A. 1966. The browning reaction. Die Stärke **18:** 319–328.
2. HERBST, R.M. & L.L. ENGEL. 1934. A reaction between α-ketonic acids and α-amino acids. J. Biol. Chemistry **107:** 505–512.
3. CRAM, D.J. & R.D. GUTHRIE. 1966. Electrophilic substitution at saturated carbon. XXVII. Carbanions as intermediates in the base-catalyzed methyleneazomethine rearrangement. J. Am. Chem. Soc. **88:** 5760–5765.
4. DAVIDEK, J., L. VELISEK & J. POKORNY, Eds. 1990. Chemical Changes During Food Processing. Elsevier. New York, NY. p. 138.
5. YAYLAYAN, V.A. & A. WNOROWSKI. 2002. The role of β-hydroxyamino acids in the Maillard reaction—transamination route to Amadori products. In Maillard Reaction in Food Chemistry and Medical Sciences: Update for the Postgenomic Era. S. Horiuchi, N. Taniguchi, F. Hayase, T. Kurata & T. Osawa, Eds.: 195–200. International Congress Series 1245. Elsevier Science. Amsterdam, the Netherlands.
6. WNOROWSKI, A. & V.A. YAYLAYAN. 2003. Monitoring carbonyl-amine reaction between pyruvic acid and α-amino alcohols by FTIR spectroscopy—a possible route to Amadori products. J. Agric. Food Chem. **51:** 6537–6543.
7. YAYLAYAN, V.A. et al. 2003. Why asparagine needs carbohydrates to generate acrylamide. J. Agric. Food Chem. **51:** 1753–1757.
8. YAYLAYAN, V.A. et al. 2004. Role of creatine in the generation of N-methylacrylamide: a new toxicant in cooked meat. J. Agric. Food Chem. **52:** 5559–5565.
9. BISHOP, J.C. et al. 1997. Prebiotic transamination. Orig. Life Evol. Biosph. **27:** 319–324.
10. CAINELLI, G. et al. 1996. Efficient transamination under mild conditions: preparation of primary amine derivatives from carbonyl compounds via imine isomerization with catalytic amounts of potassium tert-butoxide. J. Org. Chem. **61:** 5134–5139.
11. TSUGE, O. et al. 1987. Simple generation of nonstabilized azomethine ylides through decarboxylative condensation of α-amino acids with carbonyl compounds via 5-oxazolidinone intermediate. Bull. Chem. Soc. Jpn. **60:** 4079–4089.
12. AURELIO, L. et al. 2003. An efficient synthesis of N-methyl amino acids by way of intermediate 5-oxazolidinone. J. Org. Chem. **68:** 2652–2667.
13. GRANVOGL, M. et al. 2006. Formation of amines and aldehydes from parent amino acids during thermal processing of cocoa and model systems: new insights into pathways of Strecker reaction. J. Agric. Food Chem. **54:** 1730–1739.

Usefulness of Antibodies for Evaluating the Biological Significance of AGEs

RYOJI NAGAI,[a] YUKIO FUJIWARA,[a] KATSUMI MERA,[a,b] KEITA MOTOMURA,[a,c] YASUNORI IWAO,[a,b] KEIICHIRO TSURUSHIMA,[a,c] MIME NAGAI,[a] KAZUHIRO TAKEO,[a,b] MAKIKO YOSHITOMI,[a,c] MASAKI OTAGIRI,[b] AND TSUYOSHI IKEDA[c]

[a]*Department of Medical Biochemistry, Faculty of Medical and Pharmaceutical Sciences, Kumamoto University, Kumamoto 860-8556, Japan*

[b]*Department of Biopharmaceutics and* [c]*Natural Medicine, Graduate School of Pharmaceutical Sciences, Kumamoto University, Kumamoto 862-0973, Japan*

Polyclonal and monoclonal antibodies have been widely applied to demonstrate the presence of advanced glycation end products (AGEs) *in vivo*. However, our previous study showed that monoclonal anti-AGE antibody (6D12) and polyclonal anti-N^ε-(carboxymethyl)lysine (CML) antibody recognize not only CML but also N^ε-(carboxyethyl)lysine (CEL), thus indicating that we should pay attention to the specificity of the antibodies. As a result, we prepared specific monoclonal antibodies against CML, CEL, N^ω-(carboxymethyl)arginine (CMA), and S-(carboxymethyl)cysteine (CMC). Our immunochemical study using anti-CMA antibody demonstrated that the CMA content increased in a time-dependent manner when collagen was incubated with glucose, indicating that immunological quantification using the specific antibody is especially useful for measuring an acid-labile AGE structure, such as CMA. Monoclonal antibody is also applied to identify a novel biological marker in pathological lesions. We prepared antibody libraries against proteins modified with aldehydes, such as glyoxal, methylglyoxal, and glycolaldehyde (GA), and one antibody, GA5, which specifically reacts with the GA-modified protein that is recognized in human atherosclerotic lesions. Following successive high-performance liquid chromatography purification, the GA5-reactive compound was isolated and its chemical structure was found to be 3-hydroxy-4-hydroxymethyl-1-(5-amino-5-carboxypentyl) pyridinium cation, which was named GA-pyridine. Taken together, these results demonstrate that a specific antibody is a powerful tool for analyzing novel biomarkers, formation pathways, and the efficacy of AGE inhibitors.

Key words: N^ε-(carboxymethyl)lysine; N^ε-(carboxyethyl)lysine; GA-pyridine; anti-AGE antibody; diabetic complications

Introduction

Research on advanced glycation end products (AGEs) began in the field of food chemistry and, as a result of the development of polyclonal and monoclonal antibodies against AGE-modified proteins in the 1990s, has expanded to *in vivo* studies. These immunochemical approaches have greatly contributed to understanding the biological significance of AGE in the pathogenesis of age-enhanced diseases.[1] The epitope structure of monoclonal anti-AGE antibody (6D12) was identified as a N^ε-(carboxymethyl)lysine (CML)/N^ε-(carboxyethyl)lysine (CEL)-protein adduct,[2] and immunohistochemical studies using 6D12 have demonstrated the presence of CML/CEL-protein adducts in several human tissues. This suggests a potential link of CML/CEL to aging[3] and age-enhanced disease processes, such as diabetic nephropathy,[4] diabetic retinopathy,[5] diabetic neuropathy,[6] atherosclerosis,[7] hemodialysis-related amyloidosis,[8] Alzheimer's disease,[9] actinic elastosis of the skin,[10] and pulmonary fibrosis.[11] We also prepared specific monoclonal antibodies against CML,[2] CEL,[12] N^ω-(carboxymethyl)arginine (CMA),[13] and S-(carboxymethyl)cysteine (CMC). We report here the

Address for correspondence: Ryoji Nagai, Ph.D., Department of Medical Biochemistry, Faculty of Medical and Pharmaceutical Sciences, Kumamoto University, Honjo 1-1-1, Kumamoto 860-8556, Japan. Fax: +81-96-364-6940.

nagai-883@umin.ac.jp

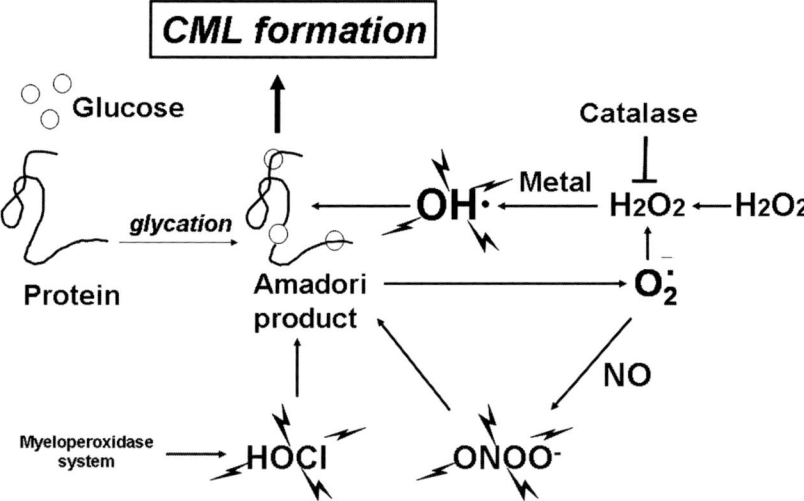

FIGURE 1. Possible pathways of N^ϵ-(carboxymethyl)lysine (CML) formation.

immunological approach for analyzing the formation pathways of AGEs and the identification of novel biomarkers.

Measurement of the Pathway for AGEs Formation by Anti-Age Antibodies

Immunological detection of AGEs is used for analyzing AGE formation pathways because immunological detection methods are easier to carry out than instrumental analysis. Using immunological detection methods, we previously demonstrated that CML is generated by the oxidative cleavage of Amadori products by a hydroxyl radical[14] and peroxynitrite.[15] CML formation was also observed when glycated-human serum albumin was incubated with activated neutrophils and was completely inhibited in the presence of a hypochlorous acid (HOCl) scavenger,[16] thus suggesting that CML is an important biological marker of oxidative stress *in vivo* (FIG. 1).

We also measured the yield of AGEs derived from 3-deoxyglucosone (3-DG).[17] Plasma 3-DG levels increase under hyperglycemic conditions and are associated with the pathogenesis of diabetic complications because of the high reactivity of 3-DG with proteins to form AGEs. To investigate potential markers for 3-DG-mediated protein modification *in vitro* and *in vivo*, we compared the yield of several 3-DG-derived AGE structures by immunochemical analysis and high-performance liquid chromatography (HPLC). When bovine serum albumin (BSA) was incubated with 3-DG at 37°C for up to 4 weeks, the amount of CML and 3-DG-imidazolone steeply increased with incubation time, whereas the level of pyrraline and pentosidine increased slightly by day 28. In contrast, a significant amount of pyrraline and pentosidine was also observed when BSA was incubated with 3-DG at 60°C to enhance AGE formation. Our results demonstrate that CML and 3-DG-imidazolone are major AGE structures in 3-DG-modified proteins and that 3-DG-imidazolone provides a better marker for protein modification by 3-DG than pyrraline.[17]

Furthermore, our recent study using the monoclonal antibody against N^ω-(carboxymethyl)arginine (CMA), an acid-labile AGE, demonstrated that a significant CMA immunoreactivity is found in atherosclerotic lesions, whereas no such immunoreactivity was observed in normal regions.[13] This suggests that the accumulation of CMA in tissue proteins may contribute to the pathophysiologies associated with aging and age-related diseases.

Identification of Novel AGE Structure In Vivo

As described above, monoclonal and polyclonal antibodies are widely used to demonstrate the presence of AGE-modified protein *in vivo*. Monoclonal antibodies are also applied to identify novel biological markers in pathological lesions. Several reactive aldehydes, such as 3-DG, glyoxal, and methylglyoxal, are formed from the Maillard reaction and the metabolic pathway *in vivo* and are considered to be precursors of AGEs. In a parallel pathway involving both enzymatic

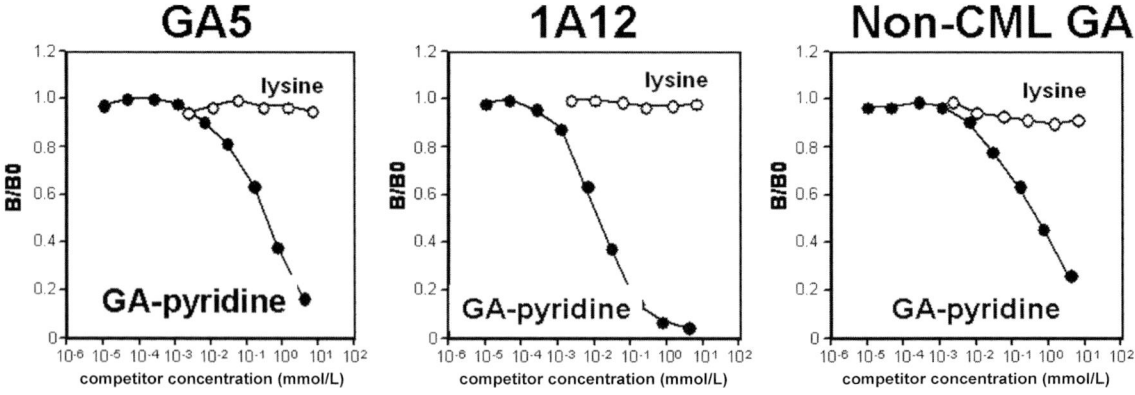

FIGURE 2. Reactivity of monoclonal and polyclonal anti-glycolaldehyde (GA)-modified protein to purified GA-pyridine. (A) GA5, (B) 1A12, and (C) rabbit polyclonal antibody (non-CML-GA).

and nonenzymatic reactions during inflammation, leukocytes are activated to secrete myeloperoxidase that mediates the formation of HOCl from H_2O_2 and chloride. A reactive aldehyde, such as glycolaldehyde (GA), which is formed by reaction of HOCl with serine, reacts to form chemical modifications in protein. GA is known to react with protein amino residues to give brown-colored cross-linked structures. Myeloperoxidase-derived GA reacts with RNase to form CML.[18] However, since CML is also formed during the Maillard reaction by oxidative cleavage of Amadori products, CML is not a specific marker for myeloperoxidase-induced protein modification *in vivo*. Therefore, to obtain a specific marker for myeloperoxidase-induced protein modification *in vivo*, a GA-protein adduct(s) other than CML must be identified. Demonstration of such a specific marker in atherosclerotic lesions would establish a role for the myeloperoxidase system in chemical modification of proteins during atherogenesis. To this end, monoclonal antibodies (GA5 and 1A12) as well as polyclonal (non-CML-GA) antibodies that are specific for GA-modified proteins have been prepared.[19] These antibodies specifically reacted with GA-modified or HOCl-modified BSA but not with BSA modified by other aldehydes, such as glucose, glyoxal, methylglyoxal, and 3-DG, indicating that the epitope structure of these antibodies could be specific for GA-modified proteins. Following successive HPLC purification procedures, the GA5-reactive compound was isolated and its chemical structure was found to be 3-hydroxy-4-hydroxymethyl-1-(5-amino-5-carboxypentyl) pyridinium cation. This compound, named GA-pyridine, could be recognized both by 1A12 and non-CML-GA (FIG. 2), indicating that GA-pyridine is an important antigenic structure in GA-modified proteins. Immunohistochemical studies with GA5 demonstrated the accumulation of GA-pyridine in the cytoplasm of foam cells and extracellularly in the central region of atheroma in human atherosclerotic lesions.[19] These results suggest that myeloperoxidase-mediated protein modification via GA may contribute to atherogenesis.

Acknowledgments

This work was supported in part by Grants-in-Aid for Scientific Research (No. 18790619 to R.N.) from the Ministry of Education, Science, Sports and Cultures of Japan.

Conflict of Interest

The authors declare no conflicts of interest.

References

1. NAGAI, R., S. HORIUCHI & Y. UNNO. 2003. Application of monoclonal antibody libraries for the measurement of glycation adducts. Biochem. Soc. Trans. **31:** 1438–1440.
2. KOITO, W. *et al.* 2004. Conventional antibody against N^ε-(carboxymethyl)lysine (CML) shows cross-reaction to N^ε-(carboxyethyl)lysine (CEL): immunochemical quantification of CML with a specific antibody. J. Biochem (Tokyo) **136:** 831–837.
3. ARAKI, N. *et al.* 1992. Immunochemical evidence for the presence of advanced glycation end products in human lens proteins and its positive correlation with aging. J. Biol. Chem. **267:** 10211–10214.
4. YAMADA, K. *et al.* 1994. Immunohistochemical study of human advanced glycosylation end-products (AGE) in chronic renal failure. Clin. Nephrol. **42:** 354–361.

5. MURATA, T. *et al.* 1997. The relationship between accumulation of advanced glycation end products and expression of vascular endothelial growth factor in human diabetic retinas. Diabetologia **40:** 764–769.
6. SUGIMOTO, K. *et al.* 1997. Localization in human diabetic peripheral nerve of N^{ε}-carboxymethyllysine-protein adducts, an advanced glycation endproduct. Diabetologia **40:** 1380–1387.
7. KUME, S. *et al.* 1995. Immunohistochemical and ultrastructural detection of advanced glycation end products in atherosclerotic lesions of human aorta with a novel specific monoclonal antibody. Am. J. Pathol. **147:** 654–667.
8. MIYATA, T. *et al.* 1993. β2-Microglobulin modified with advanced glycation end products is a major component of hemodialysis-associated amyloidosis. J. Clin. Invest. **92:** 1243–1252.
9. KIMURA, T. *et al.* 1996. Accumulation of advanced glycation end products of the Maillard reaction with age in human hippocampal neurons. Neurosci. Lett. **208:** 53–56.
10. MIZUTARI, K. *et al.* 1997. Photo-enhanced modification of human skin elastin in actinic elastosis by Ne-(carboxymethyl)lysine, one of the glycoxidation products of the Maillard reaction. J. Invest. Dermatol. **108:** 797–802.
11. MATSUSE, T. *et al.* 1998. Immunohistochemical localisation of advanced glycation end products in pulmonary fibrosis. J. Clin. Pathol. **51:** 515–519.
12. NAGAI, R. *et al.* 2008. Immunochemical detection of N^{ε}-(carboxyethyl)lysine using a specific antibody. J. Immunol. Methods In press.
13. MERA, K. *et al.* 2008. Immunological detection of N^{ω}-(carboxymethyl)arginine by a specific antibody. Ann. N.Y. Acad. Sci. The Maillard Reaction: 9th International Symposium. In press.
14. NAGAI, R. *et al.* 1997. Hydroxyl radical mediates N^{ε}-(carboxymethyl)lysine formation from Amadori product. Biochem. Biophys. Res. Commun. **234:** 167–172.
15. NAGAI, R. *et al.* 2002. Peroxynitrite induces formation of N^{ε}-(carboxymethyl) lysine by the cleavage of Amadori product and generation of glucosone and glyoxal from glucose: novel pathways for protein modification by peroxynitrite. Diabetes **51:** 2833–2839.
16. MERA, K. *et al.* Hypochlorous acid generates N^{ε}-(carboxymethyl)lysine from Amadori products. Free Radic. Res. **41:** 713–718.
17. JONO, T. *et al.* 2004. N^{ε}-(Carboxymethyl)lysine and 3-DG-imidazolone are major AGE structures in protein modification by 3-deoxyglucosone. J. Biochem (Tokyo) **136:** 351–358.
18. ANDERSON, M. M. *et al.* 1999. The myeloperoxidase system of human phagocytes generates N^{ε}-(carboxymethyl)lysine on proteins: a mechanism for producing advanced glycation end products at sites of inflammation. J. Clin. Invest. **104:** 103–113.
19. NAGAI, R. *et al.* 2002. Identification in human atherosclerotic lesions of GA-pyridine, a novel structure derived from glycolaldehyde-modified proteins. J. Biol. Chem. **277:** 48905–48912.

Receptor for Advanced Glycation End Product Expression in Experimental Diabetic Retinopathy

YUMEI WANG,[a] FRANZISKA VOM HAGEN,[a] FREDERICK PFISTER,[a] ANGELIKA BIERHAUS,[b] YUXI FENG,[a] REINHOLD GANS,[c] AND HANS-PETER HAMMES[a]

[a]*5th Medical Clinic, Medical Faculty Mannheim, University of Heidelberg, Mannheim, Germany*

[b]*Department of Internal Medicine I Heideleberg, University of Heidelberg, Heidelberg, Germany*

[c]*Department of Internal Medicine, University Medical Center, Groningen, the Netherlands*

The advanced glycation end product (AGE)–receptor for AGE (RAGE) pathway is involved in the pathogenesis of diabetic microvascular damage. The special distribution of RAGE and its engagement has an impact on the development of diabetic retinopathy. In the present study, we used immunofluorescence and confocal laser microscopy to study RAGE expression with special emphasis on Müller glia in Sprague Dawley rats. RAGE expression was low in nondiabetic retinae and was found in ganglion cells and Müller cell end feet. In diabetic retinae, upregulated RAGE was predominantly expressed in retinal glia. Since Müller cells are important in the regulation of important features of early retinal vascular damage, such as vascular permeability, homeostasis, and response to stress, RAGE appears to be a central modulator in diabetic retinopathy.

Key words: diabetic retinopathy; Müller cells; RAGE; retina

Introduction

The accelerated and increased formation of advanced glycation end products (AGEs) is one of the major biochemical abnormalities downstream of hyperglycemia-induced mitochondrial overproduction of reactive oxygen species.[1] AGEs interact with the receptor for AGE (RAGE), inducing pro-inflammatory, pro-adhesive, and growth-stimulating signals.[2] Retinal vascular damage in diabetes reflects both permanent endothelial-born alterations in transcriptional activity and consequences for the intimate cellular cross-talk with pericytes as well as the response-to-injury type of glial activation. Thus, cellular regulation of RAGE expression in the diabetic setting may have immediate consequences for cell–cell communication, including proliferative and apoptotic events.

As rodent strains differ in their propensity to develop early retinal damage in response to hyperglycemia and as RAGE expression may depend on cell types, the regulation and expression patterns in the diabetic retina gives important new insight into the pathogenetic role of RAGE in diabetic retinopathy.[3,4] We have previously observed that RAGE expression in the streptozotocin (STZ)-diabetic Wistar rat strain was absent in nondiabetic retinae and was predominantly upregulated in Müller glia in diabetic animals.[5] In the present study, we used confocal laser scanning microscopy to study RAGE expression with special emphasis on Müller glia in Sprague Dawley (SD) rats, which have been widely used to investigate experimental diabetic retinopathy and which may differ in the dynamics to develop diabetic retinopathy.

Materials and Methods

Experiments performed in this study adhered to the Association for Research in Vision and Opthalmology statement for the "Use of Animals in Ophthalmic and Vision Research."

Animals

Male SD rats (Janvier, le Genest-st.lsle, France) were randomly divided into the control (nondiabetic, $n = 3$)

Address for correspondence: Hans-Peter Hammes, M.D., 5th Medical Department, Klinikum Mannheim, University of Heidelberg, Theodor-Kutzer-Ufer 1-3, D-68167 Mannheim, Germany. Voice: +49-621-383-2663; fax: +49-621-383-2663.
hans-peter.hammes@med5.ma.uni-heidelberg.de

and the diabetic group ($n = 2$). Rats were housed in makrolon cages, were allowed free access to food and water, and were exposed to a 12 h light–dark cycle. Diabetes was induced by a single intravenous injection of STZ (50 mg/kg body weight; Roche, Mannheim, Germany) dissolved in 0.05 mol/L citrate buffer (pH 4.5). Blood glucose concentrations were measured during every third day and levels higher than 300 mg/dL were considered diabetic. Control rats were injected with vehicle alone. Glucose levels and body weight were monitored consecutively every second week, and final glycated hemoglobin concentration was determined by affinity chromatography (Glyc Affin, Akron, OH). Diabetic and nondiabetic rats were terminated under deep anesthesia 9 months after diabetes induction, and eyes were enucleated and immediately frozen at $-80°C$ for further analysis.

Immunohistochemistry

Paraffin-embedded tissue was cut to a thickness of 6 μm, dewaxed, and rehydrated. The antigens were retrieved by boiling in citrate buffer (pH 6.0–6.3) for 20 min. To permeabilize the tissue, sections were incubated with 0.5% triton x-100. Blocking of nonspecific protein interactions were achieved by incubating with 1% bovine serum albumin for 30 min at room temperature. Slides were then incubated in primary antibody (mouse anti-rat glial fibrillary acidic protein [GFAP], 1:200; Dianova, Hamburg, Germany) overnight at 4°C. After three times washing in phosphate-buffered saline solution (PBS), fluorescence-conjugated secondary antibody (goat anti-mouse) labeled with FITC (1:10) was used at room temperature for 1 h. After washing in PBS, slices were subsequently stained with rabbit anti-rat RAGE as described[5] and with TRITC-labeled swine anti-rabbit secondary antibody (1:20) (Dako Cytomation, Hamburg, Germany). For nuclei staining, Toto-3 (Invitrogen, Karlsruhe, Germany) and 4′,6-diamidino-2-phenylindole (DAPI) (Sigma, Munich, Germany) were used. Finally, the slices were mounted in 50% glycerol and photographs were taken with an Olympus (BX51) microscope (Olympus Opticals Europe, Hamburg, Germany) or with a confocal microscope (Leica TCS SP2 Confocal Microscope; Leica, Wetzlar, Germany).

Results

The metabolic data of the animals studied are summarized in TABLE 1. STZ-induced destruction of beta cells in SD rats resulted in substantial hyperglycemia and the failure to gain weight over time. Nonenzymatic

TABLE 1. Physical and metabolic parameters of the study groups

	N ($n = 3$)	D ($n = 2$)
Body weight (g)	725.7 ± 99.1	377.0 ± 86.3
Glucose level (mg/dL)	241.3 ± 9.0	574.5 ± 20.5
HbA$_1$ (%)	4.9 ± 0.01	14.7 ± 0.01

Abbreviations: N = nondiabetic; D = diabetic.

glycation of hemoglobin was also increased in diabetic rats.

RAGE expression in the nondiabetic retina was confined to ganglion cell membranes (FIG. 1B, arrows) and to the inner limiting membrane in which RAGE labeling appeared in a spotted fashion (FIG. 1B, arrowhead). Moreover, RAGE immunolabeling was found in photoreceptors. No staining was observed in vascular parts of the nondiabetic retina or in cells of the inner nuclear layer. In the nondiabetic retina, GFAP labels predominantly astrocytes. We found a staining pattern that is largely consistent with this description, including an uninterrupted staining of the inner limiting membrane (FIG. 1A, arrow). GFAP-positive filamentous bundles traversed the inner retina towards the outer half of the inner plexiform layer.

In diabetic retinae, RAGE immunolabeling was substantially stronger, occurred in more cell types, and expanded throughout the entire retina (FIG. 1F). Apart from ganglion cell labeling and a few scattered areas in the inner limiting membrane, RAGE-positive filaments were observed from the ganglion cell layer throughout the retina, including labeling of the photoreceptors. Consistent with the transdifferentiation of Müller cells in diabetic retinae, GFAP-labeling was strong throughout the retina in a filamentous staining pattern (FIG. 1E). There was a high degree of co-immunostaining of RAGE and GFAP, suggesting that most of RAGE was upregulated in Müller glia in the diabetic retina (FIG. 1G and H). Of note in vertical sections in which capillary profiles were clearly definable, RAGE-positive bundles ensheated capillary profiles (FIG. 1I–L, arrows), suggesting a close proximity of the ligand for RAGE, methylglyoxal-type AGE with RAGE as receptor. Staining patterns of vertical sections through the ganglion cell layer (FIG. 1I–L) and the inner nuclear layers confirmed that most of RAGE positivity was colocalized to GFAP-positive bundles (not shown). Vascular profiles, which were enwrapped by astrocytes in the upper parts and by Müller cells in the lower parts, did not show any distinct RAGE positivity. Thus, RAGE predominantly colocalized with glial cells in this animal model.

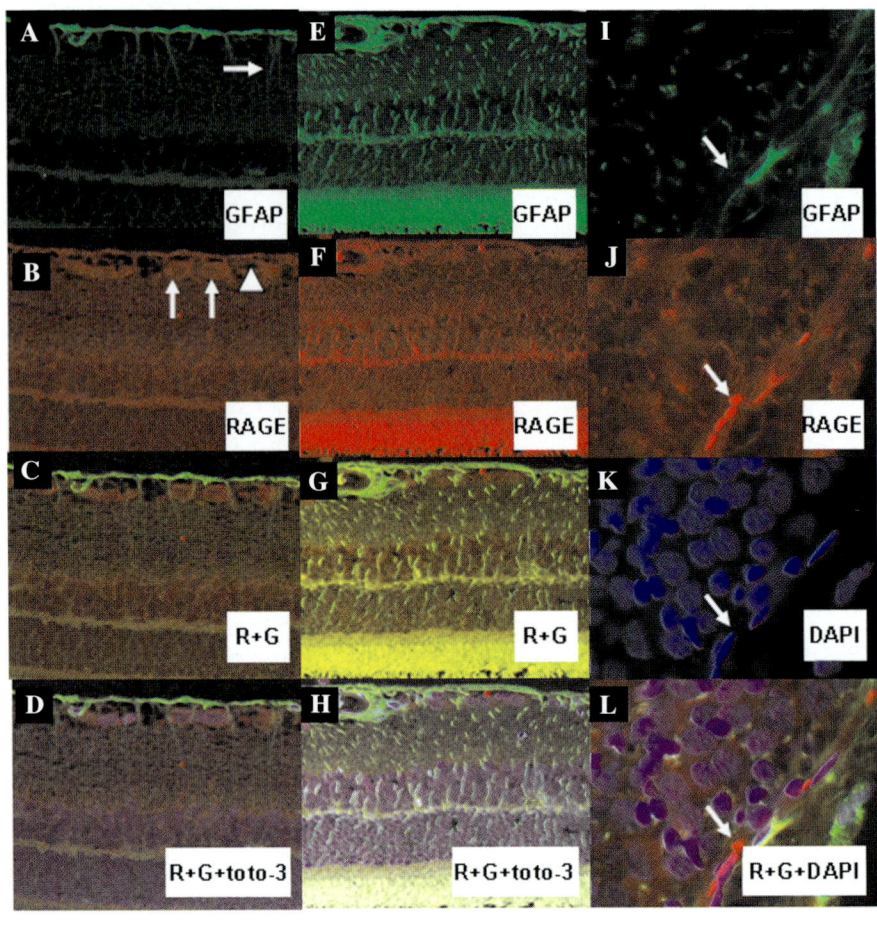

FIGURE 1. Representative photomicrographs (original magnification 400×) of (**A–D**) nondiabetic and (**E–L**) diabetic rat retinae labeled with glial fibrillary acidic protein (GFAP) (A,E,I), receptor for advanced glycation end products (RAGE) (B,F,J), and 4′,6-diamidino-2-phenylindole (DAPI) (denotes nuclear staining) (K). A–D and E–H represent vertical sections through the retina, and I–L represents a horizontal section focusing on the ganglion cell layer of a diabetic rat retina. A, arrow denotes astrocytic GFAP-labeling in the inner plexifom layer. B, arrows denote ganglion cells, arrowhead points to the inner limiting membrane. I–L, arrow indicates capillary profile. For staining protocol, see Materials and Methods section.

Discussion

In this study, we observed that the predominant cell type in which RAGE is upregulated under diabetic conditions is the retinal glia (astrocytes and Müller cells). In contrast to previous findings in other models, we did not find a prominent vascular RAGE localization.[4,6,7] However, given the proximity of retinal glia with the vasculature and the existing evidence of the interaction of Müller cells with the retinal vasculature, it is obvious that the modulation of RAGE via AGE in the diabetic setting can result in substantial vascular damage. Müller cells are of particular interest in this context. It has been recently demonstrated that glial cells in diabetic retinae are more sensitive to osmotic stress than nondiabetic cells. These alterations may be implicated in the development of retinal edema.[8] Another important inducer of increased vascular permeability in the diabetic retina is vascular endothelial growth factor (VEGF). Müller cells are a primary source of VEGF production, and VEGF expression is upregulated in diabetic rodent retina irrespective of local ischemia.[9] RAGE stimulation induces VEGF in retinal cells, suggesting a direct link between danger signaling and survival promotion by glial cells.[10]

Müller cells are also likely to play an important role in the development of pericyte loss in the early diabetic retina. Under diabetic conditions, angiopoietin-2 (Ang-2), which is a natural antagonist of the pericyte-recruiting factor Ang-1, is highly upregulated in the early diabetic retina, preceding pericyte loss. In mice with a genetic deletion of Ang-2, diabetic pericyte loss is abolished.[11] Unpublished evidence suggests the transcriptional deregulation of Ang-2 in Müller cells involves mechanisms that are similar to those in microvascular endothelial cells.[12] Thus, it is conceivable that early pericyte loss is, at least in part, controlled by transcriptional changes of Müller cell genes.

Müller cell apoptosis is evident in the diabetic rodent retina. We demonstrated earlier that neurotrophins not only prevent Müller cell apoptosis but also prevent the formation of acellular capillaries.[13] As mentioned, Müller cells are an important source for survival factors, such as VEGF.[14] The preservation of Müller cell function appears, thus, as a major contributing part in the prevention of diabetic retinopathy.

RAGE upregulation in retinal Müller cells was recently reported also in mice in which a combined retinal damage from hyperglycemia and hyperlipidemia was induced by mating db/db with ApoE-null mice.[15] In these animals, RAGE upregulation was associated with early signs of vascular damage and pericyte loss. The administration of soluble RAGE prevented both the formation of acellular capillaries and pericyte loss, underlining the importance of RAGE as a possible therapeutic target.

In summary, RAGE is a normal constituent at low levels in glial and neuronal parts of the retina. However, in the diabetic retina, RAGE is predominantly expressed in retinal glia, suggesting an important regulatory role of these cells.

Acknowledgments

This work was supported by the Deutsche Forschungsgemeinschaft (GRK 880; Y.W., F.vH., F.P.) and by a grant from the European Foundation for the Study of Diabetes (EFSD-Servier) (H.P.H., Y.F.).

Conflict of Interest

The authors declare no conflicts of interest.

References

1. BROWNLEE, M. 2001. Biochemistry and molecular cell biology of diabetic complications. Nature **414:** 813–820.
2. BIERHAUS, A., P.M. HUMPERT, M. MORCOS, et al. 2005. Understanding RAGE, the receptor for advanced glycation end products. J. Mol. Med. **83:** 876–886.
3. KERN, T.S. 2006. Differences among rat strains in the rate of development of early stages of diabetic retinopathy. Diabetes **55**(S1): 53 A.
4. BRETT, J., A.M. SCHMIDT, S.D. YAN, et al. 1993. Survey of the distribution of a newly characterized receptor for advanced glycation end products in tissues. Am. J. Pathol. **143:** 1699–1712.
5. HAMMES, H.P., A. ALT, T. NIWA, et al. 1999. Differential accumulation of advanced glycation end products in the course of diabetic retinopathy. Diabetologia **42:** 728–736.
6. NEEPER, M., A.M. SCHMIDT, J. BRETT, et al. 1992. Cloning and expression of a cell surface receptor for advanced glycosylation end products of proteins. J. Biol. Chem. **267:** 14998–15004.
7. SHOJI, T., H. KOYAMA, T. MORIOKA, et al. 2006. Receptor for advanced glycation end products is involved in impaired angiogenic response in diabetes. Diabetes **55:** 2245–2255.
8. PANNICKE, T., I. IANDIEV & A. WURM, et al. 2006. Diabetes alters osmotic swelling and membrane characteristics of glial cells in rat retina. Diabetes **55:** 633–639.
9. HAMMES, H.P., J. LIN, R.G. BRETZEL, et al. 1998. Upregulation of the VEGF/VEGF-receptor system in experimental background diabetic retinopathy of the rat. Diabetes **47:** 401–406.
10. HIRATA, C., K. NAKANO, N. NAKAMURA, et al. 1997. Advanced glycation end products induce expression of vascular endothelial growth factor by retinal Müller cells. Biochem. Biophys. Res. Commun. **236:** 712–715.
11. HAMMES, H.P., J. LIN, P. WAGNER, et al. 2004. Angiopoietin-2 causes pericyte dropout in the normal retina: evidence for involvement in diabetic retinopathy. Diabetes **53:** 1104–1110.
12. YAO, D., T. TAGUCHI, T. MATSUMURA, et al. 2007. High glucose increases angiopoietin-2 transcription in microvascular endothelial cells through methylglyoxal modification of mSin3A. J. Biol. Chem. **282:** 31038–31045.
13. HAMMES, H.P., H.J. FEDEROFF & M. BROWNLEE. 1995. Nerve growth factor (NGF) prevents both neuroretinal programmed cell death and capillary pathology in experimental diabetes. Mol. Med. **5:** 527–534.
14. STONE, J., A. ITIN, T. ALON, et al. 1995. Development of retinal vasculature is mediated by hypoxia-induced vascular endothelial growth factor (VEGF) expression by neuroglia. J. Neurosci. **15:** 4738–4747.
15. BARILE, G.R., S.I. PACHYDAKI, S.R. TARI, et al. 2005. The RAGE axis in early diabetic retinopathy. Invest. Ophthalmol. Vis. Sci. **46:** 2916–2924.

Advanced Glycation End Product Homeostasis
Exogenous Oxidants and Innate Defenses

HELEN VLASSARA,[a] JAIME URIBARRI,[b] WEIJING CAI,[a] AND GARY STRIKER[b]

[a]*Division of Experimental Diabetes and Aging, Department of Geriatrics, and the*
[b]*Division of Nephrology, Department of Medicine, Mount Sinai School of Medicine, New York, New York, USA*

Increased oxidative stress (OS) underlies many chronic diseases prevalent in aging. Data in humans confirm the hypothesis that advanced glycation end products (AGEs) and other oxidants derived from the diet may be major contributors to increased OS in normal adults as well as those with diabetes mellitus or kidney failure. Mice fed a diet with a lowered (approximately 50%) content of AGEs or a typical calorie-restricted (CR) diet, accumulated a smaller amount of AGEs, maintained normal levels of AGE receptor-1 (AGER1), and did not have increased oxidant stress or cardiac or kidney fibrosis with aging. However, the findings in mice fed a CR diet with an increased content of AGEs resembled those in mice fed a nonrestricted diet that had the usual higher content of AGEs. Thus, there was an inverse correlation between the dietary AGE content, the AGER1 to receptor for AGE (RAGE) ratio, OS, organ damage, and life span. In both humans and mice, there was an inverse correlation between the AGER1 to RAGE ratio and the levels of OS.

Key words: AGE receptor-1; RAGE; dietary AGE restriction; caloric restriction; p66^{shc}; oxidant stress

Introduction

We previously showed that a high dietary intake of glycoxidants is linked to increased oxidative stress (OS) and subclinical inflammation, a finding of particular relevance to the rising incidence of age-related chronic diseases, such as diabetes, cardiovascular disease, and renal disorders.[1–3] We now review new data, which strengthens the association between glycoxidants contained in the normal diet and pathophysiologic responses in both mice and humans.

Homeostasis of Advanced Glycation End Products

The body pool of advanced glycation end products (AGEs) reflects input from endogenous and exogenous sources, the effects of innate immunity and modifications of oxidants by cells, and excretion of oxidants by the kidneys. The endogenous formation of AGEs is markedly increased in the presence of hyperglycemia and elevated OS. The two major sources peculiar to the modern lifestyle that provide an increasing supply of exogenous AGEs are consumption of foods rich in AGEs and smoking.

AGEs are modified in the body by enzymatic degradation, innate defenses, and receptor-dependent uptake and degradation. The major degradative enzymes involved are the glyoxalase I and II system. For example, highly reactive AGE precursor molecules, such as methylglyoxal (MG), are detoxified by glyoxalase I and II at a rate proportional to the cytosolic levels of glutathione.[4] Another route of detoxification is through their reduction by aldose reductase and carbonyl reductase, which is catalyzed by aldehyde reductase.[5] There is also a group of circulating proteins that bind AGEs keeping them from causing cellular toxicity or binding to other molecules.[6] These proteins are part of the innate defense system and include lysozyme, defensins, and lactoferrin.[6,7]

There are two types of cell surface AGE receptors and molecules, those which bind AGEs and initiate cell activation[8] and those that bind, internalize, and degrade AGEs (7). The best-studied receptor that binds and initiates OS is the receptor for AGE (RAGE).[8] It recognizes AGEs as one of many ligands, is not endocytic, and is upregulated by OS. The second group of

Address for correspondence: Helen Vlassara, M.D., Mount Sinai School of Medicine, Box 1640, One Gustave Levy Place, New York, NY 10029. Voice: +-212-241-2567; fax: +1-212-241-7248.
helen.vlassara@mssm.edu

AGE receptors binds AGEs and mediates endocytosis and degradation of AGEs. This group includes the AGE receptor-1 (AGER1), AGER3, and CD36.[1,7,9] These systems help maintain normal AGE homeostasis throughout life under normal conditions, but not under conditions of chronically elevated AGEs and OS, such as in diabetes mellitus and aging.[3,10] The most extensively evaluated of the endocytic AGE receptors is AGER1, which has marked antioxidant properties modulating AGE responses via RAGE and nuclear factor kappa-B[7,9] but also via the epidermal growth factor receptor, extracellular receptor kinase, and p66^{shc}.[9] As a consequence, the AGER1 to RAGE ratio is an important element of the defense against excess OS. The importance of this ratio is emphasized by the observation that when AGER1 levels are suppressed, the levels of OS are elevated (FIGS. 1 and 2). Examples of these events include diabetes mellitus and late aging.[3,10] Thus, we hypothesize that a decrease in the level or function of AGER1 is one reason for the increased OS found in diabetes and late aging.

There have been many attempts to modify aging and its related disease manifestations by modification of OS. The role of changes in the metabolism of oxidants, such as AGEs, however, has received scant attention. Recent studies in mice provide data suggesting that both the levels of AGEs and AGE-receptor activity have a prominent effect on longevity and age-related diseases.[3,11] Young mice were randomized and pair-fed isocaloric diets that had either the regular level (Reg) or a 50% reduction (Reg$_{low}$) in the level of AGEs. A second group was pair-fed a calorie-restricted (CR) diet or a CR diet which had been heat-modified to have an elevated content of AGEs (CR$_{High}$) (TABLE 1). Both groups of mice were followed throughout their life span. Note that the regular diet (Reg) has relatively high levels of AGEs as does the CR$_{High}$ diet. The CR diet resulted in a reduced intake of both food and AGEs. Old mice maintained on the Reg or CR$_{High}$ diets developed insulin resistance, OS, decreased AGER1, increased RAGE, increased p66^{shc} (FIG. 2), and fibrosis of the heart and kidneys.[3,11] However, mice exposed to the Reg diet (which had a 50% reduction in the amount of AGEs) or the CR diet (which also resulted in a decreased intake of AGEs) had none of these changes related to increased OS but had a longer life span.[3,11] These findings suggest that avoidance of certain oxidants in the standard diet, such as AGEs, may reduce OS and delay the aging process with its related cardiovascular and kidney changes. Importantly, this can be accomplished without the need for caloric restriction, at least in mice. The AGE-receptor levels followed the pattern of lowered OS in mice on diets with a low

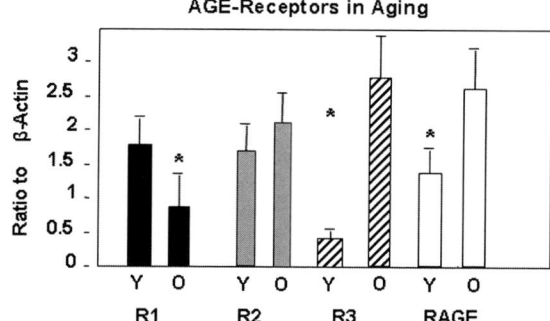

FIGURE 1. Tissue levels of advanced glycation end product receptor (AGER)1, AGER2, AGER3, and the receptor for AGE (RAGE) in 4- and 24-month mice fed diets with a regular AGE content. AGER1, AGER2, AGER3, and RAGE protein levels were assessed in spleen tissues from mice fed a regular diet (n = 6 per group) at 4 and 24 months. Data are shown as mean ±SE of three independent experiments (*P < 0.01, 4 months versus 24 months). Y: young (4 mo); O: old (24 mo). Bars: black, AGER1; gray, AGER2; striped, AGER3; white, RAGE.

content of AGEs. Namely, in CR mice there was persistence or enhancement in the expression of AGER1 (FIG. 2).[3,11] On the other hand, the level of RAGE, which promotes OS, was unchanged in the mice fed a CR diet, resulting in a high AGER1 to RAGE ratio. In contrast, RAGE levels were enhanced in mice fed CR$_{High}$ diets, and the AGER1 to RAGE ratio was decreased with a corresponding increase in OS. Because AGEs have been shown to upregulate cellular AGE uptake, partly through an AGER1-mediated mechanism, the increased expression of AGER1 in mice fed diets with a low content of AGEs suggests that AGER1 can respond to, and effectively handle, fluctuations in AGE load *in vivo*, if this load does not exceed a certain threshold. However, under conditions of chronically elevated exogenous AGEs, AGER1 levels are suppressed. As a result of decreased AGER1 levels, the capacity of the body to handle oxidants is reduced. While the time frame of the reduction in AGER1 levels is unknown, recent *in vitro* evidence revealed that sustained exposure to high levels of specific AGEs, such as MG or carboxymethyllysine, suppress AGER1 levels and there is sustained activation of proinflammatory cellular responses. This parallels data in two chronic conditions associated with high levels of AGEs, diabetes mellitus and renal failure.

A direct association between AGEs in the food and serum levels was shown by the addition of well-defined AGEs (eg., MG) to the low-AGE diet of pups born of dams that had been given a diet with a low content of AGEs during gestation (TABLE 1; FIG. 3A). Serum AGEs increased at 4 months in the mice fed

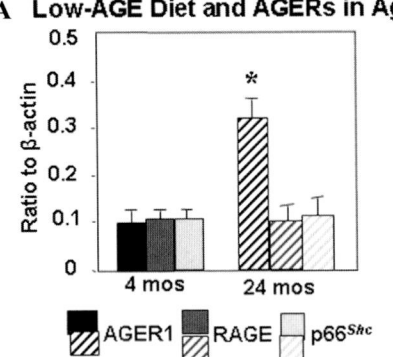 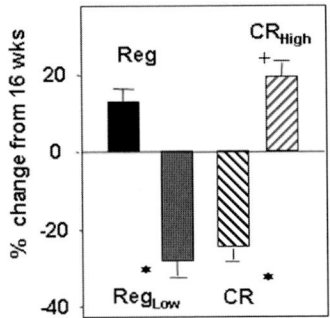

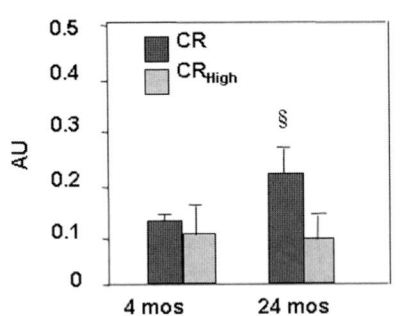 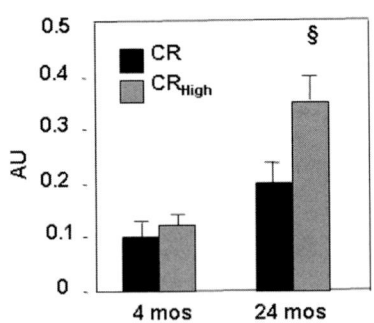

FIGURE 2. Levels of AGE receptors p66^{shc} in 4- and 24-month-old C57Bl6 mice on low-AGE or calorie-restricted (CR) diets. **(A, B)** Protein levels of AGER1, RAGE, and p66^{shc} were determined at 4 and 24 months of age in tissues (*$P < 0.01$, versus 24 months); **(C)** AGER1 levels in tissues (§$P < 0.01$, 4 months versus 24 months); **(D)** RAGE levels in tissues at 4 and 24 months (§$P < 0.01$, 4 months versus 24 months).

TABLE 1. Diets and intake

Groups	CR	CR$_{High-Age}$	Reg	Reg$_{Low}$	Reg$_{Low-AGE+MG}$[a]	Reg$_{Low-AGE}$[a]
Protein (g)	18.7	18.7	18.4	20	20	20
Fat (g)	4.4	4.4	4.47	4.5	4.5	4.5
Carbohydrate (g)	55.04	55.04	55.91	54.8	54.8	54.8
Total calories	4.02	4.02	3.95	4.02	4.02	4.02
AGE (U/g)	6.2×10^4	17.4×10^4	6.0×10^4	3.0×10^4	6.4×10^4	3.4×10^4
Food intake (g/day)	3	3	5	5	4.8	4
Calorie intake (g/day)	12.06	12.06	19.8	20.1	19.3	16.1
AGE intake (U/day)	18.6×10^4	52.5×10^4	29.5×10^4	15.5×10^4	30.7×10^4	13.6×10^4

[a]Short-term experiment: 6 months.
Abbreviations: AGE, advanced glycation end products; CR, calorie-restricted; Reg, regular.

a diet supplemented with MG and in the pups fed a regular diet. The increased serum AGE levels were accompanied by an increase in the levels of lipid peroxidation products (8-isoprostanes), which are markers of endogenous OS (FIG. 3B).

The important role of the kidneys in the metabolism and excretion of endogenous and dietary AGEs has been demonstrated in animals and humans with severe renal disease.[1,12,13] Kidney disease is associated with a sharply reduced ability to excrete an oral load of AGEs, and there is an inverse correlation between serum AGE levels and renal function. The decline in renal function noted in >15% of the "healthy" aging U.S. population[14] could contribute to increased

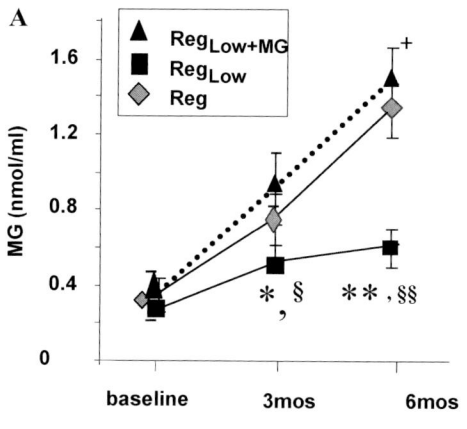 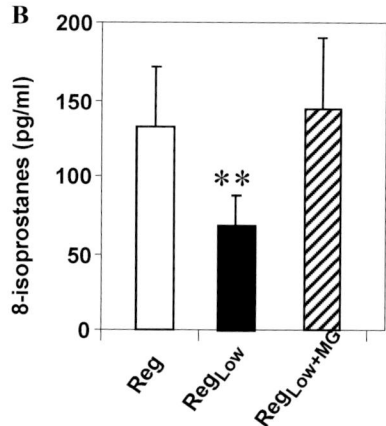

FIGURE 3. Dietary AGEs raise Serum AGEs and Oxidant Stress. Pups (F1) of dams fed a Low-AGE regular diet (Reg$_{Low}$) were maintained either on this diet, on Reg diet or on Reg$_{Low}$ diet enriched with methylglyoxal derivatives (MG) (Reg$_{Low+MG}$) for 6 months. **(A)** Serum MG levels, Reg$_{Low+MG}$ versus Reg$_{Low}$: $^+P < 0.001$, Reg$_{Low}$ versus Reg: $^*P < 0.01$, $^{**}P < 0.001$; Reg$_{Low}$ versus Reg$_{Low+MG}$: $^§P < 0.01$, $^{§§}P < 0.001$. **(B)** Plasma 8-isoprostanes: $^{**}P < 0.001$.

AGEs and OS in aging. The kidneys excrete AGEs and are also targets for AGE-induced injury. Thus, it is possible that a lifelong exposure to a high AGE diet might progressively damage the kidneys in humans, as has been documented by animal data.[3,11] Because the total load of AGEs in the tissues and OS increase with age and renal function and AGER1 levels decrease with age, we postulate that there is an interaction between these events. Namely, reduced AGER1 activity together with impaired renal function could contribute to increasing AGEs and hence OS, both of which are associated with tissue injury and organ dysfunction.

Dietary Oxidants and Advanced Glycation Homeostasis

Diet is a major source of AGEs and other oxidants that are generated during exposure of food to heat.[15] The most important determinant of the levels of AGEs in the food is the method of cooking, especially the cooking temperature.[15] Of course, AGE levels also depend on the content of proteins, lipids, and carbohydrates. Thus, meals cooked at high temperatures and under dry conditions have the highest AGE content, especially if there is a high fat content. We found a close association between the content of AGEs in the diet and serum AGE levels, OS, and inflammatory mediators across a spectrum of healthy, diabetic, and renal-failure subjects.[13,16,17]

A pathogenic role for exogenous AGEs has been documented in several *in vitro* and *in vivo* studies. Dietary AGEs have been shown to promote type 1 diabetes mellitus in nonobese diabetic (NOD) mice,[18] insulin resistance in *db/db* (+/+) mice,[19,20] diabetic nephropathy in NOD and db/db mice,[20] atherosclerosis in ApoE-deficient mice (FIG. 4),[21] impaired wound healing in NOD and *db/db* (+/+) mice,[22] and cardiovascular and/or kidney fibrosis (which is associated with a shortened life span in C57BL6 mice).[3,11]

The effect of short-term modifications in the intake of dietary AGEs on circulating and urinary AGEs was studied in a group of healthy subjects. Over a period of 9 days, these subjects were sequentially placed on isocaloric diets that varied in the content of AGEs: regular (approximately 16 AGE Eq×day 3 days), low AGE (approximately 4 AGE Eq×day 3 days), and regular (×3 days). The reduction of dietary AGE intake was associated with a 30% decrease, on average, in serum AGE levels (FIG. 5). Reinstitution of the regular diet was associated with an increase in serum AGE levels. Urinary AGE excretion directly correlated with circulating AGE levels during all three phases of the study. These data support our previous findings that variations in the AGE content in the diet were associated with corresponding changes in serum AGE levels.[12,16]

The biological effects of acute changes in dietary AGEs in humans were also studied in healthy and diabetic subjects administered a single oral load of an AGE-rich beverage.[23] Within 90 min, serum AGEs

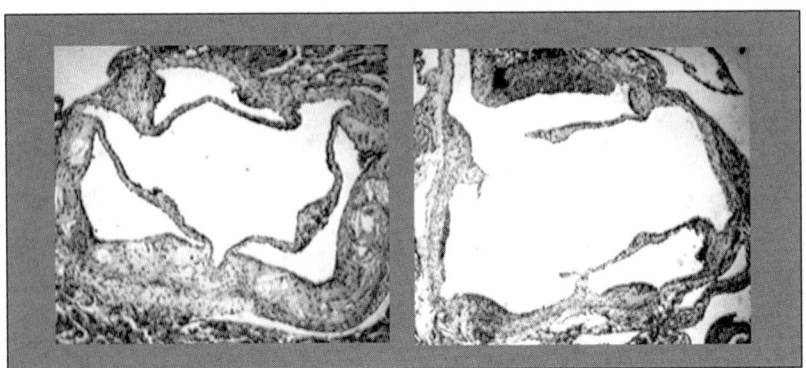

FIGURE 4. Decreased atherosclerosis in AGER1/ApoE$^{-/-}$ Mice. AGER1 immuno-staining of aortic root sections from ApoE$^{-/-}$ mice fed **(A)** a regular or **(B)** low-AGE diet for 4 months (n = 8 per group). Both groups were markedly hyperlipidemic. The atheromatous lesions are markedly reduced in the mice fed the low-AGE diet, and the levels of AGER1 in the intima and adventitia are more highly expression (red-brown staining) in these mice (B). (×200).

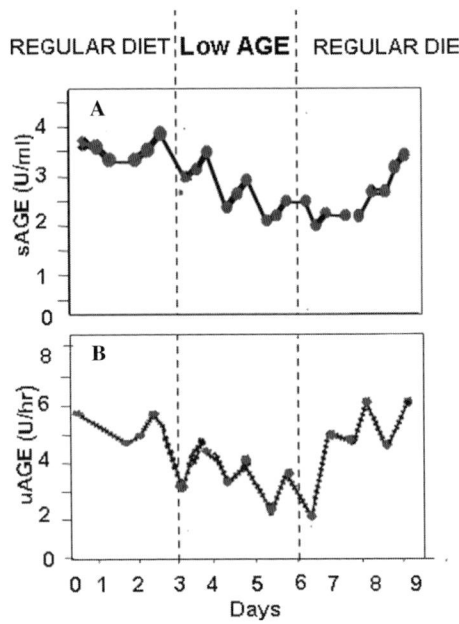

FIGURE 5. Levels of serum and urinary AGEs during serial modulation of the dietary AGE intake. Normal subjects were exposed to three consecutive periods of regular diet, low-AGE diet, and regular diet over a total of 9 days. Representative diagram depicts **(A)** serum AGE levels as U/mL and **(B)** urinary AGE as U/h. The dotted lines indicate changes in the dietary regimen: regular AGE diet (approximately 16 AGE Eq/day) versus low-AGE diet (approximately 4 AGE Eq/day) (n = 5).

levels were increased, and there was a transient impairment of flow-mediated vasodilatation. Similar results were found in diabetics who have elevated baseline levels of AGEs.[24] Comparison of flow-mediated vasodilatation following a single meal with either a high or low level of AGEs revealed markedly impaired dilation following the meal with a high-AGE content compared to an isocaloric meal with a low-AGE content. Prior treatment with very high doses of oral benfotiamine prevented the flow-mediated impairment induced by the high-AGE meal.[25]

These findings corroborated data from a larger cohort of normal subjects, which included a number of healthy elderly persons. There was an increase in the levels of serum AGEs with age, and the dietary AGE intake was an independent correlate of circulating AGEs.[17] Furthermore, these circulating AGEs correlate well with the levels of established markers of OS and inflammation, such as hsCRP. These findings support the hypothesis that AGE-mediated responses may affect OS levels in healthy adults and that circulating AGEs may now be considered as an indicator of the underlying balance between OS and innate immune responses. This is an important issue because increased OS among clinically normal individuals has been found to be important in the pathogenesis of insulin resistance and the metabolic syndrome in several large epidemiological studies.[25–27]

Thus, consumption of dietary AGEs directly influences systemic levels of AGEs, and this influence seems to contribute to a state of elevated OS and inflammation sustained throughout adulthood with significant adverse health consequences for all adults and especially for the aging population. This hypothesis should be tested in longitudinal studies and, while a number of such studies are currently in progress, large, randomized, controlled trials evaluating the effect of dietary intervention in terms of AGE restriction on health outcomes.

Conclusions

The levels of AGEs in the diet are important determinants of serum and tissue AGEs and of inflammatory markers, which result in decreased antioxidant reserves. These changes could underlie increased cardiovascular and/or renal disease and a shortened life span. The increased levels of AGEs and related inflammatory changes in both mice and humans appear to be partly a result of decreased levels of AGER1. The resultant changed AGER1 to RAGE ratio is associated with increased OS. These findings are pertinent to the normal population where consumption of large amounts of AGEs and related oxidants represents an independent factor for increased OS and inflammation. These factors can be modulated acutely, but if they remain chronically elevated, they may be associated with the development of several chronic diseases related to aging. These findings highlight the importance of consuming diets with a low-AGE content.

Acknowledgments

This work was supported by the National Institute on Aging (MERIT AG-23188 and AG-09453, H.V.) and by the National Institute of Research Resources, MO1-RR-00071, awarded to the General Clinical Research Center at Mount Sinai School of Medicine.

Conflict of Interest

The authors declare no conflicts of interest.

References

1. VLASSARA, H. 2005. Advanced glycation in health and disease. Role of modern environment. Ann. N.Y. Acad. Sci. **1043:** 452–460.
2. HUEBSCHMANN, A.G., J.G. REGENSTEINER, H. VLASSARA & J.E.B. REUSCH. 2006. Diabetes and advanced glycoxidation end products. Diabetes Care **29:** 1420–1432.
3. CAI, W., C.J. HE, L. ZHU, et al. 2007. Reduced oxidant stress and extended lifespan in mice exposed to a low glycotoxin diet. Association with increased AGER1 expression. Am. J. Pathol. **170:** 1893–1902.
4. ABORDO, E.A., H.S. MINHAS & P.J. THORNALLEY. 1999. Accumulation of alpha-oxoaldehydes during oxidative stress: a role in cytotoxicity. Biochem. Pharmacol. **58:** 641–648.
5. VANDER JAGT, D.L. & L.A. HUNSAKER. 2003. Methylglyoxal metabolism and diabetic complications: roles of aldose reductase, glyoxalase-I, betaine aldehyde dehydrogenase and 2-oxoaldehyde dehydrogenase. Chem. Biol. Interact. **143/144:** 341–351.
6. LIU, H., F. ZHENG, QI CAO, et al. 2006. Amelioration of oxidant stress by the defensin, lysozyme. Am. J. Physiol. **290:** E824–E832.
7. LU, C., J.C. HE, W. CAI, et al. 2004. AGE-Receptor-1 is a negative regulator of the inflammatory response to advanced glycation endproducts in mesangial cells. Proc. Natl. Acad. Sci. USA **101:** 11767–11772.
8. SCHMIDT, A.M., O. HORI, J. BRETT, et al. 1994. Cellular receptors for advanced glycation end products. Implications for induction of oxidant stress and cellular dysfunction in the pathogenesis of vascular lesions. Arterioscler. Thromb. **14:** 1521–1528.
9. CAI, W., J.C. HE, L. ZHU & H. VLASSARA. 2006. Advanced glycation end products (AGE) receptor-1 suppresses cell oxidant stress and activation signaling via EGF-Receptor. Proc. Natl. Acad. Sci. USA **103:** 13801–13806.
10. HE, C., T. KOSCHINSKY, C. BUENTING & H. VLASSARA. 2001. Presence of diabetic complications in Type 1 diabetic patients correlates with low expression of mononuclear AGE-receptor-1 and elevated serum AGE. Mol. Med. **7:** 159–168.
11. CAI, W., J.C. HE, L. ZHU, et al. 2007. Oral glycotoxins induce oxidant stress, insulin resistance, cardiovascular/renal disease and a shortened lifespan in calorie-restricted C57BL6 mice. In press.
12. KOSCHINSKY, T., C.J. HE, T. MITSUHASHI, et al. 1997. Orally absorbed reactive glycation products (glycotoxins): an environmental risk factor in diabetic nephropathy. Proc. Natl. Acad. Sci. USA **94:** 6474–6479.
13. URIBARRI, J., M. PEPPA, W. CAI, et al. 2003. Restriction of dietary glycotoxins reduces excessive advanced glycation end products in renal failure patients. J. Am. Soc. Nephrol. **14:** 728–731.
14. CORESH, J., B.C. ASTOR, E. GREENE, et al. 2003. Prevalence of chronic kidney disease and decreased kidney function in the adult US population: Third National Health and Nutrition Examination Survey. Am. J. Kidney Disease **41:** 1–12.
15. GOLDBERG, T., W. CAI, M. PEPPA, et al. 2004. Advanced glycoxidation end products in commonly consumed foods. J. Am. Diet. Assoc. **104:** 1287–1291.
16. VLASSARA, H., W. CAI, J. CRANDALL, et al. 2002. Inflammatory mediators are induced by dietary glycotoxins, a major risk factor for diabetic angiopathy. Proc. Natl. Acad. Sci. USA **99:** 15596–15601.
17. URIBARRI, J., W. CAI, M. PEPPA, et al. 2007. Circulating glycotoxins and dietary AGEs; two links to inflammatory response, oxidative stress and aging. J. Am. Gerontol. Sci. **62:** 427–433.
18. PEPPA, M., C. HE, C. HATTORI, et al. 2003. Fetal or neonatal low-glycotoxin environment prevents autoimmune diabetes in NOD mice. Diabetes **52:** 1441–1448.
19. HOFMANN, S.M., H.J. DONG, Z. LI, et al. 2002. Improved insulin sensitivity is associated with restricted intake of dietary glycoxidation products in the db/db mouse. Diabetes **51:** 2082–2089.
20. ZHENG, F., C. HE, W. CAI, et al. 2002. Prevention of diabetic nephropathy in mice by a diet low in glycoxidation products. Diabetes Metab. Res. Rev. **18:** 224–237.

21. LIN, R.Y., E.D. REIS, A.T. DORE, *et al.* 2002. Lowering of dietary advanced glycation endproducts (AGE) reduces neointimal formation after arterial injury in genetically hypercholesterolemic mice. Atherosclerosis **163:** 303–311.
22. PEPPA, M., P. BREM, J. EHRLICH, *et al.* 2003. Adverse effects of dietary glycotoxins on wound healing in genetically diabetic mice. Diabetes **52:** 2805–2813.
23. URIBARRI, J., A. STIRBAN, D. SANDER, *et al.* 2007. Single oral challenge by advanced glycation end products acutely impairs endothelial function in diabetic and nondiabetic subjects. Diabetes Care **30:** 2579–2582.
24. NEGREAN, M., A. STIRBAN, B. STRATMANN, *et al.* 2007. Effects of low- and high-advanced glycation endproduct meals on macro- and microvascular endothelial function and oxidative stress in patients with type 2 diabetes mellitus. Am. J. Clin. Nutr. **85:** 1236–1243.
25. STIRBAN, A., M. NEGREAN, A. STIRBAN, *et al.* 2006. Benfotiamine prevents macro- and microvascular endothelial dysfunction and oxidative stress following a meal rich in advanced glycation end products in individuals with type 2 diabetes. Diabetes Care **29:** 2064–2071.
26. COUILLARD, C., G. RUEL, W.R. ARCHER, *et al.* 2005. Circulating levels of oxidative stress markers and endothelial adhesion molecules in men with abdominal obesity. J. Clin. Endocrinol. Metab **90:** 6454–6459.
27. KEANEY, J.F., M.G. LARSON, R.S. VASAN, *et al.* 2003. Obesity and systemic oxidative stress. Clinical correlates of oxidative stress in the Framingham study. Arterioscler. Thromb. Vasc. Biol. **23:** 434–439.

Formation Mechanisms of Melanoidins and Fluorescent Pyridinium Compounds as Advanced Glycation End Products

FUMITAKA HAYASE,[a] TERUYUKI USUI,[a] YORIYUKI ONO,[a] YOSHINOBU SHIRAHASHI,[a] TOMOMI MACHIDA,[a] TAKASHI ITO,[a] NOZOMU NISHITANI,[a] KAZUHITO SHIMOHIRA,[a] AND HIROHITO WATANABE[b]

[a]*Department of Agricultural Chemistry, and*
[b]*Department of Life Science, Meiji University, Kanagawa 214-8571, Japan*

The formation mechanisms of melanoidins as advanced glycation end products (AGEs) have not been resolved. Blue and red pigments generated in the D-xylose–glycine reaction system are postulated to be intermediate oligomers in the generation of melanoidins. A novel blue pigment, designated blue-M5, was identified as a similar structure to blue-M1 except for the side chain of two dihydroxypropyl groups. Blue pigments were also generated in the D-glucose–glycine and D-xylose–β-alanine reaction systems as well as in the D-xylose–glycine reaction system. Blue pigments by the Maillard reaction might be formed by the decarboxylation of two molecules of pyrrolopyrrole-2-carbaldehydes (PPA). PPA, composed of a side chain of a dihydroxypropyl group, was identified as a precursor of blue pigments. In fact, blue-M5 was generated by the incubation of PPA alone. Blue pigments, which involved pyrrolopyrrole structures, were readily changed to brown polymers. Glyceraldehyde-derived pyridinium (GLAP) compound, a glyceraldehyde-derived fluorescent AGE, and lysyl-pyrropyridine, a 3-deoxyglucosone-derived fluorescent AGE, were detected at higher levels in the plasma proteins and the tail tendon collagen of streptozotocin-induced diabetic rats compared to normal rats. GLAP and lysyl-pyrropyridine, therefore, might be related to the progression of diabetic complications.

Key words: melanoidin; blue pigment; glyceraldehyde-derived pyridinium compound; advanced glycation end products; pyrropyridine

Introduction

The formation mechanisms of melanoidins as advanced glycation end products (AGEs) have not been resolved.[1] A major blue pigment was isolated from the reaction between D-xylose and glycine.[2] The blue pigment, designated blue-M1 (blue Maillard reaction intermediate-1) (FIG. 1), is postulated to be an intermediate oligomer in the generation of melanoidins. Blue-M1 has polymerizing activity, suggesting that it is an important Maillard reaction intermediate through the formation of melanoidins. Blue-M2 and red compounds (red-M1 and M2), which have higher molecular weights than blue-M1, were generated in D-xylose–glycine reaction systems.[3–5] In this paper, we identified other blue pigments and precursors of these pigments, which had polymerizing activity, and postulated the formation mechanisms of melanoidins. Moreover, we identified blue pigments formed from glucose–glycine and xylose–β-alanine reaction systems.

Glyceraldehyde-derived AGEs may contribute to the development of diabetic neuropathy and diabetic vascular disease.[6,7] We have reported the identification of a glyceraldehyde-derived pyridinium (GLAP) compound as a glyceraldehyde-derived AGE[8] and also identified a pyrrolopyridinium compound, which we named pyrropyridine, as a possible fluorescent lysine–lysine cross-link generated in a butylamine–3-deoxyglucosone (3DG) or glucose reaction system.[9] Here, we describe the determination of GLAP and pyrropyridine in the plasma proteins and the tail tendon collagen of streptozotocin-induced diabetic rats.

Address for correspondence: Fumitaka Hayase, Meiji University, Department of Agricultural Chemistry, Higashi-mita, Kawasaki, Kanagawa 214-8571, Japan. Voice: +81-44-934-7835; fax: +81-44-934-7902.
fumi@isc.meiji.ac.jp

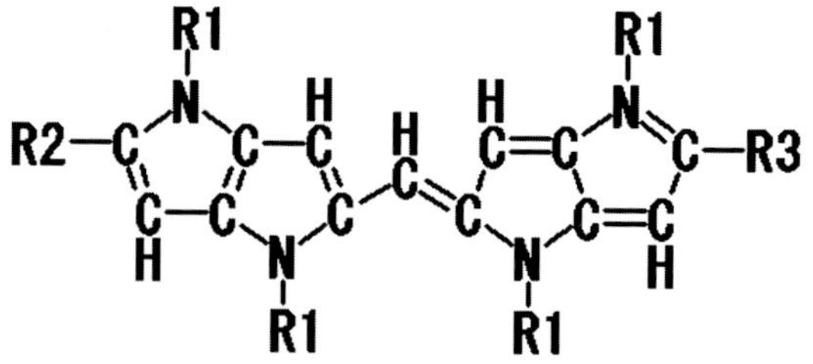

	R2	R3
Xyl-Gly(R1=CH$_2$COOH)		
Blue-M1	CH$_2$CHOHCH$_2$OH	CHOHCHOHCH$_2$OH
Blue-M5	CH$_2$CHOHCH$_2$OH	CH$_2$CHOHCH$_2$OH
Glc-Gly(R1=CH$_2$COOH)		
Blue-G1	CH$_2$(CHOH)$_2$CH$_2$OH	CH$_2$(CHOH)$_2$CH$_2$OH
Blue-G2 or G3	CH$_2$(CHOH)$_2$CH$_2$OH or	CH$_2$C(OH)=C(OH)CH$_2$OH
Blue-G4	CH$_2$(CHOH)$_2$CH$_2$OH	CH$_2$CH$_2$OH
Blue-G5	CH$_2$CHOHCH$_2$OH	CH$_2$CH$_2$OH
Blue-G6	CH$_2$CH$_2$OH	CH$_2$CH$_2$OH
Xyl-β-Ala(R1=CH$_2$CH$_2$COOH)		
Blue-B1	CH$_2$CHOHCH$_2$OH	CH$_2$CHOHCH$_2$OH

FIGURE 1. Structures of the blue pigment formed by the Maillard reaction.

Materials and Methods

Identification of Maillard Reaction Pigments

Blue-M5 was isolated from the reaction mixture of D-xylose (1 mol) and glycine (0.1 mol) and dissolved in 60% ethanol (starting pH 8.1) and NaHCO$_3$ (0.1 mol) under nitrogen at 26.5°C for 48 h or 2°C for 96 h. Pyrrolopyrrole-2-carbaldehyde (PPA), which was postulated as a precursor compound of blue pigments, was isolated from the same reaction solution. Blue compounds were isolated from the reaction mixture of D-glucose (0.7 mol) and glycine (0.3 mol) or β-alanine (0.3 mol) and dissolved in 30% ethanol (starting pH 8.0) and NaHCO$_3$ (0.1 mol) under nitrogen at 50°C for 4 days. These pigments were purified by anion exchange chromatography and reversed phase high-performance liquid chromatography (RP-HPLC) according to the elution conditions described in a previous paper.[2]

PPA was purified by anion exchange chromatography and Sephadex G-15 and was detected at 350 nm using RP-HPLC. The chemical structures of these blue pigments and PPA were determined based on NMR (500 MHz), fast atom bombardment mass spectroscopy (FAB-MS), and matrix-assisted laser desorption/ionization time-of-flight mass spectrometry (MALDI-TOF-MS) data.

Determination of Glyceraldehyde-Derived AGEs and Lysyl-pyrropyridine in Diabetic Rats

Sprague-Dawley (SD) rats (male, 6-weeks old, specific pathogen free) were obtained from Charles River Laboratories Japan (Yokohama, Japan) and were randomized into two groups (diabetic and control). Group 1 (diabetic) rats were injected with 65 mg/kg of streptozotocin. The animals were kept on a standard diet and water *ad libitum* for 12 weeks.

Subsequently, the animals were anesthetized with diethylether, and plasma samples were obtained from the abdominal vein with heparin as anticoagulant. The tail tendon was collected and defatted with acetone and used as collagen samples.

Plasma samples (0.5 mL) were combined with 0.5 mL of 10% trichloroacetic acid (TCA). The mixture was centrifuged at 1500 g for 10 min. The pellet was washed with diethylether and then dried to crude proteins. The crude protein samples and the collagen samples (1 mg) were hydrolyzed with 6 N HCl at 105°C for 20 h. Hydrolyzed samples for the determination of GLAP were treated with OASIS-MCX (Waters, Milford, MA) and Sep-Pak C18 (Waters) cartridges as described in a previous paper.[7] Hydrolyzed samples for the determination of lysyl-pyrropyridine were treated with Sep-Pak C18, CF-1 (Whatman, Kent, England), and Sep-Pak CM (Waters) cartridges. The samples were dissolved with 0.005% heptafluorobutyric acid (HFBA) for LC-MS and tandem MS (LC-MS/MS) analysis. LC-MS analysis was done by the electrospray ionization (ESI)-MS system (Mariner, Applied Biosystems, Foster City, CA) and the connecting HPLC system (Agilent, Santa Clara, CA). A 5 µL sample was placed on a Develosil C30-UG column (250 × 2.0 mm internal diameter) (Nomura Chemical, Aichi, Japan) and eluted with a linear gradient of 0.005% HFBA–60% acetonitrile from 0 to 20 min at a flow rate of 0.2 mL/min. ESI was performed under the positive polarity condition. GLAP and lysyl-pyropiridine were detected as precursor ion at m/z 255 [M^+] and 527 [M^+], respectively, by selected ion monitoring.

Results and Discussion

Identification of Maillard Reaction Pigments

A novel blue pigment (blue-M5) was isolated and purified by RP-HPLC. UV-VIS and fluorescent spectra of blue-M5 showed a maximum peak at 628 nm. FAB-MS for blue-M2 showed the presence of the M^+ ion at m/z 603. The high-resolution FAB-MS data for blue-M2 indicated the following: m/z (M^+): calculated for C27H31O12N4: 603.1939; found 603.1988. Blue-M5 was the identified compound and lacked one oxygen atom from the blue-M1 side chain as shown in FIGURE 1. MALDI-TOF-MS data by post-source decay mode (MS/MS) for a parent ion (Da = 603) of blue-M5 supported the identified structure. Blue-M5 was composed of a symmetrical structure having two dihydroxypropyl side chains and pyrrolopyrrole structures.

In addition, we investigated whether pyrrolopyrrole compounds were precursors of pigments because the identified pigments were composed of the common pyrrolopyrrole structure. We isolated a pyrrolopyrrole compound in the reaction mixture. UV-VIS and fluorescent spectra of this compound showed a maximum peak at 348 nm; FAB-MS showed the presence of the $[M+H]^+$ ion at m/z 325 and $[M+Na]^+$ ion at m/z 347. The high-resolution FAB-MS data for the pyrrolopyrrole compound indicated the following: m/z ($[M+Na]^+$) calculated for C14H16O7N2: 347.0855; found 347.0864. This compound was identified as 1,4-(dicarboxymethyl)-5-(2,3-dihydroxy propl)-pyrrolo[3,2-b]pyrrole-2-carboxaldehyde, designated PPA by NMR data as shown in FIGURE 2. PPA was composed of a dihydroxypropyl group side chain.

FIGURE 3 shows the proposed formation pathway of PPA; 1,2-enaminol would be found from D-xylose and glycine via Schiff's base, followed by 3-dexyxylosone (3-DX). 3-DX might react with glycine to give pyrrole-2-carbaldehyde. The pyrrole-2-carbaldehyde would likely react with imine to generate PPA. FIGURE 4 shows the proposed formation pathway of blue-M1 and melanoidins. Blue-M1 should be generated from PPA and an unidentified pyrrolopyrrole-2-carbaldehyde having a trihydroxypropyl group. This hypothesis was confirmed by the generation of blue-M5 when PPA alone was incubated for one day. Blue-M1 and red pigments, which involved the pyrrolopyrrole structure, were readily changed to a brown polymer as shown in FIGURE 4.[3–5] Accordingly, blue and red pigments are assumed to be generated independently and polymerized to form melanoidins.

We tried to detect blue pigments from the D-glucose–glycine reaction system. Some blue products were formed from the glucose reaction system as well as the D-xylose system. The major blue compound was blue-G1, followed by blue-G4, blue-G5, blue-G3, blue-G6, and blue-G2. FIGURE 1 shows the identified structure of these blue pigments based on NMR and MS data. Blue-G1 was composed of two pyrrolopyrrole structures, similar to blue-M5. In the case of blue-G1, the side chain was two trihydroxybutyl groups. HPLC analysis of the reaction solution obtained from D-glucose–glycine with or without D-xylose revealed that blue-G1, in the presence of D-xylose, was formed approximately eight times more than in D-glucose-glycine reaction system without D-xylose. The findings indicate that the pyrrolopyrrole structure, in part, might be formed from glycine and a C5 compound, which is generated by the retro-aldol condensation of glucose.

Moreover, we tried to detect a blue compound from the D-xylose–β-alanine reaction system. A major blue pigment (blue-B1) was identified as having a similar structure to blue-M1 except with an N-side chain as shown in FIGURE 1.

Determination of GLAP and Lysyl-pyrropyridine in Diabetic Rats

As shown in TABLE 1, body weight and plasma glucose level of the diabetic group were significantly different from the control (Student's t-test, $P < 0.05$). GLAP (FIG. 5) was detected in the plasma protein and collagen from diabetic rats but was not detected in normal rats. Identification of GLAP was verified by LC-MS and LC-MS/MS analyses. GLAP was detected as the m/z 255 (M^+) ion in the plasma protein and collagen of diabetic rats, and the fragment ion was observed at m/z 130, 126, and 84.

NMR data of PPA in D_2O

No	^{13}C δ (ppm)	^{1}H δ (ppm)
1	31.01	2.76–2.87 (m)
2	48.29	4.58 (s)
3	50.53	4.89 (s)
4	65.01	3.55–3.65 (m)
5	70.83	3.97–4.00 (m)
6	89.06	5.94 (s)
7	106.69	6.80 (s)
8	130.73	
9	132.24	
10	140.46	
11	144.99	
12	176.71	
13	176.71	
14	179.87	9.09 (s)

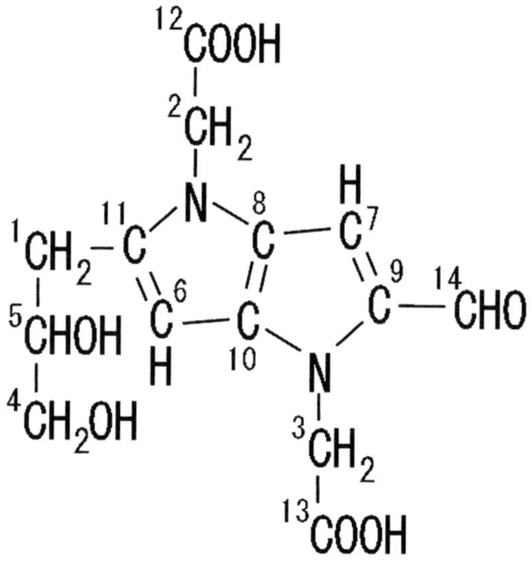

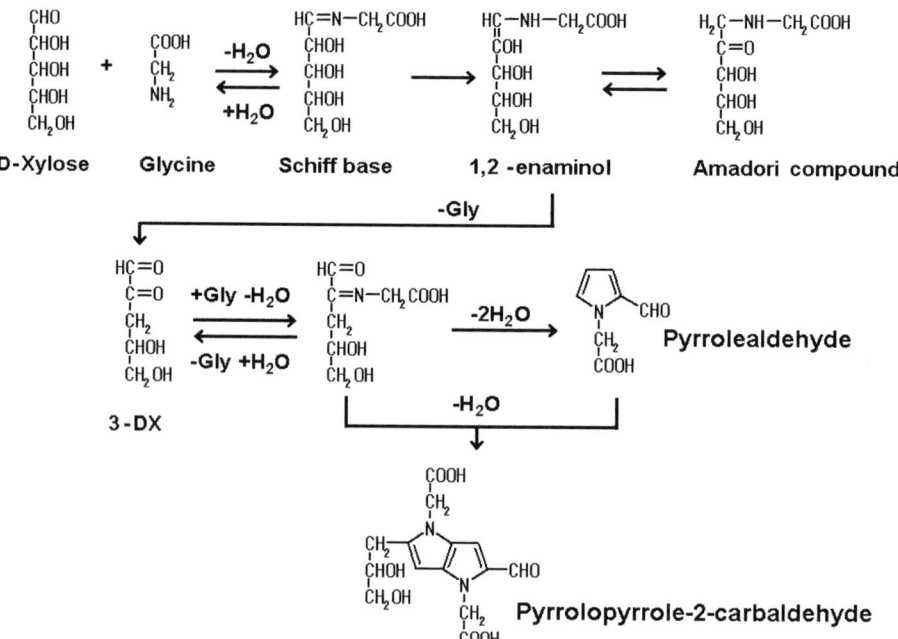

FIGURE 2. Structure of pyrrolopyrrole-2-carbaldehyde (PPA) and NMR data of PPA in D_2O.

FIGURE 3. Proposed formation pathway of PPA.

In a previous paper,[10] the GLAP level was analyzed by fluorescence HPLC in carbonyl-modified proteins using bovine serum albumin as the model protein. GLAP was detected only in the glyceraldehyde-modified albumin, not in the other carbonyls (glyceraldehyde, glycolaldehyde, methylglyoxal, glyoxal, fructose, ribose, xylose, glucose, and mannose). Glyceraldehyde was probably generated from sugars, such as glucose, by the incubation procedure, but GLAP was not detected in the sugar-modified albumin, possibly because the GLAP level might have been very low. Therefore, GLAP is considered to be a specific AGE formed by glyceraldehyde-related glycation. GLAP was also formed from glyceraldehyde-3 phosphate with lysine as well as by glyceraldehyde with lysine.[10] The degradation of glyceraldehyde-3-

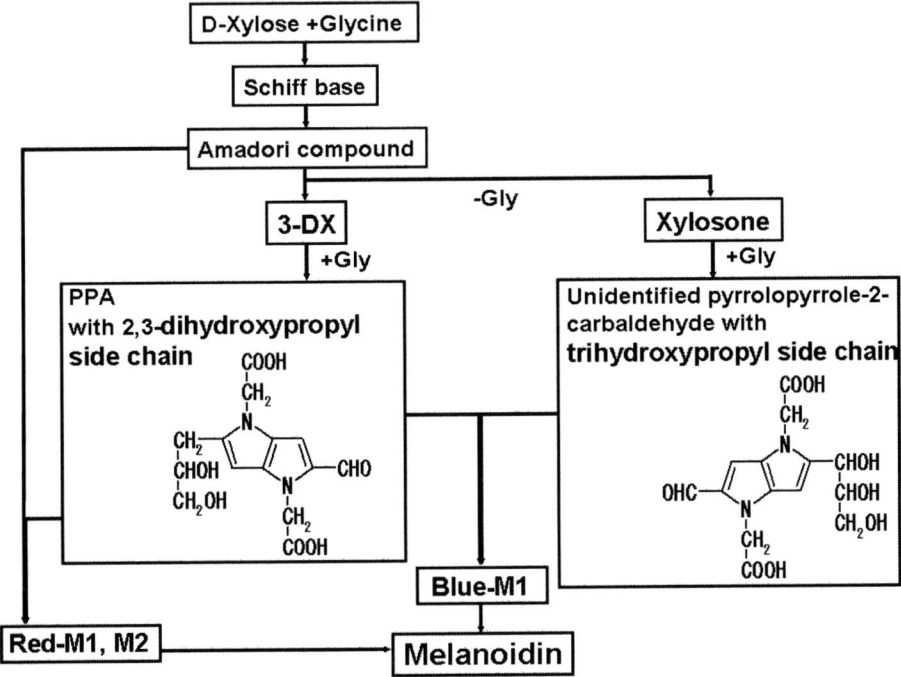

FIGURE 4. Proposed formation pathway of blue-M1 and melanoidins.

TABLE 1. Determination of glyceraldehyde-derived pyridinium (GLAP) in plasma protein and tail tendon collagen of streptozotocin-induced diabetic rats[10]

	Control	Diabetes
GLAP (pmol/mg protein)	Not detected (n=3)	71 ± 12* (n=6)
GLAP (pmol/mg collagen)	Not detected (n=3)	36 ± 14* (n=4)

Data are expressed as the means ± SD. *Differences with probability values less than 5% were considered significant.

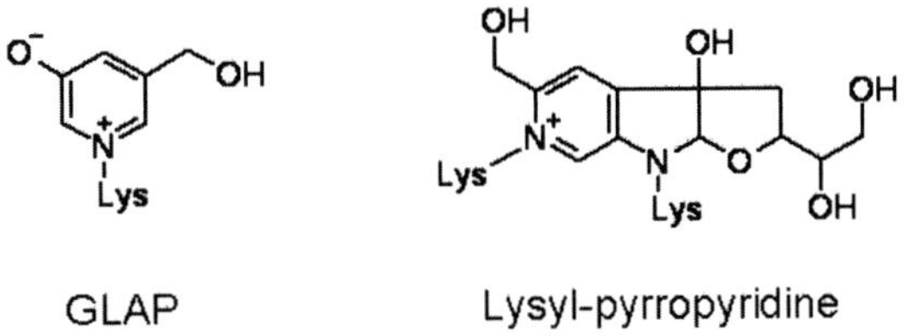

FIGURE 5. Structure of glyceraldehyde-derived pyridinium (GLAP) and lysyl-pyrropyridine.

phosphate is a key pathway in glyceraldehyde formation. Accordingly, it is suggested that an increase in the GLAP level reflects an increase in the glyceraldehyde level and the GA3P level. GLAP might be increased by reduced activity of a glyceraldehyde-related enzyme, such as GA3P dehydrogenase.

AGEs are nonenzymatic chemical modifications of *in vivo* proteins that can serve as biomarkers of carbonyl and oxidative stresses resulting from sugar and lipid oxidation. For example, pentosidine[11] and carboxymethyllysine[12] are markers of glycoxidation. Pyrraline,[13] pyrropyridine,[9] and imidazolones are from 3DG-modified proteins.[14] GLAP also induced cytotoxicity and oxidative stress for HP-60 cells.[15] Additionally, methylglyoxal-derived hydroimidazolone-1[16] serves as an index of injury to glyceraldehyde and methylglyoxal-related enzymes in glycolysis.

Since GLAP is not formed from methylglyoxal, GLAP might serve as a specific marker of reduced activity of a glyceraldehyde-related enzyme in metabolic diseases, such as diabetic complications.

Lysyl-pyrropyridine (FIG. 5) is an AGE formed from 3DG with lysine residues in protein. Identification of lysyl-pyrropyridine was verified by LC-MS analysis. Lysyl-pyrropyridine was detected as the m/z 527 (M^+) ion in the plasma protein and collagen of diabetic rats. Determination data are summarized in FIGURE 6. The lysyl-pyrropyridine level increased in the diabetic rats compared to normal rats. Therefore, lysyl-pyrropyridine might also be related to the progression of diabetic complications.

Conflict of Interest

The authors declare no conflicts of interest.

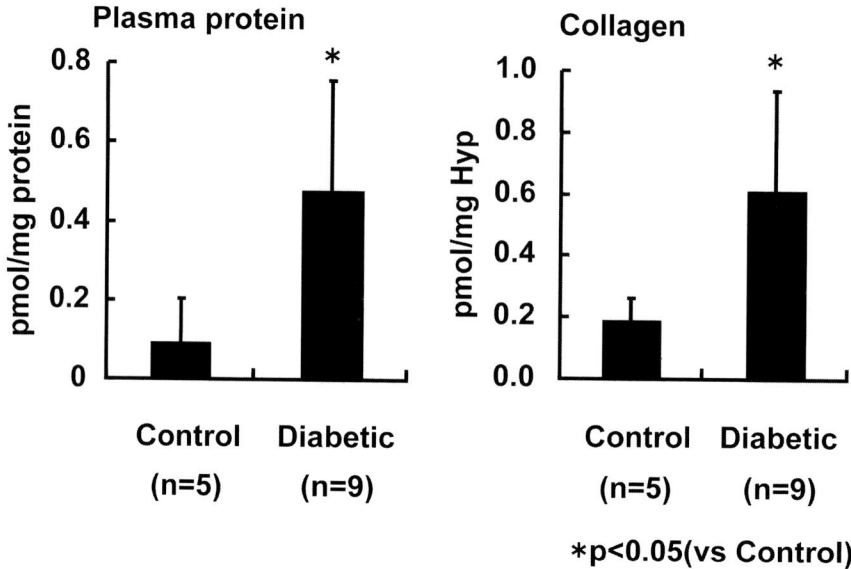

FIGURE 6. The amount of lysyl-pyrropyridine *in vivo*.

References

1. HAYASE, F. 2000. Recent development of 3-deoxyosone related Maillard reaction products. Food Sci. Technol. Res. **6:** 79–86.
2. HAYASE, F. *et al.* 1999. Identification of blue pigment formed in a D-xylose-glycine reaction system. Biosci. Biotechnol. Biochem. **63:** 1512–1514.
3. SASAKI, S. *et al.* 2006. Identification of a novel blue pigment as a melanoidin intermediate in the D-xylose-glycine reaction system. Biosci. Biotechnol. Biochem. **70:** 2529–2531.
4. HAYASE, F. *et al.* 2006. Chemistry and some biological effects of model melanoidins and pigments as Maillard intermediates. Mol. Nutr. Food Res. **50:** 1171–1179.
5. HAYASE, F. *et al.* 2005. Chemistry and biological effects of melanoidins and glyceraldehydes-derived pyrimidium as advanced glycation end products. Ann. N.Y. Acad. Sci. **1043:** 104–110.
6. TAKEUCHI, M. *et al.* 2000. Neurotoxicity of advanced glycation end-products for cultured cortical neurons. J. Neuropathol. Exp. Neurol. **59:** 1094–1105.
7. YAMAGISHI, S. *et al.* 2002. Advanced glycation end products-induced apoptosis and overexpression of vascular endothelial growth factor and monocyte chemoattractant protein-1 in human cultured mesangial cells. J. Biol. Chem. **277:** 20309–20315.
8. USUI, T. & F. HAYASE. 2003. Isolation and identification of the 3-hydroxy-5-hydroxymethyl-pyridinium compound as a novel advanced glycation end product on glyceraldehyde-related Maillard reaction. Biosci. Biotechnol. Biochem. **67:** 930–932.
9. HAYASE, F., H. HINUMA, M. ASANO, *et al.* 1994. Idetification of novel fluorescent pyrrolopyridinium compound formed from Maillard reaction of 3-deoxyglucosone and butylamine. Biosci. Biotechnol. Biochem. **58:** 1936–1937.
10. USUI, T., K. SHIMOHIRA, H. WATANABE & F. HAYASE. 2007. Detection and determination of glyceraldehyde-derived pyridinium-type advanced glycation end product in streptozotocin-induced diabetic rats. Biosci. Biotechnol. Biochem. **71:** 442–448.
11. SELL, D.R. & V.M. MONNIER. 1989. Structure elucidation of a senescence cross-link from human extracellular matrix: implication of pentoses in the aging process. J. Biol. Chem. **264:** 21597–21602.
12. AHMED, M.U. *et al.* 1997. N^ε-(carboxymethyl)lysine, a product of the chemical modification of protein by methylglyoxal, increases with age in human lens proteins. Biochem. J. **324:** 565–570.
13. HAYASE, F. *et al.* 1989. Aging of proteins: immunological detection of a glucose-derived pyrrole formed during Maillard reaction in vivo. J. Biol. Chem. **264:** 3758–3764.
14. HAYASE, F., Y. KONISHI & H. KATO. 1995. Identification of the modified structure of arginine residues in proteins with 3-deoxyglucosone, a Maillard reaction intermediate. Biosci. Biotechnol. Biochem. **59:** 1407–1411.
15. USUI, T., S. SHIZUUCHI, H. WATANABE & F. HAYASE. 2004. Cytotoxicity and oxidative stress induced by the glyceraldehyde-related Maillard reaction products for HL-60 cells. Biosci. Biotechnol. Biochem. **68:** 333–340.
16. USUI, T., H. WATANABE & F. HAYASE. 2006. Isolation and identification of 5-methyl-imidazolin-4-one derivative as glyceraldehyde-derived advanced glycation end product. Biosci. Biotechnol. Biochem. **70:** 1496–1498.

Advanced Glycation as a Basis for Understanding Retinal Aging and Noninvasive Risk Prediction

ANNA M. PAWLAK,[a] JOSEPHINE V. GLENN,[a] JAMES R. BEATTIE,[b] JOHN J. MCGARVEY,[c] AND ALAN W. STITT[a]

[a]*Centre for Vision Science, School of Biomedical Science,* [b]*Centre for Clinical Raman Spectroscopy, and* [c]*School of Chemistry and Chemical Engineering, Queen's University Belfast, Belfast, United Kingdom*

The retina is exquisitely sensitive to age-related processes, and, in many cases, these can precipitate progressive and profound loss of vision. Many asymptomatic abnormalities that accrue in the outer retina as we get older can serve as a sinister preamble to age-related macular degeneration (AMD). This condition remains the leading cause of irreversible blindness in industrialized countries, but its precise pathogenesis has yet to be completely elucidated. Over recent years, increasing evidence has suggested that the accumulation of advanced glycation end products (AGEs) and activation of the receptor for AGEs in the outer retina could play a significant role in the initiation and progression of AMD. The current review outlines this evidence and indicates how products of Maillard chemistry could be used as robust markers for outer retinal aging and susceptibility to AMD. The utility of Raman spectroscopy to measure AGE adducts in human tissues is presented. The methodology reinforces the association between AGE formation and retinal aging and provides exciting possibilities for assessing these pathogenic agents in the living eye and, perhaps, for providing a valuable index for AMD susceptibility.

Key words: retina; advanced glycation; aging; Raman spectroscopy

Basic Retinal Structure and Age-related Change

Tissue aging can be defined as a progressive decline in cellular and physiological function. As a pathogenic process, aging is associated with diminished capacity to respond to stress and concomitant susceptibility to degenerative disease.[1–3] Cellular manifestations of aging include increased chemical damage to proteins, accumulation of intracellular and/or extracellular deposits, and decreased efficiency of antioxidant defence. These processes are pronounced in long-lived post-mitotically differentiated cells, such as cardiac myocytes, central nervous system neurons, and retinal pigment epithelium (RPE).[4–7]

As a highly specialized tissue with little regenerative capacity, the retina and especially the cone photoreceptor-rich macula region is exquisitely sensitive to age-related pathology. Aging of the outer retina is associated with anatomical and functional alterations at the RPE and its underlying, highly specialized basal lamina (known as Bruch's membrane) (FIG. 1). This highly differentiated monolayer of neuroepithelial cells is located at the interface between the neural retina and the dense capillary network of the choroid (choriocapillaris). The RPE constitutes the outer blood retinal barrier and also plays a central role in normal retinal physiology.[8]

Progressive age-related dysfunction of the RPE at the Bruch's membrane axis can ultimately lead to atrophy of the outer retina. Bruch's membrane becomes progressively thicker with age, and there is remodeling of its fibrous structure, culminating in reduced hydraulic conductivity and charge selectivity.[9–11] This has implications for the RPE, which relies on this substrate for maintenance of appropriate cell function and survival, and there is a net reduction in RPE density as we age, especially in the macular region. The surviving RPE shows decreased melanin content, altered lysosomal degradative capacity, destabilization of mitochondria, and proteasome dysfunction.[12–14] Age-related change to the RPE can be readily visualized by the ophthalmologist, which is possible because of the amassing intracellular autofluorescent material (called

Address for correspondence: Alan W. Stitt, Centre for Vision Science, School of Biomedical Science, Queen's University Belfast, Royal Victoria Hospital, Belfast BT12 6BA. Voice: +011 44 (0)28-90632546; fax: +011 44 (0)28-90632699.
a.stitt@qub.ac.uk

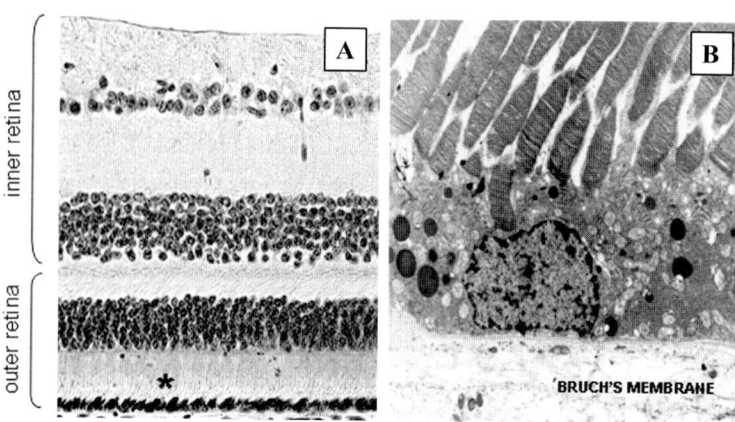

FIGURE 1. Basic anatomy of the retina and position of the retinal pigment epithelium (RPE). **(A)** histological section of the human retina. The tissue is segmented into inner and outer portions with the latter constituting the photoreceptors (including their nuclei, middle, and outer segments and the RPE (*). **(B)** An electron micrograph from the area depicted in A. It showed the RPE resting on Bruch's membrane at its basal surface and interacting with the photoreceptor outer segments on its apical surface. It has been estimated that the RPE phagocytoses and digests approximately 4000 of these phospholipid and highly oxidized membranous disks daily and engulf 300 million million disks during a 70-year life span. (In color in *Annals* online.)

lipofuscin), which is nondegradable and accumulates within the RPE cytoplasm.[15] In parallel, extracellular deposits, referred to as drusen or basal laminar deposits (BLD) (according to configuration), occur between the RPE and inner collagenous layer of Bruch's membrane.[16]

Characteristic features relating to dysfunction in the RPE are commonly referred to as early age-related maculopathy (ARM). ARM is common in people aged over approximately 55 and is characterized by deposition of sub-RPE drusen, BLD, and lipofuscin accumulation. There is no visual loss in early ARM, but a significant proportion of patients will progress to the late-stage manifestations of age-related macular degeneration (AMD). These are geographic atrophy (GA) and choroidal neovascularization (CNV), termed dry and wet AMD, respectively, and both result in severe irreversible loss of central vision. AMD constitutes the leading cause of irreversible blindness in industrialised countries[17] and its prevalence in our populations is likely to rise as a consequence of increasing longevity.[18,19]

The precise pathogenesis of ARM and AMD remains ill-defined, but changes in the chemical composition, physical structure, and hydrodynamics of Bruch's membrane during aging are important early stages in this disease.[20,65] In turn, RPE dysfunction is intimately linked to Bruch's membrane alterations, and, during aging, cells undergo inflammatory and oxidative stresses. There are strong indications that pro-inflammatory processes are linked to age-related protein modifications and influence RPE function. Indeed, the inflammation "hypothesis" has been strengthened by a recent report that, in certain susceptible individuals with polymorphisms in the complement factor H gene, there is a strong association with wet AMD.[21] This indicates that chronic pro-inflammatory factors are likely to appear in unison with cumulative protein and lipid modifications in the outer retina and are likely to be critical to the degenerative process.[22] Furthermore, in wet AMD, CNV is heavily influenced by pro-inflammatory cascades which, in combination with alterations in Bruch's membrane, can lead to upregulation of pro-angiogenic agents and endothelial cell infiltration of the subretinal space.[23,24]

Pathogenic Role of Advanced Glycation in Retinal Aging

The etiology of retinal aging is complex and multifactorial.[19] Chronic oxidative insults are strongly linked to RPE/Bruch's membrane damage leading to ARM/AMD,[25,26] and these occur in parallel with modification of proteins by advanced glycation reactions, leading to advanced glycation end products (AGEs).[27] These adducts are derived from Maillard chemistry in which ε amino acids react with glucose, lipid peroxidation products or various α-oxaloaldehydes, such as methylglyoxal and glycoaldehyde.[27] AGEs accumulate especially, but not exclusively, on long-lived structural proteins, such as collagens and lens crystallins, and are recognized as important instigators of age-related pathology by altering macromolecular structure and function.[28] Although these reactions occur during our entire life span, there is evidence that they are enhanced in pathogenic aging, leading to increased tissue levels of AGE adducts, such as pentosidine, N^ε-(carboxymethyl) lysine (CML), N^ε-(carboxyethyl)lysine, and hydroimidazolones.[29,30]

The retina is rich in polyunsaturated fatty acids (PUFA), such as docosahexaenoic acid, which are highly susceptible to lipid peroxidation.[31] This process yields lipid hydroperoxides that decompose to reactive aldehydes, such as acrolein, 4-hydroxynonenal, or malondialdehyde, which can, in turn, react with proteins to form stable adducts known as advanced lipoxidation end products (ALEs).[32] In essence, there is considerable overlap between AGEs and ALEs in terms of pathogenic potential (although the receptor interactions of ALEs remain obscure), and they should both be considered as adding to the burden of protein modifications in the aging retina.

As the outer retina has high levels of PUFAs and is constantly bathed in high oxygen and glucose levels (according to high metabolic demands), it could be anticipated that AGE and ALE formation could be comparatively high in this tissue. Evidence shows that AGEs accumulate in RPE where they can appear as free adducts or as AGE-modified proteins in lipofuscin granules.[33] Immunoreactive AGEs have also been shown in drusen and Bruch's membrane from aging eyes and at elevated levels in patients with AMD.[33-37] AGEs also occur at comparatively high levels in CNV membranes[38] where they appear to play a role in fibrous membrane formation by induction of transforming growth factor-β and platelet-derived growth factor (PDGF).[39,40] CML has been shown to promote CNV formation in cultured choroidal explants from aged rats via stimulation of growth factors, such as vascular endothelial growth factor (VEGF), tumor necrosis factor-α, and PDGF.[41] Moreover, various AGEs and ALEs can induce pro-angiogenic growth factors in RPE *in vitro*,[42] although these responses need to be confirmed *in vivo*.

Taken together, it seems likely that AGEs have an important pathogenic role in the aging retina,[43] and it is interesting that their presence is linked with chronic inflammation at the outer retinal layers in an *in vivo* model of retinal aging.[44] AGEs also alter the key complement regulatory protein CD59 and enhance inflammatory responses,[45] an important pathogenic step in age-related neurodegeneration, including Alzheimer's disease.[46] It is important to note that AGEs, in addition to their established role in damaging protein function, could act as important instigators of retinal disease by provoking pro-inflammatory cascades that are modulated by key receptors expressed by retinal cells.

The AGE–RAGE Axis in AMD

The receptor for AGEs (RAGE) is a member of the immunoglobulin super-family with a high affinity for several ligands, including AGEs, high mobility group-1 protein, β-amyloid, and S100/calgranulin. RAGE signals a range of pathophysiological responses linked to downstream transcriptional activity of nuclear factor kappa B, and receptor activation leads to induction of pro-inflammatory cytokines and oxidative stress.[47] Suppression of RAGE signaling using peptide analogues or neutralizing antibodies can prevent key pathological events in a range of cells and tissues.[48,49] In the context of the outer retina, studies have established that RAGE is expressed on RPE and that receptor levels are significantly increased in AMD (in postmortem tissue), especially on cells adjacent to drusen.[34,36] Interestingly, it has been shown that exogenous AGE-albumin and S100B modulate pro-angiogenic VEGF expression by RPE,[50] and, with prolonged exposure, these ligands may induce apoptosis.[34] *In vitro*, AGEs can induce various abnormal responses in the RPE, such as an increase in VEGF expression, a response that may be modulated by RAGE[51] and another AGE receptor known as galectin-3.[52]

Can AGEs Be Used as a Diagnostic Marker in the Aging Eye?

Traditionally, AGEs have been quantified using immunoassays,[53,54] chromatographic techniques,[55,56] and gas chromatography–mass spectroscopy (GC/MS).[57] These approaches allow identification and quantitative evaluation of defined AGE adducts in samples, but they often require considerable preprocessing of excised tissues, and their application for noninvasive characterization is limited.

AGE-linked autofluorescence can permit a degree of adduct quantification,[58] and this can also be usefully applied to *in vivo* analysis,[59] although the approach provides little information about the structural nature of the AGEs involved. Recently, Glenn et al.[53] have demonstrated age-dependent accumulation of AGEs in postmortem human Bruch's membrane/choroid. AGE adducts were quantified in homogenized tissue using standard analytical methods (high-performance liquid chromatography, GC/MS) and immunohistochemistry (FIG. 2). In addition, clinical samples were investigated using confocal Raman microscopy (CRM),

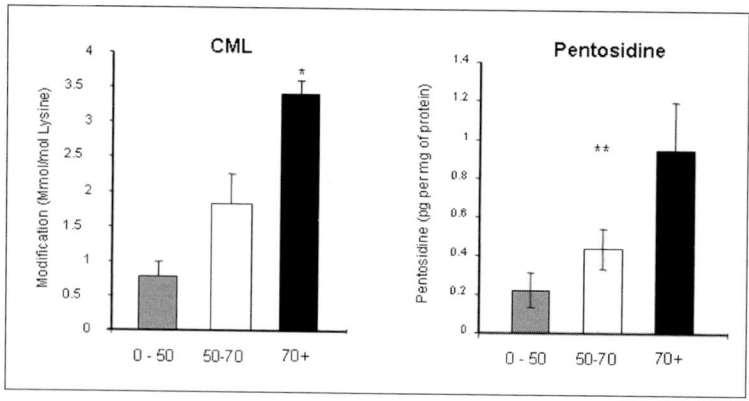

FIGURE 2. Quantification of N$^\varepsilon$-(carboxymethyl)lysine (CML) and pentosidine in human Bruch's membrane. CML was assayed using gas chromatography–mass spectroscopy from donors aged 0–50 years (grey bars), 50–70 years (white bars), and >70 years (black bars). The increase in CML in the oldest compared to the youngest cohort is statistically significant (*). Pentosidine levels were also determined using high-performance liquid chromatography analysis using "spiked" pentosidine as a standard, and levels were increased in aged donors. *$P < 0.05$; **$P < 0.005$.

and the study demonstrated how this technique can enable detection, quantification, and identification of defined AGEs in a nondestructive and noninvasive manner.[53]

Raman spectroscopy is based on the phenomenon of scattering of light by matter. Most incident monochromatic light is scattered by the sample elastically, and the frequency of both the incident and the scattered light are the same (Rayleigh scattering). However, interaction between photons and molecules causes a change of their vibrational state, leading to inelastic scattering. Thus, the light scattered by these molecules may have a lower (Stokes Raman scattering) or a higher (anti-Stokes Raman scattering) frequency than the incident light. Such differences correspond to the frequencies at which bonds of the molecules in the sample vibrate, and so each molecule possesses its own group of characteristic bond vibrations. This constitutes a unique spectrum and is, in effect, a "molecular fingerprint." Besides high specificity, the most important advantage of Raman spectroscopy is that it is nondestructive under appropriate conditions of wavelength and laser power.[60]

Raman spectroscopy has utility as a noninvasive technique to detect changes in tissue composition related to either disease or aging.[61,62] When Raman spectroscopy is combined with confocal microscopy, it can also provide spatial information about where defined molecules occur within complex tissue structures.[53,63] Raman vibrational features have narrow bandwidths, thus allowing the simultaneous identification of multiple colocalized molecules and the mapping of species distribution throughout a particular sample in three dimensions. We have uncovered Raman spectral features for various AGEs, and these fingerprints have been subsequently observed in human retina/choroid using CRM[53] (FIG. 3). The Raman data set successfully modeled defined adducts and heterogenous AGEs, and this correlated closely with accepted analytical approaches.[53] Interestingly, AGE analysis using CRM could be used to predict chronological age of the clinical samples (FIG. 3).

Conclusion

It can be concluded that AGEs and ALEs are important pathogenic factors in outer retinal aging. These adducts accumulate in the retina in association with age-related pathology, while experimental studies show that they have deleterious effects on RPE and most of their effects could be modulated by RAGE in a pro-inflammatory manner. More research is required, but with novel agents becoming increasingly available that have the ability to inhibit AGE formation or prevent RAGE activation, there is hope that this pathway could be manipulated therapeutically.

Raman spectroscopic assessment of AGEs and ALEs is an exciting development with clear therapeutic and diagnostic potential, and our group is currently assembling a comprehensive library of Raman spectra for these adducts. This should be a useful exercise because tracking pathogenic adducts in the living eye using confocal Raman spectroscopy is achievable in the near future. This has been shown by Raman-based *in situ* monitoring of macular carotenoid pigments,[64] and a similar ophthalmoscope-based approach could

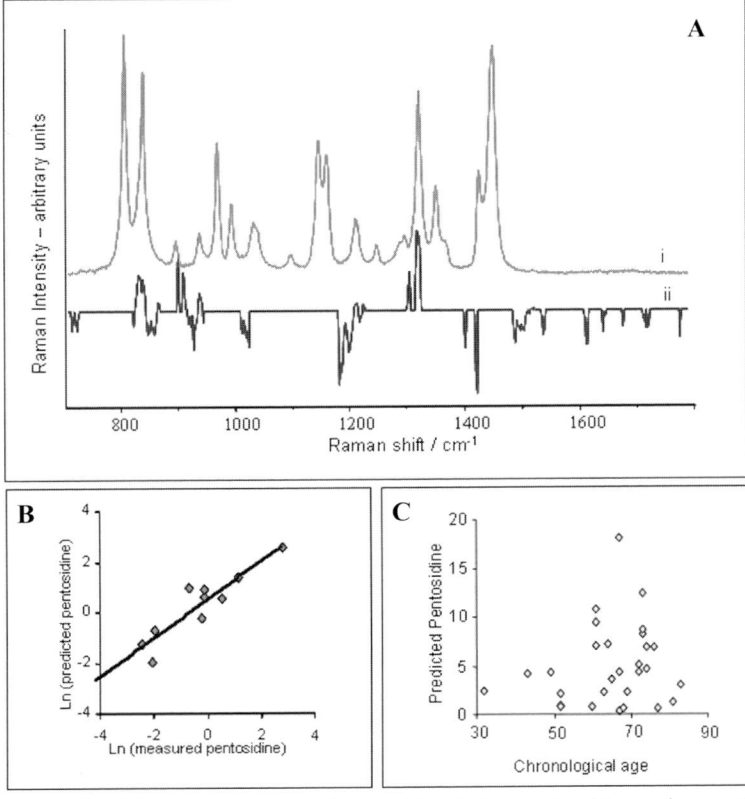

FIGURE 3. Confocal Raman microscopy of pentosidine in human Bruch's membrane. **(A)** The characteristic Raman spectrum of the pentosidine standard was recorded (i) and compared with regression coefficients selected by the uncertainty test as most correlated with pentosidine concentration (ii). The bands of the Raman signal, which positively contribute to the regression coefficients, match bands that occur in the spectrum of pentosidine. Multivariate regression analysis was used to calibrate the Raman signal against chronological age of donor and analytically quantified advanced glycation end products (AGEs). Models were constructed for each of the reference parameters and then used to predict the levels of those parameters in the remaining data set. A multivariate analysis method (Partial Least Squares [PLS]) was used. **(B)** PLS-regression plot for the prediction of pentosidine concentration from Raman spectra of Bruch's membrane. **(C)** Plot of predicted pentosidine concentration against age. The difference between samples over 60 and under 60 is significant ($P < 0.05$).

provide unique and useful diagnostic information about patient susceptibility to a range of important age-related diseases, including AMD.

Acknowledgments

J.J.M. thanks the Leverhulme Trust for the award of an Emeritus Research Fellowship (EM/2006/0049). We also acknowledge The Medical Research Council (Grant no. G0600053), The Wellcome Trust (WT066193), The Biotechnology and Biological Sciences Research Council (JREI 18471), The Research & Development Office (NI), The Belfast Association for the Blind, and Insight: The Trust for the Visually Impaired who support our research in retinal aging.

Conflict of Interest

The authors declare no conflicts of interest.

References

1. BECKMAN, K.B. & B.N. AMES. 1998. The free radical theory of aging matures. Physiol. Rev. **78:** 547–581.
2. SZWEDA, P.A. et al. 2003. Aging, lipofuscin formation, and free radical-mediated inhibition of cellular proteolytic systems. Ageing Res. Rev. **2:** 383–405.
3. YIN, D. & K. CHEN. 2005. The essential mechanisms of aging: irreparable damage accumulation of biochemical side-reactions. Exp. Gerontol. **40:** 455–465.
4. GRUNE, T. et al. 2004. Decreased proteolysis caused by protein aggregates, inclusion bodies, plaques, lipofuscin,

ceroid, and 'aggresomes' during oxidative stress, aging, and disease. Int. J. Biochem. Cell. Biol. **36:** 2519–2530.
5. LOUIE, J.L., R.J. KAPPHAHN & D.A. FERRINGTON. 2002. Proteasome function and protein oxidation in the aged retina. Exp. Eye. Res. **75:** 271–284.
6. BOULTON, M. *et al.* 2004. The photoreactivity of ocular lipofuscin. Photochem. Photobiol. Sci. **3:** 759–764.
7. TERMAN, A., B. GUSTAFSSON & U.T. BRUNK. 2007. Autophagy, organelles and ageing. J. Pathol. **211:** 134–143.
8. STRAUSS, O. 2005. The retinal pigment epithelium in visual function. Physiol. Rev. **85:** 845–881.
9. STARITA, C. *et al.* 1996. Hydrodynamics of ageing Bruch's membrane: implications for macular disease. Exp. Eye. Res. **62:** 565–572.
10. GUYMER, R., P. LUTHERT & A. BIRD. 1999. Changes in Bruch's membrane and related structures with age. Prog. Retin. Eye. Res. **18:** 59–90.
11. BINDER, S. *et al.* 2007. Transplantation of the RPE in AMD. Prog. Retin. Eye. Res. **26:** 516–554.
12. AMBATI, J. *et al.* 2003. Age-related macular degeneration: etiology, pathogenesis, and therapeutic strategies. Surv. Ophthalmol. **48:** 257–293.
13. BOULTON, M. & P. DAYHAW-BARKER. 2001. The role of the retinal pigment epithelium: topographical variation and ageing changes. Eye **15:** 384–389.
14. BOULTON, M., M. ROZANOWSKA & T. WESS. 2004. Ageing of the retinal pigment epithelium: implications for transplantation. Graefes. Arch. Clin. Exp. Ophthalmol. **242:** 76–84.
15. BIRD, A.C. *et al.* 1995. An international classification and grading system for age-related maculopathy and age-related macular degeneration. The International ARM Epidemiological Study Group. Surv. Ophthalmol. **39:** 367–374.
16. FEENEY-BURNS, L. & G.E. ELDRED. 1983. The fate of the phagosome: conversion to 'age pigment' and impact in human retinal pigment epithelium. Trans. Ophthalmol. Soc. U. K. **103**(Pt 4): 416–421.
17. KLEIN, R. *et al.* 2002. Ten-year incidence and progression of age-related maculopathy: The Beaver Dam eye study. Ophthalmology **109:** 1767–1779.
18. BEATTY, S. *et al.* 2000. The role of oxidative stress in the pathogenesis of age-related macular degeneration. Surv. Ophthalmol. **45:** 115–134.
19. ALGVERE, P.V. & S. SEREGARD. 2002. Age-related maculopathy: pathogenetic features and new treatment modalities. Acta. Ophthalmol. Scand. **80:** 136–143.
20. HJELMELAND, L.M. *et al.* 1999. Senescence of the retinal pigment epithelium. Mol. Vis. **5:** 33.
21. HAGEMAN, G.S. *et al.* 2005. A common haplotype in the complement regulatory gene factor H (HF1/CFH) predisposes individuals to age-related macular degeneration. Proc. Natl. Acad. Sci. USA **102:** 7227–7232.
22. ZARBIN, M.A. 2004. Current concepts in the pathogenesis of age-related macular degeneration. Arch. Ophthalmol. **122:** 598–614.
23. KIJLSTRA, A., E. LA HEIJ & F. HENDRIKSE. 2005. Immunological factors in the pathogenesis and treatment of age-related macular degeneration. Ocul. Immunol. Inflamm. **13:** 3–11.
24. PENFOLD, P.L. *et al.* 2001. Immunological and aetiological aspects of macular degeneration. Prog. Retin. Eye. Res. **20:** 385–414.
25. ROBERTS, J.E. 2001. Ocular phototoxicity. J. Photochem. Photobiol. B **64:** 136–143.
26. BOULTON, M., M. ROZANOWSKA & B. ROZANOWSKI. 2001. Retinal photodamage. J. Photochem. Photobiol. B **64:** 144–161.
27. BAYNES, J.W. 2001. The role of AGEs in aging: causation or correlation. Exp. Gerontol. **36:** 1527–1537.
28. MONNIER, V.M. *et al.* 1992. Maillard reaction-mediated molecular damage to extracellular matrix and other tissue proteins in diabetes, aging, and uremia. Diabetes **41**(Suppl 2): 36–41.
29. THORPE, S.R. & J.W. BAYNES. 2003. Maillard reaction products in tissue proteins: new products and new perspectives. Amino Acids **25:** 275–281.
30. THORNALLEY, P.J., A. LANGBORG & H.S. MINHAS. 1999. Formation of glyoxal, methylglyoxal and 3-deoxyglucosone in the glycation of proteins by glucose. Biochem. J. **344**(Pt 1): 109–116.
31. BAZAN, N.G. 1982. Metabolism of phospholipids in the retina. Vision Res. **22:** 1539–1548.
32. JANUSZEWSKI, A.S. *et al.* 2003. Role of lipids in chemical modification of proteins and development of complications in diabetes. Biochem. Soc. Trans. **31:** 1413–1416.
33. SCHUTT, F. *et al.* 2003. Proteins modified by malondialdehyde, 4-hydroxynonenal, or advanced glycation end products in lipofuscin of human retinal pigment epithelium. Invest. Ophthalmol. Vis. Sci. **44:** 3663–3668.
34. HOWES, K.A. *et al.* 2004. Receptor for advanced glycation end products and age-related macular degeneration. Invest. Ophthalmol. Vis. Sci. **45:** 3713–3720.
35. HANDA, J.T. *et al.* 1999. Increase in the advanced glycation end product pentosidine in Bruch's membrane with age. Invest. Ophthalmol. Vis. Sci. **40:** 775–779.
36. YAMADA, Y. *et al.* 2006. The expression of advanced glycation endproduct receptors in rpe cells associated with basal deposits in human maculas. Exp. Eye. Res. **82:** 840–848.
37. ISHIBASHI, T. *et al.* 1998. Advanced glycation end products in age-related macular degeneration. Arch. Ophthalmol. **116:** 1629–1632.
38. SWAMY-MRUTHINTI, S. *et al.* 2002. Immunolocalization and quantification of advanced glycation end products in retinal neovascular membranes and serum: a possible role in ocular neovascularization. Curr. Eye. Res. **25:** 139–145.
39. RUMBLE, J.R. *et al.* 1997. Vascular hypertrophy in experimental diabetes. Role of advanced glycation end products. J. Clin. Invest. **99:** 1016–1027.
40. HANDA, J.T. *et al.* 1998. The advanced glycation endproduct pentosidine induces the expression of PDGF-B in human retinal pigment epithelial cells. Exp. Eye. Res. **66:** 411–419.
41. KOBAYASHI, S. *et al.* 2007. Overproduction of N(epsilon)-(carboxymethyl)lysine-induced neovascularization in cultured choroidal explant of aged rat. Biol. Pharm. Bull. **30:** 133–138.
42. ZHOU, J. *et al.* 2005. Mechanisms for the induction of HNE-MDA- and AGE-adducts, RAGE and VEGF in retinal pigment epithelial cells. Exp. Eye. Res. **80:** 567–580.

43. STITT, A.W. 2001. Advanced glycation: an important pathological event in diabetic and age related ocular disease. Br. J. Ophthalmol. **85:** 746–753.
44. TIAN, J. *et al.* 2005. Advanced glycation endproduct-induced aging of the retinal pigment epithelium and choroid: a comprehensive transcriptional response. Proc. Natl. Acad. Sci. USA **102:** 11846–11851.
45. CHENG, Y. & M.H. GAO. 2005. The effect of glycation of CD59 on complement-mediated cytolysis. Cell. Mol. Immunol. **2:** 313–317.
46. MUNCH, G., J. GASIC-MILENKOVIC & T. ARENDT. 2003. Effect of advanced glycation endproducts on cell cycle and their relevance for Alzheimer's disease. J. Neural. Transm. Suppl. 63–71.
47. BIERHAUS, A. *et al.* 2005. Advanced glycation end product receptor-mediated cellular dysfunction. Ann. N.Y. Acad. Sci. **1043:** 676–680.
48. SCHMIDT, A.M. *et al.* 2000. The biology of the receptor for advanced glycation end products and its ligands. Biochim. Biophys. Acta **1498:** 99–111.
49. WAUTIER, J.L. & A.M. SCHMIDT. 2004. Protein glycation: a firm link to endothelial cell dysfunction. Circ. Res. **95:** 233–238.
50. JUSTILIEN, V. *et al.* 2007. SOD2 knockdown mouse model of early AMD. Invest. Ophthalmol. Vis. Sci. **48:** 4407–4420.
51. MA, W. *et al.* 2007. RAGE ligand upregulation of VEGF secretion in ARPE-19 cells. Invest. Ophthalmol. Vis. Sci. **48:** 1355–1361.
52. MCFARLANE, S. *et al.* 2005. Characterisation of the advanced glycation endproduct receptor complex in the retinal pigment epithelium. Br. J. Ophthalmol. **89:** 107–112.
53. GLENN, J.V. *et al.* 2007. Confocal Raman microscopy can quantify advanced glycation end product (AGE) modifications in Bruch's membrane leading to accurate, nondestructive prediction of ocular aging. Faseb J. **21:** 3542–3552.
54. TAKEUCHI, M. *et al.* 2001. Immunological detection of a novel advanced glycation end-product. Mol. Med. **7:** 783–791.
55. ODETTI, P. *et al.* 1992. Chromatographic quantitation of plasma and erythrocyte pentosidine in diabetic and uremic subjects. Diabetes **41:** 153–159.
56. PORTERO-OTIN, M., R.H. NAGARAJ & V.M. MONNIER. 1995. Chromatographic evidence for pyrraline formation during protein glycation in vitro and in vivo. Biochim. Biophys. Acta **1247:** 74–80.
57. VERZIJL, N. *et al.* 2000. Effect of collagen turnover on the accumulation of advanced glycation end products. J. Biol. Chem. **275:** 39027–39031.
58. SEBEKOVA, K., L. PODRACKA, P. BLAZICEK, *et al.* 2001. Plasma levels of advanced glycation end products in children with renal disease. Pediatr. Nephrol. **16:** 1105–1112.
59. MULDER, D.J. *et al.* 2008. Skin autofluorescence is elevated in patients with stable coronary artery disease and is associated with serum levels of neopterin and the soluble receptor for advanced glycation end products. Atherosclerosis. **197:** 217–223.
60. NOTINGHER, I., J. SELVAKUMARAN & L.L. HENCH. 2004. New detection system for toxic agents based on continuous spectroscopic monitoring of living cells. Biosens. Bioelectron. **20:** 780–789.
61. BUSCHMAN, H.P. *et al.* 2001. Raman microspectroscopy of human coronary atherosclerosis: biochemical assessment of cellular and extracellular morphologic structures in situ. Cardiovasc. Pathol. **10:** 69–82.
62. ERCKENS, R.J. *et al.* 2002. Raman spectroscopy: noninvasive determination of silicone oil in the eye: potential applications for intraocular determination of biomaterials. Retina **22:** 796–799.
63. BEATTIE, J.R. *et al.* 2007. Raman microscopy of porcine inner retinal layers from the area centralis. Mol. Vis. **13:** 1106–1113.
64. BERNSTEIN, P.S. *et al.* 2004. Resonance Raman measurement of macular carotenoids in the living human eye. Arch. Biochem. Biophys. **430:** 163–169.
65. OKUBO, A. *et al.* 1999. The relationships of age changes in retinal pigment epithelium and Bruch's membrane. Invest. Ophthalmol. Vis. Sci. **40:** 443–449.

The Aroma Side of the Maillard Reaction

CHRISTOPH CERNY

Firmenich Corporate R&D Division, CH-1217 Meyrin 2 Geneva, Switzerland

The Maillard reaction in food produces, among others, a diversity of sensory-active compounds (aroma, taste, color). The resulting key aroma compounds are often present only in trace concentrations of 1 μg/kg to 1 mg/kg. Nevertheless, they contribute to the respective flavor because of their low odor-perception thresholds. While Maillard intermediates, such as Amadori compounds and deoxyosones, are formed at percentage levels during model reactions, the yield of aroma compounds, in particular nitrogen and sulfur-containing ones, is often as low as 0.001–0.01 mol%, thus indicating their formation through chemical side reactions. The elucidation of the relevant precursors in food and the identification of previously unknown intermediates can throw light on these minor pathways. Also, model reactions with isotopically labeled precursors are of great value in gaining insight into the relevant formation mechanisms. Several examples of these studies are illustrated including work to elucidate the role of the solvent glycerol in the formation of pyrazines, trials to reveal the relative significance of 4-hydroxy-5-methyl-3(2H)-furanone as intermediate in the reaction between ribose and cysteine, and experiments to assess the proportional contribution of the precursors cysteine, xylose, and thiamine to the formation of the resulting aroma compounds in the thermal reaction.

Key words: aroma; formation pathway; intermediates; labeled precursors; Amadori compound; deoxyosone; CAMOLA

Introduction

The Maillard reaction has gained more and more attention during the past two decades by nutritionists and researchers who work in the medical and health area. As a result, today only a minority of publications deal with mechanistic aspects and formation pathways in the early phase of the Maillard reaction and still less with the sensory aspects of nonenzymatic browning and the aroma or taste compounds associated with it. The present review addresses this side of the Maillard reaction—the aroma side. It is well known that the Maillard reaction participates not only in the formation of color but also in aroma generation during cooking and thermal processing in the food industry. Examples for savory food are roasted meat, steamed fish, boiled eggs, and potatoes. Sweet and bakery aromas produced from the Maillard reaction comprise roasted cocoa and nuts, chocolate, malt, popcorn, and bread. Beverages, such as coffee, beer, whisky, and tea, owe their flavor largely to the Maillard reaction. But what are the levels of the relevant aroma compounds?

Maillard-borne Key Odorants in Food

The volatiles from the Maillard reaction that determine the flavor of a food are generally only present at trace levels. FIGURES 1–3 show the concentration ranges of selected aroma compounds in foods for which they are important odorants.[1–18] The oxygen-containing aroma compounds (FIG. 1) 2,3-butanedione, 2,3-pentanedione, methylpropanal, 3-methylbutanal, phenylacetaldehyde, 3-hydroxy-4,5-dimethyl-2(3H)furanone (sotolone), and 2,5-dimethyl-4-hydroxy-3(2H)-furanone (Furaneol®; Firmenich, Switzerland) occur in concentrations from 1 μg/kg for sotolone up to 100 mg/kg for 3-methylbutanal and Furaneol. FIGURE 2 shows 2-ethyl-3,5-dimethylpyrazine, 2,3-diethyl-5-methylpyrazine, and 2-acetyl-1-pyrroline as examples of nitrogen-containing aroma compounds. They are present in food in an order of magnitude of 0.001–10 mg/kg. On the whole, sulfur-containing Maillard odorants constitute the most powerful aroma compounds and often play, although at trace levels, a dominant role in the flavor of cooked meats and roasted coffee. FIGURE 3 shows, as examples, the concentration ranges of 2-furfurylthiol,

---

Address for correspondence: Christoph Cerny, Corporate R&D Division, Firmenich SA, Rue de la Bergère 7, 1217 Meyrin 2 Geneva, Switzerland. Voice: +41-22-780-2216; fax: +41-22-780-2734.
christoph.cerny@firmenich.com

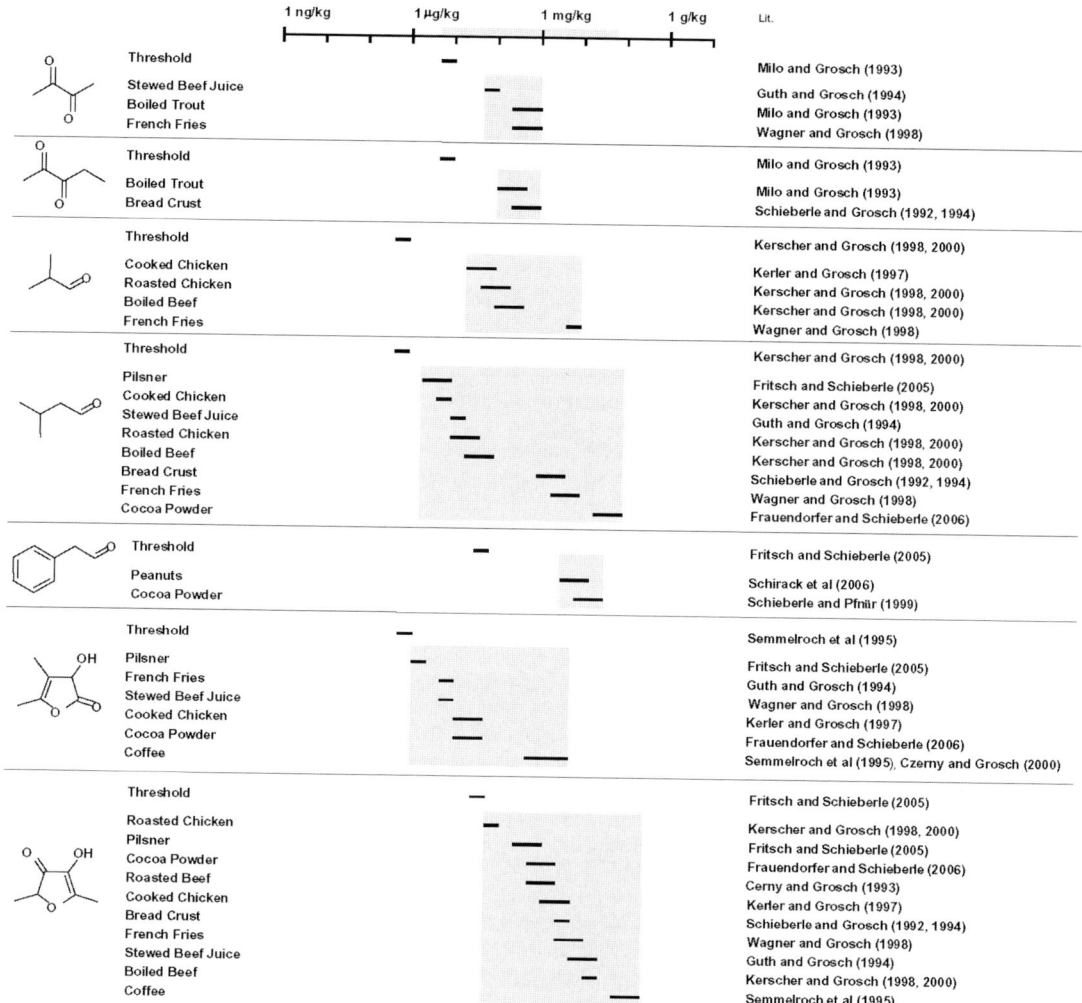

FIGURE 1. Oxygen-containing key odorants in food.[1-15]

2-methyl-3-furanthiol, 3-mercapto-2-pentanone, and methional in certain foods.

Yields of Intermediates and Odorants in Model Maillard Reactions

Several authors have investigated the early phases of the Maillard reaction in model experiments and have performed kinetic studies to follow the degradation of precursors and formation of intermediates. Davidek and co-workers[19] have analyzed the formation of the pentose-derived Amadori compound xylulosyl-glycine (Xyl-Gly) from an aqueous solution of xylose and glycine (equimolar, 1 mol/L, pH 6.0) heated at 90°C. A maximum yield of 12 mol% was observed after 20 min. On the other hand, Blank and co-workers[20] found that the reaction between glucose and proline (equimolar, 0.1 mol/L, pH 6.0, reflux) produced only a maximum yield of 0.5 mol% of the Amadori compound fructosyl-proline (Fru-Pro).

Hofmann studied the time course for the formation of 1-deoxyosone and 3-deoxyosone from alanine and reducing sugars.[21] Refluxing an equimolar solution of glucose and alanine (pH 7.0) generated a maximum of 0.2 mol% 1-deoxyosone and 0.6 mol% 3-deoxyosone. The yields were slightly higher with 0.5% and 1.0%, respectively, for the 1-deoxpentosone and the 3-deoxypentosone when glucose was replaced by xylose. The subsequent fragmentation products glyoxal, 2-oxopropanal, and 2,3-butanedione were formed in maximum yields of 0.2 mol%, 0.2 mol%, and 0.01 mol% for the glucose–alanine system and 0.1 mol%, 0.2 mol%, and 0.01 mol% for the

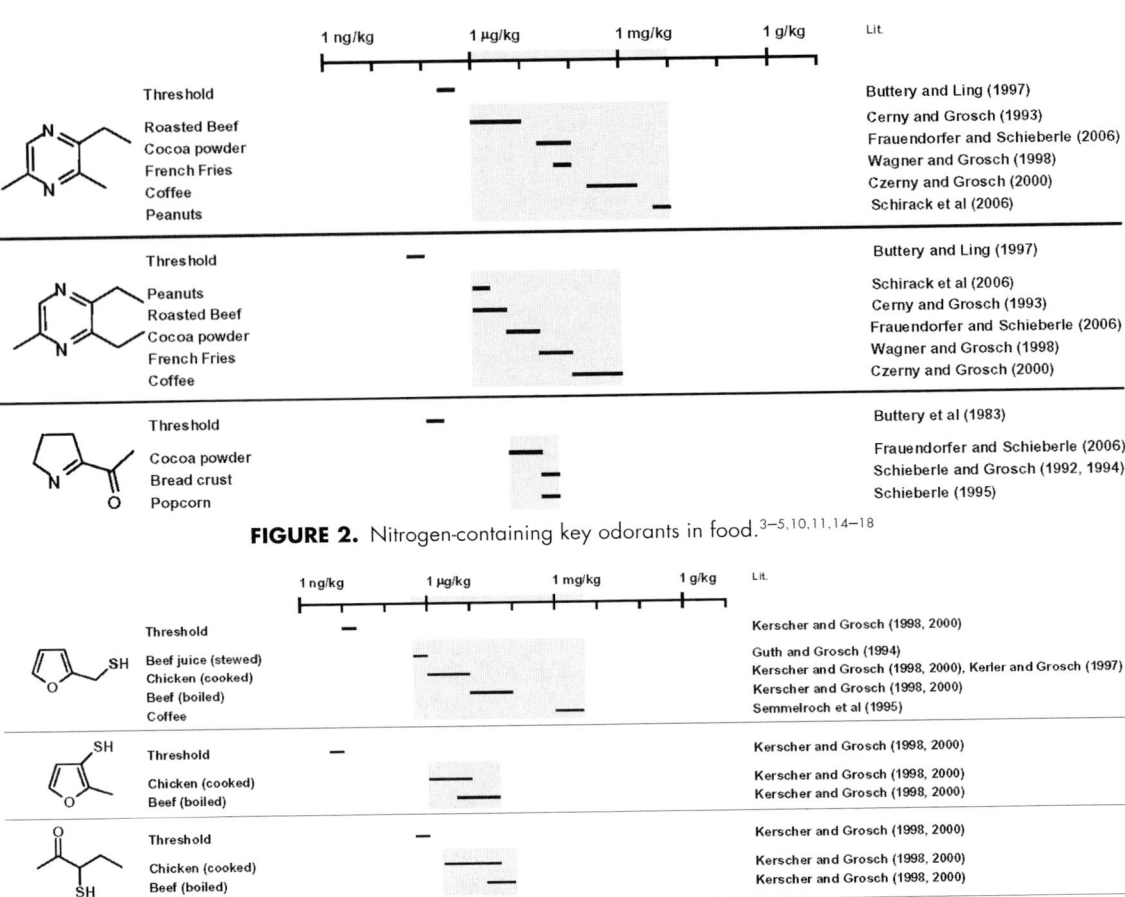

FIGURE 2. Nitrogen-containing key odorants in food.[3–5,10,11,14–18]

FIGURE 3. Sulfur-containing key odorants in food.[1–8,13–15]

xylose–alanine reaction. The author has used the derivatization agent 1,2-diaminobenzene for the quantification. However, unlike in most studies, he has not used it as an *in situ* trapping agent but added it to the reacted samples and hence avoided an overestimation of the concentration.

The yields for aroma compounds in Maillard model systems are often very low, especially for certain sulfur-containing odorants. For example, the yields from the model reaction systems cysteine–xylose and cysteine–glucose,[22] thiamine,[23] and cysteine–thiamine–xylose[24] were very low for the thiols 2-methyl-3-furanthiol (0.0005–0.0042%), 2-furfurylthiol (0.0007–0.0041%), and 3-mercapto-2-pentanone (0.0095%). Consequently, in general the yields during the Maillard reaction decrease from the Amadori compounds, deoxyosones, and their fragmentation products to the aroma compounds, as illustrated in FIGURE 4.

Elucidation of Relevant Precursors and Intermediates

Once a key aroma compound has been identified in a food, the next question is: What are the relevant precursors? In order to answer this question, a model mixture with all the potential precursors occurring in the food is reacted, under reaction conditions, which mimic those in the food. For example, a model mix of the free amino acids and sugars with the same concentrations as in beef was reacted to

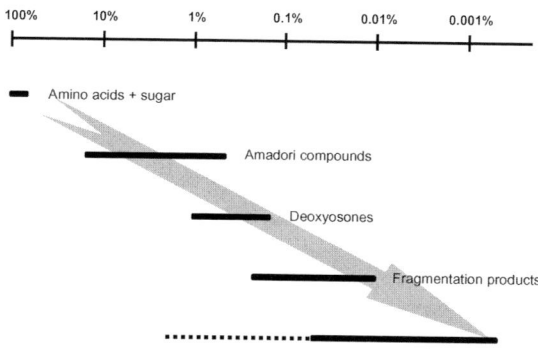

FIGURE 4. Yields in the different stages of the Maillard reaction.

study the formation of 2-ethyl-3,5-methylpyrazine and 2,3-diethyl-5-methylpyrazine in roasted beef.[25] Omission tests showed that the complete model generated both pyrazines, but, without alanine, none of them were formed. Glucose, fructose, as well as the corresponding sugar phosphates were efficient in generating them in the reaction with alanine. Hence, alanine was considered a key precursor for these ethyl-substituted pyrazines.

The identification of previously unknown or only postulated intermediates is of great help in proposing formation pathways for Maillard odorants. The dicarbonyl compounds 3-deoxyaldoketoses, 1-deoxydiketoses, and 1-amino-1,4-dideoxydiketoses are classical intermediates in the Maillard reaction cascade.[26] The structures for the hexose derivatives are shown in FIGURE 5. The identification of 3,6-dideoxyosones, like the 6-amino-3,6-dideoxy-2-hexosulose in the reaction between glucose and tBoc-lysine,[27] and the predominant formation of the 1,4-dideoxy-2,3-hexodiulose from oligosaccharides and glycine through a "peeling-off" mechanism[28] extends the number of possible deoxyosone intermediates that can play a role in formation pathways of Maillard aroma compounds.

Elucidation of Reaction Pathways Using Labeled Precursors

Model reactions with ^{13}C-labeled precursors are useful tools for obtaining information about the origin of the carbons in the generated aroma compounds. Tressl's group was one of the first to use this approach in Maillard chemistry.[29] They investigated, among others, the reaction between cysteine and [1-^{13}C]arabinose; they analyzed the resulting labeled volatiles with an intact carbon chain and determined the labeling position in the molecule. The results indicated unambiguously the relevant deoxyosone intermediates for the different compounds (e.g., the 1-deoxyosone for 3-mercapto-2-pentanone and 2-methyl-3-furanthiol and the 3-deoxyosone for 2-furfurylthiol).

Experiments, carried out by the research team of Yaylayan,[30] with differently labeled [^{13}C]glucose and [^{13}C]alanine isotopomers to study the formation of 2,3-pentanedione revealed that only 10% of the 2,3-pentanedione was formed from sugar carbons; in 90% of the molecules, alanine was involved, providing the C-2 and C-3 carbon of 2,3-pentanedione. In 50% of the 2,3-pentanedione molecules, the C1-C2-C3 carbon chain and in 40% the C-4-C-5-C-6 carbon chain from glucose was integrated in the 2,3-pentanedione skeleton.

The CAMOLA approach, introduced by Schieberle,[31] uses a mixture of unlabeled and fully [^{13}C]labeled carbohydrates to study Maillard chemistry. Schieberle demonstrated that the reaction conditions strongly influence the formation pathway of Furaneol in the reaction between glucose and proline. Under roasting conditions, the compound forms via an intact glucose C-skeleton, while heating the precursors in aqueous buffer generates Furaneol via fragmentation of the glucose and recombination of C-3 fragments.

Cerny and Davidek used the CAMOLA approach to show that the C-skeleton remains intact in the

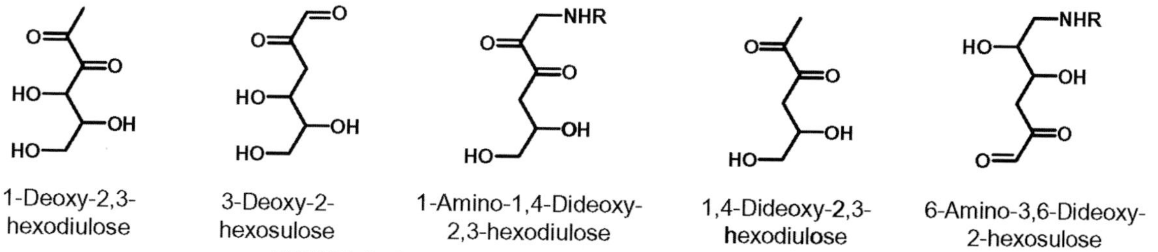

1-Deoxy-2,3-hexodiulose

3-Deoxy-2-hexosulose

1-Amino-1,4-Dideoxy-2,3-hexodiulose

1,4-Dideoxy-2,3-hexodiulose

6-Amino-3,6-Dideoxy-2-hexosulose

FIGURE 5. Deoxyosone intermediates in the Maillard reaction.

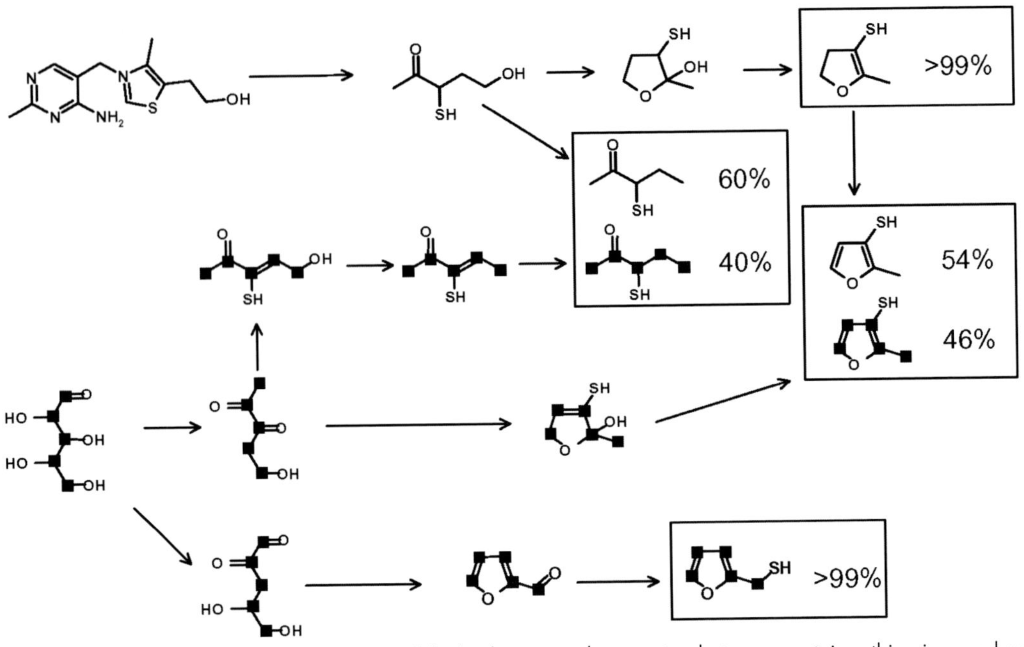

FIGURE 6. Reaction scheme for the formation of thiol odorants in the reaction between cysteine, thiamine, and xylose.

formation of 2-methyl-3-furanthiol, 2-furfurylthiol, and 3-mercapto-2-pentanone[32] during the reaction of ribose/[$^{13}C_5$]ribose with cysteine in aqueous phosphate buffer (pH 5.0). In further experiments, the labeled sugar and an unlabeled compound, which was a suspected intermediate, were reacted with cysteine. In one example, cysteine and an equimolar mix of [$^{13}C_5$]ribose and unlabeled furfural were reacted at 95 °C for 4 h. The majority of furfurylthiol (92%) that was formed was unlabeled and hence stemmed from furfural. In this vein, the role of furfural as a key intermediate for furfurylthiol was confirmed. On the other hand, only 7% of the 2-methyl-3-furanthiol, which was formed from cysteine, and an equimolar mix of [$^{13}C_5$]ribose and 4-hydroxy-5-methyl-3(2H)furanone was labeled. This indicates that ribose was substantially more efficient than 4-hydroxy-5-methyl-3(2H)furanone as a precursor. The latter compound was considered irrelevant as a key intermediate under the reaction conditions of the study.

The sulfur compounds 2-methyl-3-furanthiol, 3-mercapto-2-pentanone, furfurylthiol, and 4,5-dihydro-2-methyl-3-furanthiol were shown to be key odorants from the Maillard reaction (145 °C, 20 min, pH 5.0) between cysteine, thiamine, and xylose.[33] When [$^{13}C_5$]xylose was used instead of xylose, furfurylthiol was found completely [$^{13}C_5$]labeled, indicating that it formed only from xylose as the carbon precursor. On the other hand, 4,5-dihydro-2-methyl-3-furanthiol was virtually unlabeled, pointing to thiamine as the exclusive carbon source. Finally, 2-methyl-3-furanthiol and 3-mercapto-2-pentanone were both [$^{13}C_5$]labeled (46 and 40%, respectively and labeled) (54 and 60%, respectively). For these two compounds, different formation pathways exist starting from both xylose and thiamine (cf. FIG. 6).

Labeling experiments can also reveal the influence of the reaction solvent on the formation of aroma compounds. Cerny and Guntz-Dubini reacted fructose and alanine in [$^{13}C_3$]glycerol and found that 13% of the methylpyrazine, 26% of the 2,5-dimethylpyrazine, 27% of the 3-ethyl-2,5-dimethylpyrazine, and 20% of the 2-ethyl-3-methylpyrazine that formed was triply [^{13}C]labeled.[34] This clearly demonstrated that glycerol participates not only as a solvent in the reaction but also as a carbon precursor. The reaction probably proceeds via the supposed intermediates 1-hydroxypropanone, 2-oxopropanal, and 2,3-pentandione, which were also identified in the reacted mixture, part of it triply [^{13}C]labeled.

Conflict of Interest

The author declares no conflicts of interest.

References

1. MILO, C. & W. GROSCH. 1993. Changes in the odorants of boiled trout (Salmo fario) as affected by the storage of the raw material. J. Agric. Food Chem. **41:** 2076–2081.

2. GUTH, H. & W. GROSCH. 1994. Identification of the character impact odorants of stewed beef juice by instrumental analyses and sensory studies. J. Agric. Food Chem. **42:** 2862–2866.
3. WAGNER, R. & W. GROSCH. 1998. Key odorants of French fries. JAOCS **75:** 1385–1392.
4. SCHIEBERLE, P. & W. GROSCH. 1992. Changes in the concentrations of potent crust odorants during storage of white bread. Flavour Fragr. J. **7:** 213–218.
5. SCHIEBERLE, P. & W. GROSCH. 1994. Potent odorants of rye bread crust – differences from the crumb and from wheat bread crust. Z. Lebensm. Unters. Forsch. **198:** 292–296.
6. KERSCHER, R. & W. GROSCH. 1998. Quantification of 2-methyl-3-furanthiol, 2-furfurylthiol, 3-mercapto-2-pentanone, and 2-mercapto-3-pentanone in heated meat. J. Agric. Food Chem. **46:** 1954–1958.
7. KERSCHER, R. & W. GROSCH. 2000. Comparison of the aromas of cooked beef, pork and chicken. In Frontiers of Flavour Science. P. Schieberle & K.H. Engel, Eds.: 17–20. Deutsche Forschungsanstalt für Lebensmittelchemie. Garching.
8. KERLER, J. & W. GROSCH. 1997. Character impact odorants of boiled chicken: changes during refrigerated storage and reheating. Z. Lebensm. Unters. Forsch. **205:** 232–238.
9. FRITSCH, H. & P. SCHIEBERLE. 2005. Identification based on quantitative measurements and aroma recombination of the character impact odorants in a Bavarian Pilsner-type beer. J. Agric. Food Chem. **53:** 7544–7551.
10. FRAUENDORFER, F. & P. SCHIEBERLE. 2006. Identification of the key aroma compounds in cocoa powder based on molecular sensory correlations. J. Agric. Food Chem. **54:** 5521–5529.
11. SCHIRACK, A.V., M.A. DRAKE, T.H. SANDERS, et al. 2006. Characterization of aroma-active compounds in microwave blanched peanuts. Food Chem. Toxic. **71:** 513–520.
12. SCHIEBERLE, P. & P. PFNÜR. 1999. Characterization of key odorants in chocolate. In Flavor Chemistry: Thirty Years of Progress. R. Teranishi, L. Wick & I. Hornstein, Eds.: 147–153. Kluwer. New York.
13. SEMMELROCH, P., G. LASKAWY, I. BLANK & W. GROSCH. 1995. Determination of potent odourants in roasted coffee by stable isotope dilution assays. Flavour Fragr. J. **10:** 1–7.
14. CZERNY, M. & W. GROSCH. 2000. Potent odorants of raw arabica coffee. Their changes during roasting. J. Agric. Food Chem. **48:** 868–872.
15. CERNY, C. & W. GROSCH. 1993. Quantification of character-impact odor compounds of roasted beef. Z. Lebensm. Unters. Forsch. **196:** 417–422.
16. BUTTERY, R. & L.C. LING. 1997. 2-Ethyl-3,5-dimethylpyrazine and 2-ethyl-3,6-dimethylpyrazine: odor thresholds in water solution. Lebensm. Wiss. Technol. **30:** 109–110.
17. BUTTERY, R., L.C. LING, B.O. JULIANO, et al. 1983. Cooked rice aroma and 2-acetyl-1-pyrroline. J. Agric. Food Chem. **31:** 823–826.
18. SCHIEBERLE, P. 1995. Quantitation of important roast-smelling odorants in popcorn by stable isotope dilution assays and model studies on flavor formation during popping. J. Agric. Food Chem. **43:** 2442–2448.
19. DAVIDEK, T., K. KRAEHENBUEHL, S. DÉVAUD, et al. 2005. Analysis of Amadori compounds by high-performance cation exchange chromatography coupled to tandem mass spectrometry. Anal. Chem. **77:** 140–147.
20. BLANK, I., S. DÉVAUD, W. MATTEY-DORET, et al. 2003. Formation of odorants in Maillard model system based on L-proline as affected by pH. J. Agric. Food Chem. **51:** 3643–3650.
21. HOFMANN, T. 1999. Quantitative studies on the role of browning precursors in the Maillard reaction of pentoses and hexoses with L-alanine. Eur. Food Res. Technol. **209:** 113–121.
22. HOFMANN, T. & P. SCHIEBERLE. 1998. Quantitative model studies on the effectiveness of different precursor systems in the formation of the intense food odorants 2-furfurylthiol and 2-methyl-3-furanthiol. J. Agric. Food Chem. **46:** 234–241.
23. ZEILER, G. 1994. Model experiments on the formation of 2-methyl-3-furanthiol and 2-furfurylthiol in boiled meat (in German). Ph.D. thesis, Technical University of Munich.
24. CERNY, C. 2007. Unpublished results.
25. CERNY, C. & W. GROSCH. 1994. Precursors of ethyldimethylpyrazine isomers and 2,3-diethyl-5-methylpyrazine in roasted beef. Z. Lebensm. Unters. Forsch. **198:** 210–214.
26. LEDL, F. & E. SCHLEICHER. 1990. New aspects of the Maillard reaction in foods and the human body. Angew. Chem. Int. Ed. Engl. **29:** 565–594.
27. BIEMEL, K.M., J. CONRAD & M.O. LEDERER. 2002. Unexpected carbonyl mobility in aminoketoses: the key to major Maillard crosslinks. Angew. Chem. Int. Ed. Engl. **41:** 801–804.
28. HOLLNAGEL, A. & L.W. KROH. 2000. Degradation of oligosaccharides in nonenzymatic browning by formation of α-dicarbonyl compounds via a "peeling off" mechanism. J. Agric. Food Chem. **48:** 6219–6226.
29. TRESSL, R., B. HELAK, E. KERSTEN, et al. 1993. Formation of flavour compounds by Maillard reaction. In Recent Developments in Flavor and Fragrance Chemistry. R. Hopp & K. Mori, Eds.: 167–181. VCH. Weinheim.
30. YAYLAYAN, V.A. & A. KEYHANI. 1999. Origin of 2,3-pentanedione and 2,3-butanedione in D-Glucose/ L-Alanine Maillard model systems. J. Agric. Food Chem. **47:** 3280–3284.
31. SCHIEBERLE, P. 2005. The carbon module labelling (CAMOLA) technique—a useful tool for identifying transient intermediates in the formation of Maillard-type target molecules. Ann. N.Y. Acad. Sci. **2043:** 236–248.
32. CERNY, C. & T. DAVIDEK. 2003. Formation of aroma compounds from ribose and cysteine during the Maillard reaction. J. Agric. Food Chem. **51:** 2714–2718.
33. CERNY, C. & R. GUNTZ-DUBINI. 2007. Formation of aroma compounds in the Maillard reaction of xylose, cysteine and thiamine. Poster presentation at the 8th Wartburg Flavor Symposium. Eisenach, February 28 to March 3.
34. CERNY, C. & R. GUNTZ-DUBINI. 2006. Role of the solvent glycerol in the Maillard reaction of D-fructose and L-alanine. J. Agric. Food Chem. **54:** 574–577.

Methylglyoxal: Its Presence in Beverages and Potential Scavengers

DI TAN,[a] YU WANG,[a] CHIH-YU LO,[b] SHENGMIN SANG,[c] AND CHI-TANG HO[a,d]

[a]*Department of Food Science, Rutgers University, New Brunswick, New Jersey 08901, USA*

[b]*Department of Food Science, National Chiayi University, Chiayi, Taiwan, ROC*

[c]*Department of Chemical Biology, Ernest Mario School of Pharmacy, Rutgers University, Piscataway, New Jersey 08854, USA*

[d]*Graduate Institute of Food Science and Technology, National Taiwan University, Taipei, Taiwan, ROC*

Nonenzymic glycation, also known as the Maillard reaction, is a complex series of reactions between reducing sugars and amino compounds. Previous studies have demonstrated that reactive dicarbonyl compounds (e.g., methylglyoxal [MG] and glyoxal [GO]), formed as intermediate products of the Maillard reaction, irreversibly and progressively modify proteins over time and yield advanced glycation end products (AGEs), which are thought to contribute to the development of diabetes mellitus and its complications. Several studies have shown that higher levels of MG are present in diabetic patients' plasma than in healthy people's plasma. Thus, decreasing the levels of MG and GO will be an effective approach to reduce the formation of AGEs and the development of diabetic complications. Here, we briefly describe our effort in searching for non- or less-toxic trapping agents of reactive dicarbonyl species from dietary sources. In addition, we have discovered that commercial beverages contain extremely high levels of MG. The potential hazardous effects of dietary MG on humans remain to be explored.

Key words: methylglyoxal; diabetes; beverages; tea polyphenols; peptides

Introduction

Epidemiological and large prospective clinical studies have confirmed that hyperglycemia is the most important factor in the onset and progress of diabetic complications, both in type 1 (insulin-dependent diabetes) and type 2 diabetes mellitus.[1] Increasing evidence identifies the formation of advanced glycation end products (AGEs) as the major pathogenic link between hyperglycemic- and diabetic-related complications.[2] Nonenzymatic glycation is a complex series of reactions between reducing sugars and amino compounds. As the first step of AGEs formation, the free amino groups of proteins in the tissues react with a carbonyl group of reducing sugars, such as glucose, to form fructosamines via a Schiff base by Amadori rearrangement. Both Schiff base and Amadori product further undergo a series of reactions through dicarbonyl intermediates (e.g., glyoxal [GO], methylglyoxal [MG], and 3-deoxyglucosone), to form AGEs.[2] GO and MG, the two major α-dicarbonyl compounds found in the human body, are extremely reactive and readily modify lysine, arginine, and cysteine residues of proteins.[3] Reactive carbonyl compounds, such as GO and MG, have recently attracted much attention because of their possible clinical significance in chronic and age-related diseases.

More and more evidence indicates that the increase in reactive carbonyl intermediates is a consequence of hyperglycemia in diabetes. Carbonyl stress leads to increased modification of proteins, followed by oxidant stress and tissue damage.[4] Several studies have shown that diabetic patients have higher levels of GO and MG in their plasma than healthy individuals.[5–7] In a recent report,[7] the amount of MG from diabetic patients was found to be 16–27 μg/dL, while the amount found in normal subjects was 3.0–7.0 μg/dL. Thus, decreasing the levels of GO and MG may provide a useful approach for preventing the formation of AGEs.

Address for correspondence: Chi-Tang Ho, Department of Food Science, Rutgers University, 65 Dudley Road, New Brunswick, NJ 08901-8529. Voice: +732-932-9611 x 235; fax: +732-932-6776.
ho@aesop.rutgers.edu

Effects of Tea on Diabetes and Related Complication

The ability of preventing diabetic-related complications by using tea and its polyphenols has been tested in several studies. For instance, oral administration of tea catechins retarded the progression of functional and morphological changes in the kidney of streptozotocin-induced diabetic rats.[8] The detailed mechanisms for the prevention of diabetic complications by tea and its polyphenols require further study. Many studies have shown that those effects could partially be a result of the inhibition of AGEs formation. Rutter et al.[9] reported that green tea extract was able to delay collagen aging in C57Bl/6 mice by blocking AGEs formation and collagen cross-linking. Meanwhile, tea polyphenols may inhibit the formation of AGEs by trapping reactive dicarbonyl compounds. A recent study explored the inhibitory effect of tea polyphenols, such as catechins, epicatechin, epicatechin 3-gallate, epigallocatechin (EGC), and epigallocatechin 3-gallate (EGCG), on different stages of protein glycation, including MG-mediated protein glycation.[10] EGCG exhibited a significant inhibitory effect of 69.1% on MG-mediated protein glycation.

Tea Polyphenols as Effective Trapping Agents for Methylglyoxal

In our recent studies to find non- or less-toxic trapping agents of reactive dicarbonyl species from dietary sources, we compared the trapping effect of MG by four major catechins in green tea and three major theaflavins in black tea under physiological conditions (pH 7.4, 37 °C). Our results indicated that all the test compounds could efficiently trap MG with theaflavin 3,3′-digallate (TF-3) in theaflavins and (-)-EGC in tea catechins showing the highest trapping activity.[11]

To better understand the reaction trend of TF3 and MG, a time course study has been carried out with the same molar amount of TF3 and MG in an hour. The time course of the reaction of TF-3 and MG showed that TF-3 started the MG scavenging reaction rapidly. At the first measurement at 3 min, more than one-third of the MG was trapped. The reaction of MG with TF3 has almost been completed during the time of our experiment.

In addition, the major adduct between EGCG and MG has been identified.[11] The 1:1 adduct formation between EGCG and MG dominantly occurs at either the C8- or C6-position in the A ring of EGCG, and the gallate ring did not play an important role in the trapping of reactive dicarbonyl species.[11,12]

Formation of Pyrazinone between α-Dicarbonyl Compounds and Peptides

The formation of pyrazinone between dipeptides and GO was first proposed by van Chuyen et al.[13] who observed that a series of pyrazinones, 2-(3′-alkyl-2-oxo-pyrazin-1′-yl) alkanoic acid, were generated at the cooking condition, 100 °C and pH 5.0. Recently, the pyrazinones produced between tripeptide and GO and MG have also been reported at physiological conditions (pH 7.4, 37 °C).[14] For the reaction between MG and dipeptides, MG first adds to the primary amino group forming a Schiff base. After tautomerization, the amido nitrogen of the first peptide bond attacks the remaining carbonyl C-atom. It is finally followed by an elimination of a second molecule of water. It has been observed that 5-methylpyrazinone derivative is the main product in the reaction between Gly-Ala-Phe and MG.[14] This is supported with the observation that the aldehyde group of MG is hydrated in aqueous solution,[15] as it is the ketone group and not the aldehyde group that is targeted by the primary amino group at the first step. It was shown that the reactivity of trapping MG by N-terminal of the peptide and the guanidino function of arginine were similar under both separate and mixing incubation.[14]

We have selected dipeptides, Gly-Cys, Gly-Gly, Gly-Leu, Gly-Phe, Gly-Pro, and Gly-Ser with the N-terminal glycine and triglycine (TG) to compare their reactivity with MG. In one day incubation, the MG scavenging efficiency is in the order of Gly-Cys (53.42%) > Gly-Gly (47.01%) > Gly-Phe (16.19%) > Gly-Gly (12.19%) > Gly-Ser (11.92%) > Gly-Pro (8.92%) > Gly-Leu (5.75%). Gly-Cys has a much higher scavenging efficiency than all the other peptides we tested, which may indicate a very different mechanism for scavenging. Moreover, TG showed relatively high trapping efficiency compared to most dipeptides, including diglycine.[16] This may imply that elements other than the properties of side-chain groups influence the trapping of MG.

Current observations indicate that small peptides may serve as good scavengers for MG and other reactive carbonyl species. However, more systematic study is required to better understand the structure–activity relationship.

TABLE 1. The concentration of methylglyoxal (MG) in 13 brands of carbonated soft drinks from supermarket A, B, and C

Brand	MG concentration (μg/100 mL)		
	Supermarket A	Supermarket B	Supermarket C
A	88.3 ± 7.2	83.6 ± 6.1	78.2 ± 6.7
B	92.5 ± 8.4	89.2 ± 3.4	87.3 ± 8.8
C	54.6 ± 0.6	46.5 ± 3.1	60.0 ± 10.7
D	48.5 ± 2.1	23.6 ± 1.5	41.8 ± 1.3
E	41.1 ± 4.8	31.6 ± 2.6	31.0 ± 1.6
F	42.5 ± 2.0	38.5 ± 0.8	29.2 ± 1.2
G	62.2 ± 3.3	57.8 ± 1.0	62.1 ± 1.5
H	50.1 ± 6.1	50.2 ± 0.5	27.7 ± 0.2
I	23.5 ± 0.6	26.2 ± 1.1	26.9 ± 0.8
J	59.7 ± 4.8	43.2 ± 4.9	33.0 ± 3.2
K	139.5 ± 2.1	104.2 ± 5.9	111.2 ± 6.0
AA	7.2 ± 1.6	7.6 ± 0.7	9.3 ± 2.6
BB	7.1 ± 1.9	8.2 ± 1.2	31.5 ± 6.2

Methylglyoxal in Commercial Beverages Containing High Fructose Corn Syrup

Because of their high sugar content, soft drinks are potential sources of MG. The levels of MG in 13 brands of carbonated soft drink (CSD) (noted as: A, B, C, D, E, F, G, H, I, J, K, AA, and BB) from three different local supermarkets are measured. Two of them (AA and BB) are diet drinks, and they are the same brands as brands A and B, respectively.

TABLE 1 shows the level of MG in these carbonated beverages. It is obvious that CSDs contain significantly high levels of MG. All CSDs contain high fructose corn syrup (HFCS) as sweetener except samples AA and BB. AA and BB were both diet drinks and contained aspartame as the sweetener. Less than 10 μg/100 mL of MG, except one beverage BB has 31.5 μg/100 mL of MG from the supermarket C. However, the content of MG in beverages A, B, and all other brands that contained high fructose corn syrup were in the range of 23.5–139.5 μg/100 mL. We have also identified high fructose corn syrup as the major source of MG in CSDs.[17]

Summary

MG can be generated both *in vitro* and *in vivo*. The rate of MG formation is approximately 120 μmol/d, which is about 0.1% of the flux of glucose under normal conditions measured in *in vitro* red blood cells.[18,19] Even with such a small fraction, MG is of importance and a threat because of its high reactivity. MG presence in most foods and beverages may come from sugars and lipids. Intake of MG has been shown to induce hypertension in animal studies.[20,21] Our study on the measurement of MG levels in commercial beverages shows that regular carbonated beverages containing HFCS have astonishing high levels of MG. The health concerns about MG are thus questioned here.

Our study indicates that tea catechins are able to effectively trap MG, the most important reactive carbonyl species, *in vitro*. The key issue for future studies is to investigate whether tea catechins can trap reactive dicarbonyl species *in vivo* and thus decrease the levels of AGEs and prevent or delay the development of diabetic complications of age-related diseases.

Conflict of Interest

The authors declare no conflicts of interest.

References

1. THE DIABETES CONTROL AND COMPLICATIONS TRIAL RESEARCH GROUP. 1993. The effect of intensive treatment of diabetes on the development and progression of long-term complications in insulin-dependent diabetes mellitus. N. Engl. J. Med. **329:** 977–986.
2. SINGH, R., A. BARDEN, T. MORI & L. BEILIN. 2001. Advanced glycation end-products: a review. Diabetologia **44:** 129–146.
3. NAGARAJ, R.H., P. SARKAR, A. MALLY, et al. 2002. Effect of pyridoxamine on chemical modification of proteins by carbonyls in diabetic rats: characterization of a major product from the reaction of pyridoxamine and methylglyoxal. Arch. Biochem. Biophys. **402:** 110–119.
4. BAYNES, J.W. & S.R. THORPE. 1999. Role of oxidative stress in diabetic complications: a new perspective on an old paradigm. Diabetes **48:** 1–9.
5. ODANI, H., T. SHINZATO, Y. MATSUMOTO, et al. 1999. Increase in three alpha, beta-dicarbonyl compound levels in human uremic plasma: specific *in vivo* determination of intermediates in advanced Maillard reaction. Biochem. Biophys. Res. Commun. **256:** 89–93.
6. LAPOLLA, A., R. FLAMINI, A. DALLA VEDOVA, et al. 2003. Glyoxal and methylglyoxal levels in diabetic patients: quantitative determination by a new GC/MS method. Clin. Chem. Lab. Med. **41:** 1166–1173.
7. KHUHAWAR, M.Y., A.J. KANDHRO & F.D. KHAND. 2006. Liquid chromatographic determination of glyoxal and methylglyoxal from serum of diabetic patients using mesostilbenediamine as derivatizing agent. Anal. Lett. **39:** 2205–2215.
8. HASE, M., T. BABAZONO, S. KARIBE, et al. 2006. Renoprotective effects of tea catechin in streptozotocin-induced diabetic rats. International Urology Nephrology **38:** 693–699.

9. RUTTER, K., D.R. SELL, N. FRASER, et al. 2003. Green tea extract suppresses the age-related increase in collagen crosslinking and fluorescent products in C57BL/6 mice. Int. J. Vitam. Nutr. Res. **73:** 453–460.
10. WU, C.H. & G.C. YEN. 2005. Inhibitory effect of naturally occurring flavonoids on the formation of advanced glycation endproducts. J. Agric. Food. Chem. **53:** 3167–3173.
11. LO, C.T., S. LI, D. TAN, et al. 2006. Trapping reactions of reactive carbonyl species with tea polyphenols in simulated physiological conditions. Mol. Nutr. Food Res. **50:** 1118–1128.
12. SANG, S., X. SHAO, N. BAI, et al. 2007. Tea polyphenol (-)-epigallocatechin 3-gallate: A new trapping agent of reactive carbonyl species. Chem. Res. Toxicol. **20:** 1862–1870.
13. VAN CHUYEN, N., T. KURATA & M. FUJIMAKI. 1973. Studies on the reaction of dipeptides with glyoxal. Agric. Biol. Chem. **37:** 327–334.
14. KRAUSE, R., J. KÜHN, I. PENNDORF, et al. 2004. N-Terminal pyrazinones: a new class of peptide-bound advanced glycation end-products. Amino Acids **27:** 9–18.
15. THORNALLEY, P.J. 1996. Pharmacology of methylglyoxal: formation, modification of proteins and nucleic acids, and enzymatic detoxification- a role in pathogenesis and antiproliferative chemotherapy. Gen. Pharmac. **27:** 565–573.
16. TAN, D. 2007. Trapping of methylglyoxal by dietary compounds *in vitro*. Ph.D. Dissertation, Rutgers University, New Brunswick, New Jersey, USA.
17. LO, C.Y., S. LI, Y. WANG, et al. 2007. Reactive dicarbonyl compounds and 5-(hydroxymethyl)-2-furfural in carbonated beverages containing high fructose corn syrup. Food Chem., doi: 10.1016/j.foodchem.2007.09.028.
18. THORNALLEY, P.J. 1988. Modification of the glyoxalase system in human red blood cells by glucose *in vitro*. Biochem. J. **254:** 751–755.
19. PHILLIPS, S.A. & P.J. THORNALLEY. 1993. Formation of methylglyoxal and D-lactate in human red blood cells *in vitro*. Biochem. Soc. Trans. **21:** 163S.
20. REAVEN, G.M. 1991. Insulin resistance and compensatory hyperin-sulinemia: role in hypertension, dyslipidemia, and coronary heart disease. Am. Heart. J. **121:** 1283–1288.
21. REISER, S., A.S. POWELL, D.J. SCHOLFIELD, et al. 1989. Daylong glucose, insulin, and fructose responses of hyperinsulinemic and nonhyperinsulinemic men adapted to diets containing either fructose or high-amylose corn-starch. Am. J. Clin. Nutr. **50:** 1008–1014.

The RAGE Pathway

Activation and Perpetuation in the Pathogenesis of Diabetic Neuropathy

IVAN K. LUKIC,[a,b] PER M. HUMPERT,[a] PETER P. NAWROTH,[a] AND ANGELIKA BIERHAUS[a]

[a]*Department of Medicine I and Clinical Chemistry, University of Heidelberg, Heidelberg, Germany*

[b]*Department of Anatomy, Zagreb University School of Medicine, Zagreb, Croatia*

The molecular mechanisms underlying loss of pain perception in diabetic neuropathy are poorly understood. Experimental diabetic neuropathy models recently provided evidence that engagement of the receptor for advanced glycation end products (RAGE) and RAGE-dependent sustained activation of the proinflammatory transcription factor nuclear factor kappa B might significantly contribute to reduced nociception. Most importantly, diabetes-induced loss of pain perception is largely prevented in RAGE-deficient mice compared to RAGE-bearing wild-type mice. Identifying RAGE-dependent inflammation as one pathomechanism underlying neuronal dysfunction might provide the basis for new therapeutic approaches.

Key words: diabetic neuropathy; receptor for advanced glycation end products; nuclear factor kappa B

Introduction

Due to the increasing prevalence of diabetes, disabilities associated with its complications are constantly growing. At present, up to 25% of diabetic patients suffer from diabetic polyneuropathy.[1] Although diabetic neuropathy is one of the major late complications limiting life expectancy of affected patients,[2] pathomechanisms underlying diabetic neuropathy are only partially understood, and experimental models have not yet provided a basis for successful treatment.[3] Recent findings related to the receptor for advanced glycation end products (RAGE) might provide new solutions.

Advanced Glycation End Products and RAGE

Advanced glycation end products (AGEs) are the products of a nonenzymatic reaction of glucose, α-oxoaldehydes, and other saccharide derivates with proteins, lipids, and nucleotides. AGE formation and deposition is enhanced with concomitant hyperglycemia, oxidant stress, carbonyl stress, and a delayed macromolecular turnover. As one consequence, diabetic patients show extensive intracellular and extracellular accumulation of AGEs in perineurial basal laminae, axons, Schwann cells, and endoneurial and epineurial microvessels, as well as the perineurium.[4–7] AGE immunoreactivity also increases with duration and severity of diabetes and correlates with a reduction in myelinated fiber density.[8] A direct association between diabetes and AGE accumulation in peripheral nerves is further provided by the observation that AGEs in the sciatic nerve of diabetic rats were decreased by pancreatic islet cell transplantation.[9] The functional consequences of AGE deposition, however, are poorly understood. AGEs formation preferentially occurs on long-lived protein, such as myelin where it promotes the myelin uptake by macrophages and demyelination; likewise, AGEs increase the uptake of extracellular matrix proteins, thereby altering cellular permeability, cell interactions, and adhesion.[10] Intracellular AGE formation might further influence the assembly of the cytoskeleton (particularly vulnerable to nonenzymatic glycation in the peripheral nerve system),[11] promote protein aggregation, and modify nuclear proteins and nucleic acids. Besides changing the structure and function of the modified macromolecules,

Address for correspondence: PD Dr. Angelika Bierhaus, Ph.D., University of Heidelberg, Department of Medicine I and Clinical Chemistry, INF 410, 69120 Heidelberg, Germany. Voice: +49-6221-564752; fax: +49-6221-564754.

angelika_bierhaus@med.uni-heidelberg.de

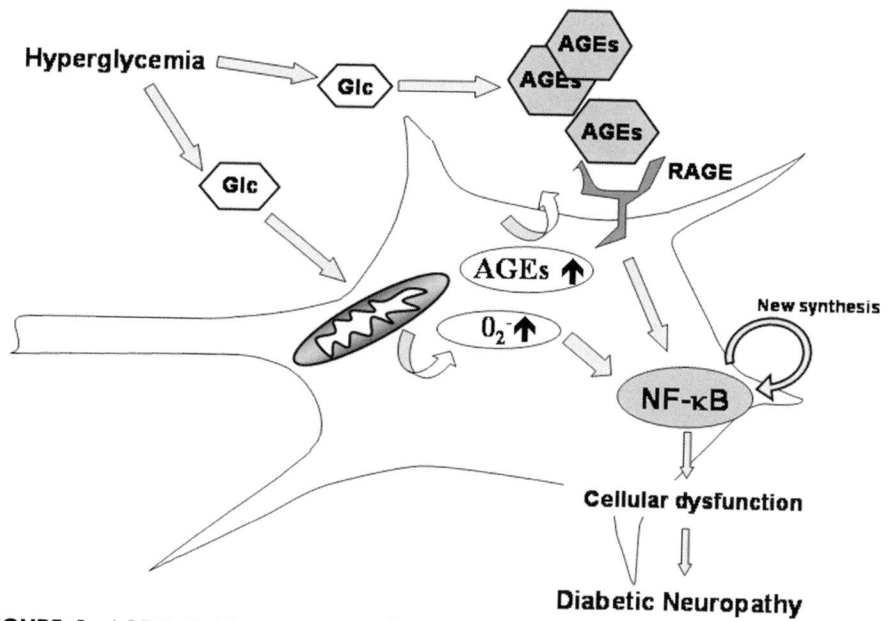

FIGURE 1. AGE–RAGE interaction contributing to the pathogenesis of diabetic neuropathy. Abbreviations: AGEs, advanced glycation end products; Glc, glucose; NF-κB, nuclear factor kappa B; O_2^-, superoxide; RAGE, receptor for AGEs.

AGE formation promotes oxidative stress,[12] depletes cellular antioxidant defense mechanisms,[13,14] and induces cellular dysfunction through binding to a number of cellular surface receptors.[15–18] Thus, the consequences of AGE accumulation in the neurovascular system are (i) irreversible structural changes of the macromolecules and (ii) uncontrolled activation of cellular metabolism and signaling pathways. The cellular activation is a consequence of binding of AGEs-modified macromolecules to several cell surface receptors, namely the scavenger receptor, galectin-3, and RAGE, which is currently the best characterized AGE receptor.[16–18]

RAGE is a multiligand member of the immunoglobulin superfamily of cell surface receptors with specificity for three-dimensional structures (e.g., β-sheets and fibrils) rather than for specific oligomere sequences.[15–18] Upon engagement of ligands, RAGE triggers intracellular signaling pathways via phosphatidylinositol-3 kinase, Ki-Ras, and mitogen-activated protein kinases, the Erk1 and Erk2.[19–24] Those pathways culminate in the activation of the transcription factor nuclear factor kappa B (NF-κB) and subsequent transcription of a number of genes, including endothelin-1, tissue factor, interleukin (IL)-1, IL-6, and tumor necrosis factor (TNF)-α.[15,16,25,26] In the context of diabetic complications, it is essential to note the unique capacity of RAGE to induce sustained NF-κB activation from increased levels of *de novo* synthesized NF-κB subunit p65 overriding endogenous negative feedback mechanisms.[27] Besides, AGE–RAGE interaction mediates activation of NADPH-oxidases and thereby increases the cellular oxidative stress that in turn also activates the redox-sensitive transcription factor NF-κB.[28,29] Since RAGE itself is regulated by NF-κB, this will further increase, amplify, and sustain cellular dysfunction.[30]

RAGE as a Mediator of Diabetic Neuropathy

Recent studies indicate that RAGE-dependent NF-κB activation might represent one mechanism significantly contributing to neuronal dysfunction and loss of function (FIG. 1).[4,31,32] Compared to healthy controls, sciatic nerves of diabetic mice showed upregulated expression of NF-κB-regulated target genes (e.g., IL-6) as well as the NF-κB subunit p65. Importantly, upregulation could be prevented by admission of soluble RAGE (sRAGE), a truncated form of RAGE comprising its extracellular ligand-binding domain (thereby acting as antagonist scavenging RAGE ligands and preventing interaction with the cell surface receptor).

Moreover, the loss of thermal pain perception observed in mice with long-standing diabetes could be prevented by treatment with sRAGE. In agreement with these observations, NF-κB activation and the loss of pain perception were largely blunted in RAGE-deficient knockout (RAGE$^{-/-}$) mice even after 6 months of diabetes, while mice overexpressing RAGE showed progressive diabetic neuropathy.[33]

Analogous pathophysiological pathways seem to exist in humans. Immunohistological stainings of sural nerve biopsies from diabetic patients showed increased expression of RAGE and colocalization with the AGE adduct N^ε-(carboxymethyl)lysine (CML)[4,5] in neurovascular compartments of endoneurium, perineurium, and epineurium. RAGE expression and accumulation of CML was overlapping with the distribution of activated NF-κBp65 and IL-6, thus implying the involvement of the AGE–RAGE–NF-κB pathway in the development of diabetic neuropathy.[4,5] Those events seem to take place at a very early stage of diabetic polyneuropathy, even before the occurrence of chronic hyperglycemia, since the sural nerve biopsies of patients with impaired glucose tolerance-related polyneuropathy revealed an identical pattern: colocalization of CML, RAGE, and NF-κB.[32]

Therefore, it can be hypothesized that the basic mechanisms by which AGEs–RAGE interaction leads to neuropathy is vascular dysfunction and consequent microangiopathy in peripheral nerves.[6] Consistently, recent clinical studies showed an association between plasma levels of sRAGE and albuminuria,[34,35] which could reflect the upregulation of the receptor and its pathophysiological role in diabetic microvascular disease. Yet, screening for peripheral and autonomic neuropathy in type 2 diabetes patients did not reveal associations of diabetic neuropathy and sRAGE.[36] In view of the strong experimental data for a role of RAGE in the development of diabetic neuropathy,[4,37] future clinical studies should focus on whether sRAGE, as a marker of diabetic vascular disease, could serve as a biomarker indicating the development of neuropathy in prospective settings. In addition, clinical data indicates that there might be a different role of the microvasculature in the development of neuropathy between patients with type 1 compared to type 2 diabetes.[38,39] These potential differences will have to be studied in suitable animal models.

In this context, it is important that activation of NF-κB and induction of cytokines, such as IL-6, can contribute neuronal plasticity and the cellular response to neurodegeneration[40] and are supposed to have initial neuroprotective effects in sensory neurons.[41] Thus, RAGE-induced NF-κB activation might play a dual role in diabetic neuropathy dependent on the compartment (endothelium versus neuronal tissue) and the time course (early stages of diabetes versus long-standing diabetes). While initial NF-κB activation in neuronal cells might be neuroprotective, perpetuated NF-κB activation is likely to cause sustained inflammation and impairment of neuronal function. The late effects of AGE–RAGE-dependent sustained NF-κB activation were shown to be of importance for the development of experimental neuropathy because NF-κB activation in dorsal root ganglia neurons is absent in RAGE$^{-/-}$ mice with long-standing diabetes and is paralleled by a partial protection from diabetic neuropathy.[4] Besides, RAGE-dependent activation of NADPH oxidases, subsequent burst of reactive oxygen species, and activation of the apoptotic cascade might be a scenario of neurodestruction.[42]

Noteworthy, deletion of RAGE does not completely protect from diabetic neuropathy since (i) NF-κB and IL-6 are not completely inhibited in diabetic RAGE$^{-/-}$ mice, (ii) pain perception is not completely restored, and (iii) diabetes-induced loss of PGP9.5-positive sensory fibers can also be observed in RAGE$^{-/-}$ mice. Thus, RAGE plays an important, but not exclusive, role in sensory deficits and loss of pain perception in diabetic neuropathy. In contrast, diabetes-induced loss of pain perception is completely reversed by sRAGE in 3-month, diabetic, wild-type mice, indicating that receptors different from RAGE but able to interact with RAGE ligands might also contribute to loss of function in diabetic neuropathy.[4]

Although the leading signs and symptoms of diabetic neuropathy relate to the dysfunction of the sensory neurons,[1] involvement of the motor neurons should not be overlooked. A recent study performed on 8-month, diabetic, wild-type mice reported the loss of axonal terminals of the motor neurons associated with decreased conduction velocity. In addition, the expression of RAGE on the perykaria of the motor neurons was elevated, implying that the AGE–RAGE is also involved in the changes of the efferent arm of the peripheral nervous system.[43]

Finally, it is noteworthy that the pathophysiological role of RAGE is not limited to the peripheral nervous system. The expression of RAGE has been shown within both gray and white matter of healthy mouse brains but was considerably higher in the brains of diabetic animals. Moreover, the pathologic changes seen in 9-month, diabetic, wild-type mice, such as leukoencephalopathy and cerebral atrophy, were largely prevented in diabetic RAGE-null mice.[44]

Concluding Remarks

Advances in recent years have pointed to the key role that RAGE plays in the development of diabetic neuropathy. Models of experimental diabetic neuropathy provided sound evidence that deletion of the RAGE gene protects animals from the detrimental effects of long-lasting diabetes, while overexpression of RAGE promotes diabetic neuropathy.[4,33,42] Thus, targeting the RAGE pathway could be a promising therapeutic strategy in clinical entities of diabetic neuropathy but also of polyneuropathies and neurodegenerative diseases of other origins.[5,31,45]

However, there is an important caveat—RAGE is not simply a "bad guy" but is involved in many physiologic processes. For instance, RAGE-mediated NF-κB activation has been shown to be involved in muscle fiber regeneration in inflammatory myopathies and limb girdle muscular dystrophy.[46] Moreover, RAGE modulates the regeneration of peripheral nerves, and the inactivation of the RAGE gene impairs the nerve repair.[47] Thus, prior to pharmacological interventions in humans, a better insight into the RAGE pathway and the physiological role of RAGE in the nervous system is urgently needed.

Acknowledgments

This study was supported by the European Foundation for the Study of Diabetes (A.B.), the Juvenile Diabetes Research Foundation (A.B., P.P.N.), the National Foundation for Science, Higher Education, and Technological Development of the Republic of Croatia (I.K.L.), and the Network Aging Research (P.M.H., A.B.).

Conflict of Interest

The authors declare no conflicts of interest.

References

1. ZIEGLER, D. & A. BIERHAUS. 2007. Therapie der diabetischen Neuropathie. Dtsch. Med. Wochenschr. **132:** 11–16.
2. ARGOFF, C.E., B.E. COLE, D.A. FISHBAIN, et al. 2006. Diabetic peripheral neuropathic pain: clinical and quality-of-life issues. Mayo Clin. Proc. **81**(4 Suppl): S3–S11.
3. ZIEGLER, D. 2006. Treatment of diabetic polyneuropathy: update 2006. Ann. N.Y. Acad. Sci. **1084:** 250–266.
4. BIERHAUS, A., K.-M. HASLBECK, P.M. HUMPERT, et al. 2004. Loss of pain perception in diabetes is dependent on a receptor of the immunoglobulin superfamily. J. Clin. Invest. **114:** 1741–1751.
5. HASLBECK, K.M., B. NEUNDOERFER, U. SCHLOETZER-SCHREHARDT, et al. 2007. Activation of the RAGE pathway: a general mechanism in the pathogenesis of polyneuropathies? Neurol. Res. **29:** 103–110.
6. WADA, R. & S. YAGIHASHI. 2005. Role of advanced glycation end products and their receptors in the development of diabetic neuropathy. Ann. N.Y. Acad. Sci. **1043:** 598–604.
7. MISUR, I., K. ZARKOVIC, A. BARADA, et al. 2004. Advanced glycation endproducts in peripheral nerve in type 2 diabetes with neuropathy. Acta Diabetol. **41:** 158–166.
8. SUGIMOTO, K., Y. NIZHIZAWA, S. HORIUCHI, et al. 1997. Localization in human diabetic peripheral nerve of Ne-carboxymethyllysine-protein adducts, an advanced glycation endproduct. Diabetologia **40:** 1380–1387.
9. SENSI, M., S. MORANO, S. MORELLI, et al. 1998. Reduction of advanced glycation-end product (AGE) levels in nervous tissue proteins of diabetic Lewis rats following islet transplants is related to different durations of poor metabolic controls. Eur. J. Neurosci. **10:** 2768–2775.
10. VLASSARA, H., M. BROWNLEE & A. CERAMI. 1985. Recognition and uptake of human diabetic peripheral nerve myelin by macrophages. Diabetes **34:** 553–557.
11. SCOTT, J.N., A.W. CLARK & D.W. ZOCHODONE. 1999. Neurofilament and tubulin gene expression in progressive experimental diabetes: failure of synthesis and export by sensory neurons. Brain **122:** 2109–2118.
12. WAUTIER, M.P., O. CHAPPEY, S. CORDA, et al. 2001. Activation of NADPH oxidase by AGE links oxidant stress to altered gene expression via RAGE. Am. J. Physiol. Endocrinol. Metab. **280:** E685–E694.
13. FELDMAN, E.L. 2003. Oxidative stress and diabetic neuropathy: a new understanding of an old problem. J. Clin. Invest. **111:** 431–433.
14. BIERHAUS, A., S. CHEVION, M. CHEVION, et al. 1997. Advanced glycation end product-induced activation of NF-kB is suppressed by a-lipoic acid in cultured endothelial cells. Diabetes **46:** 1481–1490.
15. SCHMIDT, A.M. & D.M. STERN. 2000. RAGE: a new target for the prevention and treatment of the vascular and inflammatory complications of diabetes. Trends Endocrinol. Metab. **11:** 368–375.
16. BIERHAUS, A., P.M. HUMPERT, M. MORCOS, et al. 2005. Understanding RAGE, the receptor for advanced glycation end products. **83:** 876–886.
17. MOSER, B., K.C. HEROLD & A.M. SCHMIDT. 2006. Receptor for advanced glycation endproducts and its ligands: initiators or amplifiers of joint inflammation—a bit of both? Arthritis Rheum. **54:** 14–18.
18. YAN, S.F., G.R. BARILE, V. D'AGATI, et al. 2007. The biology of RAGE and its ligands: uncovering mechanisms at the heart of diabetes and its complications. Curr. Diab. Rep. **7:** 146–153.
19. BIERHAUS, A., P.M. HUMPERT, DM. STERN, et al. 2005. Advanced glycation end product receptor–mediated cellular dysfunction. Ann. N.Y. Acad. Sci. **1043:** 676–680.
20. BARTLING, B., N. DEMLING, R.E. SILBER, et al. 2006. Proliferative stimulus of lung fibroblasts on lung cancer cells is impaired by the receptor for advanced glycation endproducts. Am. J. Respir. Cell Mol. Biol. **34:** 83–91.

21. KUNIYASU, H., Y. CHIHARA & H. KONDO. 2003. Differential effects between amphoterin and advanced glycation end products on colon cancer cells. Int. J. Cancer **104:** 722–727.
22. XU, D. & J.M. KYRIAKIS. 2003. Phosphatidylinositol 3'-kinase-dependent activation of renal mesangial cell Ki-Ras and ERK activated by advanced glycation end products. J. Biol. Chem. **278:** 39349–39355.
23. DUKIC-STEFANOVIC, S., J. GASIC-MILENKOVIC, W. DEUTHER-CONRAD, et al. 2003. Signal transduction pathways in mouse microglia N-11 cells activated by advanced glycation endproducts (AGEs). J. Neurochem. **87:** 44–55.
24. GOLDIN, A., J.A. BECKMAN, A.M. SCHMIDT, et al. 2006. Advanced glycation end products: sparking the development of diabetic vascular injury. Circulation **114:** 597–605.
25. QUEHENBERGER, P., A. BIERHAUS, P. FASCHING, et al. 2000. Endothelin 1 transcription is controlled by nuclear factor-kappaB in AGE-stimulated cultured endothelial cells. Diabetes **49:** 1561–1570.
26. BIERHAUS, A., T. ILLMER, M. KASPER, et al. 1997. Advanced glycation end product (AGE)-mediated induction of tissue factor in cultured endothelial cells is dependent on RAGE. Circulation **96:** 2262–2271.
27. BIERHAUS, A., S. SCHIEKOFER, M. SCHWANINGER, et al. 2001. Diabetes-associated sustained activation of the transcription factor nuclear factor-κB. Diabetes **50:** 2792–2808.
28. NISHIKAWA, T., D. EDELSTEIN, X.L. DU, et al. 2000. Normalizing mitochondrial superoxide production blocks three pathways of hyperglycaemic damage. Nature **404:** 787–790.
29. LIN, J., A. BIERHAUS, P. BUGERT, et al. 2006. Effect of R-(+)-alpha-lipoic acid on experimental diabetic retinopathy. Diabetologia **49:** 1089–1096.
30. LI, J. & A.M. SCHMIDT. 1997. Characterization and functional analysis of the promoter of RAGE, the receptor for advanced glycation end products. J. Biol. Chem. **272:** 16498–16506.
31. HASLBECK, K.-M., A. BIERHAUS, E. SCHLEICHER, et al. 2004. Receptor for advanced glycation endproduct (RAGE)-mediated nuclear factor κB activation in vasculitic neuropathy. Muscle Nerve **29:** 853–860.
32. HASLBECK, K.M., E. SCHLEICHER, A. BIERHAUS, et al. 2005. The AGE/RAGE/NF-(kappa)B pathway may contribute to the pathogenesis of polyneuropathy in impaired glucose tolerance. Exp. Clin. Endocrinol. Diabetes **113:** 288–291.
33. YAJIMA, N., Y. YAMAMOTO, H. YAMAMOTO, et al. 2004. Peripheral neuropathy in diabetic mice overexpressing receptor for advanced glycation endproducts (RAGE). Collected abstracts of the 8th International Symposium on the Maillard reaction (Charleston, SC): no. SXI-7: p. 55.
34. HUMPERT, P.M., S. KOPF, Z. DJURIC, et al. 2006. Plasma sRAGE is independently associated with urinary albumin excretion in type 2 diabetes. Diabetes Care **29:** 1111–1113.
35. HUMPERT, P.M., Z. DJURIC, S. KOPF, et al. 2007. Soluble RAGE but not endogenous RAGE is associated with albuminuria in patients with type 2 diabetes. Cardiovasc. Diabetol. **6:** 9.
36. HUMPERT, P.M., G. PAPADOPOULOS, K. SCHAEFER, et al. 2007. sRAGE and esRAGE are not associated with peripheral or autonomic neuropathy in type 2 diabetes. Horm. Metab. Res. **39:** 899–902.
37. CELLEK, S., W. QU, A.M. SCHMIDT, et al. 2004. Synergistic action of advanced glycation end products and endogenous nitric oxide leads to neuronal apoptosis in vitro: a new insight into selective nitrergic neuropathy in diabetes. Diabetologia **47:** 331–339.
38. KASALOVA, Z., M. PRAZNY & J. SKRHA. 2006. Relationship between peripheral diabetic neuropathy and microvascular reactivity in patients with type 1 and type 2 diabetes mellitus – neuropathy and microcirculation in diabetes. Exp. Clin. Endocrinol. Diabetes **114:** 52–57.
39. GREENMAN, R.L., S. PANASYUK, X. WANG, et al. 2005. Early changes in the skin microcirculation and mucle metabolism of the diabetic foot. Lancet **366:** 1711–1717.
40. MATTSON M.P. & S. CAMANDOLA. 2001. NF-kappaB in neuronal plasticity and neurodegenerative disorders. J. Clin. Invest. **107:** 247–254.
41. FLATTERS S.J., A.J. FOX & A.H. DICKENSON. 2004. Nerve injury alters the effects of interleukin-6 on nociceptive transmission in peripheral afferents. Eur. J. Pharmacol. **484:** 183–191.
42. VINCENT, A.M., L. PERRONE, K.A. SULLIVAN, et al. 2007. Receptor for advanced glycation end products activation injures primary sensory neurons via oxidative stress. Enodcrinology **148:** 548–558.
43. RAMJI, N., C. TOTH, J. KENNEDY, et al. 2007. Does diabetes mellitus target motor neurons? Neurobiol. Dis. **26:** 301–311.
44. TOTH, C., A.M. SCHMIDT, U.I. TUOR, et al. 2006. Diabetes, leukoencephalopathy and rage. Neurobiol. Dis. **23:** 445–461.
45. ARANCIO, O., H.P. ZHANG, X. CHEN, et al. 2004. RAGE potentiates Aβ-induced perturbation of neuronal function in transgenic mice. EMBO J. **23:** 4096–4105.
46. HASLBECK, K.M., U. FRIESS, E.D. SCHLEICHER, et al. 2005. The RAGE pathway in inflammatory myopathies and limb girdle muscular dystrophy. Acta Neuropathol. **110:** 247–254.
47. RONG, L.L., S.-F. YAN, T. WENDT, et al. 2004. RAGE modulates peripheral nerve regeneration via recruitment of both inflammatory and axonal outgrowth pathways. FASEB J. **18:** 1818–1825.

The Role of the Amadori Product in the Complications of Diabetes

VINCENT M. MONNIER,[a,b] DAVID R. SELL,[a] ZHENYU DAI,[a] INA NEMET,[a,d] FRANCOIS COLLARD,[a] AND JIANYE ZHANG[c]

Departments of [a]Pathology, [b]Biochemistry and [c]Chemistry, Case Western Reserve University, Cleveland, Ohio, USA

[d]Division of Organic Chemistry and Biochemistry, Rudjer Boskovic Institute, Zagreb, Croatia

Strong evidence has emerged in recent years in support of an association between advanced glycation and the complications of diabetes, whereby both glycoxidation products and oxoaldehydes have been implicated. In contrast, except for the fact that skin collagen-linked fructosamine (Amadori product) is a strong predictor of the risk of progression of microvascular disease in humans, Amadori products have not been associated with complications in most animal experiments. Below we develop the hypothesis that glucose-derived advanced glycation end products (AGEs), such as glucosepane, may inflict sustained damage to the extracellular matrix in diabetes and contribute to tissue stiffening and accelerated sclerosis in arteries, kidneys, and other organs as supported by immunochemical studies using a glucosepane antibody. We also hypothesize that many more structures derived from Amadori products with nucleophiles, such as primary amines and thiols, are expected. The selective prevention of Amadori-derived AGEs using deglycating enzymes would be desirable. However, x-ray diffraction studies of Amadoriase I crystals show that the active site of the enzyme is deeply embedded, explaining why this approach is unlikely to succeed *in vivo*. Preliminary experiments with nucleophiles show that aminoguanidine and other compounds block glucosepane *in vitro*.

Key words: collagen; cross-linking; Amadoriase; fructosylamino acid oxidase; Maillard reaction; glucosepane

Introduction

Research into the role of the Maillard reaction in the pathogenesis of diabetic complications is at a crossroad. Considerable evidence has accumulated showing that tissue or serum content of advanced glycation end products (AGEs) is strongly associated with the pathogenesis of microvascular or macrovascular disease, and a large number of *in vitro* experiments and intervention studies in animal models of diabetes confirm that anti-AGE treatment improves diabetic complications.[1,2] Some of the major mechanisms by which the Maillard reaction has been implicated in diabetes include triggering of the proinflammatory receptor of AGE (RAGE) signaling cascade,[3] induction of apoptosis and anoikis,[4,5] dietary AGE burden,[6] blocking of active sites in critical proteins (such as those involved in coagulation),[7] nerve function,[8] and many others.[9]

In spite of the progress achieved, reconciling mechanisms by which the Maillard reaction might be responsible for diabetic complications is facing a number of hurdles. First and foremost, clinical trials with aminoguanidine, one of the most potent AGE inhibitors in experimental diabetes, had to be stopped either because of toxicity or inability to reach the mandated clinical end point in spite of beneficial effects on proteinuria.[10] Second and more puzzling is that there is weak to no molecular evidence that aminoguanidine can indeed scavenge AGE precursors *in vivo*. On the other hand, pyridoxamine has powerful anticomplication properties in various animal models[11] as well as some beneficial effects in preliminary human trials (http://www.nephrogenex.com/clinical-programs/index.php#Pyridorin). However, it also has powerful hypolipidemic properties and is able to trap advanced lipoxidation end products (ALEs), but, surprisingly, no evidence of trapping of AGE precursors have been found![12] Further blurring the mechanistic

Address for correspondence: Vincent M. Monnier, Dept. of Pathology, CWRU, 2103 Cornell Road, Cleveland, OH 44106. Voice: 216-368-6613; fax: 216-368-1357.

vmm3@cwru.edu

picture is that several AGE inhibitors have metal chelating properties,[13] most prominently the recently discovered LR series.[14] Thus, a major, yet unanswered, question is, do AGEs act as metal chelators *in vivo* and, if so, how do chelatable metals catalyze complications?

Third, the low stoichiometric modification of cellular or extracellular proteins by glycation products, often less than a few percent, raises the question of whether the amount of detected AGEs is sufficient to impair tissue function and contribute to the pathogenesis of diabetic complications. Few studies, however, have yet attempted to identify and quantitate specific glycation targets, and it is possible that much higher levels of modification at specific sites may result in dysfunction of critical proteins. In that regard, more research on the modification of cysteine residues is needed as revealed by the recent studies on succination of thiols.[15]

The question of whether AGEs are merely markers of disease or play a causal role in complication progression may be difficult to resolve. On the one hand, drugs that have been specifically designed or tested for their *in vitro* anti-AGE activity have been subsequently found to prevent complications, as for example aminoguanidine, pyridoxamine, and the recently discovered LR90,[16] and, vice versa, several drugs with antidiabetic or renoprotective activity, such as metformin, ARB inhibitors, and hydralazine, have been found to have anti-AGE formation activity.[17,18] Thus, overall data indicate that AGE formation is intimately linked to the disease process, but existence of multiple mechanistic properties of the drugs makes it difficult to sort out causal versus innocent bystander effect.

Glucosepane: A Paradigm for Non-oxidative Age Damage to the Extracellular Matrix

Most studies described above have focused on glycoxidative mechanisms (i.e., the formation of carboxymethyllysine (CML), carboxyethyllysine (CEL), and pentosidine). Because numerous data support the existence of oxidative stress in diabetes, it is not surprising that these AGEs are associated with complications. Paradoxically, little is yet known of the actual source of their *in vivo* precursors—to what extent they form via glycoxidation or via oxoaldehydes, such as glyoxal, methylglyoxal, and 3-deoxyglucosone.[19]

While the existence of a relationship between AGEs or ALE of oxidative origin and diabetic complications is apparent, it is less clear whether the Amadori product has any significance as a source of protein damage in diabetes. The shift away from the Amadori product that occurred over the past 25 years was catalyzed, in part, by the review from Kennedy and Baynes indicating a paucity of evidence in favor of glycation (specifically Amadori product formation) in diabetic complications.[20] Whether this conclusion will have to be revised depends on whether a deleterious phenotype emerges in the diabetic fructosamine 3-kinase (FN3K) knockout mouse.[21]

The outcome of the diabetic FN3K mouse will only address phenotypes linked to intracellular glycation because the enzyme is not active in the extracellular compartment. Furthermore, the data will only be applicable to the life span of the mouse (i.e., a time frame that is often too short for most complications to develop in the human). However, while excess intracellular glycation may prove to be innocuous, excess extracellular glycation and Amadori product derived cross-links, like glucosepane, may accelerate the aging of the extracellular matrix and inflict important damage.

Glucosepane is the single major AGE cross-link and product in the aging human extracellular matrix and in diabetes discovered so far.[22] It is a seven-atom-membered protein cross-link between lysine and arginine residues that includes all six glucose carbons. Its formation from Amadori products does not require oxidation. Thus, its formation is expected to be blocked by glycoxidation and dramatically enhanced during antioxidant treatment, reflecting the trade-off of improved oxidant but worsened carbonyl stress (FIG. 1).

Such proposition was made several years ago in the context of the discovery of CML.[23] Similar paradoxical effects were also observed for CML and pentosidine,[24] suggesting that antioxidants may favor carbonyl versus oxidative pathways of "glycoxidation."

As most data show that tissue and plasma levels of CML and pentosidine tend to cosegregate with the disease process (for a review, see Ref. 25), this raises the question of whether the Amadori product and glucosepane play a role in diabetic complications. The following considerations suggest that nonoxidative glucose-derived carbonyl stress needs to be investigated. First, our data in collaboration with the Diabetes Control and Complications Trial (DCCT) showed that the risk of retinopathy progression increased from 22% to 75% if participants had not only high HbA_{1c} but in addition skin furosine (compound formed from the Amadori product during acid hydrolysis of proteins) glycation levels above the third quartile (FIG. 2).

Thus, intense modification of the extracellular matrix by Amadori products conveys an increased risk of complication progression. Second, glucosepane is the single major cross-link in skin and glomerular basement membrane collagen from diabetic individuals.[22] Unlike the Amadori product, it not only blocks

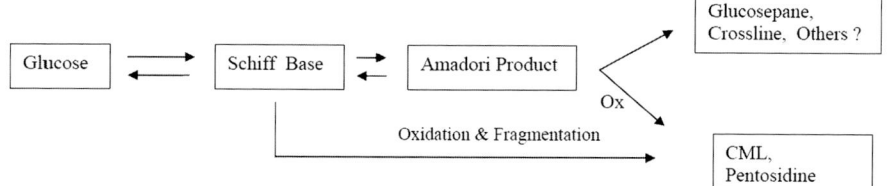

FIGURE 1. Mutual exclusivity of the pathways leading to glucosepane versus carboxymethyllysine (CML) formation from the Amadori product fructosyl-lysine.

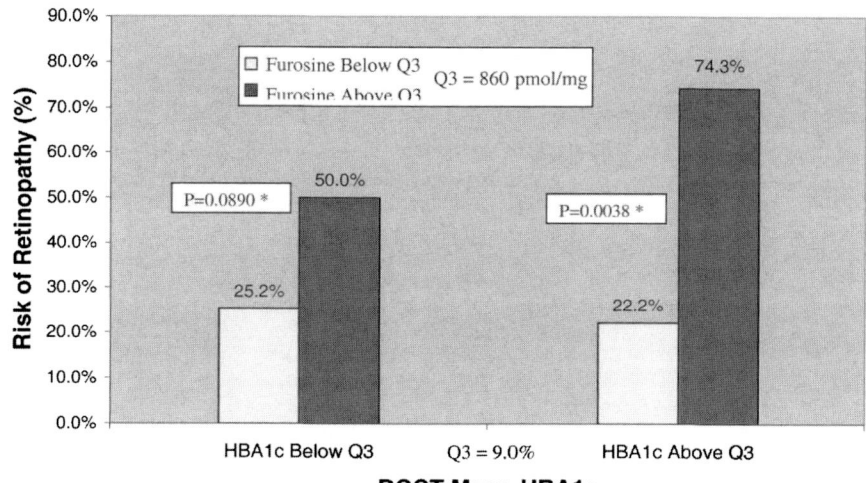

FIGURE 2. Risk of retinopathy progression by the Diabetes Control and Complications Trial (DCCT) mean A_{1C} (HbA$_{1c}$) and skin collagen furosine. Shown is the risk of progression of retinopathy in the upper quartile of furosine (>Q3) (blue bars) compared to the progression in the lower three quartiles of furosine (<Q3) (yellow bars). The comparison is made in the upper (>Q3) and lower three (<Q3) quartiles of A_{1C} separately. *P value is from a between-group furosine (>Q3 versus <Q3) comparison within each A_{1C} strata using the two-tailed test. The greatest risk of progression occurs when both furosine and A_{1C} are >Q3, but the dominance of furosine over A_{1C} as a risk factor is evident. (Reprinted with permission from Ref. 2.) (In color in *Annals* online.)

lysine but also arginine residues that might be involved in Arg-Gly-Asp (RGD) sequences needed for integrin binding. The blocking of the latter by methylglyoxal hydroimidazolones was found to result in anoikis.[5] Further damage to such residues may come from their conversion to ornithine residues by both glucosepane and methyglyoxal.[26] Thus, mapping of glucosepane formation sites in collagen will be needed to evaluate its role in anoikis.

In addition to potentially leading to anoikis, glucosepane could have deleterious properties by engaging cell surface receptors, by cross-linking proteins and increasing matrix stiffness, as well as by decreasing protein turnover rate and thus favoring tissue sclerosis and matrix accumulation in tissues, such as arteries and kidneys. In that regard, preliminary studies using a polyclonal glucosepane antibody developed in our laboratory revealed preferential localization in the thickened intima from human and apoE-null db/db aortas (not shown).

From the above, one can conclude that intensified research into the role of glucosepane in diabetic tissue dysfunction is urgently needed. While its formation may not be a major factor in the initiation of the complications, one can envisage that the accumulation of heavily modified proteins may dramatically decrease the reversibility of the lesions. As an example, tight control of glycemia had no beneficial effect on renal lesions after 5 years, but reversal of lesions was observed after 10 years.[27]

The 5,6-Dioxy-glucosone Pathway: A Rich Source of Novel AGEs?

The discovery of the 5,6-dioxoglucosone pathway suggests that the Amadori product might be a source

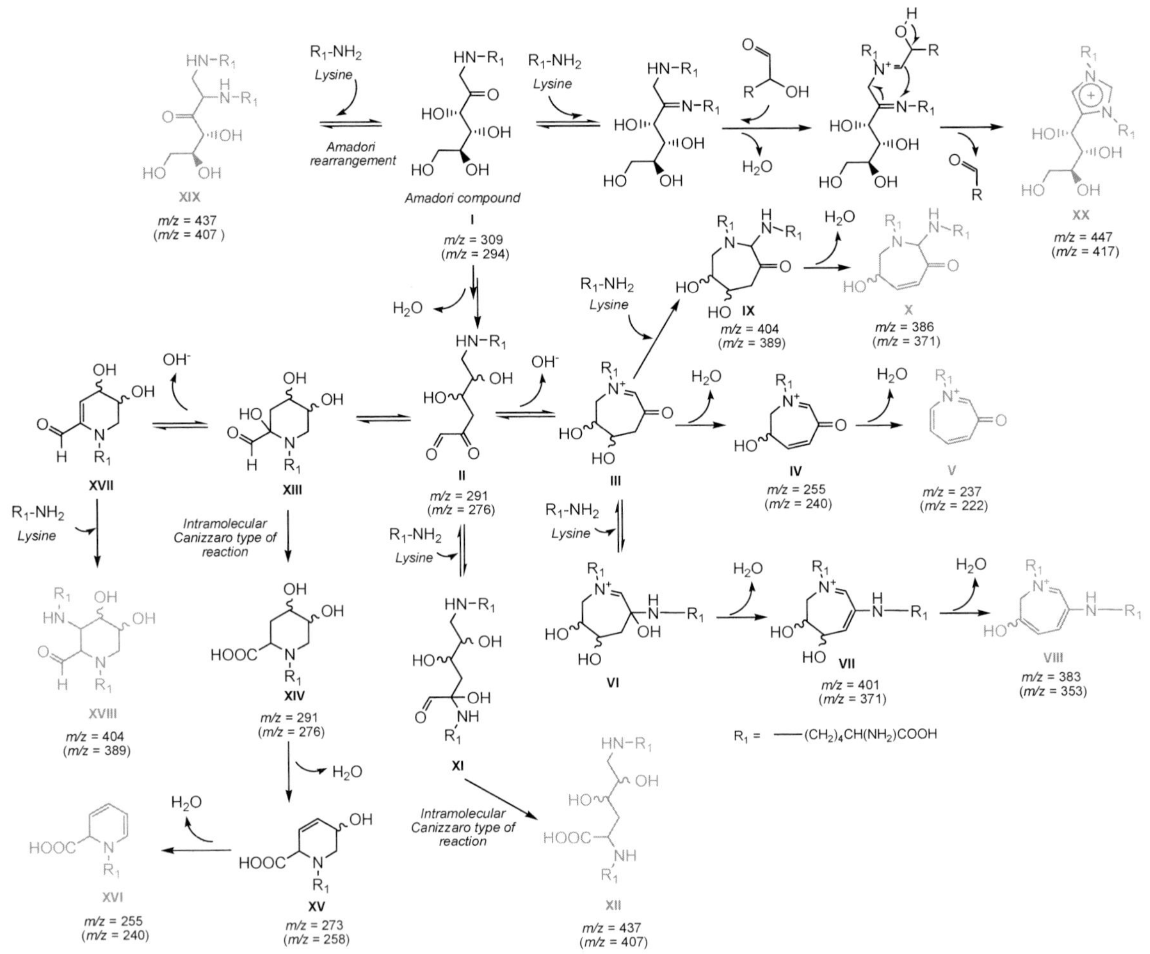

FIGURE 3. Theoretical reactivity of glucose-derived Amadori product with itself or primary amines, such as the ε-amino-lysine. Note that attack of carbon-2 would prevent formation of the 5,6-dioxoglucosone intermediate. (In color in *Annals* online.)

of many more AGE products, the structures of which can only be speculated at this time.

In FIGURE 3 we have listed a number of hypothetical structures derived from the Amadori product of glucose and lysine. We expect to find similar and other structures with cysteine SH groups, illustrating the fact that glucosepane may only be the tip of the iceberg. One exception in the scheme below is the formation of the compound with $m/z = 447$. Its formation does not proceed via the 5,6-dioxo pathway but requires attack of the keto group in position two by the nucleophilic agent. The fact that we were able to isolate it from the reaction of the Amadori product with primary amines suggests that nucleophilic attack at C-2 may drive the reactivity of the Amadori product away from both the classical 1,2- and 2,3-enolization.

Approaches toward Uncoupling of Amadori Product-derived Cross-linking from Hyperglycemia

The identification of glucosepane as the major cross-link of the extracellular matrix in human skin and glomerular basement membrane is a major step forward in our understanding of the cross-linking chemistry in aging and age-related diseases. However, many questions remain concerning glucosepane and the proposed structures above. To begin with, it is not clear where in the molecule glucosepane forms (i.e., intramolecularly versus intermolecularly). In ongoing studies, we were able to map intramolecular glucospane cross-links in glucose-incubated ribonuclease A to residues R39 and K41 and to R85 and K98.

In addition, we were able to identify an intermolecular Deoxyglucosone Dimer Imidazole Crosslink (DODIC) between K1 and R39. From these studies, we conclude that intramolecular cross-links form readily. Their significance for protein structure and enzymatic digestibility remains to be determined.

The ability of glucosepane to form intramolecular cross-links suggests that glucosepane should be immunohistochemically detectable in tissue sections. Thus we raised a polyclonal antibody to glucosepane–keyhole limpet hemocyanin (KLH) conjugate. Preliminary results indicate that glucosepane is detectable in almost all human tissues. Particularly intense staining was observed in human diabetic kidneys with sclerosed glomeruli and in the neointima of human and ApoE-null mice.

While the biological impact of the presence of glucosepane needs to be investigated, it would not be surprising that its formation contributes to the sclerosing process by rendering collagen less digestible to proteolytic enzymes. However, while *in vitro* experimentation would help clarify this question in a model system, only an approach using agents that block glucosepane formation would ultimately clarify its biomedical significance.

Enzymatic Approaches

The single most effective way to block glucose-mediated cross-linking would be either to reverse the formation of the Amadori product or to block its conversion to glucosepane and other modifications. Addressing the former approach, we have isolated and characterized two deglycating enzymes (fructosamine oxidases) from fungi named Amadoriase I and II.[28] In contrast to the FN3K enzymes that can deglycate most proteins, though only intracellularly because they are ATP dependent, Amadoriases only have activity against the free substrate or small glycated peptides. They are sterically inhibited against glycated proteins, such as bovine serum albumin. For this reason, we decided to determine the crystal structure of Amadoriase II in order to understand the problem of access to the catalytic site. These studies revealed that the depth of the pocket is 12 Å deep, i.e., too deep for access to glycated proteins (unpublished data), precluding us from producing isoforms with deglycating activity toward proteins. Intriguingly, Fogliano and colleagues[29] reported that Amadoriase I added to low-molecular weight peptides and proteins, such as insulin, could partially prevent glycation during glucose incubation. The authors suggest that the enzyme has the ability to act on unfolded more accessible states of the enzyme because little activity was observed with the glycated protein itself.

The overall outcome of these crystallographic studies is not encouraging for the prospect of engineering the enzyme for improved activity toward bulky glycated substrates. Alternative techniques, such as directed evolution, while able to improve parameters, such as thermal stability,[30] are unlikely to help solve the problem. Indeed, to our knowledge, there are no known flavin adenine dinucleotide (FAD) oxidase enzymes of any kind that can act on protein-bound substrates. In contrast, copper-bound enzymes, such as lysyl oxidase, can oxidize lysine residues in nascent procollagen chains.

Pharmacological Approaches

The dimming prospects of using Amadoriases (FAOX) for deglycation of Amadori products bound to extracellular proteins led us to explore alternative methods to prevent glucosepane formation in model systems. Incubation of ε-fructosyl aminocaproic acid with Compound NC-2 results in rapid glucosepane formation that can be quantitated using HPLC, as seen in FIGURE 4.

The two peaks correspond to the isomers resulting from the two hydroxyl groups in position 6 and 7 of the ring. Coincubation in the presence of Amadoriase II or various amounts of test compounds resulted in various degrees of inhibition as shown in TABLE 1. Amadoriase completely inhibited glucosepane formation, while the IC50 for aminoguanidine was between 0.1 and 1 mM.

Not surprisingly, salicylic acid had no blocking effect as it is not a nucleophile and glucosepane formation does not require oxidation to form. O-phenylenediamine (OPD) was also tested and found to inhibit glucosepane formation, but it also generated many new UV-active peaks, possibly as a result of autoxidation. In that regard, recent studies by Nagaraj *et al.* showed that an antibody against OPD-derived 2-methylquinoxaline-6-carboxylate proteins reacted strongly with proteins glycated with fructose and ribose, and more weakly with those glycated with glucose or ascorbic acid, providing evidence for the existence of protein-bound dideoxyglucosone intermediates.[31]

These preliminary studies suggest that inhibition of glucosepane and other Amadori product-derived modifications should be pharmacologically feasible given the proper blocking agent. Numerous previous studies have unequivocally demonstrated that the presence of aminoguanidine was effective at preventing collagen cross-linking, impaired digestibility, and other

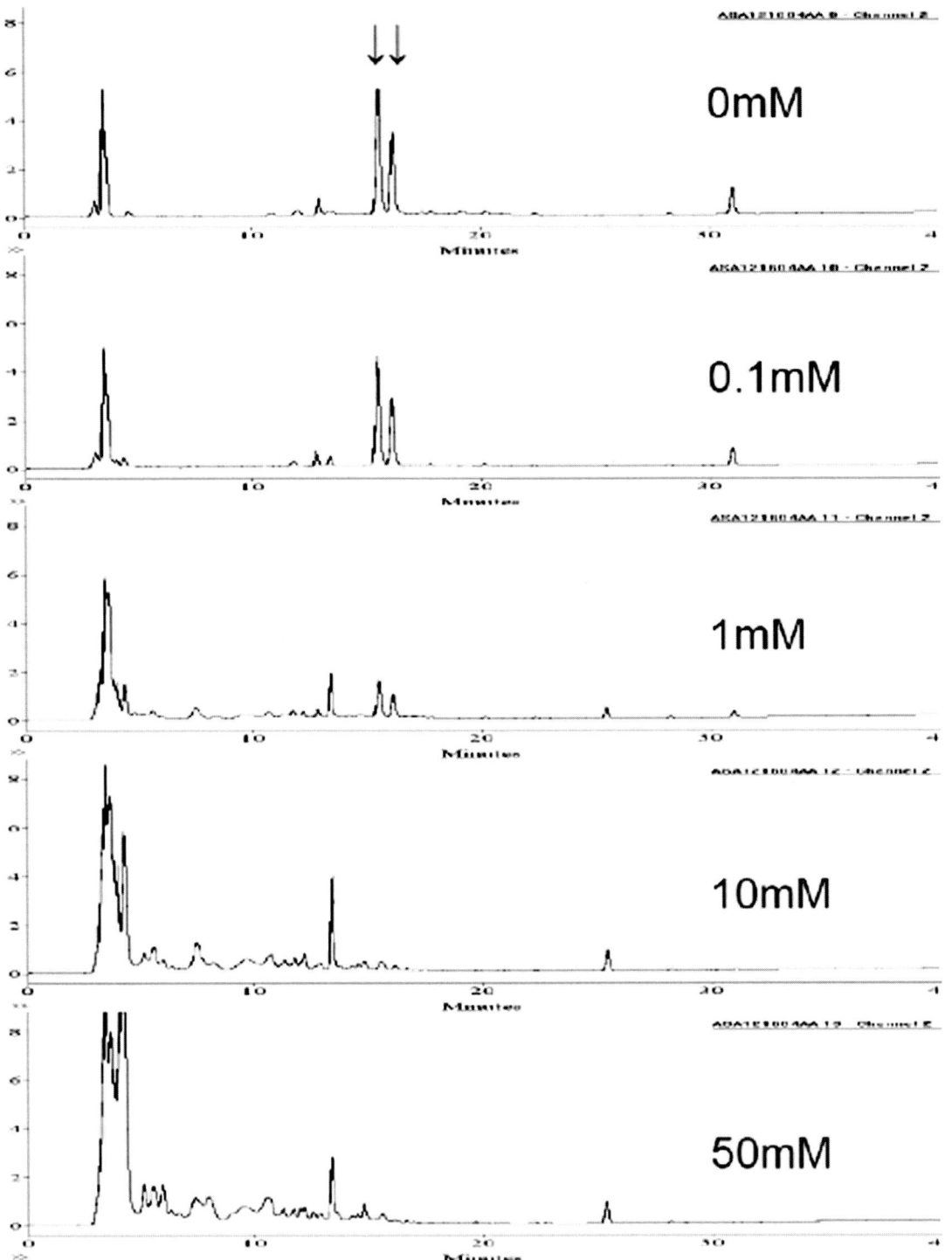

FIGURE 4. HPLC analysis of glucosepane formation from fructosyl aminocaproic acid and Compound NC-2 with or without aminoguanidine added in different concentrations. The arrows indicate the relative position of the two isomers of the glucosepane analogue.

TABLE 1. Relative inhibition of glucosepane formation by nucleophiles and other test compounds

	Glucosepane Inhibitors IC50
Amadoriase II	totally inhibit glucosepane synthesis even at 0.002 mg/mL
O-phenylenediamine	<<1 mM, but generates many other UV-active compounds
Compound NC-2	<1 mM
Aminoguanidine	<1 mM
Pyridoxamine	3.5 mM
Mercaptoethylamine	4 mM
Penicillamine	9 mM
Salicylic acid	no inhibition

properties.[32] However, because most studies were carried out under glycoxidation conditions, what is less clear is whether the reported effects stemmed from inhibition of glucosepane formation or rather from trapping of oxoaldehydes, typically glyoxal.

In other words, prolonged incubation under nonoxidative conditions will be needed in order to separate cross-linking resulting from glucosepane from that originating from oxoaldehydes.

Conclusions

We have developed the concept that molecular impairment by glucose-derived products, such as glucospane and other yet to be described structures, may result in significant consequences by increasing matrix insolubilization, stiffening, and impairment of its turnover rate, possibly leading to accumulation of AGEs with deleterious biological properties. Experimental studies in rodents showing that AGE inhibitors improved complications without affecting glycation (fructoslysine) are not adequate to address the impact of sustained hyperglycemia on collagen-rich matrix in humans with diabetes. On the other hand, the availability of mouse models, such as the db/db ApoE-null mouse, in which we found colocalization of glucosepane and collagen deposition in the thickened neointima, should allow us to approach the elusive question of how does diabetes accelerate the pathogenesis of macrovascular disease.

Acknowledgments

This work was supported by grants from the National Institute on Aging, the American Diabetes Association, and L'Oreal Research. F.C. was the recipient of a Fellowship from the Juvenile Diabetes Research Foundation International, and I.N. was the recipient of a Fellowship from the Fulbright Organization.

Conflict of Interest

The authors declare no conflicts of interest.

References

1. MONNIER, V.M. 2003. Intervention against the Maillard reaction in vivo. Arch. Biochem. Biophys. **419:** 1–15.
2. GENUTH, S. et al. 2005. Glycation and carboxymethyllysine levels in skin collagen predict the risk of future 10-year progression of diabetic retinopathy and nephropathy in the diabetes control and complications trial and epidemiology of diabetes interventions and complications participants with type 1 diabetes. Diabetes **54:** 3103–3111.
3. HEROLD, K. et al. 2007. Receptor for advanced glycation end products (RAGE) in a dash to the rescue: inflammatory signals gone awry in the primal response to stress. J. Leukoc. Biol. **82:** 204–212.
4. CHUANG, P.Y. et al. 2007. Advanced glycation endproducts induce podocyte apoptosis by activation of the FOXO4 transcription factor. Kidney Int. **72:** 965–976.
5. DOBLER, D. et al. 2006. Increased dicarbonyl metabolism in endothelial cells in hyperglycemia induces anoikis and impairs angiogenesis by RGD and GFOGER motif modification. Diabetes **55:** 1961–1969.
6. URIBARRI, J. et al. 2007. Single oral challenge by advanced glycation end products acutely impairs endothelial function in diabetic and nondiabetic subjects. Diabetes Care **30:** 2579–2582.
7. GRANT, P.J. 2007. Diabetes mellitus as a prothrombotic condition. J. Intern. Med. **262:** 157–172.
8. WILLIAMS, S.K. et al. 1982. Structural and functional consequences of increased tubulin glycosylation in diabetes mellitus. Proc. Natl. Acad. Sci. USA **79:** 6546–6550.
9. HARDING, J.J. & E. GANEA. 2006. Protection against glycation and similar post-translational modifications of proteins. Biochim. Biophys. Acta **1764:** 1436–1446.
10. BOLTON, W.K. et al. 2004. Randomized trial of an inhibitor of formation of advanced glycation end products in diabetic nephropathy. Am. J. Nephrol. **24:** 32–40.
11. METZ, T.O. et al. 2003. Pyridoxamine, an inhibitor of advanced glycation and lipoxidation reactions: a novel therapy for treatment of diabetic complications. Arch. Biochem. Biophys. **419:** 41–49.
12. METZ, T.O. et al. 2003. Pyridoxamine traps intermediates in lipid peroxidation reactions in vivo: evidence on the role of lipids in chemical modification of protein and development of diabetic complications. J. Biol. Chem. **278:** 42012–42019.
13. PRICE, D.L. et al. 2001. Chelating activity of advanced glycation end-product inhibitors. J. Biol. Chem. **276:** 48967–48972.
14. RAHBAR, S. 2007. Novel inhibitors of glycation and AGE formation. Cell. Biochem. Biophys. **48:** 147–157.
15. NAGAI, R. et al. 2007. Succination of protein thiols during adipocyte maturation—a biomarker of mitochondrial stres. J. Biol. Chem. **282:** 34219–34228.
16. FIGAROLA, J.L. et al. 2003. LR-90 a new advanced glycation endproduct inhibitor prevents progression of diabetic

nephropathy in streptozotocin-diabetic rats. Diabetologia **46:** 1140–1152.
17. RUGGIERO-LOPEZ, D. *et al.* 1999. Reaction of metformin with dicarbonyl compounds. Possible implication in the inhibition of advanced glycation end product formation. Biochem. Pharmacol. **58:** 1765–1773.
18. IZUHARA, Y. *et al.* 2005. Renoprotective properties of angiotensin receptor blockers beyond blood pressure lowering. J. Am. Soc. Nephrol. **16:** 3631–3641.
19. THORNALLEY, P.J. 2007. Endogenous alpha-oxoaldehydes and formation of protein and nucleotide advanced glycation endproducts in tissue damage. Novartis Found Symp. **285:** 229–243.
20. KENNEDY, L. & J.W. BAYNES. 1984. Non-enzymatic glycosylation and the chronic complications of diabetes: an overview. Diabetologia **26:** 93–98.
21. VEIGA DA-CUNHA, M. *et al.* 2006. Increased protein glycation in fructosamine 3-kinase-deficient mice. Biochem. J. **399:** 257–264.
22. SELL, D.R. *et al.* 2005. Glucosepane is a major protein cross-link of the senescent human extracellular matrix. Relationship with diabetes. J. Biol. Chem. **280:** 12310–12315.
23. AHMED, M.U., S.R. THORPE & J.W. BAYNES. 1986. Identification of N epsilon-carboxymethyllysine as a degradation product of fructoselysine in glycated protein. J. Biol. Chem. **261:** 4889–4894.
24. CULBERTSON, S.M. *et al.* 2003. Paradoxical impact of antioxidants on post-Amadori glycoxidation: counterintuitive increase in the yields of pentosidine and Nepsilon-carboxymethyllysine using a novel multifunctional pyridoxamine derivative. J. Biol. Chem. **278:** 38384–38394.
25. MONNIER, V.M., D.R. SELL & S. GENUTH. 2005. Glycation products as markers and predictors of the progression of diabetic complications. Ann. N.Y. Acad. Sci. **1043:** 567–581.
26. SELL, D.R. & V.M. MONNIER. 2005. Ornithine is a novel amino acid and a marker of arginine damage by oxoaldehydes in senescent proteins. Ann. N.Y. Acad. Sci. **1043:** 118–128.
27. FIORETTO, P. *et al.* 1998. Reversal of lesions of diabetic nephropathy after pancreas transplantation. N. Engl. J. Med. **339:** 69–75.
28. WU, X. *et al.* 2002. Alteration of substrate selectivity through mutation of two arginine residues in the binding site of amadoriase II from Aspergillus sp. Biochemistry **41:** 4453–4458.
29. CAPUANO, E. *et al.* 2007. Studies on the effect of Amadoriase from Aspergillus fumigatus on peptide and protein glycation in vitro. J. Agric. Food Chem. **55:** 4189–4195.
30. SAKAUE, R. & N. KAJIYAMA. 2003. Thermostabilization of bacterial fructosyl-amino acid oxidase by directed evolution. Appl. Environ. Microbiol. **69:** 139–145.
31. PUTTAIAH, S. *et al.* 2006. Detection of dideoxyosone intermediates of glycation using a monoclonal antibody: characterization of major epitope structures. Arch. Biochem. Biophys. **446:** 186–196.
32. FU, M.X. *et al.* 1994. Glycation, glycoxidation, and cross-linking of collagen by glucose. Kinetics, mechanisms, and inhibition of late stages of the Maillard reaction. Diabetes **43:** 676–683.

Mitigation Strategies to Reduce Acrylamide Formation in Fried Potato Products

FRANCISCO MORALES,[a] EDOARDO CAPUANO,[b] AND VINCENZO FOGLIANO[b]

[a]*Instituto del Frio, Consejo Superior de Investigaciones Científicas, Madrid, Spain*
[b]*Dipartimento di Scienza degli Alimenti University of Napoli "Federico II," Portici, Italy*

Potato products contain high amounts of acrylamide, which sometimes exceeds the concentration of 1 mg/L. However, many strategies for acrylamide reduction in potato products are possible. In this work, the different approaches for reducing acrylamide formation have been reviewed, keeping in mind that in the application of strategies for acrylamide formation, the main criteria to be maintained are the overall organoleptic and nutritional qualities of the final product.

Key words: acrylamide; potato; mitigation

Agronomic Strategies. Minimizing Reactants in the Potato Tuber

During the first studies of acrylamide formation in food, researchers reported that starch-derived products are very prone to the formation of acrylamide at concentrations up to 5 mg/kg.[1] One of the first lessons we learned from the systematic analysis of acrylamide in food was that there is a large variability in acrylamide content between different final products from the same food commodities, even among different lots from the same commercial brand.[1] This observation gives us some clues that there are opportunities to reduce acrylamide formation by first focusing on the raw material composition and second on the processing conditions.

The strong tendency for acrylamide to form in potatoes is because of the high levels of free asparagine (0.2–0.4% dry weight, representing 20–60% total amino acids content), compared to the amount of reducing sugars in the potato tuber, which actually represents the limiting factor. In addition, there is a significant relationship between acrylamide formation and reducing sugar (glucose, fructose) content, regardless of the potato tuber variety, under the same processing conditions.[2–7] On the other hand, sucrose content has been found to have less correlation with acrylamide content.[3]

Reduction of sugar content in the potato tuber (*Solanum tuberosum* ssp. *tuberosum* L.) was proposed as the first strategy for acrylamide minimization because sugar is the limiting reactant. The results of studies have recommended using potato cultivars with a low reducing sugar content, such as Panda, Saturna, and Lady Clare, which are less susceptible to acrylamide formation when deep frying potato crisps, or the varieties Agria, Markies, and Fontane for French fries and hash browns.[2,3,5,8,9] About 4000 cultivars are listed in the World Catalogue of Potato varieties, but only a few of them have acceptable sensory and nutritional quality after processing and should be considered for consumption. The sugar content of potatoes is determined by the genotype and several preharvest and postharvest factors. The major preharvest factors affecting sugar content are crop maturity, temperature during growth, mineral nutrition, and irrigation, while important postharvest factors are mechanical stresses and storage conditions.[10] Aspartic acid, asparagine, glutamic acid, and glutamine show no significant correlation with the acrylamide concentration in the processed products.[10]

Potatoes used for roasting or frying should contain less than 1 g/kg fresh weight of reducing sugars.[7,11] In this framework, a genetically modified potato variety was recently produced in the United States by inhibiting the expression of some specific genes that are responsible for low-temperature sweetening.[12]

However, other farming conditions could influence the sugar content of the same potato variety. A relationship between the fertilization method and the susceptibility to acrylamide

Address for correspondence: Professor Vincenzo Fogliano, Via università 100, parco gussone, ed. 84, Portici, Italy. Voice: +39 81 2539356; fax: +39 81 7754942.
fogliano@unina.it

formation has been noted.[2] Fertilization strategies influence the composition of potato tubers. Nitrogen appears to be indirectly related via its effects on dry matter and relative maturity of harvested tuber. Plants with adequate nitrogen fertilization produce tubers with lower reducing sugar concentration at harvest and accumulate less reducing sugars during storage.[10] Free asparagine content in potato tubers increases with increasing nitrogen fertilization[13] because free asparagine is a nitrogen reservoir.[14] An excess in nitrogen intake by fertilization leads to the biosynthesis of more free asparagine, thus lowering the reducing sugar content since they are used by the tuber for the biosynthesis of amino acids.[15] Consequently, decreasing nitrogen fertilization causes a reduced sugar content which reduces the potato's quality for frying.[13] However, there are other factors that must be considered regarding potato farming, and other conditions, such as type of soil, will influence nutrient uptake and metabolism by the plant.

Climatic conditions also affect the reducing sugar and asparagine content. Dry summers give rise to a lower reducing sugar content, and temperatures above 25°C result in elevated sugar levels because increased respiration has a negative effect on the rate of starch biosynthesis. Consequently, the level of reducing sugars will be lower at intermediate temperatures (15–25°C). Low temperatures during the final growing stage result in unacceptably high sugar levels,[16] and excessive rainfall in the final stages of the growing season gives rise to a lower dry-matter content and to an increase in nitrogen uptake. Finally, it has been shown that weather conditions during the growing season influence biochemical reactions as well as nutrient uptake by the roots.[17]

Inadequate potato storage conditions also represent a risk; starch can be hydrolyzed, forming high concentrations of free sugars in the tuber, and this will lead to the formation of acrylamide and an excessive Maillard reaction upon frying.[5,7,8,9,18] Senescent sweetening occurs with long-term storage after harvest, and this involves an increase in the sugar content inside the tuber. This process is enhanced at storage temperatures higher than 8°C, and it results in potato sprouting. Low temperatures mobilize sugars from starch in a process known as low-temperature sweetening.[17,19–20] Sucrose can be used as an indicator for the reducing sugar potential in enzymatic cleavage and thermal breakdown, and the concentration of reducing sugars is used as a criteria for accepting or refusing potato batches at industrial plants. The level of sucrose should be low to minimize accumulation of reducing sugars, which is an index of tuber maturity. During cold-induced sweetening in stored potatoes, starch degradation occurs primarily through the action of starch phosphorylase. Use of sprout suppressant will limit the need to use low temperatures during storage. Low-temperature sweetening is partly reversible after reconditioning the cold-stored tubers for some weeks at 15°C.[17]

Former variables are further affected if potatoes are not stored in the dark because light activates sugar metabolism.[5] Finally, low oxygen levels in the storage atmosphere inhibit the cold-induced enzymes and consequently decrease the level of reducing sugars in the product.

Processing Strategies. Controlling Variables during Deep Frying

Many stages prior to potato frying can be modified in order to reduce acrylamide formation. There are several minimization strategies related to prefrying operations and most of them focus on the elimination of reducing sugars from the reaction media. It is well known that incorporation of aqueous pretreatment steps (rinsing, soaking, blanching) of potato slices will lead to a reduction in acrylamide content. Treatment with various concentrations of citric acid before frying suppresses acrylamide in fried potatoes, but there are many concerns about sour flavors in the resulting products.

Soaking potato slices in distilled water for up to 90 min will reduce sugar content by about 30% from precursors leaching.[21] However, blanching (soaking at moderate temperatures) is more effective and has been shown to reduce glucose by 76% and asparagine by 68%, which results in a very significant reduction in acrylamide and a significant loss of texture.[21] Blanching induces gelatinization of the surface starch, reducing oil uptake, but a negative impact on the overall quality is expected.

Addition of certain additives, which reduce the pH of the water, will limit the formation of acrylamide. Immersion of potatoes in up to 2% organic acids (e.g., citric acid) decreases the pH, which reduces the formation of acrylamide at moderate temperatures of frying.[21] Surprisingly, this effect was not observed at higher temperatures of frying. Several studies have demonstrated that pH modification has the potential to reduce acrylamide formation in food and model systems. Lowering the pH of food to reduce acrylamide generation results from protonation of the alpha-amino group of asparagine, which then cannot engage the nucleophilic addition to the available carbonyls.[22–24] The acrylamide molecule contains an active double bond, which may interact with food ingredients, mainly proteins. The addition of protein-rich components from meat will reduce acrylamide by 70%, and this is likely because of the reaction of protein nucleophilic groups (-SH, NH_2) with the acrylamide double bond.[25]

Recently, an inhibition of acrylamide formation in asparagine–glucose model system by NaCl addition has been shown.[26] NaCl catalyzes the process of acrylamide polymerization and the subsequent formation of polyacryalmide. Addition of 1% NaCl (common additive in many foods) is enough to promote significant polymerization.[27]

Recently, Pedreschi et al.[28] reported that soaking solutions of potato slices in NaCl before frying dramatically reduces acrylamide formation by about 90% compared to control chips. In contrast to the experiments with the potato model system[29] but in accordance with previous reports,[28,30] Mestdagh et al.[31] confirmed that soaking potato slices in NaCl solutions significantly decreases the acrylamide content and reported that the oil content is also significantly reduced (27%) by NaCl addition (0.1%) to the blanching water. Acrylamide formation has been shown to decrease upon lowering the oil content of the potato model system, probably from a lower heat transfer from the oil to the system.[32] Therefore, decreased oil uptake seems a possible mechanism behind acrylamide reduction in the NaCl-treated crisps.

Recent studies have demonstrated that polyvalent cations also reduce acrylamide formation in thermally processed snack foods and bakery products.[33,34] Gökmen et al.[35] investigated the potential formation and degradation of acrylamide during heating in the presence of monovalent and divalent cations. Dipping potatoes into calcium chloride solution inhibits the formation of acrylamide by up to 95% during frying without adversely affecting the sensory quality of fried potato strips (in terms of color and texture). The inhibition was attributed to the presence of monovalent or divalent cations rather than to the reduction of acrylamide precursors by dipping. Gökmen and colleagues postulated that the presence of Ca^{2+} would prevent the formation of the Schiff base of asparagines, and thus of acrylamide, during heating. Similar results were found by Mestdagh et al.[31] who showed that dipping potato slices in $CaCl_2$ solution mitigates acrylamide formation, although only a marginal decrease was observed at the lowest concentration level (0.025 mol/L). As opposed to Gökmen et al.,[35] a more crispy texture and a bitter aftertaste at higher Ca^{2+} concentrations was perceived by panelists. It is interesting that several additives, such as organic acids, NaCl, and Ca^{2+}, are able to lower the absorption of oil when frying. This fits perfectly with the ongoing consumer trend to move toward healthier and low-fat products in order to counteract obesity and coronary heart diseases. Some other questions, such as the rate of oil degradation during frying when salts are added[36,37] and the consequent more rapid oil turnover rate, need to be further investigated because of the impact on process and production costs at the industrial scale.

Another mitigation strategy for acrylamide formation in foods is the use of the enzyme asparaginase (L-asparagine amidohydrolase), which catalyzes the hydrolysis of asparagine in aspartic acid and ammonia. From a theoretical point of view, this approach seems to be more useful for cereal products where asparagine is known to be the limiting factor in acrylamide formation. In fact, in potatoes, the content of free asparagine does not correlate with potential acrylamide formation,[4,9,38] whereas a highly significant correlation between the content of reducing sugars and acrylamide has been observed.[2] In raw tubers, the molar concentration of asparagine exceeds that of reducing sugars, the latter are also consumed faster during heating.[2,3,8,39] Nevertheless, when Zyzak et al.[40] evaluated the effectiveness of the asparaginase treatment in a mashed potato product that had been cooked by microwave, they found that asparaginase pretreatment results in an 88% reduction in asparagine and greater than 99% reduction in acrylamide compared to samples prepared without the enzyme. A constraint of this approach in limiting acrylamide formation is the limited availability and high cost of asparaginase, while questions remain about the contribution of free asparagine and aspartic acid to the flavor profile (e.g., pyrazines) and to nutritional aspects, as well as country-specific regulatory approval of the enzyme. Two commercial products, Acrylaway® (asparaginase form *Aspergillus oryzae*) from Novozyme (Bagsvaerd, Denmark) and PreventASe® (asparaginase form *A. niger*) from DSM (Delft, Netherlands), are on the market for food applications that have been proven to reduce acrylamide in dough-based products without influencing product taste or appearance.

Surface to volume ratio is also important because acrylamide formation is a surface phenomenon. Both home-made and industrially prepared French fries are usually cuts with different geometries. Matthäus et al.[41] demonstrated that a surface to volume ratio of 3.3:1 cm^{-1} results in a significantly lower amount of acrylamide than fine strips with a ratio of 5.4:1 cm^{-1}. Another factor to consider is the ratio of potato to oil mass, which is recommended to be less than 10%. This recommendation is important for catering services and home-frying processes where the ratio of potato to oil is not standardized and the heat load could be very different from one piece of potato to the other. Drying potatoes by microwave or hot-air treatment before frying results in a significant reduction in oil content, and, in these conditions, acrylamide is reduced by 44%.[42,43] It should be stressed that moisture content is an important quality parameter of potato crisps, which should be crispy not only after frying but also during the shelf life of the product; many investigations have not consider this aspect.

The influence of temperature on the formation of acrylamide has been repeatedly demonstrated.[44,45] Frying potatoes is based on heat transfer from the hot oil, which results in water removal and oil uptake by the potato. It has been shown that the acrylamide content of potato strips first increases exponentially with time, depending on temperature.[23] After prolonged heating, a decrease in the acrylamide content is observed because acrylamide degradation predominates. The acrylamide yield is the net result of simultaneous formation and elimination reactants and it can be modeled by consecutive first-order reactions.[46] Controlling the temperature and time of frying will have a major effect on acrylamide reduction in potato products, but this could have a negative effect on perceived product quality; it may also be difficult to introduce on a home scale. Temperature and time can be precisely controlled at industrial plants but only partly controlled at home. For instance, there are differences in the heat transfer inputs and electrical potency of domestic fryers, whereas these are exhaustively controlled at the industrial scale, and initial and end-frying temperatures of the process remain almost constant. It can be concluded that for home preparation, strategies concerning the choice of suitable potato varieties or cultivars will be more effective to reduce the acrylamide content.

Choice of the frying oil is an important aspect for deep frying. Frying oil not only acts as a heat transfer medium but also contributes to the taste and aroma of the potato fries. Some studies have focused on the effect of the type of frying oil on acrylamide formation in the product. Matthäus et al.[41] did not find differences in acrylamide formation in French fries because of the type of frying oil, and similar results were reported by Mestdasgh et al.[32] However, Gertz and Klostermann[47] found higher amounts of acrylamide in French fries fried with palmolein compared to rapeseed or sunflower oil. On the other hand, the type of oil has an influence on the shelf life and the stability of the frying medium during processing. Matthäus et al.[41] did not find a relationship between acrylamide formation and the aging of the frying oil or the addition of silicon oil as an antifoaming agent. Recently, Mestdash et al.[48] reported no evidence that the heat transfer properties of the frying oil (from oxidative or hydrolytic oil degradation) change to such an extent that acrylamide formation during the preparation of French fries would be significantly influenced. In addition, oil degradation products (glycerol, monoacylglycerols, and diacylglycerols) do not influence acrylamide formation because acrylamide formation is independent of oil oxidation and hydrolytic status. This conclusion is important since it was suggested that triacylglycerols are partially hydrolyzed during frying, followed by dehydratation of glycerol to acrolein. Acrolein may be oxidized to acrylic acid, which can finally react with ammonia to form acrylamide.[47,49]

The consumer's perception of color, flavor, and crispness should be unaffected by a reliable acrylamide mitigation procedure. Frying time should be adjusted to keep final moisture content below 3% to obtain desirable crispness. Lowering water activity at the surface of products for prefrying was proposed as a means of acrylamide reduction.[30] It has been reported that low frying temperatures result in moderate acrylamide concentrations despite the longer frying times, whereas temperatures above 170°C enhance acrylamide formation and enhance color.[6] On the other hand, frying at temperatures below 170°C will increase fat uptake, and the potato will become soft. It is suggested that the frying temperature should be below 175°C and the frying time should not be longer than necessary to obtain the right quality parameters of fried products. Therefore, a lower frying temperature will negatively influence the fat content and the moisture of crisps. Alternative heat treatments are under investigation. Granda et al.[50] performed vacuum frying experiments between 118 to 140°C at a vacuum pressure of 1333 Pa. Vacuum frying reduces acrylamide formation by 94% without important changes in the organoleptic parameters of the potato.

Finally, postfrying treatment has also been considered. Kita et al.[51] reported that applying a postdrying step after frying results in a decreased acrylamide content in crisps. When crisps were fried in oil at 185°C, it was possible to obtain a 70% decrease of acrylamide content after 2 min of frying and 75 min at 105°C of postdrying.

Acrylamide is a very polar, low mass, and volatile substance. These particular properties could be used to eliminate it after processing. A combination of adequate temperature and vacuum after the frying process could be an alternative for removing acrylamide from the matrix. However, there are many

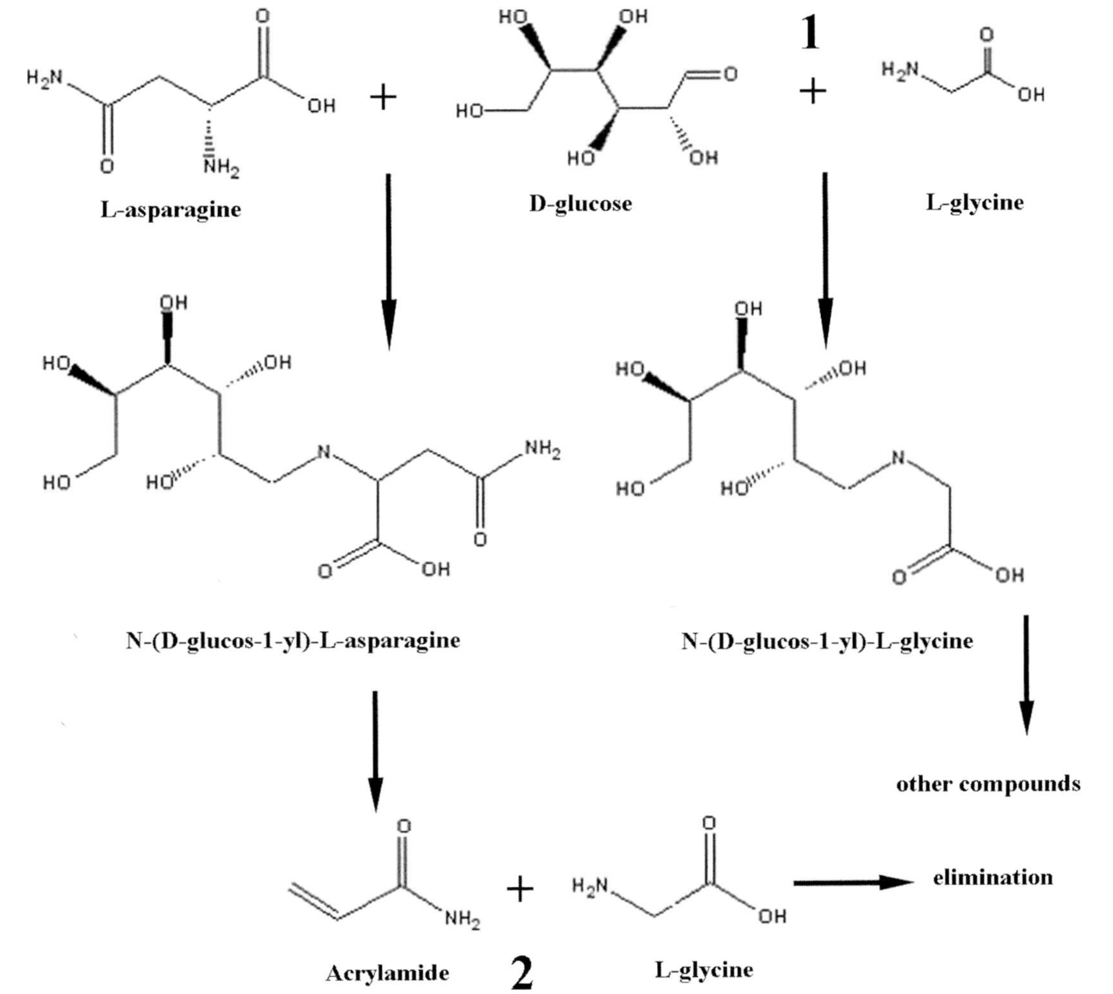

FIGURE 1. Acrylamide formation from asparagine and glucose. The R-NH2 group of asparagine participates in a nucelophilic-addition reaction with the aldehyde group of glucose to form a Schiff base, which then undergoes an Amadori rearrangement to the shown glucose–asparagine derivative (N-glycoside). The latter can then undergo decarboxylative deamination, losing the COOH and R-NH2 groups associated with asparagine to form acrylamide. Addition of glycine can reduce acrylamide level in two ways: **1)** competing effectively with asparagine in the Maillard reaction or **2)** reacting with acrylamide after its formation.

issues that should be overcome, such as loss of flavor or fragility of the potato crisp, which probably breaks. Such an alternative is possible, however, because even though the vapor pressure of acrylamide is low (i.e., 0.93 Pa at 25°C, 9.3 Pa at 50°C), it has been measured in air. On the other hand, it has been reported that acrylamide, added to foods before heating, is not completely recovered in analysis,[5] so it appears that acrylamide is eliminated either by further reactions (addition reactions or polymerization) with food components or evaporates during heating.[23,52,53]

Therefore, it is possible that adopting postfrying steps will reduce the concentration of acrylamide but may also reduce the concentration of many volatile products, thus affecting the sensorial quality of the final product.

Inhibition of the Reaction by Competitive Compounds toward Reducing Sugars Instead of Asparagine

It is well known that acrylamide is mainly formed from asparagine and reducing sugars through the Maillard reaction (FIG. 1).[44,54,55] The first step is the

TABLE 1. Effect of the addition of glycine or other amino acids on acrylamide formation in the potato model system or potato-based foodstuffs

Authors	Amino acids used	Effect on acrylamide content
Kim et al.[61]	Glycine, lysine, cysteine	Addition of 0.1 and 0.5% glycine to pallets reduced acrylamide concentration by 43% and by more than 70%, respectively; soaking potato slices in a 3% solution of either lysine or glycine reduced the formation of acrylamide by more than 80% in potato chips fried for 1.5 min at 185°C
Claeys et al.[46]	Cysteine, lysine, alanine, and glutamine	Addition of cysteine and lysine to the asparagine–glucose model system heated at 140–200°C, lowered acrylamide yield. A alanine has no effect while glutamine increased acrylamide yield
Rydberg et al.[23]	Various amino acids or protein-rich components	Addition of several amino acids or a protein-rich component to potato slurries decreased acrylamide content by more than 80%; glycine and glutamine were almost twice as effective as alanine, lysine, and glutamic acid in suppressing acrylamide levels at low concentrations
Bråthen et al.[52]	Glycine and glutamine	Addition of glycine and glutamine to a starch model system, reduced a crylamide formation by 30% compared to no addition while no effect was found on French fries
	Glycine	Addition of glycine to a model system consisting of starch, asparagine, and glucose decreased acrylamide content upon prolonged heating treatment
Low et al.[64]	Glycine	Addition of glycine to a potato model system reduced acrylamide content, increased the total volatile yield by promoting the formation of certain alkylprazines
Mestdagh et al.[29]	Glycine and lysine	Addition of glycine and lysine reduced acrylamide formation in potato slices (up to 85%) by addition to the blanching water. No effects on pH and oil content. Strong effect on browning

reaction between the alpha-amino group of the free asparagine and a carbonyl source, forming a Shiff base. Results from mechanistic studies indicate that the side-chain amide group of asparagine is incorporated in the amide bond of acrylamide.[4,40,56]

From this reaction pathway, it is clear that one possibility for reducing acrylamide formation is to use compounds able to compete with asparagine for carbonyl groups; this means, in particular, other free amino acids (e.g., glycine) or proteins. As a matter of fact, among the several patent applications for the reduction of acrylamide in heat-treated foodstuffs, applications based on the addition of amino compounds, including amino acids, have been proposed.[57,58]

Amino acids may compete effectively with asparagine in the Maillard reaction or may react with acrylamide after its formation (FIG. 1). Acrylamide has two reactive sites, the conjugated double bond and the amide group, and can thus be eliminated by reaction with numerous food constituents. In this sense, other nucleophilic compounds could work in this way and lead to a reduction in acrylamide content.

In a model system consisting of starch, asparagine, and glucose, acrylamide content decreases upon prolonged heating,[52] indicating that acrylamide reacts with amino acids. The reaction of acrylamide with the amino group of glycine was demonstrated previously,[59] and the effects of amino acids on acrylamide formation and elimination have been extensively studied (see TABLE 1).

Claeys et al.[46] reported that the addition of cysteine and/or lysine to the asparagine–glucose model system and heated at temperatures between 140 and 200°C significantly lower the acrylamide yield. According to the activation energies for acrylamide formation and elimination when these amino acids are added, Claeys and colleagues concluded that lysine would act mainly by competing with asparagines in the Maillard reaction while cysteine would increase the elimination rate of acrylamide by reacting with its SH group and forming cysteinyl-S-β-propionammide. SH groups are reported to be 100–300 times more reactive than amino groups with conjugated vinyl compounds,[59] while cysteine is classified as an amino acid with low reactivity in the Maillard reaction.[60] As such, sulfur amino acids or sulfhydryl compounds could reduce acrylamide content in heated foods. Unfortunately, their use is restricted by the unpleasant off-flavors they can generate in foods. The same authors reported that addition of alanine to the same model system has no effect on acrylamide yield, while glutamine increases the rate of acrylamide formation, thus leading to higher acrylamide levels.

The effect of soaking potato slices in a solution of amino acids in order to reduce acrylamide in fried potato slices was investigated by Kim et al.[61] They found that the longer the soaking time and the more concentrated the solution, the more effective the treatment. They reported that dipping potato slices in a 3% solution of either lysine and glycine reduces the formation of acrylamide by more than 80% in potato chips fried for 1.5 min at 185°C; however, a significant reduction (more than 40%) is obtained by using 0.1 or 0.5% lysine and glycine solutions. Of course, the overall inhibitory effect results from both the leaching of acrylamide precursors and the competitive action of amino acids. The same authors reported that the addition of lysine and glycine to potato model snacks before frying reduces the acrylamide content upon heating. In particular, the addition of 0.1 and 0.5% glycine to pallets reduces acrylamide concentration by 43% and by more than 70%, respectively.

Rydberg et al.[23] examined the effects of various amino acids on the formation of acrylamide in homogenized potatoes heated at 180°C for 25 min. The addition of glycine, alanine, lysine, glutamine, and glutamic acid at a concentration of 35 mmol/kg was found to reduce acrylamide levels by 42% to 70%. They also demonstrated that glycine and glutamine are almost twice as effective as alanine, lysine, and glutamic acid in suppressing acrylamide levels at low concentrations. Rydberg and colleagues also investigated the effect of protein addition (lean fish) to potato patties. Like free amino acids, proteins were found to significantly lower acrylamide content, as already reported by several authors working on different matrices.[5,62] Proteins could act both by inhibiting acrylamide formation reaction and eliminating acrylamide formed via reaction with amino and/or sulfhydryl groups of amino acid side chains. For that purpose, Rydberg et al.[23] added different amounts of minced cod to potato slurries and obtained a reduction of up to 70% of acrylamide content after heating.

Similar results were found by Vattem et al.[63] who used chickpea butter to coat potato chips during frying. They postulated that proteins might also be involved in complexing starch on the surface of the slices and stabilizing the complex even at higher temperatures, making the sugars in the starch less available for the Maillard reaction. The legume proteins may also be involved in delocalizing the electrons from the sugar carbonyl via their aromatic amino acids. This delocalization prevents the keto-enolization of the sugars and, therefore, the breakdown of the six-carbon chain to hydroxyacetone, which eventually may form acrylamide through a series of condensation reactions.

Bråthen et al.[52] found that the addition of glycine and glutamine during blanching of crisps reduces the amount of acrylamide by 30%, but no effect was found on French fries. Furthermore, they found that glycine is more effective than glutamine in reducing acrylamide, while Rydberg et al.[23] found comparable effects for both amino acids. This apparent discrepancy could be explained by the different systems used. Rydberg and colleagues[23] used homogenized potato slurries with amino acids added during the homogenization, which ensured that the amino acids were evenly distributed. On the other hand, in common industrial conditions, potato crisps are made from sliced potatoes and are not reconstituted. In the nonreconstituted potato crisps, the amino acid has to enter the potato tissue, and glutamine, being larger then glycine, will be transported more slowly into potato tissues. All in all, glycine is known to be more reactive in the Maillard reaction than glutamine, which may be of importance if the effect from the addition of glycine is caused by competition in the Maillard reaction. However, a reaction between glycine and the acrylamide previously formed seems to be the more likely cause of the decrease in measured acrylamide content when glycine is added.

As the Maillard reaction plays a major role in acrylamide formation, its suppression would, therefore, reduce the levels of acrylamide. However, the Maillard reaction is also a major route for the generation of desirable flavors and colors in food, ensuring the sensory quality expected by consumers. It is important to study the relationship between flavor generation, taste, and acrylamide production to develop a strategy to minimize acrylamide without adverse effects on the flavor and taste of foods. In this respect, Low et al.[64] studied the effect of treatment with citric acid or glycine on the volatile profile and acrylamide levels in a potato model system. After cooking at 180°C for 10–60 min, these treatments were found to affect the volatile profiles and, in particular, Strecker aldehydes and alkylpyrazines, key flavor compounds of cooked potato. Citric acid limits the generation of volatiles, particularly the alkylpyrazines. Glycine increases the total volatile yield by promoting the formation of certain alkylpyrazines, namely 2,3-dimethylpyrazine, trimethylpyrazine, 2-ethyl-3,5-dimethylpyrazine, tetramethylpyrazine, and 2,5-diethyl-3-methylpyrazine. However, the formation of other pyrazines and Strecker aldehydes is suppressed. The authors concluded that a combined treatment of lower levels of citric acid and glycine would have less impact on the flavor profile than a higher

TABLE 2. Effect of the addition of antioxidants on acrylamide formation in different matrices and food models

Authors	Antioxidant used	Foodstuff	Effect on acrylamide content
Becalski et al.[70]	Rosemary herb	Oil used to fry potato slices	Reduction of acrylamide formation
Fernández et al.[71]	Liquid spice mix rich in flavonoids	Potato slices before and after frying	Reduction by up to 50%
Biedermann et al.[5]	Ascorbic acid	Potato-based model system	Weak inhibition
Zhang et al.[67–69]	Different solutions of antioxidant of bamboo leaves (AOB)	Potato crisps and French fries	Reduction of 74.1% and 76.1% when AOB addition ratio was 0.1% and 0.01% (w/w); no significant effects on crispness and flavor
	Different solutions of AOB	Fried chicken wings	Nearly 57.8 and 59.0% reduction with addition ratios of 0.1 and 0.5% (w/w), respectively
	Different solutions of AOB or extract of green tea (EGT)	Fried bread sticks	Nearly 82.9% and 72.5% reduction with AOB and EGT addition levels of 1 and 0.1 g/kg, respectively; no significant effects on crispness and flavor
Tareke et al.[62]	Butylated hydroxytoluene, sesamol, Vitamin E	Meat	Enhancement of acrylamide formation
Vattem et al.[63]	Phenolic antioxidants from cranberry and oregano	Fried potato slices	Slight increase
Rydberg et al.[23]	Ascorbyl palmitate and sodium ascorbate; benzoil peroxide and hydrogen peroxide	Homogenized potato heated in an oven	Effect on acrylamide content small or nonexistent
Levine et al.[72]	Ferulic acid, ascorbic acid	Model system based on wheat flour and water; resembled crackers	Reduction of acrylamide formation; increasing acrylamide elimination

level of either treatment on its own, which would be needed to achieve the same reduction in acrylamide. Low and colleagues also suggested the possibility of using a combination of different amino acids to boost the desired Strecker aldehydes and alkylpyrazines and at the same time mitigate acrylamide levels.

In a similar way, Mestdagh et al.[29] investigated the effect of the addition of free glycine and L-lysine to potato crisp blanching water at different concentration levels. They showed that the addition of these components did not markedly change the pH of the potato at a low concentration (0.05%) as they had previously pointed out.[31] At this concentration level, these additives markedly influence the final oil content of the crisps, compared to the control. L-lysine appeared to more efficiently reduce the formation of acrylamide (by up to 85%), although the differences are less pronounced at the 0.05 and 0.025 mol/L concentration levels. However, the authors found that this treatment strongly affects the color of the final product. Darker products are obtained upon frying when glycine or L-lysine is added to the blanching water. This difference is most pronounced for L-lysine, which is not surprising because L-lysine is known to be very reactive in the Maillard reaction. Moreover, at the applied concentrations, both amino acids could counteract the inhibition of browning caused by the addition of acetic or citric acid.

Many other authors have reported an increased browning of their potato model system when glycine was added, and the same effect has been reported by several authors for bread and cereal products.[52,65,66] All in all, it is possible to conclude that the use of amino acids, and particularly of glycine, represent an attractive and promising strategy to reduce acrylamide formation in potato-based products.

Affecting Some Steps of the Reactions by Addition of Chemically Reactive Compounds that are Able to React with Intermediates

A significant part of mitigation research has consisted of studies testing the effects of the addition of chemically reactive compounds to potatoes. Most of these compounds are antioxidants, which are used in pure form, added as vegetable extracts, or present in whole spices.

It is difficult to draw firm conclusions from these studies as the reports published thus far have reported conflicting evidence. In some cases, retarding effects have been demonstrated with spice extracts, but it is clear that the effect is not necessarily a result of antioxidative properties of those additives. The main studies dealing with the effects of antioxidants on acrylamide formation are summarized in TABLE 2.

Zhang et al.[67] demonstrated that 74.1% and 76.1% of acrylamide is reduced in potato crisps and French fries, respectively, dipped into different solutions of antioxidants of bamboo (AOB), a pale brown powder extracted from bamboo leaves. The reduction depends on the AOB addition:product ratio and on immersion time. Sensory evaluation results showed that organoleptic features of products are not affected by the treatment; the crispness and flavor of the potato crisps and French fries processed by AOB solutions are not significantly different from the normal potato matrix when the AOB solution to product ratio is <0.05%. Similar results were found by the same authors in other foodstuffs, such as fried chicken wings and fried bread sticks.[68,69] Thus, the addition of plant extracts could be a possible technique for reducing acrylamide in many products. Becalski et al.[70] found that acrylamide could be reduced when adding rosemary herb to the oil used for frying potato slices. It should be noted, however, that these findings could not be confirmed by others.[41] Rosemary is known for its antioxidant content, but this effect could also be a result of many other factors. A decreasing effect of a flavonoid spice mix has also been reported by Fernández et al.[71] A liquid spice mix was added to potato slices before frying, and a powder spice mix was also added to the potato slices after frying. The acrylamide levels were reported to be reduced by up to 50% by the spice-mix treatment. Biedermann et al.[5] showed a weak inhibition effect upon addition of ascorbic acid to a potato-based model.

The ability of ferulic acid to inhibit acrylamide formation was attributed to its ability to react with acrylamide precursors or intermediates in the chemical process of its generation.[72] The same authors reported that ascorbic acid is effective in increasing acrylamide elimination and decreasing the net amount of acrylamide formed in a wheat–water model system when baked at 180°C.

In one pivotal study, Tareke et al.[62] found that the addition of antioxidants (butylated hydroxytoluene, sesamol, vitamin E) to meat before heating enhances the formation of acrylamide. This author introduced the hypothesis that antioxidants might protect the acrylamide from further radical-initiated reactions. On the other hand, studies of the effect of the antioxidant compounds present in the oil on acrylamide formation are scarce. It is possible that the minor oil components tocopherols, phenols, and sterols can influence acrylamide formation during the exposure of potatoes to these oils at high temperatures.

Vattem et al.[63] reported that formation of acrylamide in fried potato slices previously treated with phenolic antioxidants from cranberry and oregano or cooked in chickpea batter is not reduced but actually increases when exogenous phenolic is present. Based on these results, authors hypothesized a nonoxidative model for the formation of acrylamide in fried products. Rydberg et al.,[23] to investigate the effects of antioxidants on acrylamide formation, used two antioxidants, ascorbyl palmitate and sodium ascorbate, as well as oxidants, such as benzoil peroxide and hydrogen peroxide, in homogenized potato heated in an oven. They found that the effect on acrylamide content is small or nonexistent, thus suggesting that involvement of radicals or peroxidation in the formation of acrylamide may be of some, but only minor, importance; low levels of antioxidants cause a small increase, probably via protection of the acrylamide formed.

Conclusions and Future Perspectives

Many studies have attempted to find strategies to minimize the levels of acrylamide in different food commodities. This objective can be achieved either by modifying processing parameters, such as pH, temperature or time of heating, acting on precursors or key intermediates, or reducing concentration of reactants in the raw material as formerly discussed. Strategies to reduce acrylamide formation should maintain the overall organoleptic, nutritional properties, and microbiological safety of the food during its shelf life. Different approaches have been investigated to reduce the formation of acrylamide in food, but, in reality, only few of them could be implemented without any significant alteration of the food product.

It is very likely that the total elimination of acrylamide from fried products cannot be achieved. Then, the ALARA principle (as low as reasonably or technically achievable) should be applied by the different actors. And, as outlined in this review, there are many different opportunities to reduce acrylamide in fried potatoes. However, potential negative effects on the sensory characteristics of the final product have to be carefully evaluated. In this context, there will not be a unique strategy, and a combination of agronomical selection and lighter processing conditions will significantly reduce the levels of acrylamide in the final product at an industrial level. Additional strategies should be implemented on the domestic scale, mainly focused on processing conditions. We should keep in mind that the overall objective is to reduce the acrylamide intake, and many sources of acrylamide come from foods processed at home. Consumers must be aware that it is important not to fry in excess and to limit the excessive intake of over-fried products. Also, information on potato varieties particularly suitable for frying and recommendations about potato home storage should be released.

In conclusion, the reduction of the overall intake of acrylamide is not only a food company's task; in fact, companies must implement an intensive mitigation strategy for different food items. The onus also falls on the final cookers who should be aware that it is their responsibility to serve customers and/or relatives foods having a low acrylamide content.

Conflict of Interest

The authors declare no conflicts of interest.

References

1. DYBING, E., P.B. FARMER, M. ANDERSEN, et al. 2005. Human exposure and internal dose assessments of acrylamide in food. Food and Chemical Toxicology **43**: 365–410.
2. AMREIN, T.M., S. BACHMANN, A. NOTI, et al. 2003. Potential of acrylamide formation, sugars, and free asparagine in potatoes: a comparison of cultivars and farming systems. Journal of Agricultural and Food Chemistry **51**: 5556–5560.
3. AMREIN, T.M., B. SCHÖNBÄCHLER, F. ROHNER, et al. 2004. Potential for acrylamide formation in potatoes: data from the 2003 harvest. European Food Research and Technology **219**: 572–578.
4. BECALSKI, A., B.P.Y. LAU, D. LEWIS, et al. 2004. Acrylamide in French fries: influence of free amino acids and sugars. Journal of Agricultural and Food Chemistry **52**: 3801–3806.
5. BIEDERMANN, M., A. NOTI, S. BEIDERMANN-BREM, et al. 2002. Experiments on acrylamide formation and possibilities to decrease the potential of acrylamide formation in potatoes. Mitteilungen aus Lebensmitteluntersuchung und Hygiene **93**: 668–687.
6. HAASE, N.U., B. MATTHÄUS & K. VOSMANN. 2003. Acrylamide formation in foodstuffs – Minimising strategies for potato crisps. Deutsche Lebensmittel-Rundschau **99**: 87–90.
7. DE WILDE, T., B. DE MEULENAER, F. MESTDAGH, et al. 2005. Influence of storage practices on acrylamide formation during potato frying. Journal of Agricultural and Food Chemistry **53**: 6550–6557.
8. OLSSON, K., R. SVENSSON & C.A. ROSLUND. 2004. Tuber components affecting acrylamide formation and colour in fried potato: variation by variety, year, storage temperature and storage time. Journal of the Science of Food and Agriculture **84**: 447–458.
9. DE WILDE, T., B. DE MEULENAER, F. MESTDAGH, et al. 2006. Selection criteria for potato tubers to minimize acrylamide formation during frying. Journal of Agricultural and Food Chemistry **54**: 2199–2205.
10. KUMAR, D., B.P. SINGH & P. KUMAR. 2004. An overview of the factors affecting sugar content of poatoes. Ann. Appl. Biol. **145**: 247–256.
11. BIEDERMANN-BREM, S., A. NOTI, K. GROB, et al. 2003. How much reducing sugar may potatoes contain to avoid excessive acrylamide formation during roasting and baking? European Food Research and Technology **217**: 369–373.
12. ROMMENS, C.M., J. YE, C. RICHAEL & K. SWORDS. 2006. Improving potato storage and processing characteristics through all-native DNA transformation. Journal of Agricultural and Food Chemistry **54**: 9882–9887.
13. DE WILDE, T., B. DE MEULENAER, F. MESTDAGH, et al. 2006. Influence of fertilization on acrylamide formation during frying of potatoes harvested in 2003. Journal of Agricultural and Food Chemistry **54**: 404–408.
14. EPPENDORFER, W.H. & S.W. BILLE. 1996. Free and total amino acid composition of edible parts of beans, kale, spinach, cauliflower and potatoes as influenced by nitrogen fertilisation and phosphorus and potassium deficiency. Journal of the Science of Food and Agriculture **71**: 449–458.
15. KOLBE, H. 1990. Kartoffeldüngung unter differenzierten ökologischen Bedingungen. PhD Dissertation, Georg-August-Universität, Götingen, Germany.
16. GROB, K., M. BIEDERMANN, S. BIEDERMANN-BREM, et al. 2003. French fries with less than 100 μg/kg acrylamide. A collaboration between cooks and analysts. European Food Research and Technology **217**: 185–194.
17. DAVIES, H.V., R.A. JEFFERIES & L. SCOBIE. 1989. Hexose accumulation in cold-stored tubers of potato (*Solanum tuberosum* L.) – the effects of water-stress. Journal of Plant Physiology **134**: 471–475.
18. NOTI, A., S. BIEDERMANN-BREM, M. BIEDERMANN, et al. 2003. Storage of potatoes at low temperature should be avoided to prevent increased acrylamide formation during frying and roasting. Mitteilungen aus Lebensmitteluntersuchung und Hygiene **94**: 167–180.
19. BLENKINSOP, R.W., L.J. COPP, R.Y. YADA & A.G. MARANGONI. 2002. Changes in compositional parameters of tubers of potato (*Solanum tuberosum*) during low-temperature storage and their relationship to chip processing quality. Journal of Agricultural and Food Chemistry **50**: 4545–4553.
20. SOWOKINOS, J.R. 2001. Biochemical and molecular control of cold-induced sweetening in potatoes. American Journal of Potato Research **78**: 221–236.
21. PEDRESCHI, F., K. KAACK & K. GRANBY. 2004. Reduction of acrylamide formation in potato slices during frying. LWT – Food Science and Technology **37**: 679–685.
22. PEDRESCHI, F., K. KAACK & K. GRANBY. 2006. Acrylamide content and colour development in fried potato strips. Food Research International **39**: 40–46.

23. RYDBERG, P., S. ERIKSSON, E. TAREKE, et al. 2003. Investigations of factors that influence the acrylamide content of heated foodstuffs. Journal of Agricultural and Food Chemistry **51:** 7012–7018.
24. STADLER, R.H. & G. SCHOLZ. 2004. Acrylamide: an update on current knowledge in analysis, levels in food, mechanisms of formation, and potential strategies of control. Nutrition Reviews **62:** 449–467.
25. SCHOBAKER, J., T. SCHWEND & M. WINK. 2004. Reduction of acrylamide uptake by dietary proteins in a Caco-2 gut model. Journal Agricultural and Food Chemistry **52:** 4021–4025.
26. KOLEK, E., P. SIMKO & P. SIMON. 2006. Inhibition of acrylamide formation in asparagine glucose model system by NaCl addition. Eur. Food Res. Technol. **224:** 283–284.
27. KOLEK, E., P. SIMON & P. SIMKO. 2007. Non isothermal kinetics of acrylamide elimination and its acceleration by table salt – a model study. Journal of Food Science **72:** 341–344.
28. PEDRESCHI, F., O. BUSTOS, D. MERY, et al. 2007. Colour kinetics and acrylamide formation in NaCl soaked potato chips. Journal of Food Engineering **79:** 989–997.
29. MESTDAGH, F., J. MAERTENS, T. CUCU, et al. 2008. Impact of additives to lower the formation of acrylamide in a potato model system through pH reduction and other mechanisms. Food Chemistry **107**(1)**:** 26–31.
30. FRANKE, K., M. SELL & E.H. REIMERDES. 2005. Quality related minimization of acrylamide formation – An integrated approach. *In* Chemistry and Safety of Acrylamide in Food. M. Friedman & D. Mottram, Eds.: 357–369. Springer. New York, NY.
31. MESTDAGH, F., T. DE WILDE, K. DELPORTE, et al. 2007. Impact of chemical pre-treatments on the acrylamide formation and sensorial quality of potato crisps. Food Chemistry **106:** 914–922.
32. MESTDAGH, F.J., B. DE MEULENAER, C. VAN POUCKE, et al. 2005. Influence of oil type on the amounts of acrylamide generated in a model system and in french fries. Journal of Agricultural and Food Chemistry **53:** 6170–6174.
33. ELDER, V., J. FULCHER & H. LEUNG, inventors. Frito-Lay North America, Inc., assignee. 2004. Method for reducing acrylamide formation in thermally processed foods. U.S. Patent Application number: 7037540.
34. LINDSAY, R.C. & S. JANG. 2005. Model systems for evaluating factors affecting acrylamide formation in deep fried foods. Advances in Experimental Medicine and Biology **561:** 329–341.
35. GÖKMEN, V. & H.Z. ŞENYUVA. 2007. Acrylamide formation is prevented by divalent cations during the Maillard reaction. Food Chemistry **103:** 196–203.
36. MEHTA, U. & B. SWINBURN. 2001. A review of factors affecting fat absorption in hot chips. Critical Reviews in Food Science and Nutrition **41:** 133–154.
37. PADILLA, J. 2005. In Industrial frying systems. Fifth international symposium on deep-fat frying. Healthier and safer fried foods in a changing marketplace, San Francisco, CA, 20–22 February 2005.
38. WILLIAMS, J.S.E. 2005. Influence of variety and processing conditions on acrylamide levels in fried potato crisps. Food Chemistry **90:** 875–881.
39. ELMORE, J.S., G. KOUTSIDIS, A.T. DODSON, et al. 2005. Measurement of acrylamide and its precursors in potato, wheat, and rye model systems. Journal of Agricultural and Food Chemistry **53:** 1286–1293.
40. ZYZAK, D.V., R.A. SANDERS, M. STOJANOVIC, et al. 2003. Acrylamide formation mechanism in heated foods. Journal of Agricultural and Food Chemistry **51:** 4782–4787.
41. MATTHÄUS, B., N.U. HAASE & K. VOSMANN. 2004. Factors affecting the concentration of acrylamide during deep-fat frying of potatoes. European Journal of Lipid Science and Technology **106:** 793–801.
42. PEDRESCHI, F., K. KAACK, K. GRANBY & E. TRONCOSO. 2007. Acrylamide reduction under different pretreatments in French fries. Journal of Food Engineering **79:** 1287–1294.
43. PEDRESCHI, F., J. LEÓN, D. MERY, et al. 2007. Colour development and acrylamide content of pre-dried potato chips. Journal of Food Engineering **79:** 786–793.
44. MOTTRAM, D.S., B.L. WEDZICHA & A.T. DODSON. 2002. Acrylamide is formed in the Maillard reaction. Nature **419:** 448–449.
45. TAREKE, E., P. RYDBERG, P. KARLSSON, et al. 2002. Analysis of acrylamide, a carcinogen formed in heated foodstuffs. Journal of Agricultural and Food Chemistry **50:** 4998–5006.
46. CLAEYS, W.L., K. DE VLEESCHOUWER & M.E. HENDRICKX. 2005. Kinetics of acrylamide formation and elimination during heating of an asparagine-sugar model system. Journal of Agricultural and Food Chemistry **53:** 9999–10005.
47. GERTZ, C. & S. KLOSTERMANN. 2002. Analysis of acrylamide and mechanisms of its formation in deep-fried products. European Journal of Lipid Science and Technology **104:** 762–771.
48. MESTDAGH, F.J., B. DE MEULENAER & C. VAN PETEGHEM. 2007. Influence of oil degradation on the amounts of acrylamide generated in a model system and in French fries. Food Chemistry **100:** 1153–1159.
49. WEISSHAAR, R. 2004. Acrylamide in heated potato products – analytics and formation routes. European Journal of Lipid Science and Technology **106:** 786–792.
50. GRANDA, C., R.G. MOREIRA & S.E. TICHY. 2004. Reduction of acrylamide formation in potato chips by low-temperature vacumm frying. Journal of Food Science **69:** 405–411.
51. KITA, A., E. BRÅTHEN, S.H. KNUTSEN & T. WICKLUND. 2004. Effective ways of decreasing acrylamide content in potato crisps during processing. Journal of Agricultural and Food Chemistry **52:** 7011–7016.
52. BRÅTHEN, E., A. KITA, S.H. KNUTSEN & T. WICKLUND. 2005. Addition of glycine reduces the content of acrylamide in cereal and potato products. J. Agric. Food Chem. **53:** 3259–3264.
53. ERIKSSON, S., P. KARLSSON & M. TÖRNQUIST. 2007. Measurement of evaporated acrylamide during heat treatment of food and other biological materials. LWT **40:** 706–712.
54. STADLER, R.H., I. BLANK, N. VARGA, et al. 2002. Acrylamide from Maillard reaction products. Nature **419:** 449–450.
55. YAYLAYAN, V.A., A. WNOROWSKI & C.P. LOCAS. 2003. Why asparagine needs carbohydrates to generate acrylamide. J. Agric. Food. Chem. **51:** 1753–1757.

56. SANDERS, R.A., D.V. ZYZAK, M. STOJANOVIC, *et al.* 2002. An LC/MS acrylamide method and its use in investigating the role of asparagine. Acrylamide Symposium, 116th Annual AOAC International Meeting, Los Angeles, CA, Sept 26; AOAC; Gaithersburg, MD.
57. TOMODA, Y., A. HANAOKA, T. YASUDA, *et al.*, inventors. 2004. Method of decreasing acrylamide in food cooked under heat. United States Patent 20040126469.
58. PLANK, D.W. & D.J. NOVAK, inventors. 2007. Method for reducing acrylamide levels in food products and food products. US Patent application No. 7264838 B2.
59. FRIEDMAN, M. 2003. Chemistry, biochemistry, and safety of acrylamide. A review. J. Agric. Food Chem. **51:** 4504–4526.
60. ASHOOR, A. & J. ZENT. 1984. Maillard browning of common amino acids and sugars. J. Food Sci. **49:** 1206–1207.
61. KIM, C.T., E. HWANG & H.Y. LEE. 2005. Reducing acrylamide in fried snack products by adding amino acids. Journal of Food Science **70:** 354–358.
62. TAREKE, E. 2003. Identification and Origin of Potential Background Carcinogens: Endogenous Isoprene and Oxiranes, Dietary Acrylamide. Thesis, Department of Environmental Chemistry, Stockholm University.
63. VATTEM, D.A. & K. SHETTY. 2003. Acrylamide in food: a model for mechanism of formation and its reduction. Innovations Food Sci. Emerging Technol. **4:** 331–338.
64. LOW, M.Y., G. KOUTSIDIS, J.K. PARKER, *et al.* 2006. Effect of citric acid and glycine addition on acrylamide and flavor in a potato model system. Journal of Agricultural and Food Chemistry **54:** 5976–5983.
65. AMREIN, T.M., B. SCHÖNBÄCHLER, F. ESCHER & R. AMADÒ. 2004. Acrylamide in gingerbread: critical factors for formation and possible ways for reduction. Journal of Agricultural and Food Chemistry **52:** 4282–4288.
66. FINK, M., R. ANDERSSON, J. ROSÉN & P. AMAN. 2006. Effect of added asparagine and glycine on acrylamide content yeast-leavened bread. Cereal Chemistry **83:** 218–222.
67. ZHANG, Y., J. CHEN, X. ZHANG, *et al.* 2007. Addition of antioxidant of bamboo leaves (AOB) effectively reduces acrylamide formation in potato crisps and french fries. J. Agric. Food Chem. **55:** 523–528.
68. ZHANG, Y. & Y. ZHANG. 2007. Study on reduction of acrylamide in fried bread sticks by addition of antioxidant of bamboo leaves and extract of green tea. Asia Pac. J. Clin. Nutr. **16**(Suppl 1): 131–136.
69. ZHANG, Y., W. Xu, X. WU, *et al.* 2007. Addition of antioxidant from bamboo leaves as an effective way to reduce the formation of acrylamide in fried chicken wings. Food Additives and Contaminants **24:** 242–251.
70. BECALSKI, A., B.P-Y. LAU, D. LEWIS & S. SEAMAN. 2002. Acrylamide in foods: occurrence and sources. [Abstracts]; 116th Annual AOAC International Meeting, Los Angeles, CA, September 26, 2002; AOAC: Gaithersburg, MD.
71. FERNÁNDEZ, S., L. KURPPA & L. HYVONEN. 2003. Content of acrylamide decreased in potato chips with addition of a proprietary flovoniod spicemix (Flavomare®) in frying. Innovation Food Technol.: 24–26.
72. LEVINE R.A. & R.E. SMITH. 2005. Sources of variability of acrylamide levels in a cracker model. Journal of Agricultural and Food Chemistry **53:** 4410–4416.

Therapeutic Interruption of Advanced Glycation in Diabetic Nephropathy

Do All Roads Lead to Rome?

Karly C. Sourris, Josephine M. Forbes, and Mark E. Cooper

JDRF Albert Einstein Centre for Diabetes Complications, Diabetes and Metabolism Division, Baker Heart Research Institute, Melbourne, Victoria, Australia

A major common feature of the chemically disparate compounds that inhibit advanced glycation end product (AGE) accumulation or signaling is their ability to show end-organ protection in experimental models of diabetes complications. The mechanisms by which these AGE-lowering therapies confer their benefits remain unsolved. Is it the reduction in tissue AGE levels *per se* or the inhibition of downstream signal transduction (as has been described with the soluble receptor for AGE)? Possible modes of action that need to be investigated include the ability of some of these agents to stimulate antioxidant defenses, to lower cholesterol and other lipid levels, and to inhibit low-grade inflammation. To understand these novel mechanisms, further examination of the advanced glycation pathway and, in particular, the diverse action of these agents in ameliorating the development of diabetic complications is needed.

Key words: advanced glycation end products; diabetic nephropathy; reactive oxygen species; receptor for advanced glycation end products

Introduction

In recent times there has been ongoing identification of novel inhibitors of nonenzymatic glycation as well as increased recognition that existing therapies generally considered to act on other pathways can also interrupt this biochemical process. Currently available therapies that have been noted to influence the advanced glycation pathway include thiazolidinediones,[1] blockers of the renin–angiotensin system,[2] and high-dose aspirin.[3]

Early inhibitors of advanced glycation, such as aminoguanidine (pimagedine)[4] and OPB-9195,[5] relied on direct scavenging of advanced glycation end product (AGE) precursors, such as pyridoxal acting to trap reactive carbonyl groups. The antidiabetic agent metformin can also trap reactive carbonyls in addition to lowering glucose (because of its guanidine moiety) and lowering circulating levels of reducing sugars.[6] Although carnosine[7] and aspirin[3] trap reactive carbonyls to reduce AGE formation, they also chelate copper and other transition metals, as has been reported for angiotensin-converting enzyme (ACE) inhibitors (ACEi) and angiotensin type 1 receptor (AT1R) antagonists,[8] and this chelation activity could possibly influence their ability to act as AGE inhibitors. Subsequently, putative AGE cross-link breakers, such as the prototype phenacylthiazolium bromide (PTB)[9] and a more stable derivative, alagebrium chloride,[10] were reported to cleave preformed AGEs. Furthermore, therapeutic benefits of lowering dietary intake of AGEs are also seen in diabetes complications.[11] Finally, B-group vitamins and derivatives (such as thiamine, benfotiamine,[12] and pyridoxamine[13]), which are potent inhibitors of advanced glycation, show many of the inhibitory mechanisms of action listed above.

These compounds appear to have diverse mechanisms of action and yet have many similarities, particularly with respect to their effects on downstream pathways. This review provides an overview of some of the pathways implicated in diabetic nephropathy, focusing on those agents that clearly lower AGE levels, particularly in tissues susceptible to diabetes-related injury (TABLE 1).

Pathways to End-organ Damage in Diabetic Nephropathy

It is likely that the damage seen in the diabetic kidney is the result of an interaction between hemodynamic and metabolic abnormalities,[14] as evidenced by the major clinical determinants of diabetic nephropathy,

Address for correspondence: Associate Professor Josephine Forbes, JDRF Albert Einstein Centre for Diabetes Complications, Baker Heart Research Institute, PO Box 6492, St Kilda Road Central, Melbourne, 8008, Australia. Voice: +61 3 8532 1456; fax: +61 3 8532 1288.
Josephine.forbes@baker.edu.au

TABLE 1. A summary of advanced glycation end product (AGE)-lowering therapies with diverse mechanisms of action

Therapy	Tissue AGEs	Circulating AGEs
Carnosine	√	√
Benfotiamine	√	√
Thiamine	√	ND
Pyridoxamine	√	ND
Aminoguanidine	√	√
OPB-9195	√	√
ACE inhibitors	√	√
AT1 antagonists	√	X
Aspirin	√	ND
Metformin	√	√
Thiazolidinediones	ND	√
sRAGE	ND	ND
ALT-711 (alagebrium)	√	√

ND, not determined ACE, angiotensin-converting enzyme; AT1, angiotensin type 1 receptor; SRAGE, soluble RAGE.

hyperglycemia,[15] and hypertension.[16] Furthermore, these hemodynamic and metabolic pathways have been shown to interact in diabetic nephropathy.[17,18] FIGURE 1 represents a theoretical cascade of events which would likely result in the end-organ damage seen in the diabetic kidney.

Metabolic Links to AGE Inhibitors

A range of metabolic abnormalities, in addition to hyperglycemia, are seen in the diabetic kidney. However, it is obvious from studies in diabetic patients that glucose is the predominant metabolic abnormality in type 1 diabetes and strict glycemic control remains the critical, but often unattainable, strategy to retard the progression of nephropathy.[15,19] Both metformin and thiazolidinediones, which are agents that are widely used to improve glycemic control in type 2 diabetes, also influence tissue and circulating levels of AGEs.[20] Interestingly, a number of agents that may influence AGEs, such as ACEi and thiamine derivatives, have been reported to directly influence intracellular glucose uptake. Whether this ultimately leads to reduced intracellular AGE accumulation has not been extensively examined but should be considered. There is increasing evidence that intracellular AGEs and potentially AGE-binding proteins that are predominantly intracellular in location, such as the ezrin–radixin–moesin (ERM) proteins, may play a pivotal role in diabetic complications, such as nephropathy.[21,22] Indeed, the influence of other AGE inhibitors on glycemic control and cellular uptake of glucose has not been previously defined. For example, low AGE-containing diets have been reported to improve insulin sensitivity in models of type 2 diabetes[23] and in insulin-resistance states as occur with high fat feeding.[24]

Another metabolic abnormality characteristic of patients with type 1 and type 2 diabetes is hyperlipidemia, including hypertriglyceridemia and increased oxidized low-density lipoproteins.[25] The dyslipidemia in type 1 diabetes is not as overt as that seen in type 2 diabetes and most likely involves abnormal chemical modifications of lipoproteins with subsequent changes in biological function, rather than significant changes in lipoprotein levels. Interestingly, a number of the AGE inhibitors assessed in this review (TABLE 1) have been reported as improving lipid profiles in diabetic patients[26] and in experimental models of diabetic complications.[12,13,27] The relevance of this effect of end-organ proteins conferred by these agents remains to be elucidated, but reducing atherogenic lipids is likely to be clinically desirable in individuals with or at risk of diabetic complications.

The renin–angiotensin system (RAS) and in particular its hormonal vasoactivator peptide angiotensin II play a critical role not only in the regulation of systemic and glomerular hemodynamics but also in glomerular hypertrophy and ultimately glomerulosclerosis. Indeed, the therapeutic blockade of the renin–angiotensin system with either ACEi and AT1R antagonists remains a major component of therapies in both type 1 and type 2 diabetic patients with complications.[28,29] It should be appreciated that in addition to the agents that interrupt the RAS, other AGE-reducing agents, including pyridoxamine,[13] OPB-9195,[30] thiazolidinediones (mild), and carnosine,[7] have been shown to have direct hemodynamic effects, including reductions in systemic blood pressure.[16] This may not be a surprising finding since we have previously reported direct interactions between the advanced glycation pathway and the RAS. Specifically, the administration of exogenous AGE-BSA modulates the expression of various intrarenal components of the RAS with a pattern similar to that seen in the diabetic kidney.[18] Furthermore, both ACEi[17] and AT1R[31] decrease tissue accumulation of AGEs.

Downstream Effectors of Metabolic and Hemodynamic Pathways

There are four main downstream pathways that have been hypothesized to explain how glucose, through excess generation of reactive oxygen species (ROS), leads to the development of diabetic complications, including nephropathy. These also include increased flux via the polyol and hexosamine pathways, accumulation of AGEs, activation of protein kinase

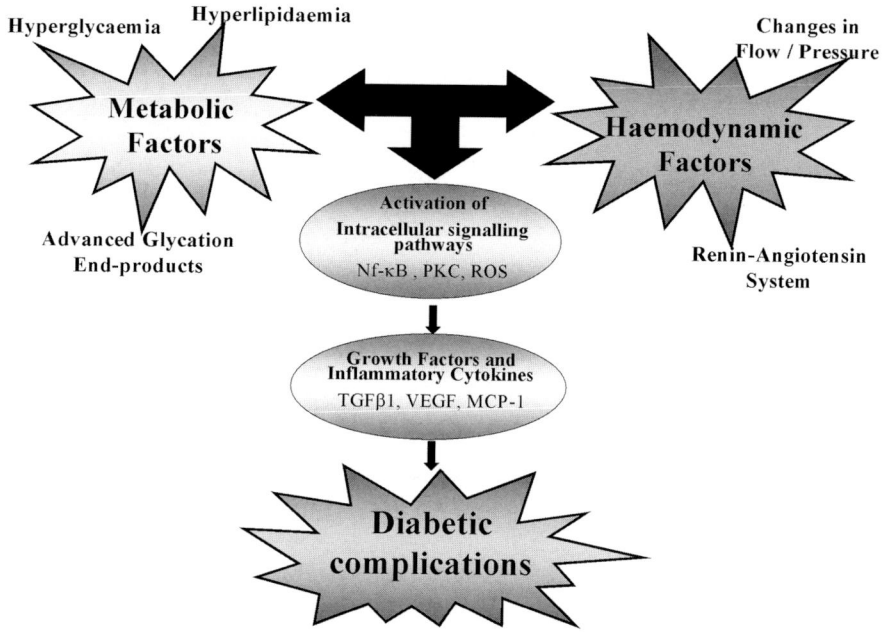

FIGURE 1. A theoretical cascade for the pathogenesis of diabetic complications: the downstream consequences of interactions between hemodynamic and metabolic pathways. (Adapted from Diabetologia Cooper 2001.[14])

C (PKC), or translocation of nuclear factor kappa B (NF-κB).[21] Indeed, therapeutic manipulation of each of these individual pathways has been shown to confer benefits in experimental models of diabetic complications. However, their individual contributions of modulating each of these pathways for the clinical management of diabetic nephropathy remains to be determined.

The excess generation of ROS as a result of hyperglycemia appears to enhance the progression of diabetic complications, with both cytosolic and mitochondrial sources of ROS being implicated. Indeed, the overexpression of cellular antioxidants, such as copper- or zinc-containing superoxide disumtase, protect against end-organ damage in models of type 2 diabetic nephropathy.[32] In addition, there have been many studies suggesting direct modulatory effects of AGEs on oxidative stress. The therapeutic approaches for reducing AGE accumulation and/or signaling discussed within this review have, in general, been reported to decrease ROS generation within complication-prone organs or in relevant cell culture experiments. It is likely, however, that the exact mechanism whereby each AGE inhibitor results in decreasing ROS generation may differ.[33] Furthermore, the involvement of specific cellular compartments, in particular the relative contribution of the mitochondrial versus cytosolic sources of ROS, remains to be determined.

Activation of PKC is considered to specifically activate a number of downstream signaling pathways in the pathogenesis of diabetic complications. Of particular relevance is the renoprotection afforded by blockade of the PKC-β1 isoform with LY333531[34] in experimental diabetic nephropathy. Furthermore, another PCK isoform, PCK-α, may also be involved since the genetic deletion of the PKC-α isoform in diabetic mice completely abrogates renal functional abnormalities.[35] Indeed, the majority of therapeutic agents described in this review (TABLE 1) prevent diabetes-induced activation and often membrane translocation of PKC-β1[12,36] or -α.[27] It should be noted that effects on PKC have not been determined in all the anti-AGE agents described in this review, and, therefore, it is possible that PKC inhibition is a common feature of agents which retard renal AGE accumulation.

The nuclear translocation of NF-κB by hyperglycemia has been demonstrated on many occasions in acute tissue culture experiments[37] and has also been confirmed in more experimental models of chronic diabetic complications.[38,39] However, it should be noted that when examined, AGE inhibitors do not appear to significantly influence diabetes-induced NF-κB translocation within renal tissues.[2,12,40]

The specific contribution of low-grade inflammation to a chronic disease, such as diabetic nephropathy, is increasingly being delineated. What is clear is that the blockade of specific cytokines and chemokines involved in processes such as the recruitment of inflammatory cells, including monocyte chemoattractant molecule (MCP-1), appears to be a valid therapeutic strategy in models of diabetic nephropathy.[41] For example, blockade of the chemokine MCP-1 has been shown to attenuate not only macrophage infiltration but also progressive renal injury in *db/db* mice, a model of type 2 diabetic nephropathy. The link between AGEs and these chemokines has been evaluated by examining the reduction tissue expression of MCP-1 with a range of AGE-modifying drugs including, AT1 antagonists, aminoguanidine, aspirin,[42] soluble RAGE (sRAGE),[43] and thiazolidinediones.[1] Interestingly, each of the approaches described above that inhibit AGE accumulation or signaling of AGE-dependant pathways appears to be, in general, anti-inflammatory, although the specific cytokines that they modulate appear to vary among the different drugs.

Several *in vitro* and *in vivo* studies have implicated transforming growth factor-β (TGF-β), a fibrogenic cytokine, as a key effector molecule in promoting diabetic renal disease. To date, several anti-AGE therapies, including alagebrium,[10] ACEi, AT1 antagonists,[44] sRAGE,[45] aminoguanidine,[10] OPB-9195,[46] and aspirin,[42] have been shown to ameliorate diabetes-induced increases in either TGF-β1 or another profibrotic cytokine, connective tissue growth factor. The utility of TGF-β1 as a target for therapeutic intervention in diabetic nephropathy is hampered by its vital role in inflammatory and immune processes, and it may be preferable to suppress renal TGF-β1 levels by an alternative approach, such as therapies that focus on the advanced glycation pathway.

Other growth factors have been implicated in the progression of diabetic nephropathy, including angiogenic cytokines, such as vascular endothelial growth factor (VEGF).[47] One must, however, be cautious in the interpretation of all data since the suppression of VEGF or its receptors, particularly in the kidney, remains controversial with some studies suggesting that VEGF blockade will result in less albuminuria[48] and recent studies, albeit in a nondiabetic context, suggesting that VEGF is a critical renal survival factor and that blockade may promote renal damage.[49] This is best demonstrated by the diverse effects seen with anti-VEGF antibodies.[48] To date, within renal tissues, a specific decrease in VEGF expression in conjunction with improved renal functional and structural parameters has been seen with a range of AGE inhibitors, including alagebrium,[27] ACEi,[47] sRAGE,[45] and OPB-9195.[46]

Conclusion

Despite diverse chemical structures and a variety of different mechanisms of action, each of the strategies presented in this review that reduce accumulation of AGEs in tissues and/or relevant downstream signaling pathways, appear to confer their benefits via a number of common pathways. These anti-AGE therapies reduce cellular oxidative stress, decrease inflammation and macrophage infiltration, lower renal cytokine expression, and often alter serum lipid levels. These diverse biological actions were observed in the context of providing end-organ protection in a variety of models of diabetic complications. In addition, many of these agents reduce blood pressure and exhibit PKC activation (some 60%). Interestingly, only a few of these agents appear to have direct glucose-lowering effects, and effects on NF-κB activation are not generally observed. Importantly, however, current treatment strategies in clinical practice have little effect on lipid profiles, full-length RAGE expression, and cellular glucose uptake and compartmentalized mitochondrial production of superoxide. In conclusion, this review summarizes our understanding of the relative importance of multiple pathways of tissue damage implicated in the pathogenesis of diabetic complications and, in particular, the relationship of these pathways to AGE-lowering therapies that are currently being used, albeit for other reasons, or are in ongoing preclinical or clinical development.

Conflict of Interest

The authors declare no conflicts of interest.

References

1. MARX, N. *et al.* 2004. Thiazolidinediones reduce endothelial expression of receptors for advanced glycation end products. Diabetes **53:** 2662–2668.
2. FORBES, J.M. *et al.* 2002. Reduction of the accumulation of advanced glycation end products by ACE inhibition in experimental diabetic nephropathy. Diabetes **51:** 3274–3282.
3. URIOS, P., A.M. GRIGOROVA-BORSOS & M. STERNBERG. 2007. Aspirin inhibits the formation of pentosidine, a cross-linking advanced glycation end product, in collagen. Diabetes Res. Clin. Pract. **77:** 337–340.
4. BROWNLEE, M. *et al.* 1986. Aminoguanidine prevents diabetes-induced arterial wall protein cross-linking. Science **232:** 1629–1632.
5. MIYATA, T. *et al.* 2000. Mechanism of the inhibitory effect of OPB-9195 [(+/−)-2-isopropylidenehydrazono-4-oxo-thiazolidin-5-yla cetanilide] on advanced glycation

end product and advanced lipoxidation end product formation. J. Am. Soc. Nephrol. **11:** 1719–1725.
6. BEISSWENGER, P. & D. RUGGIERO-LOPEZ. 2003. Metformin inhibition of glycation processes. Diabetes Metab. **29:** 6S95–103.
7. PRICE, D.L. et al. 2001. Chelating activity of advanced glycation end-product inhibitors. J. Biol. Chem. **276:** 48967–48972.
8. MIYATA, T. et al. 2002. Angiotensin II receptor antagonists and angiotensin-converting enzyme inhibitors lower in vitro the formation of advanced glycation end products: biochemical mechanisms. J. Am. Soc. Nephrol. **13:** 2478–2487.
9. VASAN, S. et al. 1996. An agent cleaving glucose-derived protein crosslinks in vitro and in vivo. Nature. **382:** 275–278.
10. FORBES, J.M. et al. 2003. The breakdown of preexisting advanced glycation end products is associated with reduced renal fibrosis in experimental diabetes. FASEB J. **17:** 1762–1764.
11. ZHENG, F. et al. 2002. Prevention of diabetic nephropathy in mice by a diet low in glycoxidation products. Diabetes Metab Res. Rev. **18:** 224–237.
12. BABAEI-JADIDI, R. et al. 2003. Prevention of incipient diabetic nephropathy by high-dose thiamine and benfotiamine. Diabetes **52:** 2110–2120.
13. DEGENHARDT, T.P. et al. 2002. Pyridoxamine inhibits early renal disease and dyslipidemia in the streptozotocin-diabetic rat. Kidney Int. **61:** 939–950.
14. COOPER, M.E. 2001. Interaction of metabolic and haemodynamic factors in mediating experimental diabetic nephropathy. Diabetologia **44:** 1957–1972.
15. 1998. Effect of intensive blood-glucose control with metformin on complications in overweight patients with type 2 diabetes (UKPDS 34). UK Prospective Diabetes Study (UKPDS) Group. Lancet **352:** 854–865.
16. ADLER, A.I. et al. 2000. Association of systolic blood pressure with macrovascular and microvascular complications of type 2 diabetes (UKPDS 36): prospective observational study. BMJ **321:** 412–419.
17. FORBES, J.M. et al. 2005. Modulation of soluble receptor for advanced glycation end products by angiotensin-converting enzyme-1 inhibition in diabetic nephropathy. J. Am. Soc. Nephrol. **16:** 2363–2372.
18. THOMAS, M.C. et al. 2005. Interactions between renin angiotensin system and advanced glycation in the kidney. J. Am. Soc. Nephrol. **16:** 2976–2984.
19. Writing Team for the Diabetes Control and Complications Trial /Epidemiology of Diabetes Interventions and Complications Research Group. 2002. Effect of intensive therapy on the microvascular complications of type 1 diabetes mellitus. JAMA **287:** 2563–2569.
20. RAHBAR, S. et al. 2000. Evidence that pioglitazone, metformin and pentoxifylline are inhibitors of glycation. Clin. Chim. Acta. **301:** 65–77.
21. BROWNLEE, M. 2001. Biochemistry and molecular cell biology of diabetic complications. Nature **414:** 813–820.
22. MCROBERT, E.A. et al. 2003. The amino-terminal domains of the ezrin, radixin, and moesin (ERM) proteins bind advanced glycation end products, an interaction that may play a role in the development of diabetic complications. J. Biol. Chem. **278:** 25783–25789.
23. HOFMANN, S.M. et al. 2002. Improved insulin sensitivity is associated with restricted intake of dietary glycoxidation products in the db/db mouse. Diabetes **51:** 2082–2089.
24. SANDU, O. et al. 2005. Insulin resistance and type 2 diabetes in high-fat-fed mice are linked to high glycotoxin intake. Diabetes **54:** 2314–2319.
25. CHAIT, A. & E.L. BIERMAN. 1994. Joslin's Diabetes Mellitus. Lea & Febiger. Philadelphia.
26. NAGI, D.K. & J.S. YUDKIN. 1993. Effects of metformin on insulin resistance, risk factors for cardiovascular disease, and plasminogen activator inhibitor in NIDDM subjects. A study of two ethnic groups. Diabetes Care **16:** 621–629.
27. THALLAS-BONKE, V. et al. 2004. Attenuation of extracellular matrix accumulation in diabetic nephropathy by the advanced glycation end product cross-link breaker ALT-711 via a protein kinase C-alpha-dependent pathway. Diabetes **53:** 2921–2930.
28. BRENNER, B.M. et al. 2001. Effects of losartan on renal and cardiovascular outcomes in patients with type 2 diabetes and nephropathy. N. Engl. J. Med. **345:** 861–869.
29. LEWIS, E.J. et al. 1993. The effect of angiotensin-converting-enzyme inhibition on diabetic nephropathy. The Collaborative Study Group. N. Engl. J. Med. **329:** 1456–1462.
30. MIZUTANI, K. et al. 2002. Inhibitor for advanced glycation end products formation attenuates hypertension and oxidative damage in genetic hypertensive rats. J. Hypertens. **20:** 1607–1614.
31. FORBES, J.M. et al. 2004. The effects of valsartan on the accumulation of circulating and renal advanced glycation end products in experimental diabetes. Kidney Int. **Suppl:** S105–107.
32. DERUBERTIS, F.R., P.A. CRAVEN & M.F. MELHEM. 2007. Acceleration of diabetic renal injury in the superoxide dismutase knockout mouse: effects of tempol. Metabolism. **56:** 1256–1264.
33. BAYNES, J.W. & S.R. THORPE. 1999. Role of oxidative stress in diabetic complications: a new perspective on an old paradigm. Diabetes **48:** 1–9.
34. KOYA, D. et al. 1997. Characterization of protein kinase C beta isoform activation on the gene expression of transforming growth factor-beta, extracellular matrix components, and prostanoids in the glomeruli of diabetic rats. J. Clin. Invest. **100:** 115–126.
35. MENNE, J. et al. 2004. Diminished loss of proteoglycans and lack of albuminuria in protein kinase C-alpha-deficient diabetic mice. Diabetes **53**(8)**:** 2101–2109.
36. OSICKA, T.M. et al. 2000. Prevention of albuminuria by aminoguanidine or ramipril in streptozotocin-induced diabetic rats is associated with the normalization of glomerular protein kinase C. Diabetes **49:** 87–93.
37. NISHIKAWA, T. et al. 2000. Normalizing mitochondrial superoxide production blocks three pathways of hyperglycaemic damage. Nature **404:** 787–790.

38. BIERHAUS, A. *et al.* 2001. Diabetes-associated sustained activation of the transcription factor nuclear factor-kappaB. Diabetes **50:** 2792–2808.
39. LEE, F.T. *et al.* 2004. Interactions between angiotensin II and NF-{kappa}B-dependent pathways in modulating macrophage infiltration in experimental diabetic nephropathy. J. Am. Soc. Nephrol. **15:** 2139–2151.
40. COUGHLAN, M.T. *et al.* 2007. Combination therapy with the advanced glycation end product cross-link breaker, alagebrium, and angiotensin converting enzyme inhibitors in diabetes: synergy or redundancy? Endocrinology **148:** 886–895.
41. CHOW, F.Y. *et al.* 2007. Monocyte chemoattractant protein-1-induced tissue inflammation is critical for the development of renal injury but not type 2 diabetes in obese db/db mice. Diabetologia **50:** 471–480.
42. MAKINO, H. *et al.* 2003. Roles of connective tissue growth factor and prostanoids in early streptozotocin-induced diabetic rat kidney: the effect of aspirin treatment. Clin. Exp. Nephrol. **7:** 33–40.
43. GU, L. *et al.* 2006. Role of receptor for advanced glycation end-products and signalling events in advanced glycation end-product-induced monocyte chemoattractant protein-1 expression in differentiated mouse podocytes. Nephrol Dial Transplant. **21:** 299–313.
44. CAO, Z. *et al.* 2001. Additive hypotensive and anti-albuminuric effects of angiotensin-converting enzyme inhibition and angiotensin receptor antagonism in diabetic spontaneously hypertensive rats. Clin. Sci. (Colch.) **100:** 591–599.
45. WENDT, T.M. *et al.* 2003. RAGE drives the development of glomerulosclerosis and implicates podocyte activation in the pathogenesis of diabetic nephropathy. Am. J. Pathol. **162:** 1123–1137.
46. WADA, R. *et al.* 2001. Effects of OPB-9195, anti-glycation agent, on experimental diabetic neuropathy. Eur. J. Clin. Invest. **31:** 513–520.
47. RIZKALLA, B. *et al.* 2003. Increased renal vascular endothelial growth factor and angiopoietins by angiotensin II infusion is mediated by both AT1 and AT2 receptors. J. Am. Soc. Nephrol. **14:** 3061–3071.
48. DE VRIESE, A.S. *et al.* 2001. Vascular endothelial growth factor is essential for hyperglycemia-induced structural and functional alterations of the peritoneal membrane. J. Am. Soc. Nephrol. **12:** 1734–1741.
49. ADVANI, A. *et al.* 2007. Role of VEGF in maintaining renal structure and function under normotensive and hypertensive conditions. Proc. Natl. Acad. Sci. USA **104:** 14448–14453.

The Other Side of the Maillard Reaction

RAM H. NAGARAJ,[a] ASHIS BISWAS,[a,b] ANTONIA MILLER,[a,c] TOMOKO OYA-ITO,[a,d] AND MANJUNATHA BHAT[e]

[a]*Department of Ophthalmology and Visual Sciences, Case Western Reserve University, Cleveland, Ohio, USA*

[b]*Present address: Department of Pathobiology, Lerner Research Institute, The Cleveland Clinic Foundation, Cleveland, Ohio, USA*

[c]*Present address: Monash University, Alfred Medical Research and Education Precinct, Victoria, Australia*

[d]*Present address: Department of Medical Proteomics, Kyoto Prefectural University of Medicine, Kyoto, Japan*

[e]*Department of Anesthesiology, The Cleveland Clinic Foundation, Cleveland, Ohio, USA*

The Maillard reaction plays an important role in eye lens aging and cataract formation. Methylglyoxal (MGO) is a metabolic dicarbonyl compound present in the lens. It reacts with arginine residues in lens proteins to form advanced glycation end products (AGEs), such as hydroimidazolones and argpyrimidine. α-Crystallin, comprising αA- and αB-crystallin, is a major protein of the lens and it functions as a chaperone protein. We have found that upon reaction with MGO, human αA-crystallin becomes a more effective chaperone. Modification of specific arginine residues to AGEs appears to be the reason. Mutation of these arginine residues to alanine mirrors the effect of MGO, suggesting neutralization of the positive charge on arginine residues as a cause for improved chaperone function. Reaction with MGO also blocks the loss of the chaperone function of αA-crystallin caused by nonenzymatic glycation by ascorbate and ribose. These findings suggest that low levels of MGO might help the lens remain transparent during aging.

Key words: methylglyoxal; α-crystallin; chaperone function; argpyrimidine; hydroimidazolone

The eye lens is a transparent tissue that functions by focusing light onto the retina. The crystallins are the major proteins of the lens, and there are three major classes of crystallins: α, β, and γ. Of the three crystallins, α-crystallin is the most abundant in the adult lens, making up nearly 50% of the total lens protein. There are two α-crystallins, αA and αB, and although these proteins are products of two different genes, they display considerable amino acid sequence homology. The α-crystallins are approximately 20 kDa proteins with αA-crystallin having 173 amino acids and αB-crystallin having 175 amino acids. α-Crystallin exists as large aggregates in the lens; usually 40 subunits compose a single aggregate. As the lens ages, α-crystallin tends to aggregate further through covalent and noncovalent interactions, and in the adult lens, there is a wide array of α-crystallin aggregates with molecular masses ranging from 800 kDa to more than 1 million kDa.

Both αA- and αB-crystallins belong to the small heat shock-protein family. They share considerable sequence homology with other members of this family, the most conspicuous similarity being the presence of an α-crystallin domain of 80 to 100 amino acids common to all members. This domain is flanked by a highly variable N-terminal domain and a short flexible C-terminal extension. The N-terminal domain is believed to be necessary for protein oligomer formation, while the C-terminal extension imparts solubility.[1]

Both αA- and αB-crystallins exhibit chaperone-like functions. Numerous studies have shown that α-crystallin inhibits protein aggregation caused by thermal, chemical, or UV-light stress.[2] Mild structural perturbation in the target proteins is believed to be necessary for α-crystallin binding, and several recent studies have mapped the hydrophobic target protein-binding sites in the α-crystallins.[3–5]

Address for correspondence: Ram H. Nagaraj, Ph.D., Department of Ophthalmology and Visual Sciences, 2085 Adelbert Road, Room 311, Case Western Reserve University, Cleveland, OH 44106. Voice: 216-368-2089; fax: 216-368-0743.
ram.nagaraj@case.edu

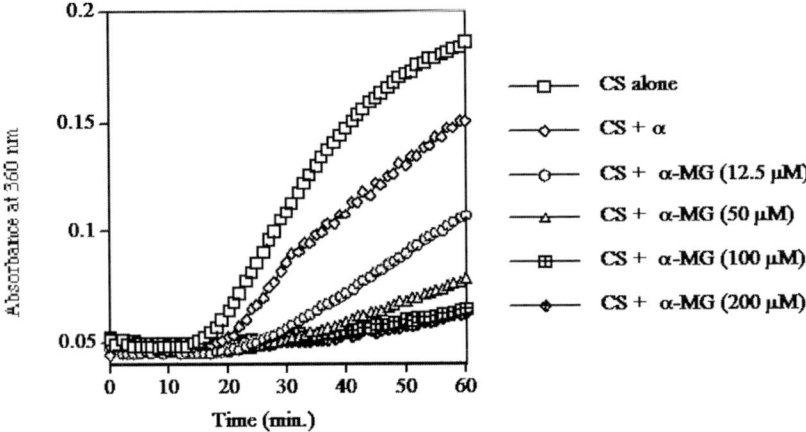

FIGURE 1. Low concentrations of methylglyoxal (MG) affect chaperone function of α-crystallin. α-Crystallin was incubated with 12.5 – 200 mM MG and assayed for chaperone function by citrate synthase (CS) aggregation. CS (40 mg) in 0.1 M HEPES buffer (pH 7.4) was incubated at 43 °C in the presence or absence of 5 mg of α-crystallin. Chaperone assays were done in triplicate and in a microwell plate; representative data are shown. Total volume in each assay was 250 mL. (Reprinted from Nagaraj et al.[24])

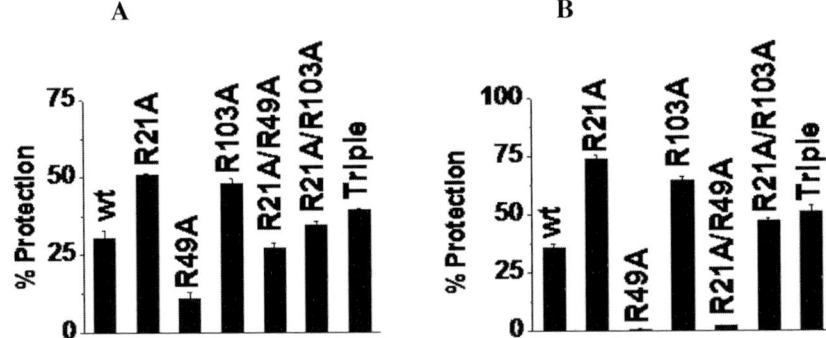

FIGURE 2. Chaperone function of human αA-crystallin. Dithriothreitol-induced aggregation of 0.32 mg/mL insulin at 25 °C (**A**) and thermal aggregation of 0.06 mg/mL citrate synthase (CS) at 43 °C (**B**) in the presence or absence of different αA-crystallin preparations. Data show the percent protection ability of different αA-crystallin against protein aggregation. The chaperone:substrate ratio (w/w) was 1:5 and 1:15 for insulin and CS aggregation assays, respectively. Data are mean ± standard deviation from triplicate determinations. (Reprinted from Biswas et al.[27])

A number of mutations, such as R116C in αA-crystallin and R120G in αB-crystallin, detected either in humans and animals or generated *in vitro* have been found to cause a loss in α-crystallin chaperone function. Overexpression of R116C in the mouse lens results in cataract formation.[6] Furthermore, knocking out α-crystallin from the mouse lens results in developmental and morphological changes and opacity of the lens.[7,8] Together these observations strongly suggest that the chaperone function of α-crystallin is necessary to prevent protein aggregation and consequently maintain transparency of the aging lens. In addition to its role in lens transparency, α-crystallin also seems to play a role in other diseases, such as autoimmune demyelination, desmin-related myopathy, and neurological disorders, such as Parkinson's and Alzheimer's disease.[9,10]

Nonenzymatic glycation is the reaction of carbonyl compounds with free amino groups in proteins. This reaction is also known as the Maillard reaction, and it proceeds through the formation of a Schiff's base to form structurally diverse stable adducts on proteins. Such adducts are collectively known as advanced glycation end products or AGEs.[11] Many AGEs are amino acid cross-linking adducts, although others are noncross-linking. Some AGEs are chromophores, and others are fluorophores. Many aged proteins, especially

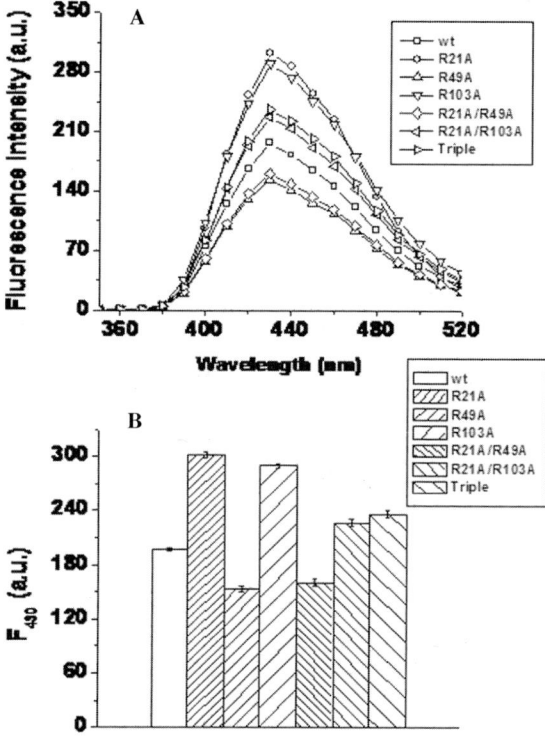

FIGURE 3. Fluorescence spectra of 2-p-toluidinyl-naphthalene-6-sulfonate (TNS)-bound αA-crystallin. Protein concentration was 0.1 mg/mL and TNS concentration was 100 μmol/L. **(A)** The fluorescence spectrum of different samples at 25°C was recorded from 350–520 nm. The excitation wavelength was 320 nm. **(B)** Fluorescence intensity at 430 nm (emission maxima) for wt and mutant αA-crystallin at 25°C. (Reprinted from Biswas et al.[27])

those that have a long half-life, exhibit cross-linking and fluorescence that are very similar to that seen in proteins modified by AGEs *in vitro*. Such similarities have led to many investigations over the past 25 years to identify AGEs in aging tissues. These investigations have resulted in detection and quantification of several AGEs in human as well as animal tissues.[11–13]

Lens proteins are unique in that they have a very limited turnover, and therefore they accumulate postsynthetic modifications, including AGEs, during normal aging. The AGEs that have been detected in the lens are derived from ascorbic acid, sugars, and glyoxal and methylglyoxal (MGO).[14–21] Some of these studies have shown that AGEs are closely associated with lens aging and cataract development, which point to a causal role for AGEs in lens aging and cataract formation.

Because α-crystallins are major proteins in the lens and AGEs are produced in lens proteins, it was of interest to determine the effect of AGE formation on the chaperone-like function of α-crystallin. *In vitro* incubation of α-crystallin with sugars resulted in the loss of α-crystallin's chaperone function.[22,23] Thus, we predicted that modification of α-crystallin by MGO would also make it a poor chaperone. To our surprise, contrary to the effects of sugars and ascorbate, we found that upon MGO modification αA-crystallin became a more effective chaperone (FIG. 1).[24] Moreover, we demonstrated that the enhancement of chaperone function occurred when the MGO concentration was close to the physiological level and the protein concentration was 5 mg/mL. These initial experiments also indicated that MGO modified specific arginine residues in human αA-crystallin by converting residues R21, R49, and R103 to argpyrimidine. Subsequent immunoprecipitation of argpyrimidine-bearing αA-crystallin with a monoclonal antibody revealed that the remaining unprecipitated αA-crystallin was a less efficient chaperone compared to the argpyrimidine-bearing αA-crystallin, suggesting that argpyrimidine formation was responsible for the enhancement of the chaperone-like function.[24] It is believed that hydrophobic pockets in α-crystallin are responsible for the interaction with client proteins during the chaperone function.[25,26] The hydrophobicity as assessed by 1,1′-Bis (1-anilinonaphthalene 8-sulfonate) binding increased in αA-crystallin upon modification by 2–100 μM MGO but decreased in samples modified with higher concentrations of MGO. These observations suggest that the chaperone function improved as a result of MGO modification of specific arginine residues in the protein. It is very likely that formation of MGO-derived hydroimidazolone adducts will also have an effect similar to that of argpyrimidine on αA-crystallin. Our studies also suggested that neutralization of the positive charge on these arginine residues might be responsible for enhancement of the chaperone function.

To determine whether neutralization of the positive charge on arginine improved chaperone function after MGO modification, we performed site-directed mutagenesis of argpyrimidine-modifiable arginine residues and replaced them with alanine. We found that R21A and R103A mutations enhanced the chaperone function against thermal denaturation of citrate synthase and chemical denaturation of insulin, but R49A decreased it (FIG. 2).[27] In a triple mutant in which all three arginine residues were replaced with alanine, we found a gain in chaperone function when compared to the wild-type protein. The increase in function coincided with an increase in protein hydrophobicity as determined using the hydrophobic probe, 2-(p-toludino) naphthalene-6-sulfonic acid, sodium salt

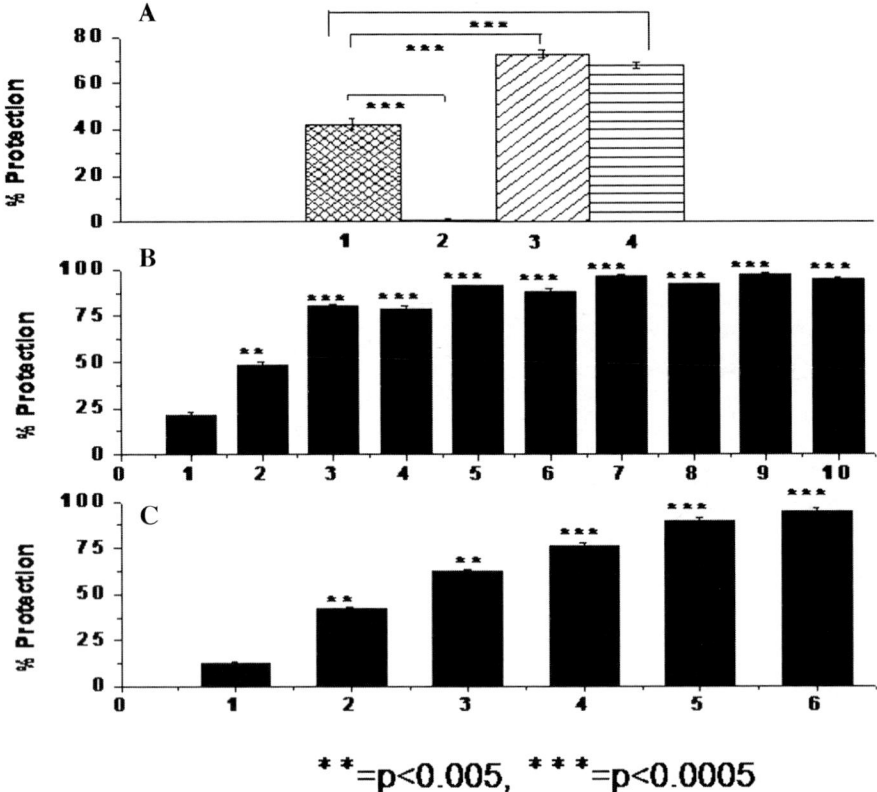

FIGURE 4. Inhibition of glycation-mediated loss of α-crystallin's chaperone function by MGO. **(A)** Chaperone function was measured by the CS aggregation assay at 43°C. Chaperone function of the control (bar 1) and 10 mM ascorbate-incubated α-crystallin (bar 2); chaperone function of α-crystallin (5 mg/mL) incubated with 5 mM MGO alone for 96 h (bar 3); chaperone function of α-crystallin (5 mg/mL) incubated with 5 mM MGO for 24 h followed by incubation with 10 mM ascorbate for 72 h (bar 4). **(B)** Chaperone function of α-crystallin control (bar 2) and α-crystallin incubated with various concentrations of MGO—0.1 mM (bars 3 and 4), 0.5 mM (bars 5 and 6), 2.5 mM (bars 7 and 8), and 5 mM (bars 9 and 10) for 24 h followed by incubation with 10 mM ribose (bars 4, 6, 8, 10) at 37°C for 72 h. The chaperone function of α-crystallin incubated with 10 mM ribose for 72 h is shown in bar 1. **(C)** α-Crystallin was co-incubated with 10 mM ribose and one of the following concentrations of MGO—0.1, 0.5, 2.5, 5 mM (bars 3 to 6) at 37°C for 96 h. Chaperone function of the α-crystallin control is in bar 2 and that of the 10 mM ribose-incubated sample is in bar 1. (Reprinted from Puttaiah et al.[28])

(FIG. 3). The mutant proteins showed minor changes in tertiary structure as revealed by near-UV circular dichroism spectroscopy. We found that the substitution of positively charged arginine residues with alanine residues at R21, R49, and R103 resulted in higher molecular aggregates when compared to the wild-type protein. We also found that the R21A mutant had a lower molar mass and polydispersity when compared to R49A and the wild type. Both R21A and R103A had higher protein-binding constants and lower Kd (as measured using carbonic anhydrase) when compared with R49A and the wild-type protein. The R21A and R103A mutant proteins also were better than the wild type and R49A in refolding guanidine hydrochloride-inactivated malate dehydrogenase. These observations confirmed our hypothesis that MGO modification of specific arginine residues neutralizes the positive charge and enhances chaperone function. Thus, in physiological conditions, MGO modification may improve the chaperone function of αA-crystallin. We also investigated the effect of MGO modification on glycation-mediated inactivation of the chaperone function of α-crystallin. Glycation with either 10 mM ribose or 10 mM ascorbate of 5 mg/mL α-crystallin (for 4 days at 37°C) resulted in a substantial loss in the chaperone function (FIG. 4).[28] However, prior incubation with 10 mM MGO protected against such a loss in the chaperone function. What

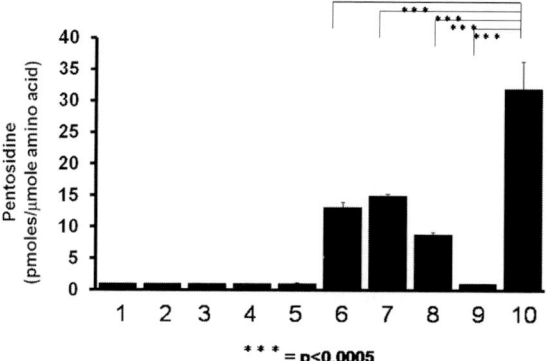

FIGURE 5. Effect of various concentrations of methylglyoxal (MGO) on pentosidine synthesis. Pentosidine in α-crystallin was incubated with MGO for 96 h (0.1, 0.5, 2.5, 5 mM, bars 2 to 5) or first with 0.1, 0.5, 2.5, 5 mM MGO for 24 h followed by incubation with 10 mM ribose for 72 h at 37°C (bars 6 to 9). Pentosidine in α-crystallin incubated with ribose only is shown in bar 10; control incubation without MGO or ribose is bar 1. (Reprinted from Puttaiah et al.[28])

is even more striking was that such incubation enhanced the chaperone function beyond what was seen with the unmodified protein. Such an increase occurred even when α-crystallin was not covalently cross-linked.

Pentosidine is a lysine–arginine cross-linking AGE. Its formation has been linked to protein cross-linking during normal aging and in diabetic complications.[29] We found that prior incubation of α-crystallin with MGO prevented ribose-mediated pentosidine synthesis (FIG. 5). These observations strongly suggest that MGO protects α-crystallin against pentosidine formation and restores or preserves α-crystallin's ability to act as a chaperone even when the sugar concentration is high.

In summary, our results show that the Maillard reaction, by low concentrations of MGO may benefit the lens by enhancing the chaperone function of α-crystallin and preventing glycation-mediated loss of chaperone function. In addition, low concentrations of MGO might also prevent the formation of the protein cross-linker pentosidine, and possibly other AGEs, in α-crystallin. Together, these effects may help to maintain lens transparency during aging. However, we caution that high concentrations of MGO, prolonged exposure, or both might lead to cross-linking of α-crystallin, resulting in loss of chaperone function. Furthermore, during cataract formation, many other post-translational modifications, such as oxidation and kynurenine-modification, deamidation, and truncation, may have overriding effects that might mask the beneficial effects of MGO on α-crystallin.

Acknowledgments

Our studies were supported by National Institute of Health Grants R01EY-016219 and R01EY-09912, P30EY-11373 (Visual Sciences Research Center of Case Western Reserve University); the Carl F. Asseff, M.D. Professorship (R.H.N); and Research to Prevent Blindness, New York and Ohio Lions Eye Research Foundation.

Conflict of Interest

The authors declare no conflict of interest.

References

1. SUN, Y. & T.H. MACRAE. 2005. Small heat shock proteins: molecular structure and chaperone function. Cell. Mol. Life Sci. **62:** 2460–2476.
2. ANDLEY, U.P. 2007. Crystallins in the eye: function and pathology. Prog. Retin Eye Res. **26:** 78–98.
3. SHARMA, K.K., H. KAUR & K. KESTER. 1997. Functional elements in molecular chaperone alpha-crystallin: identification of binding sites in alpha B-crystallin. Biochem. Biophys. Res. Commun. **239:** 217–222.
4. SHARMA, K.K. et al. 2000. Synthesis and characterization of a peptide identified as a functional element in alphaA-crystallin. J. Biol. Chem. **275:** 3767–3771.
5. GHOSH, J.G., M.R. ESTRADA & J.I. CLARK. 2005. Interactive domains for chaperone activity in the small heat shock protein, human alphaB crystallin. Biochemistry **44:** 14854–14869.
6. HSU, C.D., S. KYMES & J.M. PETRASH. 2006. A transgenic mouse model for human autosomal dominant cataract. Invest. Ophthalmol. Vis. Sci. **47:** 2036–2044.
7. BRADY, J.P. et al. 1997. Targeted disruption of the mouse alpha A-crystallin gene induces cataract and cytoplasmic inclusion bodies containing the small heat shock protein alpha B-crystallin. Proc. Natl. Acad. Sci. USA **94:** 884–889.
8. BOYLE, D.L. et al. 2003. Morphological characterization of the Alpha A- and Alpha B-crystallin double knockout mouse lens. BMC Ophthalmol. **3:** 3.
9. SUN, Y. & T.H. MACRAE. 2005. The small heat shock proteins and their role in human disease. FEBS J. **272:** 2613–2627.
10. OUSMAN, S.S. et al. 2007. Protective and therapeutic role for alphaB-crystallin in autoimmune demyelination. Nature **448:** 474–479.
11. AHMED, N. & P.J. THORNALLEY. 2007. Advanced glycation endproducts: what is their relevance to diabetic complications? Diabetes Obes. Metab. **9:** 233–245.

12. PEPPA, M. & H. VLASSARA. 2005. Advanced glycation end products and diabetic complications: a general overview. Hormones (Athens). **4:** 28–37.
13. STITT, A.W. 2005. The maillard reaction in eye diseases. Ann. N.Y. Acad. Sci. **1043:** 582–597.
14. CHENG, R. et al. 2005. K2P–a novel cross-link from human lens protein. Ann. N.Y. Acad. Sci. **1043:** 184–194.
15. ARGIROV, O.K., B. LIN & B.J. ORTWERTH. 2004. 2-ammonio-6-(3-oxidopyridinium-1-yl)hexanoate (OP-lysine) is a newly identified advanced glycation end product in cataractous and aged human lenses. J. Biol. Chem. **279:** 6487–6495.
16. NAGARAJ, R.H. et al. 1991. High correlation between pentosidine protein crosslinks and pigmentation implicates ascorbate oxidation in human lens senescence and cataractogenesis. Proc. Natl. Acad. Sci. USA **88:** 10257–10261.
17. NAGARAJ, R.H. & C. SADY. 1996. The presence of a glucose-derived Maillard reaction product in the human lens. FEBS Lett. **382:** 234–238.
18. WILKER, S.C. et al. 2001. Chromatographic quantification of argpyrimidine, a methylglyoxal-derived product in tissue proteins: comparison with pentosidine. Anal. Biochem. **290:** 353–358.
19. AHMED, M.U. et al. 1997. N-epsilon-(carboxyethyl)lysine, a product of the chemical modification of proteins by methylglyoxal, increases with age in human lens proteins. Biochem J. **324:** 565–570.
20. FRYE, E.B. et al. 1998. Role of the Maillard reaction in aging of tissue proteins. Advanced glycation end product-dependent increase in imidazolium cross-links in human lens proteins. J. Biol. Chem. **273:** 18714–18719.
21. AHMED, N. et al. 2003. Methylglyoxal-derived hydroimidazolone advanced glycation end-products of human lens proteins. Invest. Ophthalmol. Vis. Sci. **44:** 5287–5292.
22. CHERIAN, M. & E.C. ABRAHAM. 1995. Decreased molecular chaperone property of alpha-crystallins due to posttranslational modifications. Biochem. Biophys. Res. Commun. **208:** 675–679.
23. VAN BOEKEL, M.A. et al. 1996. The influence of some post-translational modifications on the chaperone-like activity of alpha-crystallin. Ophthalmic Res. 28 Suppl. **1:** 32–38.
24. NAGARAJ, R.H. et al. 2003. Enhancement of chaperone function of alpha-crystallin by methylglyoxal modification. Biochemistry **42:** 10746–10755.
25. DAS, K.P. & W.K. SUREWICZ. 1995. Temperature-induced exposure of hydrophobic surfaces and its effect on the chaperone activity of alpha-crystallin. FEBS Lett. **369:** 321–325.
26. RAMAN, B., T. RAMAKRISHNA & C.M. RAO. 1995. Temperature dependent chaperone-like activity of alpha-crystallin. FEBS Lett. **365:** 133–136.
27. BISWAS, A. et al. 2006. Effect of site-directed mutagenesis of methylglyoxal-modifiable arginine residues on the structure and chaperone function of human alphaA-crystallin. Biochemistry **45:** 4569–4577.
28. PUTTAIAH, S. et al. 2007. Methylglyoxal inhibits glycation-mediated loss in chaperone function and synthesis of pentosidine in alpha-crystallin. Exp. Eye Res. **84:** 914–921.
29. MONNIER, V.M., D.R. SELL & S. GENUTH. 2005. Glycation products as markers and predictors of the progression of diabetic complications. Ann. N.Y. Acad. Sci. **1043:** 567–581.

Dicarbonyls Stimulate Cellular Protection Systems in Primary Rat Hepatocytes and Show Anti-inflammatory Properties

TIMO M. BUETLER, HÉLIA LATADO, ALEXANDRA BAUMEYER, AND THIERRY DELATOUR

Nestlé Research Center, Lausanne 26, Switzerland

Advanced glycation endproducts (AGEs) and their precursor dicarbonyls are generally perceived as having adverse health effects. They are also considered to be initiators and promoters of disease and aging. However, proof for a causal relationship is lacking. On the other hand, it is known that AGEs and melanoidins possess beneficial properties, such as antioxidant and metal-chelating activities. Furthermore, some AGEs may stimulate the cellular detoxification system, generally known as the phase II drug metabolizing system. We show here that several reactive dicarbonyl intermediates have the capability to stimulate the cellular phase II detoxification systems in both a reporter cell line and primary rat hepatocytes. In addition, we demonstrate that dicarbonyls can attenuate the inflammatory signaling induced by tumor necrosis factor-α in a reporter cell system.

Key words: AGEs; glycation; dicarbonyls; phase II enzymes; hepatocytes; protection; detoxification; anti-inflammatory

Introduction

Advanced glycation endproducts (AGEs) are often perceived as having adverse health effects. However, there is no proof for a causal relationship between AGEs and disease occurrence or progression. In addition, many studies have only studied the effects of heated diets that contain AGEs on physiological parameters[1] and did not consider other heat-induced alterations of food, such as lipid and protein oxidations or thermal decomposition of vitamins and polyphenols.

Recently, a few studies have shown that AGEs can also have beneficial effects.[2] The antioxidant properties of AGEs have long been known to food chemists,[3] but their *in vivo* antioxidant effects are difficult to assess. The Somoza group has shown that some AGEs or Maillard reaction products possess the ability to stimulate cellular detoxification systems[4–6]; this was also confirmed in a rat study.[7] The major compound able to stimulate phase II detoxification enzymes in bread crust was identified as pronyllysine.[6,8]

Phase II drug-metabolizing enzymes are important detoxification systems[9] expressed in most tissues of the body with the highest expression in the liver, followed by the intestine, kidney, and lung. Many of these enzymes possess potent conjugating activities and serve to eliminate reactive metabolic intermediates and carcinogens. Their induction and expression are associated with protection against many types of cancer[10] as well as diabetes[11] and cardiovascular diseases.[12] Phase II enzymes appear to be inducible by a common transcription factor, Nrf-2,[13] which recognizes and binds to the antioxidant response element (ARE). More recently, targeting Nrf-2 as a way to stimulate these protective mechanisms has become a focus of pharmacological interest.[14]

We have investigated the potential ability of AGEs and dicarbonyls to stimulate phase II enzyme expression in cellular systems.

Materials and Methods

Cultures of Reporter Cell Lines

The AREc32 cell line[15] was maintained in Dulbecco's modified Eagle's medium (DMEM) supplemented with 10% heat-inactivated fetal calf serum (FCS), 50 U/mL penicillin/streptomycin and 0.8 mg/mL geneticin, all from Invitrogen (Basel, Switzerland). The cells were plated at 0.8×10^4 cells/well in 96 well plates. After 24 hours (40–50% confluence), cells were rinsed with phosphate-buffered saline solution (PBS) and pre-incubated with medium without FCS for 30 min before the addition of the various test

Address for correspondence: Timo M. Buetler, Nestlé Research Center, Vers-chez-les-Blanc, 1000 Lausanne 26, Switzerland. Voice: +4121 7859223; fax: +4121 785 8553.

Timo.Buetler@rdls.nestle.com

compounds in serum-free medium. The expression of the reporter gene product luciferase was measured with the Luciferase Assay System (Promega, Wallisellen, Switzerland).

The HT29c34 cell line[16] was maintained in high-glucose DMEM (4.5 g/L) supplemented with glutamine (Amimed, BioConcept, Basel, Switzerland), 10% heat-inactivated FCS (Amimed), 0.5 mg/mL geneticin (Invitrogen), and 1% penicillin/streptomycin (Invitrogen). The cells were plated at 2×10^4 cells/well in 96 well plates. After 24 hours (40–50% confluence), cells were rinsed with PBS and pre-incubated with medium without FCS for 30 min. After treatment with compounds for 24 h in serum-free medium, the medium was replaced by medium containing the compounds plus 1 ng/mL tumor necrosis factor-α (TNFα). The expression of the reporter gene product, secreted alkaline phosphatase (sAP), was measured with the Phospha-Light kit (Applied Biosystems, Rotkreuz, Switzerland) and cell viability was determined with the CellTiter-Glo Luminescent Cell Viability Assay (Promega).

Primary Hepatocyte Cultures

Hepatocytes were isolated and cultured as described by Cavin et al.[17] Cells were maintained in serum-free William's E medium supplemented with 25 nM dexamethasone and a Matrigel overlay (233 μg/mL). Treatment durations were 16 h and 24 for qPCR and Western blot analyses, respectively.

Real-time, Quantitative PCR

RNA was isolated from one 60-mm dish using the Trizol method (Invitrogen). One microgram of total RNA was reverse-transcribed into cDNA using random hexamers and Multiscribe RT (Applied Biosystems), and 20 ng of cDNA were used for real-time PCR employing TaqMan genes. Real-time PCR was performed using an ABI Prism 7000 Sequence Detection System (Applied Biosystems) and analyzed with SDS software version 1.1 (Applied Biosystems). The mRNA expression for each target gene was normalized to the mRNA expression of TATA box binding protein (TBP).

Western Blot

Hepatocytes were treated in 60-mm dishes for 24 h at 37 °C in a 5% CO_2 incubator. Cells were washed twice with PBS and collected in RIPA buffer (Sigma-Aldrich, Buchs, Switzerland) containing protease inhibitor cocktail and phosphatase inhibitor cocktails (Sigma). The homogenates were collected by centrifugation at 8000g for 10 min at 4 °C, and the protein concentration of the supernatant was determined with a Bio-Rad Protein assay kit (Bio-Rad, Reinach, Switzerland). Ten micrograms of protein were loaded onto NuPAGE® NOVEX Bis-Tris Pre-Cast Gel 4–12% System (Invitrogen). After electrophoresis, proteins were transferred onto polyvinylidene fluoride membranes (Invitrogen) and probed with antibodies according to the manufacturer's recommendations.

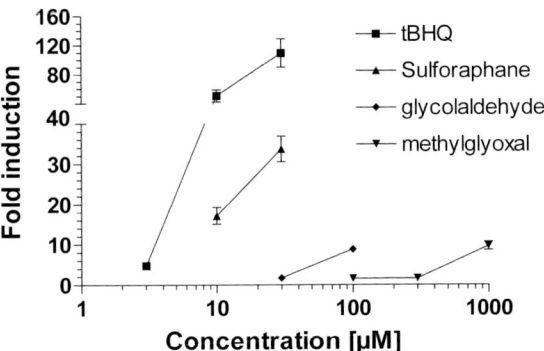

FIGURE 1. Induction of the Nrf-2 reporter gene in AREc32 cells. Data are expressed relative to untreated controls (means ± SEM, n = 3).

Results

We first investigated whether AGEs (glucose-, lactose-, carboxymethyllysine- and glycolaldehyde-modified beta lactoglobulin) would be able to stimulate the Nrf-2 pathway using the ARE-reporter cell line AREc32.[15] However, none of the AGE preparations were able to stimulate the reporter gene. We then investigated whether the reactive dicarbonyls, generally formed from glucose and Amadori product breakdown, could stimulate the Nrf-2 responsive transgene in AREc32 cells. FIGURE 1 shows that both glycolaldehyde (8-fold induction at 0.1 mM) and methylglyoxal (13-fold induction at 1 mM) were able to stimulate the luciferase reporter gene. However, the concentrations needed to stimulate the reporter gene were 1–2 orders of magnitude higher than those required for induction by the reference Nrf-2-stimulating chemicals, *tert*-butylhydroquinone (tBHQ)[18] and sulforaphane.[19] The fold-induction was also significantly lower than for the established Nrf-2 inducers. Higher dicarbonyl concentrations could not be tested in this system because of toxicity. Glyoxal was also tested but only a three-fold induction of the reporter was observed at 1 mM (not shown).

We then tested whether dicarbonyls could stimulate phase II enzyme genes in primary cultures of rat hepatocytes. FIGURE 2 shows that glyoxal, glycoladehyde,

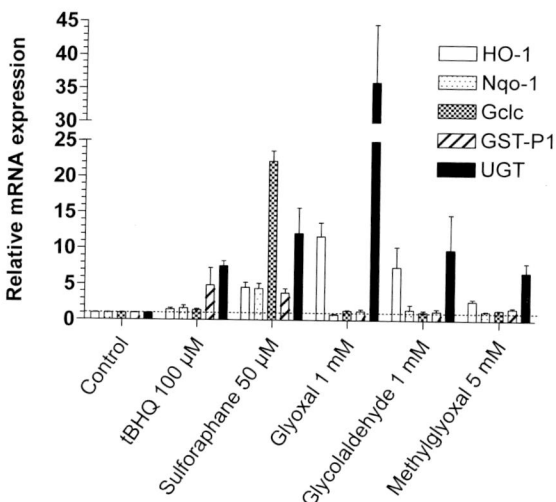

FIGURE 2. Induction of phase II enzyme mRNAs in primary rat hepatocytes. Messenger RNA expression, normalized to TBP mRNA expression, is shown relative to untreated controls. tBHQ and sulforaphane were used as positive controls (means ± SEM, n = 3). Abbreviations: HO-1, heme oxygenase 1; Nqo-1, NAD(P)H:quinone oxidoreductase 1; Gclc, γ-glutmalycysteine ligase catalytic subunit; GST-P1, glutathione S-transferase P; UGT, UDP-glucuronosyl transferase.

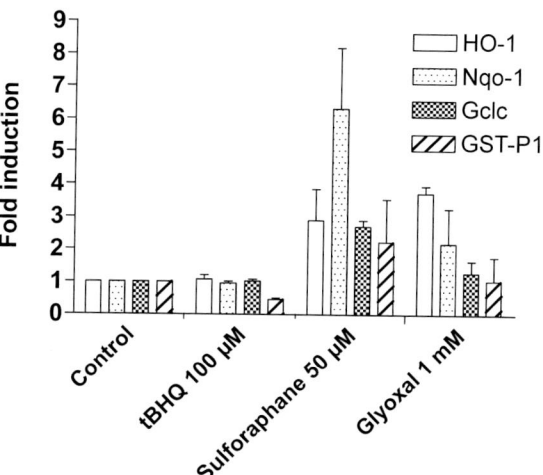

FIGURE 3. Quantitation of Western blots of phase II protein induction by glyoxal in primary rat hepatocytes. tBHQ and sulforaphane served as positive controls. Values are means and SEM of two Western blots. Abbreviations: HO-1, heme oxygenase 1; Nqo-1, NAD(P)H:quinone oxidoreductase 1; Gclc, γ-glutmalycysteine ligase catalytic subunit; GST-P1, glutathione S-transferase P.

and methylglyoxal were able to stimulate several phase II enzyme genes in primary hepatocytes. The relative potencies of the carbonyls tested in hepatocytes differed to those observed using the AREc32 reporter cells. For methylglyoxal, a concentration of 5 mM or higher was required for induction of phase II enzyme genes, whereas glycoladehyde was less potent in hepatocytes compared with the AREc32 cells. Glyoxal, which only yielded very low induction activity in the reporter cells, was the best inducer in hepatocytes.

Western blot analyses (FIG. 3) confirmed the induction of phase II enzymes by glyoxal in primary hepatocytes.

In a last set of experiments we tested whether dicarbonyls could influence the TNFα-mediated induction of the pro-inflammatory transcription factor NF-κB using the HT29c34 cell line stably expressing a sAP reporter construct under the control of NF-κB.[16] Preliminary experiments have shown that a 24-h dicarbonyl pre-incubation resulted in the optimal effect. After a 24-h pre-incubation with dicarbonyls, cells were treated with TNFα for 24 h in the presence of dicarbonyls. TNFα induced a large increase in the sAP reporter gene product that was determined in the cell culture supernatant. The pre-incubation with dicarbonyls attenuated this pro-inflammatory response up to 100% for 0.5 mM glycolaldehyde (FIG. 4). It is worth noting that the dicarbonyl concentrations used did not induce any cell toxicity as measured by a cell viability test.

Discussion

Our data show that dicarbonyls at low millimolar and sub-millimolar concentrations can stimulate the expression of the phase II detoxification system in both a reporter cell line and primary rat hepatocytes. In addition, our results provide evidence that they can also attenuate the pro-inflammatory response to TNFα exposure in a reporter cell line. These novel findings suggest that dicarbonyls can have beneficial effects in cellular systems when used at low concentrations. At higher concentrations, dicarbonyls showed some toxicity, depending on the cell system used.

Although high dicarbonyl concentrations were needed to stimulate phase II enzyme gene expression in hepatocytes, the concentration of dicarbonyls that actually reached the cytoplasm to exert their action is not known. Extracellularly added dicarbonyls could react with proteins on the cell surface or membrane lipids, and it is likely that only a fraction of the original concentration may have reached the cytoplasm. Intracellular steady-state dicarbonyl levels were estimated to be in the low micromolar range. Because the biologically relevant dicarbonyls are most likely formed intracellularly from glucose metabolism and breakdown, it

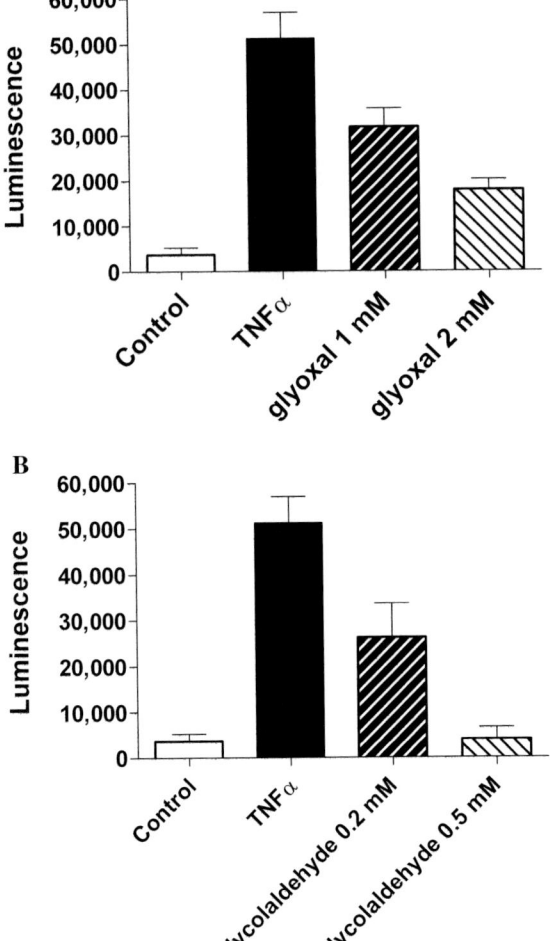

FIGURE 4. Interference with TNFα-induced stimulation of inflammatory signaling by dicarbonyls. Panel A shows the effect of glyoxal, panel B that of glycolaldehyde. Data are expressed as luminescence [RLU] after induction with 1 ng/mL TNFα (means ± SEM, n = 5). RLU = relative luminescence unit.

appears realistic to suggest that these dicarbonyls could induce protective responses to toxic insults, especially under conditions of physiological stress. The concept that toxicants at low concentrations could stimulate protection whereas high concentrations are toxic is often referred to as hormesis.[20]

Although the mechanism by which dicarbonyls stimulate Nrf-2 activation and NF-κB inhibition remains obscure, one possibility may be interference with ubiquitination by dicarbonyls via modification of lysine and arginine residues.[21] Indeed, several reports have shown that glycation can interfere with ubiquitination.[22–24] The difference between Nrf-2 and NF-κB is that interference with ubiquitination of Nrf-2 is expected to result in its activation, whereas interference with ubiquitination of NF-κB is expected to result in the inhibition of its activation. In the inactive state, Nrf-2 is constantly ubiquitinated and degraded by the proteasome system. Activation of Nrf-2 prevents ubiquitination and results in the stabilization of Nrf-2. Indeed, it has been shown that interference with the proteasomal breakdown of Nrf-2 can result in the upregulation of some phase II enzymes.[25] NF-κB is activated by ubiquitination and degradation of the NF-κB repressor IκB which liberates active NF-κB.

In summary, our results show that the reactive dicarbonyl intermediates of glucose metabolism may have some beneficial effects by inducing the cell protective phase II enzymes and by interfering with the pro-inflammatory cell signaling mediated by NF-κB.

Conflict of Interest

The authors declare no conflicts of interest.

References

1. VLASSARA, H. & G. STRIKER. 2007. Glycotoxins in the diet promote diabetes and diabetic complications. Curr. Diab. Rep. **7:** 235–241.
2. SOMOZA, V. 2005. Five years of research on health risks and benefits of Maillard reaction products: an update. Mol. Nutr. Food Res. **49:** 663–672.
3. MASTROCOLA, D. *et al*. 2000. Interaction between Maillard reaction products and lipid oxidation in starch-based model systems. J. Sci. Food Agr. **80:** 684–690.
4. HOFMANN, T. *et al*. 2001. Determination of the molecular weight distribution of non-enzymatic browning products formed by roasting of glucose and glycine and studies on their effects on NADPH-cytochrome c-reductase and glutathione-S-transferase in Caco-2 cells. Nahrung. **45:** 189–194.
5. FAIST, V. *et al*. 2002. Influence of molecular weight fractions isolated from roasted malt on the enzyme activities of NADPH-cytochrome c-reductase and glutathione-S-transferase in Caco-2 cells. J. Agric. Food Chem. **50:** 602–606.
6. LINDENMEIER, M. *et al*. 2002. Structural and functional characterization of pronyl-lysine, a novel protein modification in bread crust melanoidins showing in vitro antioxidative and phase I/II enzyme modulating activity. J. Agric. Food Chem. **50:** 6997–7006.
7. WENZEL, E. *et al*. 2002. Effect of heat-treated proteins on selected parameters of the biotransformation system in the rat. Ann. Nutr. Metab. **46:** 9–16.
8. HOFMANN, T. *et al*. 2005. Pronyl-lysine-A novel protein modification in bread crust melanoidins showing *in vitro* antioxidative and phase I/II enzyme modulating activity. Ann. N.Y. Acad. Sci. **1043:** 887.

9. HAYES, J.D. et al. 2005. Glutathione transferases. Annu. Rev. Pharmacol. Toxicol. **45:** 51–88.
10. ZHANG, Y. & G.B. GORDON. 2004. A strategy for cancer prevention: stimulation of the Nrf2-ARE signaling pathway. Mol. Cancer Ther. **3:** 885–893.
11. VAN DAM, R.M. 2006. Coffee and type 2 diabetes: from beans to beta-cells. Nutr. Metab. Cardiovasc. Dis. **16:** 69–77.
12. RAHMAN, K. & G.M. LOWE. 2006. Garlic and cardiovascular disease: a critical review. J. Nutr. **136:** 736S–740S.
13. HAYES, J.D. & M. MCMAHON. 2001. Molecular basis for the contribution of the antioxidant responsive element to cancer chemoprevention. Cancer Lett. **174:** 103–113.
14. JEONG, W.S. et al. 2006. Nrf2: a potential molecular target for cancer chemoprevention by natural compounds. Antioxid. Redox. Signal. **8:** 99–106.
15. WANG, X.J. et al. 2006. Generation of a stable antioxidant response element-driven reporter gene cell line and its use to show redox-dependent activation of Nrf2 by cancer chemotherapeutic agents. Cancer Res. **66:** 10983–10994.
16. RIEDEL, C.U. et al. 2006. Anti-inflammatory effects of bifidobacteria by inhibition of LPS-induced NF-kappaB activation. World J. Gastroenterol. **12:** 3729–3735.
17. CAVIN, C. et al. 2007. Reduction in antioxidant defenses may contribute to ochratoxin A toxicity and carcinogenicity. Toxicol. Sci. **96:** 30–39.
18. NAKAMURA, Y. et al. 2003. Pivotal role of electrophilicity in glutathione S-transferase induction by tert-butylhydroquinone. Biochemistry **42:** 4300–4309.
19. THIMMULAPPA, R.K. et al. 2002. Identification of Nrf2-regulated genes induced by the chemopreventive agent sulforaphane by oligonucleotide microarray. Cancer Res. **62:** 5196–5203.
20. CALABRESE, E.J. & L.A. BALDWIN. 2003. Hormesis: the dose-response revolution. Annu. Rev. Pharmacol. Toxicol. **43:** 175–197.
21. HAGLUND, K. & I. DIKIC. 2005. Ubiquitylation and cell signaling. EMBO J. **24:** 3353–3359.
22. TAKIZAWA, N. et al. 1993. Inhibitory effect of nonenzymatic glycation on ubiquitination and ubiquitin-mediated degradation of lysozyme. Biochem. Biophys. Res. Commun. **192:** 700–706.
23. BULTEAU, A.L. et al. 2001. Proteasome inhibition in glyoxal-treated fibroblasts and resistance of glycated glucose-6-phosphate dehydrogenase to 20 S proteasome degradation in vitro. J. Biol. Chem. **276:** 45662–45668.
24. DU, J. et al. 2006. Methylglyoxal downregulates Raf-1 protein through a ubiquitination-mediated mechanism. Int. J. Biochem. Cell Biol. **38:** 1084–1091.
25. YAMAMOTO, N. et al. 2006. Proteasome inhibition induces glutathione synthesis and protects cells from oxidative stress: relevance to Parkinson disease. J. Biol. Chem. **282:** 4364–4372.

N-terminal Glycation of Proteins and Peptides in Foods and in Vivo

Evaluation of N-(2-Furoylmethyl)Valine in Acid Hydrolyzates of Human Hemoglobin

ILKA PENNDORF, CHANGHAO LI, UWE SCHWARZENBOLZ, AND THOMAS HENLE

Institute of Food Chemistry, Technische Universität Dresden, D-01062 Dresden, Germany

Specific determination of N-(2-furoylmethyl)valine (FM-Val) together with furosine in acid hydrolyzates of human hemoglobin of healthy volunteers ($n = 6$) and diabetic patients ($n = 14$) by means of reversed-phase HPLC with electrospray ionization–time-of-flight mass spectroscopy is reported. Whereas FM-Val is formed during acid hydrolysis of the N-terminal hemoglobin adduct N-fructosylvaline, furosine results from acid degradation of lysine residues glycated at the ε-amino group. Quantification was based on the use of synthesized isotopomers, namely N-[2-($^{13}C_6$)furoylmethyl]valine and N-ε-[2-($^{13}C_6$)furoylmethyl]lysine, thus enabling interference-free detection and calibration. Taking the conversion factors into account, the amount of N-terminally bound N-fructosylvaline in human hemoglobin was between 518 and 774 pmol/mg protein for healthy volunteers and between 586 and 1426 pmol/mg protein for diabetic patients. Derivatization at the side chain of peptide-bound lysine residues to N-ε-fructosyllysine was from 1156 to 1753 pmol/mg protein for healthy controls and from 1191 to 2409 pmol/mg protein for diabetics. For these patients, the amount of N-fructosylvaline showed good correlation with the values for HbA_{1c}. The significantly higher relative extent of glycation at the N terminus compared to side-chain glycation points to a specific and intraindividual capacity for enzymatic deglycation in human erythrocytes, which can be assessed using the proposed method.

Key words: glycation; Amadori compounds; hemoglobin HbA_{1c}; diabetes; N-fructosylvaline; furosine

Introduction

Analysis of protein-bound reaction products resulting from the complex Maillard reaction (also referred to as nonenzymatic browning or glycation) is of particular importance in understanding and controlling the impact of processing on the nutritional and functional properties of food proteins.[1] Furthermore, glycated amino acids in heated food and *in vivo* are discussed with respect to pathophysiological processes, such as diabetes or uremia.[2] From a chemical point of view, analysis of glycation compounds has mainly focused on the side chains of peptide-bound lysine and arginine residues as main targets for a derivatization by carbohydrates or their degradation products.[3] A putative glycation at the N-termini of peptides and proteins in foods or *in vivo*, respectively, has had limited attention to date. It may be noteworthy in this context that an N-terminally glycated hemoglobin variant (HbA_{1c}) was the first protein identified to result from the Maillard reaction *in vivo*.[4,5] In HbA_{1c}, predominantly the N-terminal valine of the β-chain is modified by glucose to give the Amadori product N-fructosylvaline (FIG. 1). Today, HbA_{1c} is widely used as a tool to monitor long-term glycation in diabetic patients.[6] Several methods for the assessment of HbA_{1c} have been published and are in use by clinical laboratories; these include ion-exchange chromatography, affinity chromatography, electrophoresis and immunoassays which generally express the extent of glycation as percentage of total hemoglobin.[7] Recently, a reference method for the measurement of HbA_{1c} has been suggested based on cleavage of hemoglobin to peptides by an endoproteinase and quantification of the ratio of the glycated and nonglycated N-terminal hexapeptides by HPLC and mass spectrometry.[8] Very little, however, is known about the extent of formation of individual glycation compounds present at the

---

Address for correspondence: T. Henle, Institute of Food Chemistry, Technische Universität Dresden, D-01062 Dresden, Germany. Fax: +49 351 463 34138.
Thomas.Henle@chemie.tu-dresden.de

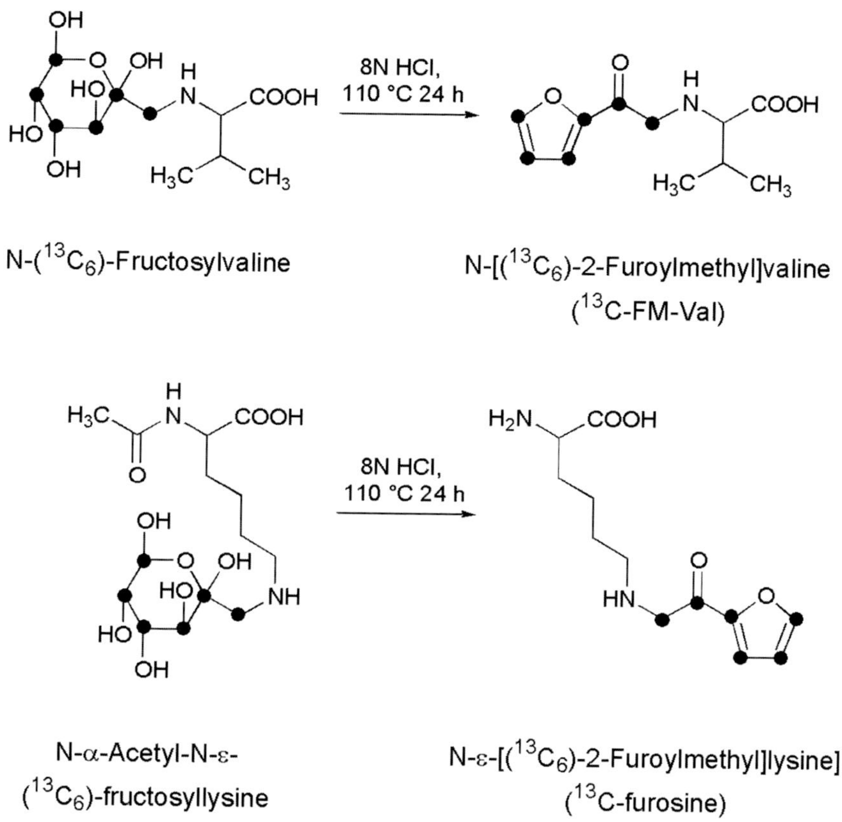

FIGURE 1. Synthesized compounds (see text for details).

N-termini of hemoglobin or the side chains of lysine or arginine moieties. Direct quantification of the N-terminal-bound fructosylvaline has not been performed yet, which probably is because the Amadori compounds are degraded during acid hydrolysis and, therefore, cannot be determined directly using routine amino acid analysis. As proposed recently, information about the extent of the early Maillard reaction between the N-termini of peptides or proteins and reducing sugars can be obtained by measuring α-N-(2-furoylmethyl) amino acids (FMAAs).[9] FMAAs are formed together with the well-known lysine derivative furosine [ε-N-(2-furoylmethyl)lysine] during acid hydrolysis of Amadori products.[10] Using reversed-phase (RP)-HPLC with UV detection, FMAAs could be detected in acid hydrolyzates of hypoallergenic infant formulas. Taking the factors of conversion of the Amadori products to the corresponding FMAAs into account, modification of N-terminal amino acids by reducing carbohydrates in such peptide-containing foods was between 0.3 and 8.4%, which has to be considered within the discussion concerning the nutritional quality of peptide-containing foods.[9] The purpose of the present study was to develop an isotope dilution assay for the sensitive quantification of N-(2-furoylmethyl)valine together with furosine as indicators for the quantification of the extent of the early Maillard reaction in human hemoglobin. After synthesis of N-[2-($^{13}C_6$)furoylmethyl]valine (^{13}C-FMVal) and N-ε-[2-($^{13}C_6$)furoylmethyl]lysine (^{13}C-furosine) (FIG. 1), quantification was achieved in acid hydrolysates by means of liquid chromatography (LC)–electrospray ionization (ESI)–time-of-flight (TOF) mass spectroscopy (MS).

Materials and Methods

Synthesis of Isotopomers

We synthesized ^{13}C-FMVal and ^{13}C-furosine starting from ^{13}C-D-glucose and valine or N-α-acetyllysine (all from Merck, Darmstadt, Germany), following the procedure described for the preparation of various FMAAs as reference material.[9] Briefly, the mixtures consisting of 2 mol/L amino acid and 4 mol/L glucose in methanol were heated under reflux for 8 h followed by acid hydrolysis with 7.95 mol/L HCl for 23 h at 110°C and semipreparative RP-HPLC. Identity of the isolated FMAAs was verified by LC–ESI–TOF-MS, nuclear magnetic resonance, and UV spectroscopy. The concentration of ^{13}C-FMVal and ^{13}C-furosine in solutions used as standard samples were calculated using the theoretical nitrogen content of the corresponding compound as measured by RP-HPLC with chemiluminescence nitrogen detection.[9]

Sample Preparation

Blood samples of 14 diabetic patients (HbA_{1c} between 5.1 and 12.1%) and six healthy controls were drawn into vacuum tubes containing EDTA as an anticoagulant. Erythrocytes were isolated as previously published.[11] For lysis of erythrocytes, 1.5 mL erythrocyte isolate was mixed with 4.5 mL of a hypotonic lysis buffer containing 5 mol/L 4-(2-hydroxyethyl)-piperazine-1-propanesulfonic acid, pH = 7.5. After storage (4°C, 10 min), the erythrocyte lysate was centrifuged (2000 g, 4°C, 10 min) and the membrane pellet was discarded. Determination of hemoglobin concentration was performed using the alkaline hematin D-575 method as described previously.[11] We mixed 2 mL of lysed erythrocytes with 40 mL acetone containing 2% concentrated HCl. After incubation for 10 min at −20°C, globin was obtained by centrifugation. Precipitated globin was washed twice with 40 mL acetone and evaporated to dryness. The dried protein (40 mg) was hydrolyzed in the presence of 4 mL of 7.95 mol/L HCl for 23 h at 110°C. A defined amount of isotopically labeled standards was added to a 1.0 mL aliquot of the hydrolyzate, and the mixture was evaporated to dryness and reconstituted in 500 μL of 0.2 mol/L HCl. After membrane (0.2 μm) filtration, 50 μL was subjected to LC–ESI–TOF-MS.

Liquid Chromatography–Electrospray Ionization–Time-of-Flight Mass Spectrometry

An Agilent 1100 series HPLC system (Agilent Technologies, Palo Alto, CA) to a PerSeptive Biosystems Mariner TOF-MS equipped with an ESI source working in the positive mode (Applied Biosystems, Stafford, TX) was used. The column was an Eurospher 100-C18 column (5 μm, 4.6 mm × 250 mm; Knauer, Berlin, Germany) protected by a guard column (8 × 5 mm) containing the same material. The injection volume was 50 μL, the flow rate was set at 0.7 mL/min, and the temperature was set at 25°C. A gradient was applied with water (solvent A) and methanol (solvent B), each containing 2% formic acid. Gradient elution was as follows: 0% solvent B for 10 min, 0% to 2% solvent B in 5 min, 2% to 6% solvent B in 19 min, 6% to 80% solvent B in 11 min, 80% solvent B for 5 min, 80% to 0% solvent B in 5 min, 0% solvent B for 10 min. The detection wavelength was 280 nm. MS conditions were as described.[9]

Results and Discussion

The purpose of our study was to establish a chromatographic assay for the quantification of the extent of "early" N-terminal glycation of hemoglobin. As direct determination of the Amadori product N-fructosylvaline after acid hydrolysis is not possible, we decided to use the corresponding hydrolysis product, namely α-N-(2-furoylmethyl)valine (FM-Val) as a suitable parameter. FMAAs are known as characteristic degradation products of amino ketoses.[12] The determination of furosine, which is formed during acid hydrolysis of N-ε-fructosyllysine, is a widely used tool to assess the extent of the Maillard reaction in protein-containing foods.[13–15] Prerequisite for the use of FMAAs as indicators for the early Maillard reaction is their formation during acid hydrolysis in constant and reproducible amounts. Recently, we could demonstrate that 6.3% FM-Val is formed from fructosylvaline during hydrolysis in the presence of 7.95 mol/L HCl, for 23 h at 110°C. Direct quantification of FM-Val in acid hydrolyzates of hemoglobin by RP-HPLC with UV detection, however, was not possible because of numerous co-eluting substances (FIG. 2). Therefore, an isotope dilution assay was established. Sufficient amounts of ^{13}C-labeled FM-Val and ^{13}C-furosine were prepared to high purity by semipreparative RP-HPLC after incubation of the amino acids with ^{13}C-glucose and subsequent acid hydrolysis (FIG. 1). Using RP-HPLC coupled to ESI–TOF-MS, FM-Val could be detected and quantified in the chromatograms by extracted ion monitoring at a retention time of 27.5 min using the nominal m/z values [M+1] of 226 and 232 for FM-Val and ^{13}C-FMVal, respectively (FIG. 4). Furosine eluted at 5.6 min and was recorded at m/z values of 255 and 261 for furosine and ^{13}C-furosine, respectively (FIG. 3). Internal calibration was performed using the labeled compounds as references. Calibration was linear in the working range from 20 to 1200 pmol FMAAs per mg protein. The limits of quantification were not determined as they are of only little importance in the quantification of these FMAAs in blood samples. The reproducibility of the method was determined by repeated measurements of the blood samples. Standard deviation from different analysis ($n = 3$) was typically less than 5%.

Values for FM-Val in acid hydrolyzates of blood samples ranged from 32.6 to 48.7 pmol/mg protein

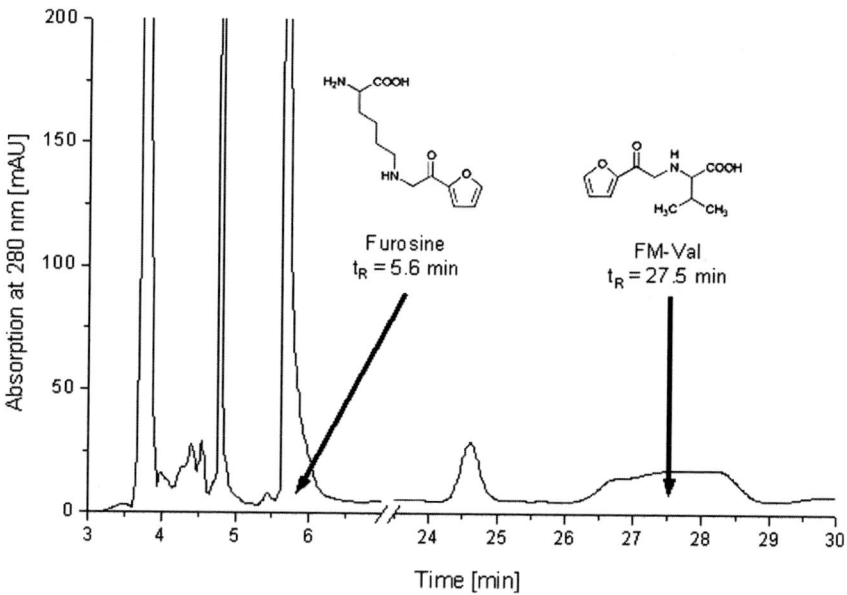

FIGURE 2. Reversed-phase HPLC with UV detection of an acid hydrolysate of a diabetic erythrocyte lysate showing elution position of furosine and FM-Val.

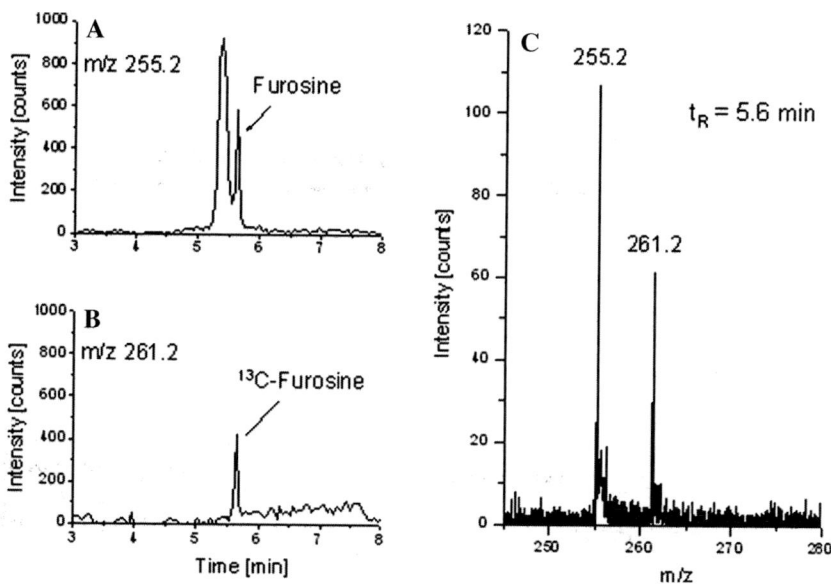

FIGURE 3. Determination of furosine in acid-hydrolyzed hemoglobin: Extracted ion profiles obtained for the chromatogram shown in FIGURE 2 at m/z values of **A)** 255.2 and **B)** 261.2. **C)** Mass spectra at a retention time of 5.6 min.

for healthy controls and from 36.9 to 89.7 pmol/mg protein for diabetic subjects. Taking the conversion factor of 15.9 into account,[9] it was possible to calculate the amount of N-terminal fructosylvaline present in the samples before hydrolysis (FIG. 5). Analogously, the amount of N-ε-fructosyllysine was calculated from the furosine values, which ranged from 578 to 868 pmol/mg protein in healthy controls and 596 to 1205 pmol/mg protein in diabetics, using a conversion factor of 2.0 (FIG. 5). The amount of N-fructosylvaline and of N-ε-fructosyllysine in the blood of diabetic patients should largely depend on glycemic control, which may explain why a broader range of data were found from diabetic samples compared to

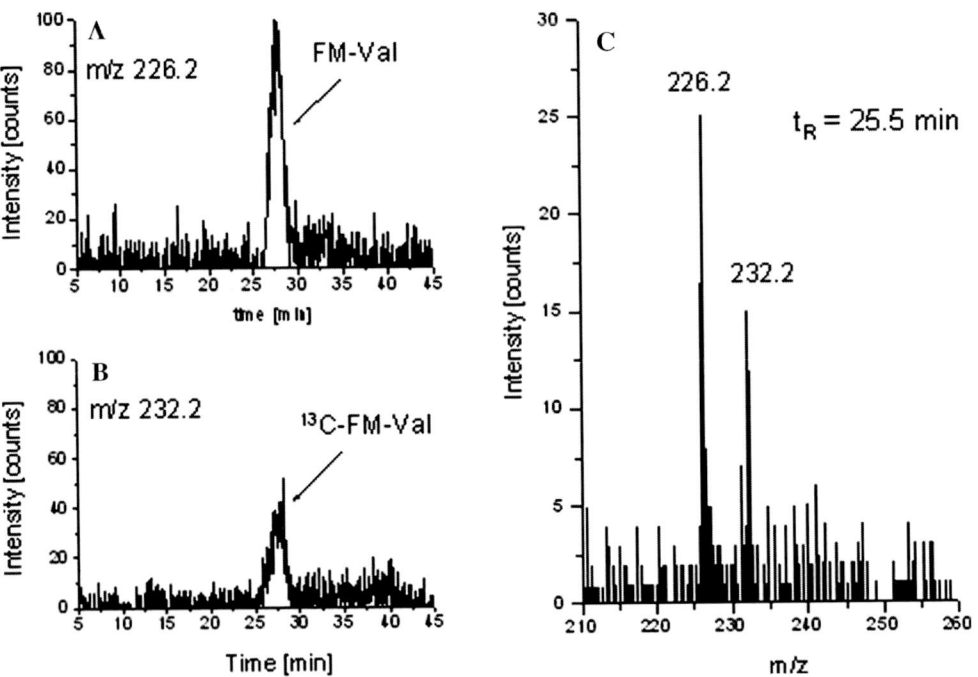

FIGURE 4. Determination of FM-Val in acid-hydrolyzed hemoglobin: Extracted ion profiles obtained for the chromatogram shown in FIGURE 2 at m/z values of **A)** 226.2 and **B)** 232.2. **C)** Mass spectra at a retention time of 25.5 min.

healthy controls. The amounts of fructosylvaline and, to a lesser extent, N-ε-fructosyllysine correlated well with the corresponding data for HbA_{1c} found for the diabetic samples (FIGS. 6 and 7).

The less pronounced correlation found for the lysine derivative and HbA_{1c} and the fact that the corresponding regression line does not pass through the origin (parameter $a = 533$ in the regression equation for N-ε-fructosyllysine compared to a value of 62 for N-fructosylvaline) may be explained by examining the activity of the deglycating enzyme fructosamine-3-kinase (FN3K) in erythrocytes. FN3K was found to be predominantly active towards a conversion of N-ε-fructosyllysine, whereas N-fructosylvaline was a poorly recognized substrate for the enzyme.[16] This means that high side-chain glycation is avoided by the activity of the enzyme. Furthermore, low amounts of N-ε-fructosyllysine accompanying high amounts of fructosylvaline, as was found for several samples, may reflect high activity of the deglycating enzyme in the erythrocytes. Analogously, relatively high side-chain glycation together with low N-terminal modification may be an indicator of impaired deglycation. These observations point to intraindividual differences in deglycating capacities, which may represent a constitutional factor to be taken into account in diabetes control.

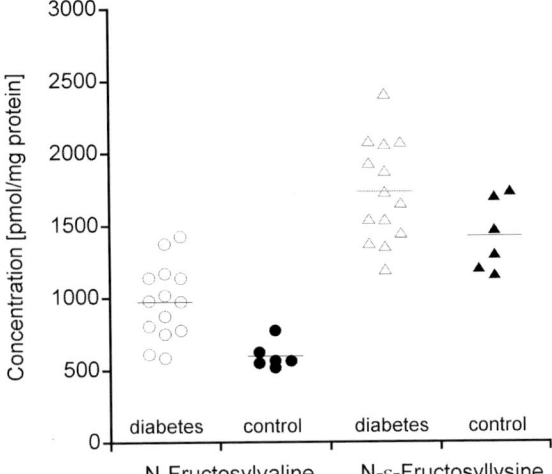

FIGURE 5. N-Fructosylvaline and N-ε-fructosyllysine in blood samples of diabetic patients ($n = 14$) and healthy controls ($n = 6$).

In summary, to the best of our knowledge, this is the first report on the direct and simultaneous quantification of the N-terminally bound Amadori product N-fructosylvaline in human hemoglobin as the corresponding FM-Val together with furosine using RP-HPLC with mass spectrometric detection. This

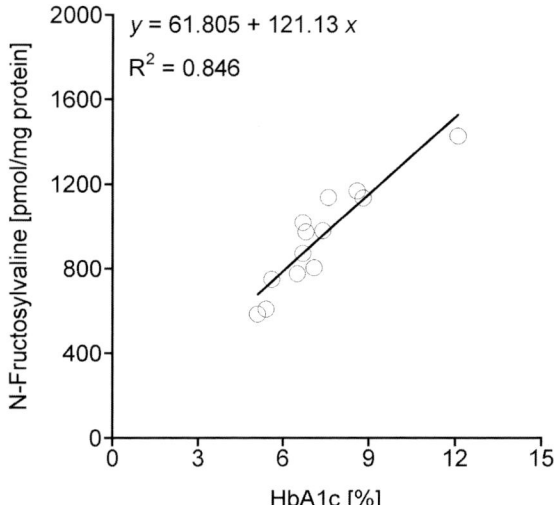

FIGURE 6. Correlation between hemoglobin HbA1c and N-fructosylvaline in blood samples of diabetic patients ($n = 14$).

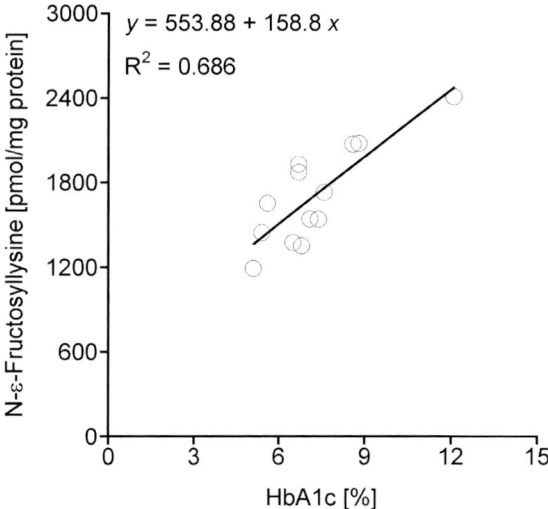

FIGURE 7. Correlation between hemoglobin HbA1c and N-ε-fructosyllysine in blood samples of diabetic patients ($n = 14$).

method may serve as an interesting alternative for the evaluation of an overall early glycation of hemoglobin in clinical diabetes management. Furthermore, as the method allows the simultaneous determination of N-terminal and side-chain glycation, this technique may be of interest for investigations of glycation and deglycation mechanisms in erythrocytes.

Acknowledgments

This paper is dedicated to Prof. Dr. H. Klostermeyer on the occasion of his 75th birthday. We thank Dr. K. Herbrig, Nephrology, Department of Internal Medicine III, Technische Universität Dresden, for acquiring the blood samples.

Conflict of Interest

The authors declare no conflict of interest.

References

1. FRIEDMAN, M. 1996. Food browning and ist prevention: an overview. J. Agric. Food Chem. **44:** 631–653.
2. HENLE, T. & T. MIYATA. 2003. Advanced glycation end products in uremia. Adv. Ren. Repl. Ther. **10:** 321–331.
3. HENLE, T. 2005. Protein-bound advanced glycation end-products (AGEs) as bioactive amino acid derivatives in foods. Amino Acids **29:** 313–322.
4. RAHBAR, S., O. BLUMENFELD & H. RANNEY. 1969. Hemoglobin studies of the unusual hemoglobin in patients with diabetes mellitus. Biochem. Biophys. Res. Comm. **36:** 838–843.
5. RAHBAR, S. 2005. The discovery of glycated hemoglobin: a major event in the study of nonenzymatic chemistry in biological systems. Ann. N.Y. Acad. Sci. **1043:** 9–19.
6. JEFFCOATE, S.L. 2003. Diabetes control and complications: the role of glycated haemoglobin, 25 years on. Diab. Med. **21:** 657–665.
7. JOHN, W.G. 2003. Haemoglobin A_{1c}: analysis and standardization. Clin. Chem. Lab. Med. **41:** 1199–1212.
8. JEPPSON, J.-O., U. KOBOLD, J. BARR, et al. 2002. Approved IFCC reference method for the measurement of HbA_{1c} in human blood. Clin. Chem. Lab. Med. **40:** 78–89.
9. PENNDORF, I., D. BIEDERMANN, S.V. MAURER & T. HENLE. 2007. Studies on N-terminal glycation of peptides in hypoallergenic infant formulas: quantification of α-N-(2-furoylmethyl) amino acids. J. Agric. Food Chem. **55:** 723–727.
10. DOLORES DEL CASTILLO, M., N. CORZO & A. OLANO. 1999. Early stages of Maillard reaction in dehydrated orange juice. J. Agric. Food Chem. **47:** 4388–4390.
11. KRAUSE, R., A. OEHME, K. WOLF & T. HENLE. 2006. A convenient HPLC assay for the determination of fructosamine-3–kinase activity in erythrocytes. Anal. Bioanal. Chem. **386:** 2019–2025.
12. HEYNS, K., J. HEUKESHOVEN & K.H. BROSE. 1968. Degradation of fructose amino acids to N-(2-furoylmethyl)amino acids. Intermediates in browning reactions. Angew. Chemie Int. Ed. Engl. **7:** 528–529.
13. FINOT, P.A., R. BRICOUT, R. VIANI & J. MAURON. 1968. Identification of a new lysine derivative obtained upon hydrolysis of heated milk. Experientia **24:** 1097–1099.
14. ERBERSDOBLER, H.F. & V. SOMOZA. 2007. Forty years of furosine - forty years of using Maillard reaction products as indicators of the nutritional quality of foods. Mol. Nutr. Food Res. **51:** 423–430.
15. HENLE, T., G. ZEHETNER & H. KLOSTERMEYER. 1995. Fast and sensitive determination of furosine. Z. Lebensm. Unters. Forsch. **200:** 235–238.
16. DELPIERRE, G. & E. VAN SCHAFTINGEN. 2003. Fructosamine 3-kinase, an enzyme involved in protein deglycation. Biochem. Soc. Trans. 2003. **31:** 1354–1357.

The Dicarbonyl Proteome

Proteins Susceptible to Dicarbonyl Glycation at Functional Sites in Health, Aging, and Disease

NAILA RABBANI AND PAUL J. THORNALLEY

Clinical Sciences Research Institute, Warwick Medical School, University of Warwick, Coventry CV2 2DX, United Kingdom

Reactive, physiological, dicarbonyl, glycating agents, glyoxal and methylglyoxal, are arginine-directed glycating agents forming mainly hydroimidazolone residues. Arginine residues have high-frequency occurrence in sites of protein–protein, enzyme substrate and protein–nucleotide binding sites. There is emerging evidence that functionally important arginine residues in proteins are often activated toward dicarbonyl glycation—leading to functional impairment. When uncontrolled, this is associated with aging, degenerative diseases, and metabolic disorders where dicarbonyl glycation may be viewed as damage to the proteome. The glyoxalase system, particularly glyoxalase 1, is the vanguard against dicarbonyl glycation in physiological systems. Functional regulation of glyoxalase 1 suggests a role for dicarbonyl glycation in cell signaling. Although extents of modification are usually low, the dicarbonyl proteome is a critical feature of the impact of glycation on physiological function—particularly in mitochondrial dysfunction, vascular disease, and potentially in disorders of lipoprotein metabolism.

Key words: glycation; glyoxal; methylglyoxal; glyoxalase; arginine; hydroimidazolone; bioinformatics

Introduction

Glyoxal and methylglyoxal are potent arginine-directed glycating agents. They react with proteins to form major and important advanced glycation end product (AGE) residues in cellular and extracellular proteins. Important AGEs quantitatively are hydroimidazolones derived from arginine residues modified by glyoxal and methylglyoxal—N_δ-(5-hydro-4-imidazolon-2-yl)ornithine (G-H1) and N_δ-(5-hydro-5-methyl-4-imidazolon-2-yl)-ornithine (MG-H1).[1,2] Glyoxal also forms a related ring-opened AGE, N_ω-carboxymethyl-arginine (CMA)[3] (FIG. 1). For methylglyoxal, there is a trace fluorescent adduct formed called argpyrimidine.[4,5] Hydroimidazolones have relatively moderate half-lives (2–6 weeks) and slow dynamic reversibility; therefore, protein content of hydroimidazolones can be decreased if the concentrations of the precursor α-oxoaldehydes are decreased.[1,6] Where hydroimidazolones accumulate with donor age, the increase is related to decreased expression and activity of the major enzyme involved in the metabolism of glyoxal and methylglyoxal, glyoxalase 1 (Glo1).[7,8] Proteins in mammalian tissues, plasma, and extracellular matrix *in vivo* have relatively high MG-H1 residue content (0.1–15 mmol/mol amino acid modified). Hydroimidazolone AGEs are found in highest concentrations in lens protein of elderly human subjects; MG-H1 residue content was 1–2% of total arginine residues.[9]

The physiological importance of protein glycation remains under intensive investigation. Particularly damaging effects are produced by covalent cross-linking of proteins, which confers resistance to proteolysis.[10] Protein modification is also damaging when amino acid residues are located in sites of protein–protein interaction, enzyme–substrate interaction, and protein–DNA interaction (for transcription factors). A bioinformatics analysis of receptor binding domains indicated that arginine residues have the highest probability of being located in such sites (19.6%).[11] The major modification of proteins by glyoxal and methylglyoxal is on arginine residues. Formation of hydroimidazolones causes structural distortion, loss of side chain

---

Address for correspondence: Naila Rabbani, Clinical Sciences Research Institute, Warwick Medical School, University of Warwick, Coventry CV2 2DX, UK. Voice: +44-24 7696 8593; fax: +44-24 7696 8595.
N.Rabbani@warwick.ac.uk

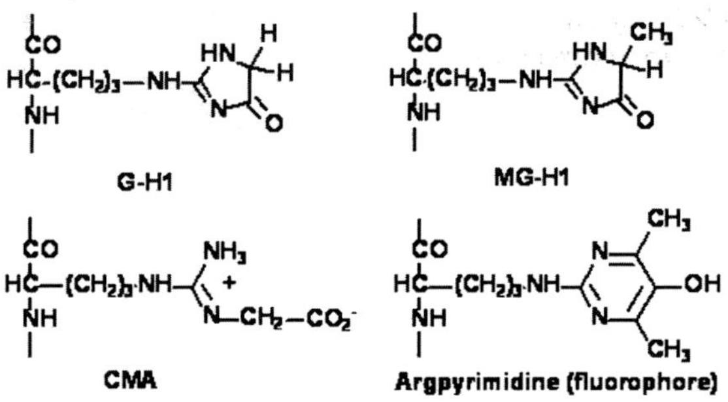

FIGURE 1. Arginine-derived advanced glycation end products formed from glyoxal and methylglyoxal in physiological systems.

charge, and functional impairment.[5] Two examples we studied recently are methylglyoxal modification of human serum albumin and vascular type IV collagen.

Modification of human serum albumin by methylglyoxal produced hotspot modification of arg-410 with hydroimidazolone MG-H1 residue formation. Modification of arg-410 by methylglyoxal was found in albumin glycated *in vivo*. Arg-410 is located in drug-binding site II and the active site of albumin-associated esterase activity. Hydroimidazolone formation at arg-410 inhibited drug binding and esterase activity. Molecular dynamics and modeling studies indicated that hydroimidazolone formation caused structural distortion leading to disruption of arginine-directed hydrogen bonding and loss of electrostatic interaction.[5] MG-H1 residue formation in albumin and other plasma proteins was increased in clinical diabetes and end-stage renal disease.[2,12]

Modification of vascular basement membrane type IV collagen by methylglyoxal formed MG-H1 residues at hotspot modification sites in Arg-Gly-Asp (RGD) and Gly-Phe-hydroxypro-Gly-Glu-Arg (GFOGER) integrin-binding sites of collagen, causing endothelial cell detachment, anoikis, and inhibition of angiogenesis. Endothelial cells incubated in model hyperglycemia *in vitro* and experimental diabetes *in vivo* produced the same modifications of vascular collagen, inducing similar responses. Increased methylglyoxal formed from triosephosphate degradation within endothelial cell crossed the plasma membrane and modified type IV collagen.[13] This may contribute to increased shedding of vascular endothelial cells and increased numbers of circulating endothelial cells in diabetes and uremia where plasma levels of methylglyoxal and other α-oxoaldehydes are abnormally high.[14,15] The number of circulating endothelial cells, indicative of damage to the endothelium, is prognostic for vascular disease.[16]

Physiological Regulation of Glyoxalase 1 and Glycation by Glyoxal and Methylglyoxal

The modifications of proteins by glyoxal and methylglyoxal, may be viewed as damage to the proteome in physiological systems. This is suppressed by the glyoxalase system.[17] There are examples, however, of regulation of Glo1 activity in physiological systems that suggests Glo1 substrates and glycation may have a functional role in cell signaling.

First, the high expression and activity of Glo1 in many tumor cells and some micro-organisms.[18–20] This is likely to be linked to the high glycolytic activity of many tumors and some micro-organisms.[21] A high Glo1 activity is required, concomitant with a high flux of methylglyoxal associated with high flux of triosephosphates and anaerobic glycolytic activity, to protect the proteome and genome from functional impairment and mutation, respectively. Second, activation of the receptor for advanced glycation end products (RAGE) by S100 proteins, which are found in increased plasma concentration in inflammatory disorders, was associated with decreased expression of Glo1. Induction of diabetes in wild-type mice decreased the expression of Glo1, whereas induction of diabetes in RAGE (−/−) mice did not.[22] Decreased Glo1 expression leads to increased protein glycation.[17] AGE residues in tissue protein were found colocalized with RAGE. This may be because of RAGE activation by S100 proteins, decreasing the local expression of Glo1 and thereby increasing the local formation of AGE residues. This may be part of an inflammatory response leading to labeling proteins with dicarbonyl-derived hydroimidazolone AGE residues and directing these proteins to the proteasome for destruction. Methylglyoxal modification of proteins is thought to target proteins for proteasomal destruction.[23] If correct, this will link the formation of glyoxal and methylglyoxal-derived AGEs to removal and destruction of proteins, at least for intracellular proteins—a re-interpretation of the role originally attributed to AGEs by Cerami as "signals" on extracellular proteins to target them for cellular uptake and proteolysis.[24] Third, Glo1 was shown to be phosphorylated, and glycation of proteins increased in apoptosis induced by tumor necrosis factor.[25] This, taken together with the overexpression of Glo1 in multidrug resistance in cancer chemotherapy, suggests glyoxal and methylglyoxal may have a role in apoptosis.

These examples of regulation of the Glo1 and cellular concentrations of methylglyoxal-derived AGEs are suggestive of a signaling role for methylglyoxal glycation of proteins and nucleotides in physiological systems. The role appears to include influences in malignant transformation,[26] cell death (apoptosis, anoikis [detachment-stimulated apoptosis], and eryptosis [erythrocyte-programmed cell death]),[13,27,28] and inflammation.[29] There may also be a role of dicarbonyl glycation in chaperone function of some proteins.[30] Functionally important arginine residues in lipoproteins suggest that dicarbonyl glycation of low-density lipoprotein and high-density lipoprotein may also be a target for dicarbonyl glycation and components of the dicarbonyl proteome.[31,32]

Concluding Remarks: Emerging Evidence that Glyoxalase 1 and Glycation by Glyoxal and Methylglyoxal Play a Critical Role in Disease and Aging

Glycation by glyoxal and methylglyoxal and the related influence of Glo1 are now emerging as playing a critical role in aging[33] and disease processes, such as vascular complications associated with diabetes,[2,14] renal failure,[12,34] Alzheimer's disease,[35,36] and tumorigenesis and multidrug resistance in cancer chemotherapy.[37]

Acknowledgments

We thank the Wellcome Trust, British Heart Foundation, Cancer Research UK, and Diabetes UK for support for our research.

Conflict of Interest

The authors declare no conflicts of interest.

References

1. THORNALLEY, P.J., S. BATTAH, N. AHMED, et al. 2003. Quantitative screening of advanced glycation endproducts in cellular and extracellular proteins by tandem mass spectrometry. Biochem. J. **375:** 581–592.
2. AHMED, N., R. BABAEI-JADIDI, S.K. HOWELL, et al. 2005. Degradation products of proteins damaged by glycation, oxidation and nitration in clinical type 1 diabetes. Diabetologia **48:** 1590–1603.
3. ODANI, H., K. IIJIMA, M. NAKATA, et al. 2001. Identification of Nw-carboxymethylarginine, a new advanced glycation endproduct in serum proteins of diabetic patients: possibly a new marker of aging and diabetes. Biochem. Biophys. Res. Comm. **285:** 1232–1236.
4. SHIPANOVA, I.N., M.A. GLOMB & R.H. NAGARAJ. 1997. Protein modification by methylglyoxal: chemical nature and synthetic mechanism of a major fluorescent adduct. Arch. Biochem. Biophys. **344:** 29–36.
5. AHMED, N., D. DOBLER, M. DEAN & P.J. THORNALLEY. 2005. Peptide mapping identifies hotspot site of modification in human serum albumin by methylglyoxal involved in ligand binding and esterase activity. J. Biol. Chem. **280:** 5724–5732.
6. AHMED, N., O.K. ARGIROV, H.S. MINHAS, et al. 2002. Assay of advanced glycation endproducts (AGEs): surveying AGEs by chromatographic assay with derivatisation by aminoquinolyl-N-hydroxysuccimidyl-carbamate and application to Nε-carboxymethyl-lysine- and Nε-(1-carboxyethyl)lysine-modified albumin Biochem. J. **364:** 1–14.
7. AHMED, N., P.J. THORNALLEY, J. DAWCZYNSKI, et al. 2003. Methylglyoxal-derived hydroimidazolone advanced glycation endproducts of human lens proteins. Invest. Ophthalmol. Vis. Sci. **44:** 5287–5292.
8. HAIK, J.R., G.M., T.W.C. LO & P.J. THORNALLEY. 1994. Methylglyoxal concentration and glyoxalase activities in human lens. Exp. Eye Res **59:** 497–500.
9. AHMED, M.U., E. BRINKMANN FRYE, T.P. DEGENHARDT, et al. 1997. Nε-(Carboxyethyl)lysine, a product of chemical modification of proteins by methylglyoxal, increases with age in human lens proteins. Biochem. J. **324:** 565–570.
10. DEGROOT, J., N. VERZIJL, M.J.G. WENTING-VAN WIJK, et al. 2001. Age-related decrease in susceptibility of human articular cartilage to matrix metalloproteinase-mediated degradation—The role of advanced glycation end products. Arthritis and Rheumatism **44:** 2562–2571.
11. GALLET, X., B. CHARLOTEAUX, A. THOMAS & R. BRASEUR. 2000. A fast method to predict protein interaction sites from sequences. J. Mol. Biol. **302:** 917–926.
12. AGALOU, S., N. AHMED, R. BABAEI-JADIDI, et al. 2005. Profound mishandling of protein glycation degradation products in uremia and dialysis. J. Am. Soc. Nephrol. **16:** 1471–1485.
13. DOBLER, D., N. AHMED, L.J. SONG, et al. 2006. Increased dicarbonyl metabolism in endothelial cells in hyperglycemia induces anoikis and impairs angiogenesis by RGD and GFOGER motif modification. Diabetes **55:** 1961–1969.
14. MCLELLAN, A.C., P.J. THORNALLEY, J. BENN & P.H. SONKSEN. 1994. The glyoxalase system in clinical diabetes mellitus and correlation with diabetic complications. Clin. Sci. **87:** 21–29.
15. MANN, V.M., B. TUCKER, P.J. THORNALLEY & A. DAWNAY. 1999. Elevated plasma methylglyoxal and glyoxal in uraemia: implications for advanced glycation endproduct formation. Kidney Internat. **55:** 2582.
16. SEGAL, M.S., A. BIHORAC & M. KOC. 2002. Circulating endothelial cells: tea leaves for renal disease. Am. J. Physiol. Renal Physiol. **283:** F11–F19.
17. THORNALLEY, P.J. 2003. Glyoxalase I—structure, function and a critical role in the enzymatic defence against glycation. Biochem. Soc. Trans. **31:** 1343–1348.
18. AYOUB, F.M., M.A. ZAMAN, P.J. THORNALLEY & J.R.W. MASTERS. 1993. Glyoxalase activity in human tumour cell lines *in vitro*. Anticancer Res. **13:** 151–156.

19. THORNALLEY, P.J. 1995. Advances in glyoxalase research. Crit. Rev. Oncol. Haematol. **20:** 99–128.
20. VANDER JAGT, D.L. 1990. D-Lactate production in erythrocytes infected with *Plasmodium falciparum*. Mol. Biochem. Parasitol. **42:** 277–284.
21. DILLS, W.L. 1993. Nutritional and physiological consequences of tumour glycolysis. Parasitology **107:** S177–S186.
22. BIERHAUS, A., S. STOYANOV, G.M. HAAG, *et al*. 2006. RAGE-deficiency reduces diabetes-associated impairment of glyoxalase-1 in neuronal cells. Diabetes **55:** A511.
23. DU, J., J. ZENG, X. OU, *et al*. 2006. Methylglyoxal downregulates Raf-1 protein through a ubiquitination-mediated mechanism. Internat. J. Biochem. Cell Biol. **38:** 1084–1091.
24. CERAMI, A. 1986. Aging of proteins and nucleic acids: what is the role of glucose? TIBS **11:** 311–314.
25. VAN HERREWEGHE, F., J. MAO, F.W.R. CHAPLEN, *et al*. 2002. Tumour necrosis factor-induced modulation of glyoxalase I activities through phosphorylation by PKA results in cell death and is accompanied by the formation of a specific methylglyoxal-derived AGE. Proc. Natl. Acad. Sci. USA **99:** 949–954.
26. THORNALLEY, P.J. 2003. Protecting the genome: defence against nucleotide glycation and emerging role of glyoxalase I over expression in multidrug resistance in cancer chemotherapy. Biochem. Soc. Trans. **31:** 1372–1377.
27. KANG, Y., L.G. EDWARDS & P.J. THORNALLEY. 1996. Effect of methylglyoxal on human leukaemia 60 cell growth: modification of DNA, G_1 growth arrest and induction of apoptosis. Leuk. Res. **20:** 397–405.
28. STRZINEK, R.A., V.E. SCHOLES & S.J. NORTON. 1972. Purification and Characterization of Liver Glyoxalase-I from Normal Mice and from Mice Bearing A Lymphosarcoma. Cancer Res. **32:** 2359–2364.
29. BIERHAUS, A., I. KONRADE, G.M. HAAG, *et al*. 2005. The receptor RAGE regulates glyoxalase-1 transcription, expression and activity. Diabetologia **48:** A90.
30. BISWAS, A., A. MILLER, T. OYA-ITO, *et al*. 2006. Effect of Site-Directed Mutagenesis of Methylglyoxal-Modifiable Arginine Residues on the Structure and Chaperone Function of Human aA-Crystallin. Biochemistry **45:** 4569–4577.
31. BOREN, J., U. EKSTROM, B. AGREN, *et al*. 2001. The molecular mechanism for the genetic disorder familial defective apolipoprotein B100. J. Biol. Chem. **276:** 9214–9218.
32. ROOSBEEK, S., B. VANLOO, N. DUVERGER, *et al*. 2001. Three arginine residues in apolipoprotein A-I are critical for activation of lecithin:cholesterol acyltransferase J. Lipid Res. **42:** 31–40.
33. MORCOS, M., X.L. DU, A.A.S.H. HUTTER, *et al*. 2005. Life extension in Caenorhabditis elegans by overexpression of glyoxalase I – The connection to protein damage by glycation, oxidation and nitration. Free Radic. Res. **39:** S43.
34. RABBANI, N., K. SEBEKOVA, K. SEBEKOVA, JR., *et al*. 2007. Protein glycation, oxidation and nitration free adduct accumulation after bilateral nephrectomy and ureteral ligation Kidney Int. In press.
35. CHEN, F., M.A. WOLLMER, F. HOERNDLI, *et al*. 2004. Role for glyoxalase I in Alzheimer's disease. Proc. Natl. Acad. Sci. USA **101:** 7687–7692.
36. AHMED, N., U. AHMED, P.J. THORNALLEY, *et al*. 2004. Protein glycation, oxidation and nitration marker residues and free adducts of cerebrospinal fluid in Alzheimer's disease and link to cognitive impairment. J. Neurochem. **92:** 255–263.
37. SAKAMOTO, H., T. MASHIMA, S. SATO, *et al*. 2001. Selective activation of apoptosis program by S-p-bromobenzylglutathione cyclopentyl diester in glyoxalase I-overexpressing human lung cancer cells. Clin. Cancer Res. **7:** 2513–2518.

Peroxyl Radicals Are Essential Reagents in the Oxidation Steps of the Maillard Reaction Leading to Generation of Advanced Glycation End Products

GERHARD SPITELLER

Institute of Organic Chemistry, University of Bayreuth, Bayreuth, Germany

Polyunsaturated fatty acids (PUFAs) are incorporated in all membranes of mammalian and plant cells and are extremely sensitive to oxygen. This property is used in nature to respond to any changes in cell membrane structure. In the first step of a response, lipid hydroperoxide molecules are generated. An increasing impact switches the enzymatic reaction to a nonenzymatic one by generation of lipid peroxyl radicals, which attack sugars by oxidation. In the course of these reactions, hydrogen peroxyl radicals are generated, resembling lipid peroxyl radicals in their reactivity. The reactions induced by these radicals are not under genetic control, they attack nearly all types of biological molecules (such as proteins, lipids, and sugars), and are responsible for the deleterious cell alterations in aging and age-related diseases (such as diabetes, Alzheimer's disease, or atherosclerosis) and probably also in autoimmune diseases, which involve sugars at the cell membranes. Lipid peroxidation processes are induced by heating fats, meat, and other nutritional products. The oxidation products generated by consumption of heated food cause damage of mammalian cells. The deleterious reactions can be partly reduced by consumption of plants and/or algae. These contain, among other well-known antioxidants, furan fatty acids, which are important scavengers of peroxyl radicals.

Key words: advanced glycation end products; aging; carboxymethyllysine; Maillard reaction; membrane; lipid peroxidation; peroxyl radicals

Introduction

Polyunsaturated fatty acids (PUFAs), such as linoleic, linolenic, or arachidonic acid, are characterized by the presence of one or several —CH=CH-CH$_2$-CH=CH— groups. The C-H bonds of the methylene groups between the double bonds are extremely sensitive to oxidation.[1] Nevertheless, PUFAs are incorporated in the outermost layer of mammalian and plant cells in the form of phospholipids where they are exposed to wounding (e.g., by an attack of microorganisms).

Mammals and plants have lipoxygenases (LOX) incorporated in their tissues[2]; these further reduce the energy necessary for oxidation of PUFAs. When a cell membrane is altered (stressed),[3,4] its ion channels are also altered, causing the influx of Ca^{2+}-ions. These contribute to the activation of phospholipases[5] that cleave the phosphoric ester bonds in phospholipids and liberate PUFAs. The free PUFAs are substrates for LOX,[5] and these transform the PUFAs to lipid hydroperoxide molecules (LOOHs). LOOHs are reduced to corresponding alcohols, which form protein complexes[6] that induce an adequate gene response. These physiological processes switch to nonenzymatic liquid peroxidation (LPO) when the level of free PUFAs exceeds a certain limit. LOX contain iron ions which catalyze a two-step electron transfer reaction.[7] The radicals derived from PUFAs within the enzyme never leave the complex before the reaction is completed. When the amount of liberated PUFAs exceeds a certain limit, the radicals produced within the LOX attack the enzyme, resulting in liberation of iron ions that react with previously generated LOOH molecules in a Fenton reaction by generation of LO• radicals and OH$^-$ anions. The LO• radicals attack adjacent PUFAs still bound in phospholipids and transform these to corresponding peroxyl radicals, thus inducing a chain reaction. The switch[3] from the enzymatic reaction, in which exclusively LOOH molecules are generated,

Address for correspondence: Prof. Gerhard Spiteller, Institute of Organic Chemistry, University of Bayreuth, Universitätsstr. 30; D95440 Bayreuth, Germany. Voice: +0049 921 552692; fax: +0049 921 552671.
Gerhard.spiteller@uni-bayreuth.de

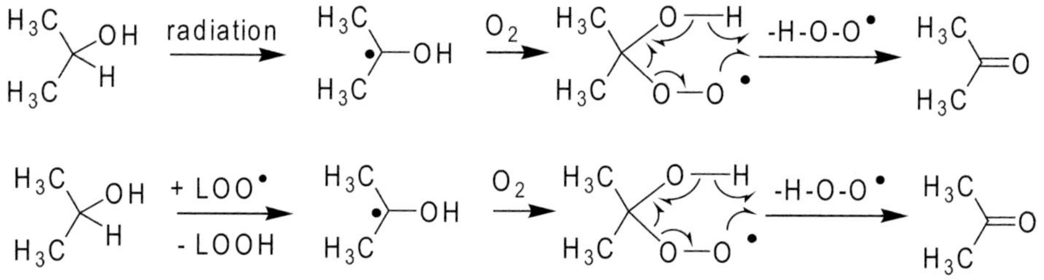

FIGURE 1. Generation of hydrogen peroxyl (HOO•) radicals either by radiation or by attack of lipid peroxyl (LOO•) radicals on secondary alcohols.

to the nonenzymatic reaction, in which lipid peroxyl (LOO•) radicals are produced, has not been sufficiently recognized in the past to produce species of very different reactivity. LOOH molecules are rather inactive, while LOO• radicals attack nearly all types of biological molecules, including primary and secondary alcohols, such as sugars. These radicals represent the agents inducing the oxidation reactions observed in the generation of advanced glycation end products (AGEs).

Oxidation of Alcoholic Groups by LOO• Radicals

About 30 years ago, Schulte-Frohlinde et al.[8] revealed the mechanism of acetone production by irradiation of isopropanol. In the first step of the reaction, the C-H bond of the secondary alcoholic group is cleaved to generate a carbon-centered radical (FIG. 1). This radical adds oxygen instantly to produce a peroxyl radical. The latter releases a hydrogen peroxyl (HOO•) radical and generates acetone (FIG. 1) in a five-membered hydrogen rearrangement reaction. Although the HOO• radical may decay to $O_2•^-$ and a proton, it is also able to attack another isopropanol molecule by hydrogen removal. Thus, two products are generated, acetone and hydrogen peroxide (H_2O_2). Each of the secondary CH(OH) groups of a sugar (and also their primary ones) are subjected to an identical attack by radiation; thus sugars are transformed to their oxo derivatives and H_2O_2.[9]

Identical reaction products are obtained when radiation peroxyl radicals derived from LPO of PUFAs are used as inducing reagents (FIG. 1). Moreover, radicals can be generated artificially when iron ions and H_2O_2 are added to a sugar solution because then •OH radicals are produced in a Fenton reaction.[10]

The reaction outlined above explains the generation of oxo derivatives of sugars but not their cleavage products, such as glyoxal, glycerine aldehyde, and erythrose or glycolic acid. Radical-induced hydrogen shift reactions occur not only via five- but also via six-membered, intramolecular, rearrangement reactions that are preferred for steric reasons. Such a six-membered hydrogen rearrangement involving the hydrogen of a hydroxylic group at the carbon adjacent to that carrying the peroxyl group leads to chain cleavage by formation of an aldehyde and a new carbon-centered radical (FIG. 2). The latter again adds oxygen, and the new peroxyl radical finally abstracts from another molecule, a hydrogen radical, to produce after hydrolysis an acid and H_2O_2 (FIG. 2).

Besides glyoxal, formaldehyde, glycerine aldehyde, and erythrose are also generated, depending on the site of attack by the LOO• radical.[11] Previously, H_2O_2 had been regarded to be a byproduct of the Maillard reaction.[12] In fact, it is a main product. Generation of H_2O_2 seems to be a much more important marker compound for the induction of deleterious processes than any of the known AGE markers.

Many of the aldehydes produced by oxidation of sugars were recognized to be involved in the generation of AGEs.[13–16] Likewise, peroxyl radicals are involved in the generation of N^ϵ-(carboxymethyl)lysine, which is started by formation of a Schiff base between a sugar molecule and the basic amino group of the lysyl residue of proteins (FIG. 3).[17] The oxidation step in the reaction is started by an enolization reaction. The hydrogen in position 4 of the endiol is activated more than the other hydrogens in the molecule and therefore prone to abstraction of a hydrogen radical by an attacking radical. It seems reasonable to assume that the generated radical reacts as other carbon-centered radicals do by addition of oxygen. The peroxyl radical attacks the double bond forming an endoperoxide, resembling analogous reactions from similar radicals formed from PUFAs.[1] The new carbon-centered radical adds oxygen, followed by a six-membered hydrogen shift, causing chain cleavage. The generated radical adds oxygen to form another peroxyl radical that picks up a hydrogen atom from another molecule. Finally, the resulting molecule is cleaved to two H_2O_2 molecules and N^ϵ-(carboxymethyl)lysine (FIG. 3).

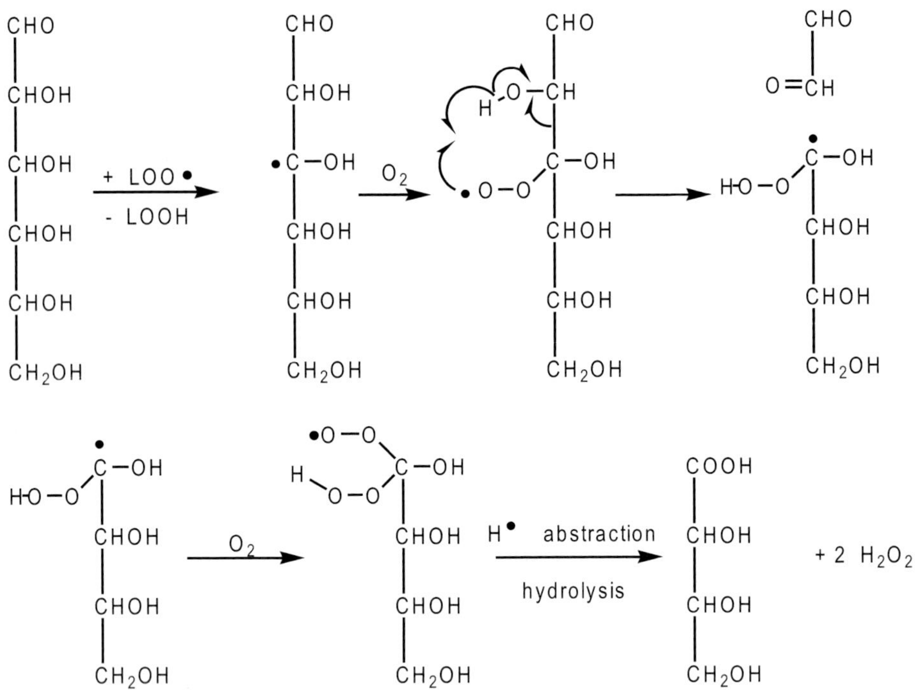

FIGURE 2. Generation of aldehydes and acids from sugar molecules by a six-membered hydrogen shift reaction induced by peroxyl radicals.

The generation of LOO• radicals is not under genetic control as cells do not have specific receptors for radicals. Instead, the high reactivity of HOO• and LOO• radicals allows them to attack not only alcoholic groups but also amino groups, for instance those in lysyl residues of proteins, all types of double bonds by epoxidation (e.g., those of sterols, terpenes, PUFAs, plasmalogens), phenolic groups (e.g., those of tyrosine and adrenaline), CH bonds of imidazole rings (such as in histidine), and many more types of bonds.[3] Consequently, the detected AGE products represent only a small fraction of all their reaction products. The spectrum of reaction products is expanded by considering that many of the primarily obtained oxidation products still contain activated C-H bonds and double bonds prone to further attack. Nevertheless, identification of easily detectible marker compounds, such as N^ϵ-(carboxymethyl)lysine or pentosidine,[16] is important because these marker compounds prove the induction of LPO reactions.

Peroxyl Radicals as Inducers of Aging and Age-related Diseases

Although each of the identified marker compounds of the Maillard reaction and AGEs is produced in extremely low amounts, which are probably insufficient to cause severe damage, the detection of pentosidine and N^ϵ-(carboxymethyl)lysine and other markers unquestionably prove the onset and progression of LPO reactions.

Enrichment of AGEs was first detected in diabetes[18,19] and later in related diseases, such as cataract formation[20,21]; it was also detected in brain tissues of Alzheimer's disease patients,[22] in atherosclerotic tissues,[23,24] and even in loss of bone mass.[25] Moreover, AGEs are produced in enriched amounts in aging.[26,27] Since sugar molecules easily suffer LPO-induced oxidation—as demonstrated above—and since they are localized at the outer site of cell membranes where they are responsible for immune reactions, it is expected that AGEs should also be enriched in autoimmune diseases (e.g., Parkinsons's disease).

Previously it was postulated that all these diseases are related to the generation of reactive oxygen species (ROS). ROS were assumed to be generated by a "leakage" process of mitochondria. Induction of ROS generation by mitochondria requires a preceding "stimulation" which is induced, for instance, by adding ionophores. Ionophores change the phospholipid layer of mitochondria, and I suspect that it is the generation of LOO• radicals, not a leakage process, that induces the generation of ROS. Another source of

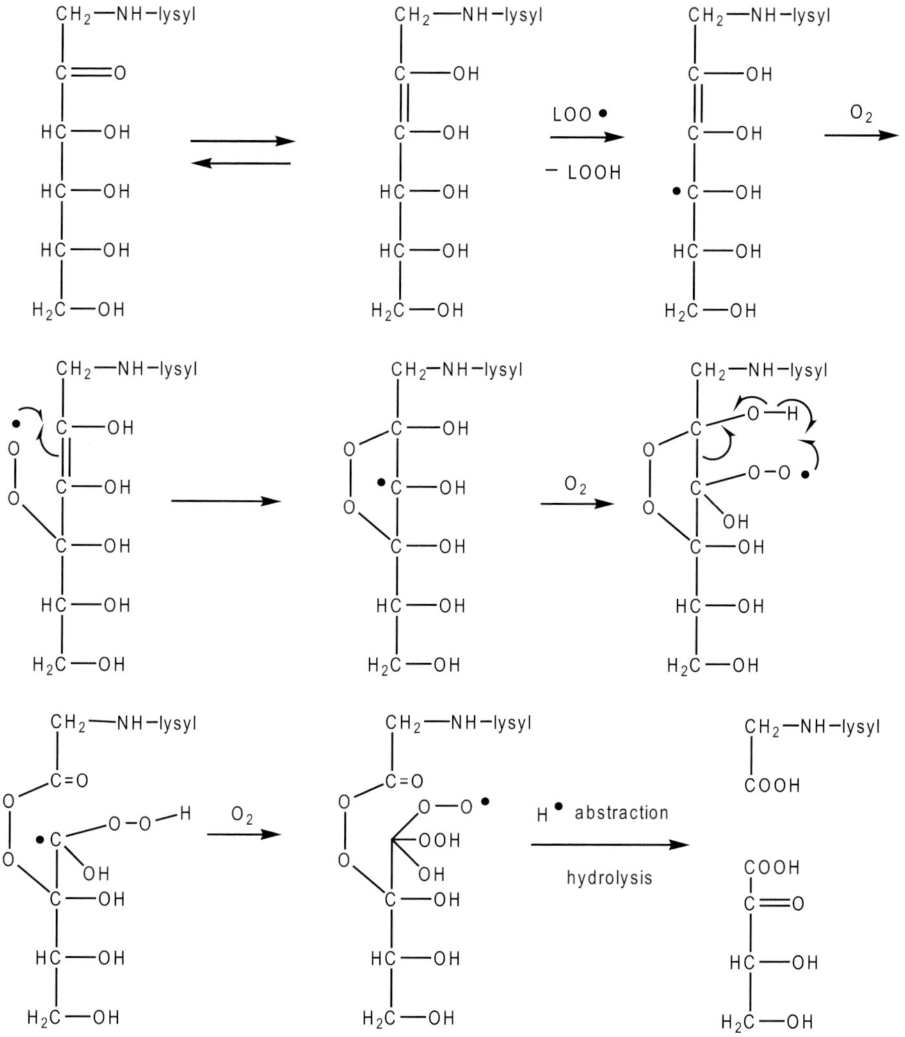

FIGURE 3. Generation of N^ε-(carboxymethyl)lysine.

ROS was seen in the activation of redox enzymes, such as $NADPH^+$ oxidase or catalase. These enzymes suffer deactivation, apparently like LOX by liberation of metal ions which react with LOOHs to form LOO• radicals. Therefore, I assume that all biological sources of radicals are based on changes in cell membrane structures.

Are We Able to Influence the Generation of Peroxyl Radicals?

LPO is induced by frying meat or fats, for instance, by preparation of chips in boiling oil that has been used for a long time. Although the generated peroxyl radicals react instantly with other molecules and are therefore not transferred to the body, the transformed products are partly toxic, for instance cholesterol linoleate is oxidized via peroxyl radicals to toxic oxidation products of cholesterol.[3] The toxic products are incorporated into our cells. LPO products injure the cells and start new LPO processes that spread like an infection from cell to cell,[3] except that the radicals are scavenged by antioxidants.

Many compounds, such as plasmalogens, are considered to be scavengers, although only their disappearance was recognized in contact with radicals. Radicals can not "disappear." If they react with a molecule by hydrogen abstraction, they generate from the molecule a new radical and the reaction is carried on. Therefore "disappearance" of a compound after reaction with a radical does not qualify it to be a radical scavenger. The

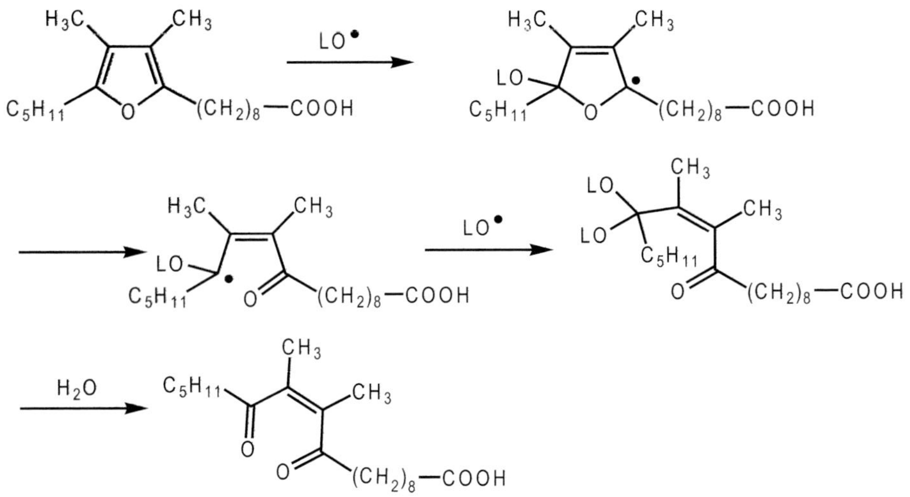

FIGURE 4. The action of furan fatty acids as scavengers of radicals.

chain reaction is only stopped if a radical has a sufficiently long lifetime to be able to react with a second radical.

Plants and algae are exposed to UV radiation, which induces generation of LOO• radicals. Therefore, plants and algae produce long-living radicals, such as vitamin E or flavonoids. Radicals produced from these compounds have an extended mesomeric system that stabilizes the radical and enhances their lifetime. Nevertheless, the consumption of flavonoids seems to have little effect because flavonoids are not adsorbed in the intestine and, therefore, are not incorporated into our tissues. As a consequence, I doubt that drinking red wine, which contains flavonoids, protects against radicals. Plants and especially algae produce furan fatty acids. These not only are perfect radical scavengers but also are incorporated, instead of PUFAs, in phospholipids (FIG. 4).[28]

Conclusions

It seems that any change in cell membrane structure induces degrading enzymes, inducing liberation of PUFAs and their oxidation to LOOH molecules. Increase of free PUFAs is combined with suicide of LOX and a switch from enzymatic to nonenzymatic LPO. The switch leads to the production of LOO• radicals. These attack proteins, sugars, and other lipids. Attacked sugar molecules release HOO• radicals, resembling LOO• radicals in their reactivity. Radical formation and radical action are not under genetic control. Radical-induced processes can only be stopped by scavengers. The lifetime of the formed radical intermediates determines the scavenger qualities of the radicals. Furan fatty acids, produced by plants and algae, are apparently some of the best scavengers of HOO• radicals.

Conflict of Interest

The author declares no conflicts of interest.

References

1. FRANKEL, E.N. 2005. Lipid Oxidation, 2nd ed. pp. 165–186. Oily Press. Bridgewater, UK.
2. YAMAMOTO, S., H. SUZUKI, N. UEDA, et al. 2004. Mammalian lipoxygenases. In Eicosanoids. P. Curtis-Prior, Ed.: 53–59. Wiley. Chichester, UK.
3. SPITELLER, G. 2006. Peroxyl radicals: inductors of neurodegenerative and other inflammatory diseases. Their origin and how they transform cholesterol, phospholipids, plasmalogens, polyunsaturated fatty acids, sugars, and proteins into deleterious products. Free Radic. Biol. Med. **41:** 362–387.
4. SPITELLER, G. 2003. The relation between changes in the cell wall, lipid peroxidation, proliferation senescence and cell death. Physiol. Plant. **119:** 5–18.
5. GALLIARD, T. 1970. Aspects of lipid metabolism in higher plants. III. Enzymic break down of lipids in potato tuber by phospholipids- galactolipid-acyl hydrolase activities and by lipoxygenase. Phytochemistry **9:** 1725–1734.
6. NAGY, L. & A. SZANTO. 2005. Roles for lipid-activated transcription factor in atherosclerosis. Mol. Nutr. Food Res. **49:** 1072–1074.
7. DE GROOT, J.J.M.C., G.A. VELDINK, J.F.G. VLIEGENTHART, et al. 1975. Demonstration by EPR spectroscopy of the functional role of iron in soybean lipoxygenase-1. Biochem. Biophys. Acta **377:** 71–79.

8. BOTHE, E., G. BEHRENDS & D. SCHULTE-FROHLINDE. 1977. Mechanism of the first order decay of 2-hydroxy-propyl-2-peroxyl radicals. Formation in aqueous solution. Z. Naturforsch. **32B:** 886–889.
9. BOTHE, E., M.N. SCHUCHMANN & D. SCHULTE-FROHLINDE, et al. 1978. $HO_2^\bullet$ Elimination from α-hydroxyalkylperoxyl radicals in aqueous solution. Photochem. Photobiol. **28:** 639–644.
10. JAHN, M. & G. SPITELLER; 1996. Oxidation of D-(-)-ribose with H_2O_2 and lipidhydroperoxides. Z. Naturforsch. **51C:** 870–876.
11. THORNALLEY, P., S. WOLFF, J. CRABBE, et al. 1984. The autoxidation of glyceraldehyde and other simple monosaccharides under physiological conditions catalysed by buffer ions. Biochem. Biophys. Acta **797:** 276–287.
12. CHO, S.-J., G. ROMAN, F. YEBOAH, et al. 2007. The road to advanced glycation end products: A mechanistic perspective. Chem. Med. Chemistry **14:** 1653–1671.
13. GLOMB, M.A. & V.M. MONNIER. 1995. Mechanism of protein modification by glyoxal and glycolaldehyde, reactive intermediates of the Maillard reaction. J. Biol. Chem. **270:** 10017–10026.
14. WELLS-KNECHT, K.J., D.V. ZYZAK, J.E. LITCHFIELD, et al. 1995. Mechanism of autoxidative glycosylation: Identification of glyoxal and arabinose as intermediates in the autoxidative modification of proteins by glucose. Biochemistry **34:** 3702–3709.
15. AHMED, M. U., S.R. THORPE & J.W. BAYNES. 1986. Identification of N^ε-carboxymethyllysine as a degradation product of fructoselysine in glycated protein. J. Biol. Chem. **261:** 4889–4894.
16. GRANDHEE, S. K. & V.M. MONNIER. 1991. Mechanism of formation of the Maillard protein cross-link pentosidine. J. Biol. Chem. **266:** 11649–11653.
17. LEDL, F. & E. SCHLEICHER. 1990. The Maillard reaction in food and in the human body—new results in chemistry, biochemistry and medicine. Angew. Chem. **102:** 597–626.
18. DAY, J. F., S.R. THORPE & J.W. BAYNES. 1979. Nonenzymatically glucosylated albumin. In vitro preparation and isolation from normal human serum. J. Biol. Chem. **254:** 595–597.
19. DOLHOFER, R. & O.H. WIELAND. 1979. Glycosylation of serum albumin: elevated glycosyl-albumin in diabetic patients. FEBS Lett. **103:** 282–286.
20. MONNIER, V. M. & A. CERAMI. 1981. Nonenzymatic browning in vivo: possible process for aging of long-lived proteins. Science **211:** 491–493.
21. CHENG, R., Q. FENG, O.K. ARGIROV, et al. 2004. Structure elucidation of a novel yellow chromophore from human lens protein. J. Biol. Chem. **279:** 45441–45449.
22. PRAKASH REDDY, V., M.E. OBRENOVICH, C.S. ATWOOD, et al. 2002. Involvment of Maillard reactions in Alzheimer disease. Neurotox. Res. **4:** 191–209.
23. Zhang M., A.L. KHO, N.A. KUMAR, et al. 2006. Glycated Proteins stimulate reactive oxygen species production in cardiac mycocytes. Involvement of Nox2 (gp91phox)-containing NADPH oxidase. Circulations **113:** 1235–1243.
24. SCHLEICHER, E. & U. FRIESS. 2007. Oxidative stress, AGE, and atherosclerosis. Kidney Int. **72**(S106): S17–S26.
25. ODETTI, P., S. ROSSI, F. MONACELLI, et al. 2005. Advanced glycation end products and bone loss during aging. Ann. N.Y. Acad. Sci. **1043:** 710–717.
26. LAMBERT, A.J., M. PORTERO-OTIN, R. PAMPLONA, et al. 2004. Effect of ageing and caloric restriction on specific markers of protein oxidative damage and membrane peroxidizability in rat liver mitochondria. Mech. Ageing Dev. **125:** 529–538.
27. BRINKMANN FRYE, E., T.D. DEGENHARDT, S.R. THORPE, et al. 1998. Role of the Maillard reaction in aging of tissue proteins. J. Biol. Chem. **273:** 18714–18719.
28. SPITELLER, G. 2005. Furan fatty acids: occurrence, synthesis and reactions. Are furan fatty acids responsible for the cardio-protective effects of a fish diet? Lipids **40:** 755–771.

Application of Mass Spectrometry for the Detection of Glycation and Oxidation Products in Milk Proteins

JASMIN MELTRETTER AND MONIKA PISCHETSRIEDER

Department of Chemistry and Pharmacy, University of Erlangen-Nuremberg, 91052 Erlangen, Germany

Protein mass spectometry techniques, such as electrospray ionization mass spectrometry or matrix-assisted laser desorption/ionization time-of-flight mass spectrometry (MALDI-TOF-MS), are effective methods to screen for protein modifications derived from the Maillard reaction. The analysis of the intact proteins reveals the major modification, most commonly the Amadori product, whereas partial enzymatic hydrolysis prior to mass spectrometry additionally allows the detection of minor adducts. Therefore, a mass spectrometric method was developed for the analysis of whey protein modifications occurring during heat treatment. The two main whey proteins, α-lactalbumin and β-lactoglobulin, were incubated with lactose in a milk model and modifications were recorded using MALDI-TOF-MS. The analysis of the intact proteins revealed protein species with 0–4 lactulosyl residues. Partial enzymatic hydrolysis with endoproteinase AspN prior to mass spectrometric analysis enabled the detection of further modifications and their localization in the amino acid sequence. Detected modifications were lactulosyllysine, N^ε-(carboxymethyl)lysine, lysine aldehyde, methionine sulfoxide, cyclization of N-terminal glutamic acid to a pyrrolidone, and oxidation of cysteine or tryptophan. Protein modifications in heated milk and commercially available dairy products can be analyzed after the separation of the milk proteins using one-dimensional SDS-PAGE.

Key words: advanced glycation end products; α-lactalbumin; β-lactoglobulin; Maillard reaction; matrix-assisted laser desorption/ionization time-of-flight mass spectrometry; milk products; oxidation

Dietary Advanced Glycation End Products: From Physiology to Analytical Chemistry

One of the hot topics these days in the field of Maillard reaction is the question of the physiological role of dietary advanced glycation end products (AGEs).[1] The results of intervention studies with humans and animals are still controversial. Several recent studies suggest that a diet rich in Maillard products promotes low-grade inflammatory reactions in animals as well as in humans.[2] Furthermore, it is well established that Maillard products, which are formed during the heat treatment of foodstuffs, such as heterocyclic aromatic amines or acrylamide, have mutagenic and probably also cancerogenic properties.[3] On the other hand, it has often been documented that Maillard products have strong antioxidative activity *in vitro*, suggesting that these compounds also improve the antioxidative status *in vivo* after resorption.[4,5] Finally, there is evidence that AGEs are only poorly absorbed and, after ingestion, readily excreted by the urine so that their impact on human health is only limited.[6,7] The high diversity of results on the nutritional properties of Maillard products clearly indicates that it is crucial to differentiate among this heterogeneous group of compounds and relate physiological effects to substances with defined chemical structures. With different food, we ingest Maillard products of very different chemical structures which have, most likely, very different nutritional properties, ranging from toxic to health-promoting effects. For addressing the question of the health impact of Maillard products, one primary goal of research will be to systematically analyze which products are formed in which foodstuffs during processing.

Address for correspondence: Monika Pischetsrieder, Department of Chemistry and Pharmacy, University of Erlangen-Nuremberg, Schuhstr. 19, 91052 Erlangen, Germany. Voice: +49 9131 8524102; fax: +49 9131 8522587.

pischetsrieder@lmchemie.uni-erlangen.de

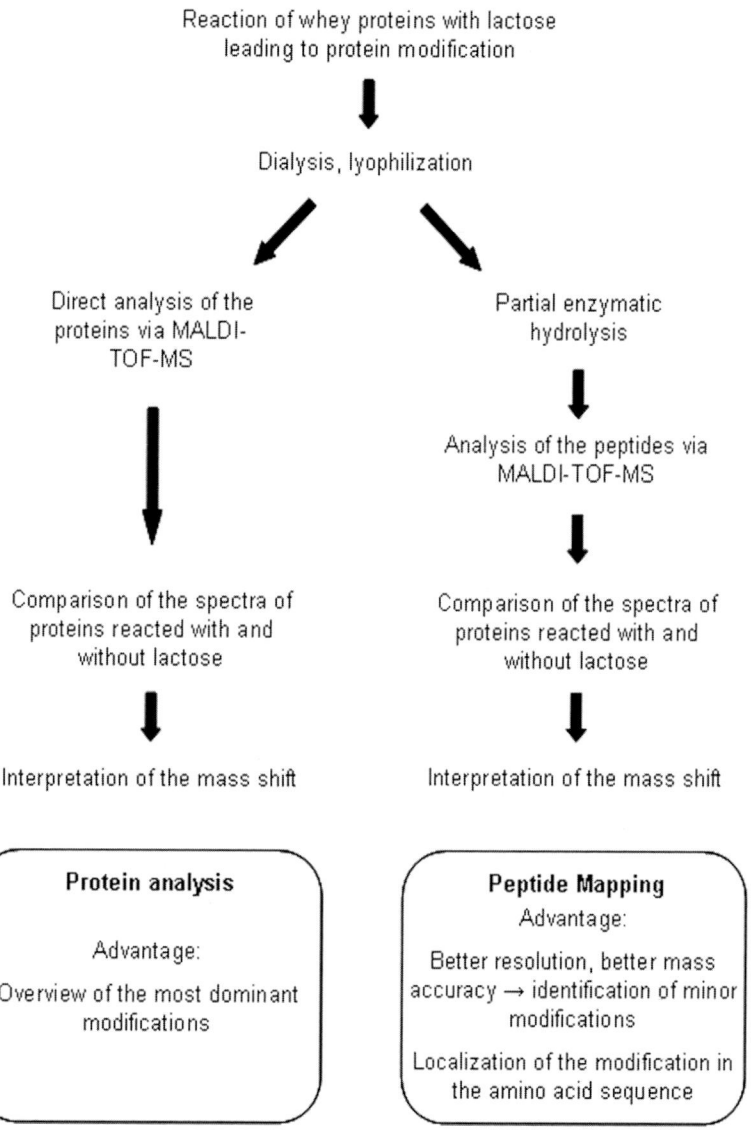

FIGURE 1. Principle and advantages of the presented matrix-assisted laser desorption/ionization time-of-flight mass spectrometry (MALDI-TOF-MS) method for the identification and localization of modifications of whey proteins in heated milk models.

Mass Spectrometry—An Effective Method for the Systematical Analysis of Protein-bound AGEs

During food processing and *in vivo*, proteins are modified, for example by glycation, glycoxidation, and oxidation reactions. Among others, lysine, arginine, methionine, cysteine, or tryptophan can be affected. As a consequence, not only nutritional but also biochemical and technological protein properties can be dramatically changed. Therefore, it is important to identify and monitor the major protein modifications that occur during food processing and *in vivo*. During the last decade, mass spectrometry has become an important method for the analysis of protein adducts. The advantages of mass spectrometry are (i) the potential to analyze intact proteins without major sample preparation, thus avoiding artifact formation, and (ii) the simultaneous detection of all modification types independent of their structures. For mass spectrometric analysis of protein glycation, matrix-assisted laser desorption/ionization mass spectrometry with time-of-flight detection (MALDI-TOF-MS) or electrospray ionization mass spectrometry (ESI-MS) have been successfully applied.[8,9] With these methods, it is possible to detect Amadori adducts of proteins glycated in models, in milk samples, or *in vivo* in diabetic patients.[10,11] After partial enzymatic hydrolysis, the modification site can be determined (peptide mapping; see FIG. 1). For this purpose, the endoproteases GluC, AspN, or trypsin are most commonly used.[12,13] The structures of the adducts are identified by the mass difference between the new signal(s) and the parent signal, whereas the binding site can be assigned by protein database analysis. Furthermore, partial hydrolysis prior to mass spectrometry yields a better mass resolution so that glycation, glycoxidation, and oxidation products, which are formed as minor products, can be detected in addition to the Amadori products.[14]

MALDI-TOF-MS Analysis of Intact Whey Proteins in a Heated Milk Model

In a recent study, MALDI-TOF-MS with and without peptide mapping was applied for the systematical screening of whey protein modifications in heated milk models as well as in dairy products.[15] The analysis of the intact proteins revealed four new signals in the spectra of α-lactalbumin and β-lactoglobulin after incubation with lactose (data not shown).[15] The mass differences of 324 Da or multiples thereof correspond with the addition of up to four lactose molecules to the proteins, leading to the Amadori product lactulosyllysine. Lactulosyllysine has been identified as the most common modification in milk proteins[16] and has been detected in a variety of dairy products.[17,18] Further modifications could not be detected because of insufficient resolution of the MS analysis in the mass range of the intact proteins.

MALDI-TOF-MS Analysis of β-Lactoglobulin in a Heated Milk Model after Peptide Mapping

To achieve a better resolution and to get a more detailed insight into the nature and the site of further modifications, a partial enzymatic hydrolysis was carried out prior to mass spectrometric analysis by using the endoproteinase AspN. The latter specifically cleaves peptide bonds N-terminally to aspartic acid and, with lower affinity, to glutamic acid. The resulting peptides were analyzed using MALDI-TOF-MS and assigned to the corresponding amino acid sequence

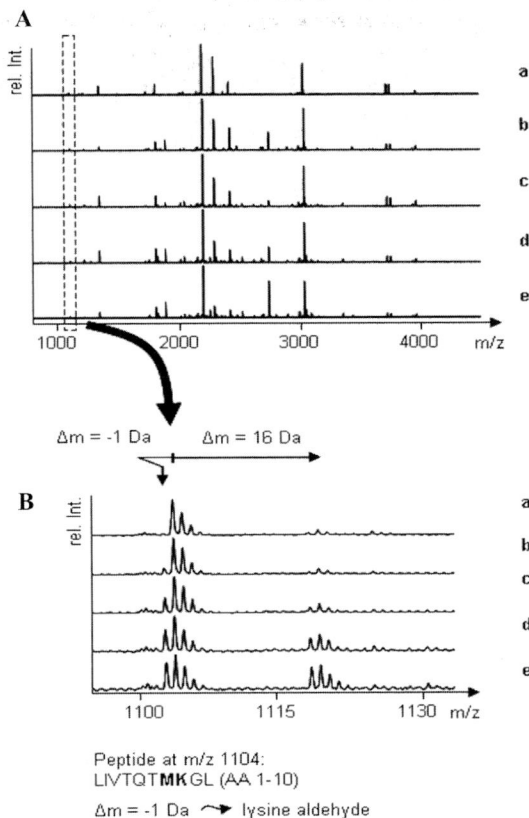

FIGURE 2. (A) MALDI-TOF-MS spectra of β-lactoglobulin incubated with lactose at 60 °C for 3 (c), 7 (d), and 14 days (e). The commercial standard (a) and the protein heated without lactose (b) were used as controls. **(B)** The spectrum of peptide at m/z 1104 is zoomed in. Mass shifts of newly formed signals are shown with the possible interpretation of the modification in the amino acid sequence. AA: amino acid.

by means of the software Peptide Mass (available on www.expasy.org). New signals resulting from the incubation with lactose were interpreted by their mass shift in respect to the native peptide. In FIGURE 2, the results are exemplified by the mass spectra obtained for the peptide m/z 1104. The principle of this method and its advantages are shown in FIGURE 1. For an overview of the detected modifications see FIGURES 3 and 4.

Heating of β-lactoglobulin with lactose led to the addition of lactose at lysine 47 and either lysine 138 or 141, which subsequently reacted to N^{ε}-(carboxymethyl)lysine (CML). The lysine residues 138 and 141 are located on the same peptide fragment and, therefore, cannot be distinguished by this method. The observed lactolysation sites are in accordance with previous studies that showed site-specific formation of the

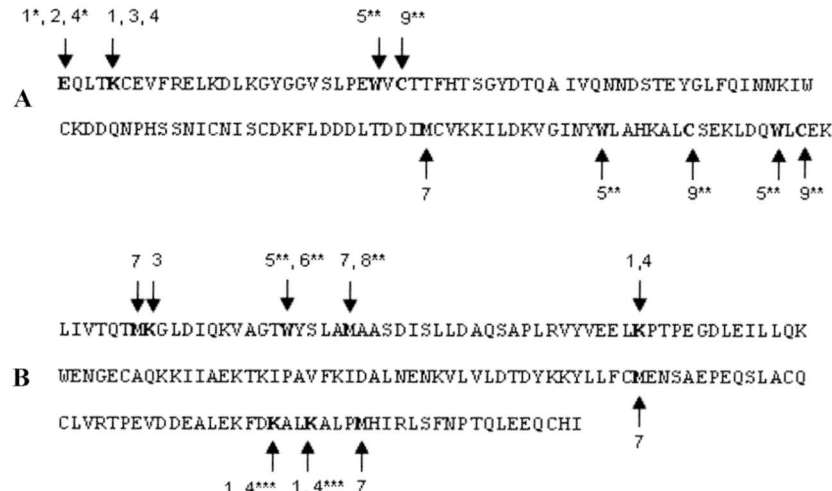

FIGURE 3. Amino acid sequences of α-lactalbumin **(A)** and β-lactoglobulin **(B)**. Postulated modification sites are indicated with numbers referring to the structures of the modifications depicted in FIGURE 4. *Definite identification of this modification site is not possible; **clear differentiation of the modifications is not possible; ***clear differentiation of the modification site is not possible.

Amadori product in model solutions or dairy products.[19] CML is an approved marker for the heat treatment of milk,[20] but to our knowledge, an identification of its binding sites in the protein has not been possible yet. Lactolysation and CML formation on other lysine residues were not observed. The strict concurrence of the Amadori product and CML suggests the formation of the latter by oxidative degradation of lactulosyllysine.

Lysine 8 was selectively modified to lysine aldehyde (FIG. 2B), a well-known oxidation product. It has been reported that its formation is promoted during heat treatment by the presence of sugars, which are degraded to dicarbonyl compounds. The latter induce a Strecker-type oxidation of the amino group to the aldehyde (FIG. 4).[21]

Furthermore, the oxidation of methionine—the amino acid most susceptible to oxidation—to methionine sulfoxide was observed at position 7, 24, and 145 with a mass shift of 16 Da (FIG. 2B). Mass increments of 16 Da have already been detected in ESI-MS spectra of whey, although the modification site was not determined.[22] Since methionine sulfoxide was not detected in the unheated samples, artificial formation during sample preparation could be excluded.

A mass shift of 32 Da at a peptide containing methionine and tryptophan as amino acids, which are susceptible to oxidation, could be explained by several modifications: Either methionine at position 24 was oxidized to the sulfone, or tryptophan 19 reacted to N-formylkynurenine, or a parallel oxidation to methionine sulfoxide at position 24 and to hydroxytryptophan at position 19 (both with a mass shift of 16 Da) occurred. Further structural analyses are required for the unambiguous assignment of this new signal.

MALDI-TOF-MS Analysis of α-Lactalbumin in a Heated Milk Model after Peptide Mapping

Upon incubation of α-lactalbumin with lactose, lactulosyllysine and its degradation product CML were identified at the N-terminal peptide containing lysine 5 and the N-terminal amino group. Since the modifications were also present in the condensed form where the α-amino group is blocked (see below), lactosylation and CML formation at lysine 5 could be clearly demonstrated. These findings are in accordance with results from Marvin *et al.*[16] Glycation of the N-terminal amino group could also occur.

Oxidation of lysine to lysine aldehyde was detected at position 5. Furthermore, mass shifts of 16 Da indicated the formation of cysteine sulfenic acid at cysteine 28, 111, and 120 or of hydroxytryptophan at tryptophan 26, 104, and 118. Cysteine sulfenic acid is stable in proteins under certain structural circumstances[23] but has not been detected in milk proteins yet, whereas oxidation products of tryptophan have been demonstrated in human milk.[24]

As a major modification of α-lactalbumin, condensation of the N-terminal glutamic acid to a pyrrolidone

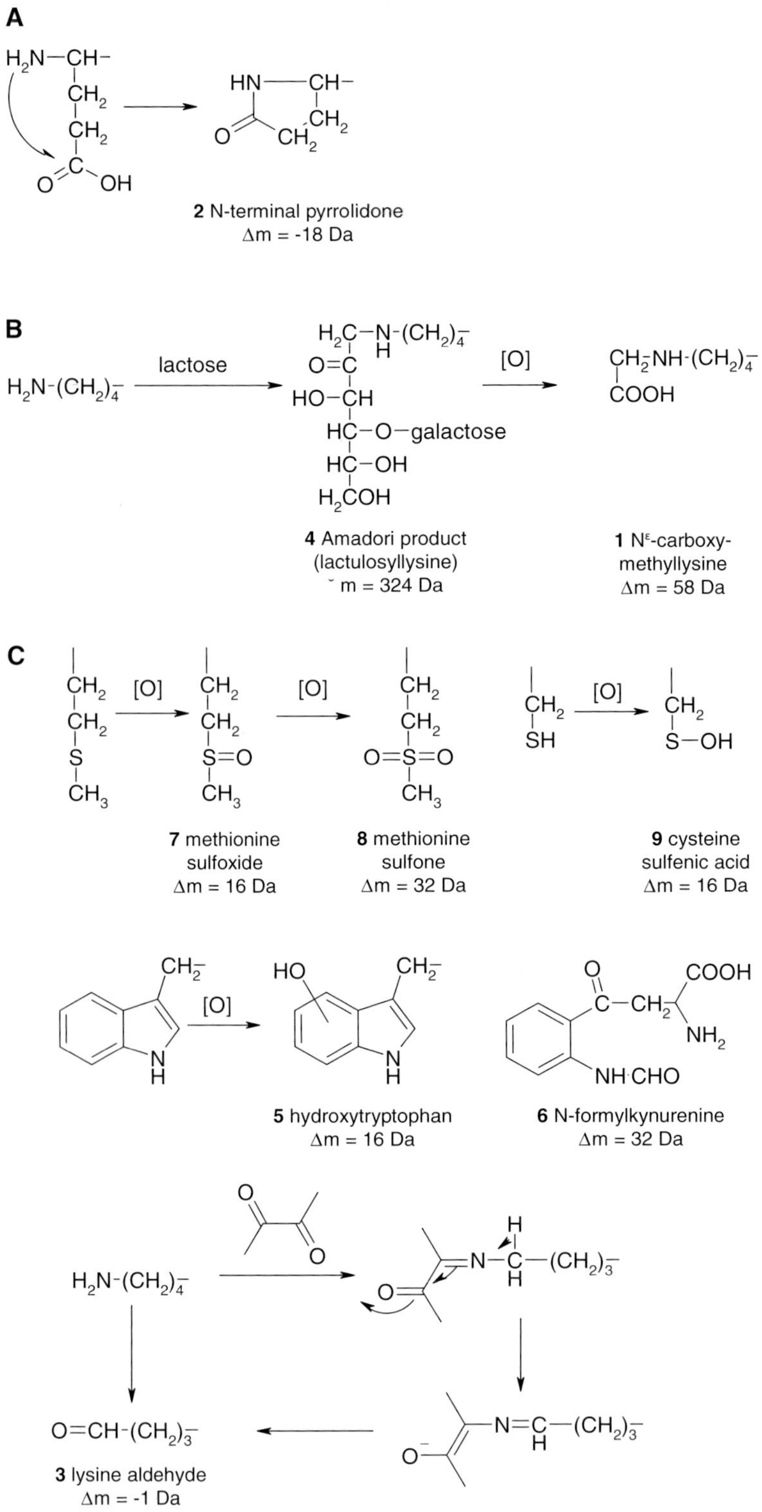

FIGURE 4. Postulated modifications of α-lactalbumin and β-lactoglobulin in a heated milk model based on MALDI-TOF-MS analysis after partial enzymatic hydrolysis; **(A)** condensation reaction; **(B)** glycation and glycoxidation; **(C)** oxidation.

derivative was detected. The pyrrolidone was the only modification which was also present in α-lactalbumin heated in the absence of lactose, indicating that its formation is independent of the Maillard reaction. Condensation of N-terminal glutamic acid has already been shown to occur under mild conditions[25] and was obviously enhanced by the heat process.

MALDI-TOF-MS Analysis of Oxidized Proteins after Peptide Mapping

Furthermore, oxidized whey proteins were prepared by incubating solutions of α-lactalbumin and β-lactoglobulin at 120°C in the presence of hydrogen peroxide. The goal of this experiment was to confirm the assumption that protein oxidation occurred in the model solutions with lactose by detecting analogous mass differences in the oxidized proteins.

In the spectra of the hydrogen peroxide-treated β-lactoglobulin, the formation of methionine sulfoxide at methionine 107 was observed besides oxidation of methionine 7, 24, and 145. When the lactosylated β-lactoglobulin was analyzed, the peptide containing methionine 107 was not covered by the method. It seems, however, very likely that this amino acid was also oxidized in the milk model.

After partial enzymatic hydrolysis of oxidized α-lactalbumin with endoproteinase AspN, no differences were visible in comparison with the native protein. As the only methionine in the sequence at position 90, which was suspected to be the most readily oxidized amino acid, was not covered by this method, hydrolysis was repeated with trypsin. This approach allowed the detection of additional sequences. Hydrolysis by trypsin revealed oxidation of methionine 90 to the sulfoxide, suggesting that this amino acid was also modified when the protein was incubated with lactose. None of the cystein or tryptophan oxidation products were detected in the whey proteins treated with hydrogen peroxide.

These data underline the importance of the Maillard reaction for protein oxidation. Incubation of the whey proteins in the milk models led to more severe protein oxidation in comparison to the reaction with hydrogen peroxide. When the whey proteins were heated under conditions similar to the milk models but in the absence of lactose, none of the oxidation products were detectable.

Analysis of Whey Protein Modifications in Heated Milk and Dairy Products

For the analysis of protein modifications in heated milk, it is necessary to perform a protein separation prior to mass spectrometry. Otherwise, the spectra resulting from protein mixture after enzymatic digestion would not allow the detection of minor new signals because of the complexity of the spectra. For this purpose, one-dimensional SDS-PAGE can be used as a standard technique for the combination of protein separation and structural analyses by MALDI-TOF-MS. An advantage of this method is the possibility to perform the partial enzymatic hydrolysis directly in the destained and washed gel. This method allows the determination of major modifications of β-lactoglobulin, which are formed during heating or industrial processing of milk.

Conclusion

MALDI-TOF-MS is a useful method for the systematic elucidation of glycation and oxidation reactions in heated whey proteins. Furthermore, it allows the identification of the modification sites in the amino acid sequence. By applying a protein separation technique prior to partial enzymatic hydrolysis, the analysis of modifications in dairy products can be achieved. In addition to the glycation products lactulosyllysine and CML, mainly protein oxidation products and a condensation product of N-terminal glutamic acid were formed in the milk models. As a consequence, these products should be considered as important protein modifications when the physiological, nutritional, and technological properties of processed dairy food are studied.

Conflict of Interest

The authors declare no conflicts of interest.

References

1. PISCHETSRIEDER, M. 2007. Are dietary AGEs/ALEs a risk to human health and, if so, what is the mechanism of action? Mol. Nutr. Food Res. **51:** 1169–1170.
2. SEBEKOVA, K. & V. SOMOZA. 2007. Dietary advanced glycation endproducts (AGEs) and their health effects – PRO. Mol. Nutr. Food Res. **51:** 1079–1084.
3. JAGERSTAD, M. & K. SKOG. 2005. Genotoxicity of heat-processed foods. Mutat. Res. **574:** 156–172.

4. GOYA, L. et al. 2007. Effect of coffee melanoidin on human hepatoma HepG2 cells. Protection against oxidative stress induced by tert-butylhydroperoxide. Mol. Nutr. Food Res. **51:** 536–545.
5. DITTRICH, R. et al. 2003. Maillard reaction products inhibit oxidation of human low-density lipoproteins in vitro. J. Agric. Food Chem. **51:** 3900–3904.
6. AMES, J.M. 2007. Evidence against dietary advanced glycation endproducts being a risk to human health. Mol. Nutr. Food Res. **51:** 1085–1090.
7. BAYNES, J.W. 2007. Dietary ALEs are a risk to human health – NOT! Mol. Nutr. Food Res. **51:** 1102–1106.
8. AHMED, N. et al. 2005. Peptide mapping identifies hotspot site of modification in human serum albumin by methylglyoxal involved in ligand binding and esterase activity. J. Biol. Chem. **280:** 5724–5732.
9. KISLINGER, T., A. HUMENY & M. PISCHETSRIEDER. 2004. Analysis of protein glycation products by matrix-assisted laser desorption ionization time-of-flight mass spectrometry. Curr. Med. Chem. **11:** 2185–2193.
10. TRALDI, P. et al. 1997. Mass spectrometry in the study of nonenzymatic glyco-oxidation of proteins. Rapid Commun. Mass Spectrom. **11:** 673–678.
11. SICILIANO, R. et al. 2000. Modern mass spectrometric methodologies in monitoring milk quality. Anal. Chem. **72:** 408–415.
12. KISLINGER, T. et al. 2005. Analysis of protein glycation products by MALDI-TOF/MS. Ann. N.Y. Acad. Sci. **1043:** 249–259.
13. COTHAM, W.E. et al. 2004. Proteomic analysis of arginine adducts on glyoxal-modified ribonuclease. Mol. Cell. Proteomics. **3:** 1145–1153.
14. HUMENY, A. et al. 2002. Qualitative determination of specific protein glycation products by matrix-assisted laser desorption/ionization mass spectrometry peptide mapping. J. Agric. Food Chem. **50:** 2153–2160.
15. MELTRETTER, J. et al. 2007. Site-specific formation of Maillard, oxidation, and condensation products from whey proteins during reaction with lactose. J. Agric. Food Chem. **55:** 6096–6103.
16. MARVIN, L.F. et al. 2002. Characterization of lactosylated proteins of infant formula powders using two-dimensional gel electrophoresis and nanoelectrospray mass spectrometry. Electrophoresis **23:** 2505–2512.
17. GALVANI, M., M. HAMDAN & P.G. RIGHETTI. 2001. Two-dimensional gel electrophoresis/matrix-assisted laser desorption/ionisation mass spectrometry of commercial bovine milk. Rapid Commun. Mass Spectrom. **15:** 258–264.
18. GALVANI, M., M. HAMDAN & P.G. RIGHETTI. 2000. Two-dimensional gel electrophoresis/matrix-assisted laser desorption/ionisation mass spectrometry of a milk powder. Rapid Commun. Mass Spectrom. **14:** 1889–1897.
19. LEONIL, J. et al. 1997. Characterization by ionization mass spectrometry of lactosyl beta-lactoglobulin conjugates formed during heat treatment of milk and whey and identification of one lactose-binding site. J. Dairy Sci. **80:** 2270–2281.
20. HEWEDY, M.H. et al. 1994. Effects of UHT heating of milk in an experimental plant on several indicators of heat treatment. J. Dairy Res. **61:** 305–309.
21. AKAGAWA, M. et al. 2005. Formation of alpha-aminoadipic and gamma-glutamic semialdehydes in proteins by the maillard reaction. Ann. N.Y. Acad. Sci. **1043:** 129–134.
22. HAU, J. & L. BOVETTO. 2001. Characterisation of modified whey protein in milk ingredients by liquid chromatography coupled to electrospray ionisation mass spectrometry. J. Chromatogr. A **926:** 105–112.
23. CLAIBORNE, A. et al. 1993. Protein-sulfenic acid stabilization and function in enzyme catalysis and gene regulation. FASEB J. **7:** 1483–1490.
24. DAZZI, C. et al. 2001. New high-performance liquid chromatographic method for the detection of picolinic acid in biological fluids. J. Chromatogr. B Biomed. Sci. Appl. **751:** 61–68.
25. CHELIUS, D. et al. 2006. Formation of pyroglutamic acid from N-terminal glutamic acid in immunoglobulin gamma antibodies. Anal. Chem. **78:** 2370–2376.

Inhibition of Advanced Glycation End Products

An Implicit Goal in Clinical Medicine for the Treatment of Diabetic Nephropathy?

TOSHIO MIYATA AND YUKO IZUHARA

Institute of Medical Sciences and Division of Nephrology, Hypertension and Metabolism, Tokai University School of Medicine, Kanagawa, Japan

Several factors have been incriminated in the genesis of diabetic nephropathy. To elucidate their interplay, we have used a hypertensive, obese, diabetic rat model with nephropathy (SHR/NDmcr-cp) that mimics human type 2 diabetes. This model is characterized by hypertension, obesity with the metabolic syndrome, diabetes with insulin resistance, and intrarenal advanced glycation end product (AGE) accumulation. In order to achieve renoprotection, which was evaluated by histology and albuminuria, various therapeutic approaches were used: caloric restriction, antihypertensive agents (angiotensin II receptor blocker [ARB] and calcium channel blocker), lipid- (bezafibrate) or glucose-lowering (insulin and pioglitazone) agents, and cobalt chloride (a hypoxia-inducible factor activator). Altogether, renoprotection is not necessarily associated with blood pressure or glycemic control. By contrast, it is almost always associated with decreased AGE formation, with the exception of insulin, which induces hyperinsulinemia, eventually leading to an overproduction of transforming growth factor-β. AGE formation is reduced directly by *in vitro* active compounds (e.g., ARBs) or indirectly by *in vitro* inactive compounds (e.g., pioglitazone and cobalt). In the latter cases, AGE reduction may reflect a decreased oxidative stress as it is concomitant with a marked reduction of oxidative stress markers. It remains to be seen whether the renoprotection offered by these various approaches may be additive.

Key words: diabetic nephropathy; blood pressure; renin–angiotensin system; oxidative stress; advanced glycation; chronic hypoxia

Diabetic nephropathy has become one of the main causes of end-stage renal disease. The molecular mechanisms responsible for its development are, as yet, incompletely understood. Hemodynamic and metabolic factors have been incriminated in the genesis of diabetic nephropathy, including hypertension, obesity, hyperglycemia, hyperinsulinemia, and hyperlipidemia.

To elucidate their interplays, we use a unique diabetic rat model SHR/NDmcr-cp (SHR/ND). Its genetic background is the same as that of the spontaneously hypertensive rats (SHR) and original Wistar–Kyoto rat strain. Like SHR, SHR/ND rats become hypertensive, but systemic blood pressure is significantly lower in SHR/ND rats than SHR.[1,2] The uniqueness of SHR/ND rats is the metabolic derangement accompanying hypertension. In SHR, metabolic markers, such as body weight and plasma levels of glucose, insulin, and lipids, are normal. In SHR/ND rats, by contrast, the genetic mutation of the leptin receptor generates obesity, hyperglycemia, hyperinsulinemia, and hyperlipidemia, all characteristics of human type 2 diabetes.[1,2]

As might be expected, this obese diabetic rat develops significant renal injury.[1,2] Despite severe hypertension, SHR, a strain with hypertension alone, has only a slight increase in proteinuria and glomerular sclerosis. In SHR/ND rats, by contrast, despite milder hypertension, proteinuria and glomerular sclerosis develop and worsen over time. In agreement with our common clinical experience, these observations point to the critical role played not only by hypertension but also by metabolic derangements in diabetic renal damage.

Several questions then arise: Are these renal disorders preventable? Which factors contribute most to the genesis of diabetic renal damage? Which therapeutic interventions are most effective and easy to achieve? In order to address these issues, we have undertaken to test several therapeutic approaches in this model, including caloric restriction, antihypertensive

Address for correspondence: Toshio Miyata, M.D., Ph.D., Institute of Medical Sciences and Division of Nephrology, Hypertension and Metabolism, Tokai University School of Medicine, Isehara, Kanagawa 259-1193, Japan. Voice: +81 463 93 1936; fax: +81 463 93 1938.
t-miyata@is.icc.u-tokai.ac.jp

agents, glucose-lowering agents, and cobalt chloride (a hypoxia-inducible factor [HIF] activator).

Caloric Restriction

In order to investigate the benefits of the correction on the metabolic syndrome, we restricted the caloric intake of SHR/ND rats by 30%.[2] Both obesity and hyperlipidemia were significantly ameliorated. Still, hyperglycemia and especially hyperinsulinemia persisted. Nevertheless, 20 weeks of a low-calorie diet prevented proteinuria and histological abnormalities of the kidney. Prevention was independent of systemic blood pressure as SHR-derived hypertension remained unchanged.

Thus, renoprotection occurred in association with a significant correction of obesity and hyperlipidemia, despite unaltered hypertension, hyperinsulinemia, and a limited partial (at best) correction of hyperglycemia. In our model, renal damage was correlated well with body weight and with the renal content of an advanced glycation end product (AGE), pentosidine.[2]

Antihypertensive Agents

In the second approach, we relied on the correction of hypertension with various antihypertensive agents.[1,3] We evaluated the renoprotection offered by several types of antihypertensive agents with different capacities to inhibit the renin–angiotensin system (RAS).

Three different types of antihypertensive agents, olmesartan (angiotensin II receptor blocker [ARB]), nifedipine (calcium channel blocker), and atenolol (beta blocker), were given for 20 weeks to SHR/ND rats. These agents all normalized systolic blood pressure to the same extent. However, olmesartan successfully decreased proteinuria in contrast with nifedipine and atenolol, which were ineffective. This finding fits with our clinical experience that agents able to inhibit the RAS afford a better renoprotection than other types of antihypertensive drugs.[4] Renal benefits of ARB in this model appear independent of decreases in systemic blood pressure and of changes in metabolic abnormalities.

The benefits of ARB beyond lowering blood pressure were also confirmed in another model of mouse diabetic nephropathy. Recently, we generated a novel mouse model of diabetic nephropathy (i.e., megsin/RAGE/iNOS triple transgenic mice[5–7]). Although normotensive, these mice progressively develop severe diabetic nephropathy, mimicking that observed in humans. We tested whether olmesartan (ARB) achieves a better renoprotection than amlodipine (calcium channel blocker) in the transgenic mice. Drug treatment was initiated at aged 6 weeks and lasted for 12 weeks. Despite both drugs lowering blood pressure equally, only olmesartan suppressed the progression of albuminuria, supporting benefits of ARBs beyond lowering blood pressure (S. Ohtomo and T. Miyata, unpublished observations, Tokai University).

The optimal doses of ARB needed for lowering blood pressure and for renoprotection were further defined. We gave SHR/ND rats supramaximal doses of olmesartan, exceeding the level necessary for angiotensin II receptor saturation. Beyond 50 mg/kg/d, olmesartan failed to further reduce blood pressure, suggesting a complete blockade of angiotensin II receptor. Nevertheless, renoprotection, as indicated by proteinuria, progressed continuously in a dose-dependent manner (N. Tominaga and T. Miyata, unpublished observations, Tokai University), providing evidence that renoprotection is independent of reduced blood pressure.

Renal AGE content is markedly elevated in hypertensive diabetic SHR/ND rats.[1] Impressively, olmesartan, but not nifedipine or atenolol, markedly reduces renal AGE content without modification of the concomitant metabolic syndrome.[1,3] Renal AGE content significantly correlates with proteinuria in our diabetic rats, regardless of the antihypertensive agent.[1,3]

Renoprotection by ARB is associated with iron metabolism.[3] Prussian blue staining revealed abnormal iron deposition in SHR/ND rats but not in nondiabetic control rats. ARB inhibits this abnormal iron deposition in diabetic kidneys.

Mediators that are effectively modulated by ARBs are tentatively integrated in the hypothetical scheme (FIG. 1). Clearly, the interrelationship among these various elements precludes the identification of a single culprit in the genesis of diabetic kidney lesions. Whatever the hypothesis, ARBs and probably angiotensin-converting enzyme inhibitors (ACEIs) have unique renoprotective properties, including the inhibition of RAS, oxidative stress, AGEs, and iron deposition and the correction of chronic hypoxia.

Glucose-lowering Agents

The third approach concentrates on the correction of hyperglycemia or hyperinsulinemia by glucose-lowering agents. Strict glycemic control of diabetic patients with nephropathy is associated with reduction

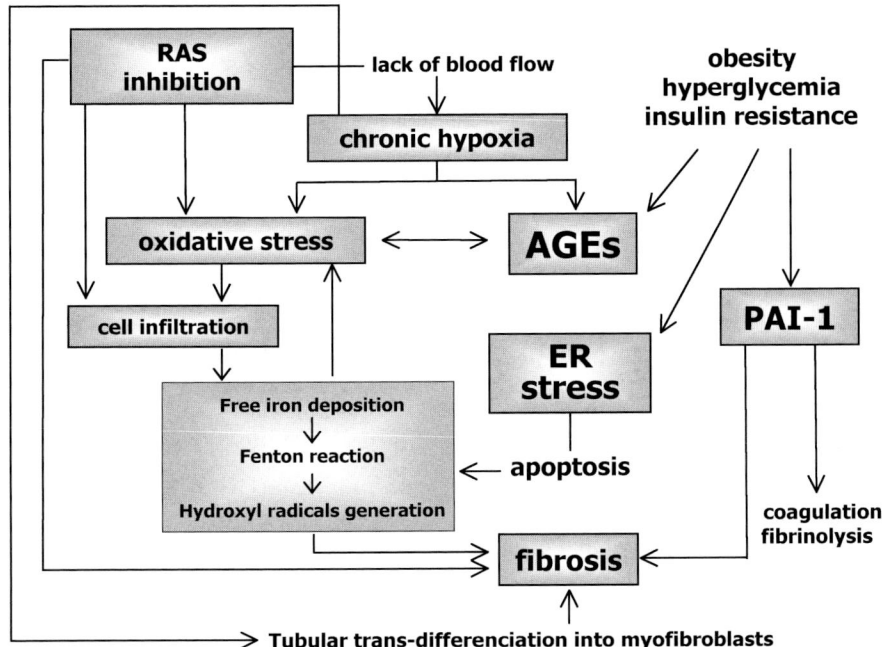

FIGURE 1. Hypothetical scheme for diabetic renal injury. Among three antihypertensive agents used in this study, only ARB possess all properties beneficial for renoprotection (i.e., inhibition of the renin–angiotensin system (RAS), prevention of abnormal iron deposition in the interstitium, correction of chronic hypoxia, hydroxyl radical scavenging, reduction of expressions of heme oxygenase (HO)-1 and NADPH oxidase, amelioration of inflammatory cell infiltration, and inhibition of pentosidine formation. These interrelated benefits of ARB may contribute to renoprotection (i.e., reduction of proteinuria and improvement of glomerular and tubulointerstitial damage). PAI-1, plasminogen activator inhibitor-1; ER, endoplasmic reticulum; AGEs, advanced glycation end products.

of microalbuminuria.[8] Recent studies have implicated insulin resistance (IR) or hyperinsulinemia in the development of type 2 diabetic nephropathy.[9] Insulin sensitizer is a new class of agents that lowers IR and thus blood glucose.[10]

We therefore asked a clinically important question: Does pioglitazone and its attendant correction of hyperinsulinemia provide better renoprotection than insulin at similar glucose levels? SHR/ND rats were given pioglitazone or insulin for 20 weeks.[11] Neither treatment modified hypertension. Pioglitazone aggravated obesity and provided a poorer glycemic control than insulin. Still, pioglitazone decreased plasma insulin levels significantly in contrast with insulin. Renoprotection was markedly better with pioglitazone than with insulin treatment as shown by proteinuria reduction.

Deeper insights into biological features of diabetic nephropathy disclosed that both pioglitazone and insulin reduced the renal accumulation of pentosidine and markers of oxidative stress. Again, renal pentosidine correlated with proteinuria in SHR/ND rats given pioglitazone. Of note, pioglitazone, but not insulin, reduced renal expression of TGF-β in SHR/ND rats. This was ascribed to insulin-induced TGF-β expression as shown by experiments in cultured rat proximal tubular cells.

Our results thus suggest that hyperinsulinemia and the attendant increase of TGF-β expression are potential therapeutic targets, independent of glycemic control. Such conclusions agree with clinical evidence that pioglitazone enhances renoprotection in obese diabetic patients with nephropathy.

Hypoxia-inducible Factor Activator (Cobalt Chloride)

The fourth therapeutic approach concerns renal hypoxia and relies on the activation of HIF.[12] The peritubular capillary plexus, fed by glomerular efferent arterioles, is known to supply oxygen and nutrients

TABLE 1. Summary of the animal experiments in hypertensive type 2 diabetic rats SHR/NDmcr-cp

Treatment	Renoprotection	AGE inhibition	BP lowering	RAS inhibition	Obesity correction	Glycemic control	Lipid lowering	Hyperinsulinemia correction
Caloric restriction	+	+	−	−	+	+	+	−
ARB	+	+	+	+	−	−	+	−
CCB	−	−	+	−	−	−	−	−
β-blocker	−	−	+	−	−	−	−	−
Pioglitazone	+	+	−	−	worsening	+	+	+
Insulin	−	+	−	−	−	+	+	worsening
Cobalt	+	+	−	−	−	−	−	−

AGE, advanced glycation end product; BP, blood pressure; RAS, renin-angiotensin system; ARB, angiotensin II receptor blocker; CCB, calcium channel blocker.

to tubular and interstitial cells. Diabetic glomerular damage decreases the number of peritubular capillaries and thus oxygen diffusion to tubulointerstitial cells, leading to tubular dysfunction and fibrosis.[13] In human type 2 diabetic patients, a study of intrarenal hemodynamics has indeed shown a correlation between a decreased peritubular capillary flow and tubulointerstitial dysfunction, supporting a pathogenic role of chronic hypoxia of the diabetic kidney.

Pimonidazole selectively stains hypoxic cells. Pimonidazole staining was indeed detected in diabetic kidneys but not in control kidneys, a finding confirming tubulointerstitial chronic hypoxia.[3] Of note, the renoprotection provided by ARB is partially associated with improved tubulointerstitial chronic hypoxia.[3]

We know that HIF is a crucial molecule in the defense against hypoxia.[14] Its activation might lead to the protection of hypoxic tissues as it induces the expression of a broad range of genes, erythropoietin, vascular endothelial growth factor (VEGF), heme oxygenase (HO)-1, and glucose transporters, all of which are critically involved in the defense against chronic hypoxia. HIF stability is known to be drastically reduced by the oxygen-dependant hydroxylation of HIF proline residues by prolyl hydroxylase (PHD). Hydroxylated HIF recruits the von Hippel Lindau protein, which in turn tags HIF with ubiquitin and targets it for degradation within the proteasome. Under hypoxic conditions, HIF is not hydroxylated and is thus able to activate a host of genes involved in the adaptation to hypoxia. We were interested in the ability of cobalt to inhibit HIF degradation by PHDs.[15] Cobalt does so by replacing iron, an essential element for PHD activity, so that prolylhydroxylase function is inhibited.

Cobalt was given to SHR/ND rats.[12] Treatment was initiated at the age of 13 weeks and continued for 26 weeks. Cobalt did not correct hypertension and metabolic abnormalities, but it reduced proteinuria as well as histological kidney injury. Cobalt upregulated renal HIF expression and increased the expression of HIF-regulated genes, including those for erythropoietin, VEFG, and HO-1. The renal expression of TGF-β was also significantly reduced by cobalt.

Of note, cobalt reduced the renal content of pentosidine and expression of NADPH oxidase, a marker of oxidative stress. Again, renal pentosidine correlated with proteinuria in SHR/ND rats given cobalt.

Cobalt thus achieved renal protection independently of metabolic status and blood pressure through the upregulation of HIF and HIF-regulated genes, and it mitigated advanced glycation and oxidative stress.

Renoprotection

In summary (TABLE 1), these various animal experiments demonstrate that renoprotection does not necessarily rely on blood pressure or glycemic control. Interestingly, renoprotection is associated with decreased AGE formation and decreased oxidative stress. We should not forget, however, that insulin treatment prevents advanced glycation and oxidative stress without renoprotection, a fact probably a result of persistent hyperinsulinemia and its attendant overproduction of TGF-β.

Molecular mechanisms for decreasing AGE *in vivo* remain elusive. Some agents, such as ARBs, directly inhibit AGE formation *in vitro*.[16,17] Others lower AGE formation indirectly *in vivo* by the reduction of prevailing glucose levels (e.g., insulin, pioglitazone, caloric restriction). Of interest, despite its inability in lowering AGE *in vitro*, cobalt significantly lowered *in vivo* AGE formation, most probably through decreased oxidative stress.

Renoprotection remains a major therapeutic goal for nephrologists. Our results suggest that medical agents that interfere with oxidative stress, advanced glycation, and TGF-β expression may contribute to this goal.

TABLE 2. Summary of the characteristics of TM2002 and previously reported AGE-inhibitory compounds

	Carbonyl compound entrapment	Pyridoxal entrapment	Metal chelation[a]	Radical scavenging[b]	Blood pressure lowering
Aminoguanidine	o	o	—	—	—
OPB-9195	o	o	o	ND	—
Pyridoxamine	o	—	—	o	—
LR-90[c]	o	NR	o	NR	—
ARBs (ACEIs)	—	—	o	o	o
Edaravone	o	o	o	o	—
TM2002	—	—	o	o	—

[a] Assessed by the copper-catalyzed oxidation of ascorbic acid.[19]
[b] Assessed by inhibition of hydroxyl radical-mediated phenylalanine modification.
[c] Reference 21.
ACEIs, angiotensin-converting enzyme inhibitors; ND, not determined because of the low solubility in the reaction mixture; NR, not reported.

AGE Inhibitors

We screened a large chemical library (approximately 1300 compounds), including edaravone,[18] a drug used to treat cerebral infarction, for the *in vitro* AGE inhibitory activity. Unfortunately, like most AGE inhibitors, edaravone also traps pyridoxal, limiting its clinical usefulness. We therefore synthesized a novel AGE inhibitor, TM2002, which does not trap pyridoxal.[19] *In vitro*, TM2002 showed powerful AGE inhibitory activity. Markers of oxidation (i.e., *o*-tyrosine formation and transition metal chelation) were efficiently inhibited by TM2002, like ARBs. TM2002 did not bind to the angiotensin II–type 1 receptor, and it was readily bioavailable and nontoxic. *In vivo*, TM2002, given acutely or for 8 weeks, had no adverse effects. In four different rat models of renal injury (anti-Thy1 and ischemia–reperfusion) and cardiovascular injury (carotid artery balloon injury and angiotensin II–induced cardiac fibrosis), TM2002 improved renal and cardiovascular lesions without modification of blood pressure.[19]

TM2002 also significantly decreased infarct volume in both transient and permanent focal ischemia rat models.[20] Inhibition of advanced glycation and oxidative stress was confirmed by a significantly reduced number of cells positive for AGEs and HO-1. TM2002 reduced the levels of protein carbonyl formation in ischemic caudate.

Our compound, TM2002, belongs to a novel class of AGE inhibitors (TABLE 2). Like ARBs, but unlike aminoguanidine, OPB-9195, pyridoxamine, and LR-90, TM2002 does not trap carbonyl precursors. R147176 chelates metals and reduces oxidative stress. Unlike ARBs, it lacks specific binding affinity to the human angiotensin II–type 1 receptor *in vitro* and therefore does not alter blood pressure *in vivo*.

Conclusion

A growing body of evidence demonstrated the significance of AGEs as surrogate markers for diabetic renal injury. Indeed, several inhibitors have been proved as protective in experimental diabetic models; some are being tested in human clinical studies (e.g., pyridoxamine). The issue of importance at this stage is clinical evidence in human trials.

Acknowledgments

This study was supported by a grant from the Program for Promotion of Fundamental Studies in Health Sciences of the Pharmaceuticals and Medical Devices Agency (PMDA).

Conflict of Interest

The authors declare no conflicts of interest.

References

1. NANGAKU, M., T. MIYATA, T. SADA, *et al*. 2003. Antihypertensive agents inhibit in vivo the formation of advanced glycation end products and improve renal damage in a type 2 diabetic nephropathy rat model. J. Am. Soc. Nephrol. **14:** 1212–1222.

2. NANGAKU, M., Y. IZUHARA, N. USUDA, et al. 2005. In a type 2 diabetic nephropathy rat model, the improvement of obesity by a low calorie diet reduces oxidative/carbonyl stress and prevents diabetic nephropathy. Nephrol. Dial. Transplant. **20:** 2661–2669.

3. IZUHARA, Y., M. NANGAKU, R. INAGI, et al. 2005. Renoprotective properties of angiotensin receptor blockers beyond blood pressure lowering. J. Am. Soc. Nephrol. **16:** 3631–3641.

4. WOLF, G., E. RITZ. 2005. Combination therapy with ACE inhibitors and angiotensin II receptor blockers to halt progression of chronic renal disease: Pathophysiology and indications. Kidney Int. **67:** 799–812.

5. MIYATA, T., M. NANGAKU, D. SUZUKI, et al. 1998. A mesangium-predominant gene, megsin, is a new serpin up-regulated in IgA nephropathy. J. Clin. Invest. **102:** 828–836.

6. MIYATA, T., R. INAGI, M. NANGAKU, et al. 2002. Overexpression of the serpin megsin induces progressive mesangial cell proliferation and expansion. J. Clin. Invest. **109:** 585–593.

7. INAGI, R., Y. YAMAMOTO, M. NANGAKU, et al. 2006. A severe diabetic nephropathy model with early development of nodular-like lesions induced by megsin overexpression in RAGE/iNOS transgenic mice. Diabetes **55:** 356–366.

8. THE EPIDEMIOLOGY OF DIABETES INTERVENTIONS AND COMPLICATIONS (EDIC) STUDY. 2003. Sustained effect of intensive treatment of type 1 diabetes mellitus on development and progression of diabetic nephropathy. JAMA **290:** 2159–2167.

9. SARAFIDIS, P.A., L.M. RUILOPE. 2006. Insulin resistance, hyperinsulinemia, and renal injury: mechanisms and implications. Am. J. Nephrol. **26:** 232–234.

10. PISTROSCH, F., K. HERBRIG, B. KINDEL, et al. 2005. Rosiglitazone improves glomerular hyperfiltration, renal endothelial dysfunction, and microalbuminuria of incipient diabetic nephropathy in patients. Diabetes **54:** 2206–2211.

11. OHTOMO, S., Y. IZUHARA, S. TAKIZAWA, et al. 2007. Thiazolidinediones protects the kidney better than insulin in an obese, hypertensive type 2 diabetes Rat Model. Kidney Int. **72:** 1512–1519.

12. OHTOMO, S., M. NANGAKU, Y. IZUHARA, et al. 2008. Cobalt ameliorates renal injury in an obese, hypertensive type 2 diabetes rat model. Nephrol. Dial. Transplant. In Press.

13. NANGAKU, M. 2006. Chronic hypoxia and tubulointerstitial injury: a final common pathway to end-stage renal failure. J. Am. Soc. Nephrol. **17:** 17–25.

14. SCHOFIELD, C.J., P.J. RATCLIFFE. 2004. Oxygen sensing by HIF hydroxylases. Nat. Rev. Mol. Cell Biol. **5:** 343–354.

15. SALNIKOW, K., S.P. DONALD, R.K. BRUICK, et al. 2004. Depletion of intracellular ascorbate by the carcinogenic metals nickel and cobalt results in the induction of hypoxic stress. J. Biol. Chem. **279:** 40337–40344.

16. MIYATA, T., C. VAN YPERSELE DESTRIHOU, Y. UEDA, et al. 2002. Angiotensin II receptor antagonists and angiotensin-converting enzyme inhibitors lower in vitro the formation of advanced glycation end products: biochemical mechanisms. J. Am. Soc. Nephrol. **13:** 2478–2487.

17. FORBES, J.M., M.E. COOPER, V. THALLAS, et al. 2002. Reduction of the accumulation of advanced glycation end products by ACE inhibition in experimental diabetic nephropathy. Diabetes **51:** 3274–3282.

18. TSUJITA, K., H. SHIMOMURA, H. KAWANO, et al. 2004. Effects of edaravone on reperfusion injury in patients with acute myocardial infarction. Am. J. Cardiol. **94:** 481–484.

19. IZUHARA, Y., M. NANGAKU, S. TAKIZAWA, et al. 2008. A novel class of advanced glycation inhibitors ameliorates renal and cardiovascular damage in experimental rat models. Nephrol. Dial. Transplant. In Press.

20. TAKIZAWA, S., Y. IZUHARA, Y. KITAO, et al. 2007. A novel inhibitor of advanced glycation and endoplasmic reticulum stress reduces infarct volume in rat focal cerebral ischemia. Brain Res. **1183:** 124–137.

21. FIGAROLA, J.L., S. SCOTT, S. LOERA, et al. 2006. LR-90 a new advanced glycation endproduct inhibitor prevents progression of diabetic nephropathy in streptozotocin-diabetic rats. Diabetologia **46:** 1140–1152.

Inflammation and the Redox-sensitive AGE–RAGE Pathway as a Therapeutic Target in Alzheimer's Disease

ANNETTE MACZUREK, KIRUBAKARAN SHANMUGAM, AND GERALD MÜNCH

Dept. of Biochemistry and Molecular Biology/Comparative Genomics Centre, James Cook University, Townsville, Australia

Alzheimer's disease (AD) is the most common cause of dementia. Neuritic amyloid plaques and concomitant chronic inflammation are prominent pathological features of AD. β-amyloid peptide (Aβ), the major component of plaques, and advanced glycation end products (AGEs), post-translational protein modifications, are key activators of plaque-associated inflammation. Aβ, AGEs, S100b, and amphoterin bind to the receptor for AGEs (RAGE), which transmits the signal from RAGE via redox-sensitive pathways to nuclear factor kappa-B (NF-κB)-regulated cytokines. RAGE-mediated inflammation caused by glial cells and subsequent changes in neuronal glucose metabolism are likely to be important contributors to neurodegeneration in AD. As long as the neuronal damage is reversible, drugs interfering with the Aβ and AGE–RAGE pathways might be interesting novel therapeutics for the treatment of AD.

Key words: Alzheimer's disease; advanced glycation end products; antioxidants; redox-sensitive signaling

Alzheimer's Disease: Epidemiology and Costs

Alzheimer's disease (AD) is the most common cause of dementia. AD affects about 160,000 Australians with numbers expected to rise to 500,000 by 2040. In 2002, the estimated cost of dementia in Australia was about 1% of the gross domestic product (GDP); by 2050, this is expected to increase to about 3%. The development of neuroprotective pharmacological strategies is an important task for the community.

Inflammation as a Relevant Pathogenic Factor in AD

AD patients suffer from a progressive decline in cognitive functions and develop a defect in their ability to store newly acquired information. Conversion from short-term to long-term memory, requiring energy consuming steps for local synaptic protein synthesis, appears to be particularly impaired in these patients. Among the histological hallmarks, neuritic amyloid plaques are the most obvious pathological features of AD. They accumulate in the cerebral cortex and hippocampus in a stage- and region-specific manner. In histochemical studies, activation of glial cells and increased levels of radicals and proinflammatory cytokines (interleukin [IL]-1β, IL-6, and tumor necrosis factor [TNF]-α) can be observed in microglial or astroglial cells adjacent to plaques.[1] In imaging studies, glial activation can be measured with position emission tomography (PET). Patients with AD show significantly increased binding of [11C](R)-PK11195, a specific ligand for glial activation, in the entorhinal, temporoparietal, and cingulate cortex.[2] Analysis of cerebrospinal fluid (CSF) shows that AD patients have increased cytokine levels (IL-1β and TNF) in the CSF with TNF being a good predictor for the progression from mild cognitive impairment to AD.[3] Genetic studies show evidence of an association of AD with certain polymorphisms in cytokine genes, such as IL-6, IL-1, and TNF.[4] Epidemiological evidence suggests a protective effect of nonsteroidal anti-inflammatory drugs (NSAIDs), such as indomethacin, ibuprofen, and acetylsalicylic acid, in AD.[5] A recently identified target of NSAIDs is the peroxisome proliferator-activated receptor γ (PPAR-γ), which is involved in the down-regulation of proinflammatory gene expression. In a prospective study (phase II trial), the insulin-sensitizer

---

Address for correspondence: Dr. Gerald Münch, Department of Biochemistry and Molecular Biology/Comparative Genomics Centre, James Cook University, Townsville, Australia. Voice: +61 7 4781 4709; fax: +61 7 4781 6078.

gerald.muench@jcu.edu.au

and PPAR-γ agonist rosiglitazone improved cognitive performance in AD patients lacking ApoE-ε4.[6] In another trial, the TNF-α inhibitor etanercept, administered by perispinal extrathecal route, showed significant improvement: Mini-mental State Examination (MMSE) increased by 2.13 ± 2.23 (mean ± SD) and Alzheimer's Disease Assessment Scale-Cognitive Subscale (ADAS-Cog) improved by 5.48 ± 5.08 (mean ± SD).[7]

Pro-inflammatory Stimuli in AD

β-amyloid peptid (Aβ) was proposed to be the key activator of plaque-associated inflammation. In cell culture experiments, fibrillar Aβ induces the expression of various cytokines and chemokines, such as TNF or monocyte chemotactic protein-1.[8] Results from a further study, using rat astrocyte cultures, suggest that Aβ oligomers induce a profound early-inflammatory response whereas fibrillar Aβ showed less increase of proinflammatory molecules, consistent with a more chronic form of inflammation.[9] However, the Tg2576 mouse, an animal model of AD, demonstrates an inflammatory response less pronounced than in AD patients despite high levels of soluble and fibrillar amyloid.[10] In particular, IL-6 is rarely, and inducible nitric oxide (NO) synthase never, detected around plaques in the Tg2576 mouse.[11] One of the reasons for the "mild inflammation" in these mice might be the lack of a costimulus priming the inflammatory response to Aβ. We have proposed that advanced glycation end products (AGEs) might be one of the physiological costimuli which is lacking in the Tg2576 mouse.[10] In diabetes, high levels of extracellular glucose and high levels of glucose metabolites in insulin-insensitive cells, such as neurons and mesangial cells, favor the formation of both intracellular and extracellular AGEs. At normal extracellular glucose concentrations, such as in the AD brain, AGEs accumulate on plaques because of the extremely long half-life (of up to 30 years) of the amyloid peptide.[12] A further contributor to increased AGE formation in AD might be the higher levels of reactive carbonyl compound, such as hydroxynonenal and acrolein derived from lipid peroxidation, which might contribute to protein cross-linking and deposition.[13] We have previously demonstrated that AGE accumulation in AD is age and stage dependent.[14] The auditory association area of the superior temporal gyrus (Brodmann area 22) was used; AGE immunoreactivity was detected in neurons and glia cells of young and old nondemented control patients and compared with early- and late-stage AD. Both the percentage of AGE-positive neurons and astroglia were found to increase with age. The age-related rise in AGE deposition was found to be dramatically increased in AD patients and correlated with the progression of the disease defined by the Braak stages, which classifies the density and distribution of neurofibrillary tangles in the brain.[14] Cell culture studies also indicate that the expression of cytokines induced by Aβ is strongly enhanced by a second proinflammatory stimulus, such as IFN-γ or Il-1β.[15] We have confirmed these data and shown that a variety of proinflammatory stimuli, such as lipopolysaccharide (LPS) AGEs or IFN-γ, increase the Aβ-induced production of NO, IL-6, TNF-α, and macrophage colony stimulating factor.[16]

Impaired Glucose Metabolism in AD

A further characteristic observed in AD patients is a decreased, regional, cerebral metabolic rate for glucose, demonstrated by PET using [18F]-fluorodeoxyglucose, in the parietal, temporal, occipital, frontal, and posterior cingulate cortex, which progressively worsens during the course of the disease.[17] Glucose is an indispensable substrate for the brain as it is necessary for the synthesis of acetylcholine, glutamate, aspartate, γ-aminobutyric acid, and glycine as well as for the production of ATP. Thus, limited energy supply can be (at least partially) responsible for impaired neurotransmitter production (in addition to loss of membrane potential) and impaired cognitive function in AD patients.

Links between Inflammation and Impaired Glucose Metabolism

Release of reactive oxygen and nitrogen species from activated glia cells has been extensively discussed as a contributor to neuronal damage and energy metabolism disorders. However, the influence of cytokines and chemokines on energy metabolism appears to be a potentially interesting field of research. Evidence for an influence of cytokines on cellular energy production comes from conditions, such as cancer-associated cachexia. In cachexia patients, cytokines, including TNF, IL-6, and IFN-γ, are released by the tumor or unaffected tissue of the patient, leading to increased lactate production and decreased ATP levels by various mechanisms such as:

a) *Inhibition of pyruvate dehydrogenase activity and mitochondrial function:* TNF and IL-1 inhibit both pyruvate dehydrogenase and mitochondrial complex I and

II activity in cardiomyocytes, leading to loss of energy and increased lactate production.[18]

b) *Activation of a futile phosphofructokinase–fructose bisphosphate phosphatase cycle*: TNF induces an increase in lactate production and glucose metabolism in cultured myocytes but presumably induces an ATP-depleting cellular process to account for the lack of feedback inhibition on glycolysis by the ATP produced. This observation led to the identification of a futile substrate cycle between fructose 6-phosphate and fructose 1,6-bisphosphate as an energy dissipation that is activated by TNF.[19]

c) *Increased ATP consumption*: IEC-6 rat enterocytes, incubated with TNF, IL-1ß, and IFN-γ, showed an ATP consumption rate approximately threefold higher than control cells.[20]

Anti-inflammatory Medications Stabilize Glucose Metabolism in AD Patients with Positive Effects on Cognition

A very interesting twin study describes findings in 90-year-old monozygotic female twins who have remained heterogenous for AD for at least 7 years. The twins' lifestyles were similar, except for continuous NSAID use by the unaffected twin. Regional cerebral glucose metabolic rates and cognitive parameters of the affected twin were much lower than in the unaffected twin.[21] A further indication that anti-inflammatory medication normalizes neuronal energy metabolism comes from a double-blind placebo-controlled trial with the phosphodiesterase inhibitor propentofylline, which is known to downregulate cytokine expression.[22] In this study, 30 AD patients underwent [18F]-fluorodeoxyglucose-PET during stimulation with an auditory memory paradigm. Only in the propentofylline treatment group was a significant increase of cerebral glucose metabolism to the memory task observed.[23]

The AGE–RAGE Pathway as the Main Inflammatory Pathway in AD

The receptor for AGE (RAGE), which binds a variety of proinflammatory ligands, such as Aβ, AGEs (both components of amyloid plaques), S100b, and amphoterin, transmits the signal from the ligand to NF-κB-regulated cytokines via redox-sensitive pathways.[24] These pathways are considered interesting drug targets for the treatment of AD.[25] Three strategies aiming at the receptor level might be promising:

a) *β-sheet breakers*: The RAGE-binding properties of Aβ are suggested to be dependent on its oligomerization, aggregation and formation of β-sheet structures. β-sheet breakers, such RS-0466, have been shown to decrease the direct neurotoxic properties of Aβ.[26]

b) *RAGE-antagonist:* RAGE antagonists have been shown to inhibit both binding of AGEs and Aβ to RAGE and AGE-induced nuclear translocation of NF-κB (United States Patent 6,613,801).

c) *soluble RAGE:* The N-terminal domain of RAGE, by interacting with Aβ, is a powerful inhibitor of Aβ polymerization and Aβ binding to RAGE.[27]

It is known that many signal transduction pathways, including that of RAGE responsible for cytokine production, involve reactive oxygen species (ROS) as second messengers or can be inhibited by agonists of the PPAR-γ.[28] Intracellular pathways downstream of RAGE might provide interesting drug targets as well, and the following drug classes might become novel antidementia drugs:

a) *NSAIDs with PPAR-γ activity:* A recently characterized target of NSAIDs is PPAR-γ.[29] COX-inhibitors (which also have PPAR-γ agonist activity), such as indomethacin and ibuprofen, and specific designed PPAR-γ agonists (also known as insulin-sensitizers), such as rosiglitazone and pioglitazone, might be interesting drug candidates.

b) *Membrane permeable antioxidants:* Proinflammatory intracellular signaling uses ROS as second messengers ("redox-active" signaling). Thiol antioxidants, such as α-lipoic acid, as well as plant polyphenols, such as apigenin and diosmetin, possess radical scavenging activity. We have previously shown that these antioxidants and also other inhibitors of NF-κB activation downregulate NO production in N-11 microglia, which is activated by AGEs and LPS.[28]

c) *Statins:* Some statins slow the progression of AD.[30] The p21^{ras}–MAP kinase pathway has been identified as a trigger for NF-κB activation in the RAGE signaling pathway.[31] By inhibition of 3-hydroxy-3-methyl-glutaryl-CoA reductase statins also inhibit the isoprenylation of small G-proteins and their localization to the membrane, which then might cause downregulation of the cytokine response induced by RAGE ligands.

In summary, RAGE-mediated inflammation caused by glial cells and subsequent changes in neuronal glucose metabolism are likely to be important contributors to neurodegeneration in AD. As long as the neuronal damage is reversible, anti-inflammatory medication (e.g., medications interfering with the Aβ and AGE–RAGE pathways) might be interesting therapeutics for AD treatment.

Acknowledgments

This work was supported by the National Health and Medical Research Council, Alzheimer's Australia, and the J.O. and J.R Wicking Foundation.

Conflict of Interest

The authors declare no conflicts of interest.

References

1. WONG, A. et al. 2001. Advanced glycation endproducts co-localize with inducible nitric oxide synthase in Alzheimer's disease. Brain Res. **920:** 32–40.
2. CAGNIN, A. et al. 2007. Positron emission tomography imaging of neuroinflammation. Neurotherapeutics **4:** 443–452.
3. TARKOWSKI, E. et al. 2003. Cerebral pattern of pro- and anti-inflammatory cytokines in dementias. Brain Res. Bull. **61:** 255–260.
4. ALVAREZ, V. et al. 2002. Association between the TNFalpha-308 A/G polymorphism and the onset-age of Alzheimer disease. Am. J. Med. Genet. **114:** 574–577.
5. SZEKELY, C.A. et al. 2004. Nonsteroidal anti-inflammatory drugs for the prevention of Alzheimer's disease: a systematic review. Neuroepidemiology **23:** 159–169.
6. RISNER, M.E. et al. 2006. Efficacy of rosiglitazone in a genetically defined population with mild-to-moderate Alzheimer's disease. Pharmacogenomics J. **6:** 246–254.
7. TOBINICK, E. et al. 2006. TNF-alpha modulation for treatment of Alzheimer's disease: a 6-month pilot study. MedGenMed. **8:** 25.
8. SZCZEPANIK, A.M., D. RAMPE & G.E. RINGHEIM. 2001. Amyloid-beta peptide fragments p3 and p4 induce pro-inflammatory cytokine and chemokine production in vitro and in vivo. J. Neurochem. **77:** 304–317.
9. WHITE, J.A. et al. 2005. Differential effects of oligomeric and fibrillar amyloid-beta 1–42 on astrocyte-mediated inflammation. Neurobiol. Dis. **18:** 459–465.
10. MÜNCH, G. et al. 2003. Advanced glycation endproducts and pro-inflammatory cytokines in transgenic Tg2576 mice with amyloid plaque pathology. J. Neurochem. **86:** 283–289.
11. APELT, J. & R. SCHLIEBS. 2001. Beta-amyloid-induced glial expression of both pro- and anti-inflammatory cytokines in cerebral cortex of aged transgenic Tg2576 mice with Alzheimer plaque pathology. Brain Res. **894:** 21–30.
12. MÜNCH, G. et al. 1997. Advanced glycation endproducts in ageing and Alzheimer's disease. Brain Res. Brain Res. Rev. **23:** 134–143.
13. KUHLA, B. et al. 2007. Effect of pseudophosphorylation and cross-linking by lipid peroxidation and advanced glycation end product precursors on tau aggregation and filament formation. J. Biol. Chem. **282:** 6984–6991.
14. LÜTH, H.J. et al. 2005. Age- and stage-dependent accumulation of advanced glycation end products in intracellular deposits in normal and Alzheimer's disease brains. Cereb Cortex. **15:** 211–220.
15. HOLMLUND, L., V. CORTES TORO & K. IVERFELDT. 2002. Additive effects of amyloid beta fragment and interleukin-1beta on interleukin-6 secretion in rat primary glial cultures. Int. J. Mol. Med. **10:** 245–250.
16. GASIC-MILENKOVIC, J. et al. 2003. Beta-amyloid peptide potentiates inflammatory responses induced by lipopolysaccharide, interferon -gamma and 'advanced glycation endproducts' in a murine microglia cell line. Eur. J. Neurosci. **17:** 813–821.
17. ALEXANDER, G.E. et al. 2002. Longitudinal PET Evaluation of Cerebral Metabolic Decline in Dementia: A Potential Outcome Measure in Alzheimer's Disease Treatment Studies. Am. J. Psychiatry **159:** 738–745.
18. ZELL, R. et al. 1997. TNF-alpha and IL-1 alpha inhibit both pyruvate dehydrogenase activity and mitochondrial function in cardiomyocytes: evidence for primary impairment of mitochondrial function. Mol Cell Biochem. **177:** 61–67.
19. ZENTELLA, A., K. MANOGUE & A. CERAMI. 1993. Cachectin/TNF-mediated lactate production in cultured myocytes is linked to activation of a futile substrate cycle. Cytokine **5:** 436–447.
20. BERG, S. et al. 2003. Proinflammatory cytokines increase the rate of glycolysis and adenosine-5′-triphosphate turnover in cultured rat enterocytes. Crit. Care Med. **31:** 1203–1212.
21. JARVENPAA, T. et al. 2003. A 90-year-old monozygotic female twin pair discordant for Alzheimer's disease. Neurobiol Aging. **24:** 941–945.
22. YOSHIKAWA, M. et al. 1999. Effects of phosphodiesterase inhibitors on cytokine production by microglia. Mult. Scler. **5:** 126–133.
23. MIELKE, R. et al. 1998. Propentofylline enhances cerebral metabolic response to auditory memory stimulation in Alzheimer's disease. J. Neurol. Sci. **154:** 76–82.
24. SAJITHLAL, G. et al. 2002. Receptor for advanced glycation end products plays a more important role in cellular survival than in neurite outgrowth during retinoic acid-induced differentiation of neuroblastoma cells. J. Biol. Chem. **277:** 6888–6897.
25. LUE, L.F. et al. 2005. Preventing activation of receptor for advanced glycation endproducts in Alzheimer's disease. Curr Drug Targets CNS Neurol. Disord. **4:** 249–266.
26. NAKAGAMI, Y. et al. 2002. A novel beta-sheet breaker, RS-0406, reverses amyloid beta-induced cytotoxicity and impairment of long-term potentiation in vitro. Br. J. Pharmacol. **137:** 676–682.
27. CHANEY, M.O. et al. 2005. RAGE and amyloid beta interactions: atomic force microscopy and molecular modeling. Biochim. Biophys. Acta **1741:** 199–205.
28. WONG, A. et al. 2001. Anti-inflammatory antioxidants attenuate the expression of inducible nitric oxide synthase mediated by advanced glycation endproducts in murine microglia. Eur. J. Neurosci. **14:** 1961–1967.

29. HENEKA, M.T. & G.E. LANDRETH. 2007. PPARs in the brain. Biochim. Biophys. Acta **1771:** 1031–1045.
30. SPARKS, D.L. *et al*. 2006. Statin therapy in Alzheimer's disease. Acta Neurol. Scand. Suppl. **185:** 78–86.
31. LANDER, H.M. *et al*. 1997. Activation of the receptor for advanced glycation end products triggers a p21(ras)-dependent mitogen-activated protein kinase pathway regulated by oxidant stress. J Biol. Chem. **272:** 17810–17814.

Some Natural Compounds Enhance N$^\varepsilon$-(Carboxymethyl)lysine Formation

YUKIO FUJIWARA,[a,c] NAOKO KIYOTA,[a] KEITA MOTOMURA,[a,b] KATSUMI MERA,[a] MOTOHIRO TAKEYA,[c] TSUYOSHI IKEDA,[b] AND RYOJI NAGAI[a]

[a]*Department of Medical Biochemistry,*
[b]*Natural Medicine, and*
[c]*Cellular Pathology, Faculty of Medical and Pharmaceutical Sciences, Kumamoto University, Kumamoto 860-8556, Japan*

Since pyridoxamine, which traps intermediates in the Maillard reaction and lipid peroxidation reaction, significantly inhibits the development of retinopathy and neuropathy in the streptozotocin-induced diabetic rat, treatment with advanced glycation end product inhibitors and antioxidants may be a potential strategy for the prevention of clinical diabetic complications. However, the paradoxical effect of green tea has been reported; although plasma hydroperoxide levels were ameliorated, the level of N$^\varepsilon$-(carboxyethyl)lysine (CML) in tendon and plasma was increased by the oral administration of green tea to diabetic rats. In the present study, we measured the effect of natural compounds on CML formation by enzyme-linked immunosorbent assay. A significant amount of CML was observed when bovine serum albumin was incubated with ribose for 7 days. Under the same conditions, natural compounds, such as desgalactotigonin, showed inhibitory effects, whereas quercetin and acteoside enhanced CML formation, indicating that natural compounds contain both inhibitors and enhancers for CML formation.

Key words: N$^\varepsilon$-(carboxymethyl)lysine; anti-AGE antibody; natural compounds

Introduction

Immunological studies using anti-advanced glycation end product (AGE) antibodies, particularly the anti-N$^\varepsilon$-(carboxyethyl)lysine (CML) antibody, have greatly contributed to our understanding of AGE-modified proteins *in vivo*. Monoclonal anti-AGE antibody (6D12), which recognizes CML and N$^\varepsilon$-(carboxyethyl)lysine (CEL), has successfully demonstrated the presence of CML and CEL in several human tissues, suggesting a potential link of AGEs to aging and age-enhanced disease processes, such as diabetic nephropathy, diabetic retinopathy and diabetic neuropathy, atherosclerosis,[1] hemodialysis-related amyloidosis, Alzheimer's disease, actinic elastosis of the skin, and pulmonary fibrosis. We previously demonstrated that CML is generated by the oxidative cleavage of Amadori products by the hydroxyl radical and peroxynitrite.[2] Furthermore, CML formation was also observed when glycated-human serum albumin was incubated with activated neutrophils and was completely inhibited in the presence of a hypochlorous acid scavenger[3]; this suggests that CML is an important biological marker of oxidative stress *in vivo*.

Pyridoxamine, which traps intermediates in the Maillard reaction and lipid peroxidation reaction, significantly inhibits CML formation and the development of retinopathy and neuropathy in the streptozotocin-induced diabetic rat,[4] indicating that treatment with AGE inhibitors may be a potential strategy for the prevention of clinical diabetic complications. For this reason, we have developed the assay system for CML formation by enzyme-linked immunosorbent assay (ELISA) and measured the inhibitory effects of natural compounds on CML formation.

Materials and Methods

Inhibitory Effect of Natural Compounds on CML Formation

Natural compounds, such as desgalactotigonin, neoaspidistrin, acteoside, and quercetin 3-sambubioside, were prepared as described previously.[5,6] Bovine serum albumin (BSA) (2 mg/mL)

Address for correspondence: Ryoji Nagai, Ph.D., Department of Medical Biochemistry, Faculty of Medical and Pharmaceutical Sciences, Kumamoto University, Honjo 1-1-1, Kumamoto 860-8556, Japan. Fax: +81 96 364 6940.

nagai-883@umin.ac.jp

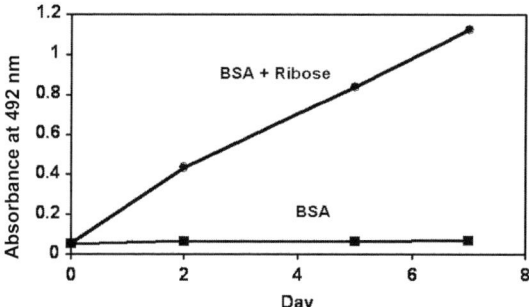

FIGURE 1. Formation of N^{ε}-(carboxymethyl)lysine (CML) during the incubation of bovine serum albumin (BSA) with ribose. The samples (10 μg/mL) were coated on the enzyme-linked immunosorbent assay (ELISA) plate and incubated for 2 h. The wells were washed and blocked with gelatin, followed by reaction with 6D12. The antibodies bound to the wells were detected by horseradish peroxidase-conjugated anti-mouse immunoglobulin G antibody.

was incubated with 33 mmol/L ribose in phosphate-buffered saline solution (PBS) at 37°C for 7 days in the presence (1 mg/mL) or absence of the natural compounds. The yield of CML was measured by ELISA.

ELISA

ELISA was performed as described previously.[3] Briefly, each well of a 96-well microtiter plate was coated with 100 μL of the indicated concentration of sample in PBS and incubated for 2 h. The wells were washed three times with PBS containing 0.05% Tween 20 (washing buffer). The wells were then blocked with 0.5% gelatin in PBS for 1 h. After washing three times, the wells were incubated for 1 h with 100 μL of the indicated concentration of monoclonal anti-CML antibody (6D12). After triplicate washing, the wells were incubated with horseradish peroxidase-conjugated anti-mouse immunoglobulin G, followed by reaction with 1,2-phenylenediamine dihydrochloride. The reaction was terminated with 100 μL of 1.0 mol/L sulfuric acid, and then the absorbance was read at 492 nm with a micro-ELISA plate reader.

Results and Discussion

Hammes *et al.*[7] reported that thiamine and benfotiamine prevent intracellular AGE formation by reducing the concentration of methylglyoxal, a strong AGE-precursor, and hence inhibit diabetic retinopathy. Furthermore, Babaei-Jadidi *et al.*[8] also demonstrated that administration of thiamine and benfotiamine resulted in reduction of intracellular methylglyoxal concentration by increasing transketolase expression and prevented the development of diabetic nephropathy in diabetic rats. Taken together, these findings suggest that compounds that inhibit intracellular AGE formation could be potentially useful agents for the treatment of diabetic complications and atherosclerosis. CML is now widely accepted as one of the most important structures generated through post-translational modifications that contribute to the pathogeneses of age-related disorders.

In the present study, we developed the assay system for CML formation by ELISA and measured the inhibitory effect of natural compounds on CML

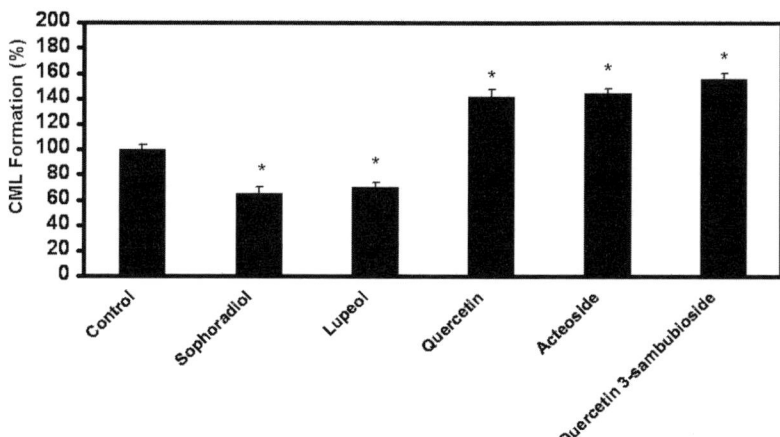

FIGURE 2. Enhancing effect of natural compounds on CML formation. BSA (2 mg/mL) was incubated with 33 mmol/L ribose in phosphate-buffered saline solution at 37°C for 7 days in the presence (1 mg/mL) or absence of a natural compound. The yield of CML was measured by ELISA. Bar indicates mean ± SD; *$P < 0.005$ versus control.

formation. As shown in FIGURE 1, incubation of BSA with ribose for 7 days resulted in the generation of CML. Under the same conditions, natural compounds, such as desgalactotigonin, showed inhibitory effects, whereas quercetin and acteoside enhanced CML formation (FIG. 2).

CML is recognized by the receptor for AGE (RAGE), and CML–RAGE interaction activates cell signaling pathways, such as nuclear factor kappa-B (NF-κB), and enhances the expression of vascular cell adhesion molecule-1 in human umbilical vein endothelial cells. Furthermore, Alikhani et al.[9] demonstrated that CML-collagen is recognized by RAGE and induces fibroblast apoptosis via the expression of caspase-3, -8, and -9. As described in their summary, the oral administration of green tea to diabetic rats enhances the level of CML in tendon. These reports demonstrate that the excessive administration of crude natural products may cause aggravation of diabetic complications.

Acknowledgments

This work was supported in part by Grants-in-Aid for Scientific Research (No. 18790619 to R.N.) from the Ministry of Education, Science, Sports and Cultures of Japan.

Conflict of Interest

The authors declare no conflicts of interest.

References

1. KUME, S. et al. 1995. Immunohistochemical and ultrastructural detection of advanced glycation end products in atherosclerotic lesions of human aorta with a novel specific monoclonal antibody. Am. J. Pathol. **147:** 654–667.
2. NAGAI, R. et al. 2002. Peroxynitrite induces formation of N(epsilon)-(carboxymethyl) lysine by the cleavage of Amadori product and generation of glucosone and glyoxal from glucose: novel pathways for protein modification by peroxynitrite. Diabetes **51:** 2833–2839.
3. MERA, K. et al. 2007. Hypochlorous acid generates N epsilon-(carboxymethyl)lysine from Amadori products. Free Radic Res. **41:** 713–718.
4. METZ, T.O. et al. 2003. Pyridoxamine, an inhibitor of advanced glycation and lipoxidation reactions: a novel therapy for treatment of diabetic complications. Arch. Biochem. Biophys. **419:** 41–49.
5. YAN, W. et al. 1996. Steroidal saponins from fruits of Tribulus terrestris. Phytochemistry **42:** 1417–1422.
6. TAKEDA, Y. et al. 1998. An acyclic monoterpene glucosyl ester from Lantana lilacia. Planta Med. **64:** 78–79.
7. HAMMES, H.P. et al. 2003. Benfotiamine blocks three major pathways of hyperglycemic damage and prevents experimental diabetic retinopathy. Nat. Med. **9:** 294–299.
8. BABAEI-JADIDI, R. et al. 2004. High-dose thiamine therapy counters dyslipidaemia in streptozotocin-induced diabetic rats. Diabetologia **47:** 2235–2246.
9. ALIKHANI, Z. et al. 2005. Advanced glycation end products enhance expression of pro-apoptotic genes and stimulate fibroblast apoptosis through cytoplasmic and mitochondrial pathways. J. Biol. Chem. **280:** 12087–12095.

Immunological Detection of N$^\omega$-(Carboxymethyl)arginine by a Specific Antibody

KATSUMI MERA,[a,b] YUKIO FUJIWARA,[b] MASAKI OTAGIRI,[a] NORIYUKI SAKATA,[c] AND RYOJI NAGAI[b]

[a]*Department of Biopharmaceutics, Graduate School of Pharmaceutical Sciences, Kumamoto University, Kumamoto 862-0973, Japan*

[b]*Department of Medical Biochemistry, Faculty of Medical and Pharmaceutical Sciences, Kumamoto University, Kumamoto 860-8556, Japan*

[c]*Department of Pathology, School of Medicine, Fukuoka University, Fukuoka 814-0180, Japan*

N$^\omega$-(carboxymethyl)arginine (CMA) is an acid-labile advanced glycation end product (AGE) that was discovered in enzymatic hydrolysate of glycated collagen. Subsequently, CMA was also detected in human serum, and its level in patients with diabetes was found to be higher than in people without the disease. However, the histological localization of CMA and its pathophysiological significance remains poorly understood. Here, to address this issue, we developed a monoclonal antibody specific for CMA. This antibody reacted with CMA and CMA-protein adduct, whereas it did not cross-react with its analogues, such as N$^\varepsilon$-(carboxymethyl)lysine and S-(carboxymethyl)cysteine, indicating that the antibody specifically recognizes CMA. Upon immunohistochemical analysis, a significant CMA immnoreactivity was found in atherosclerotic lesions, whereas no such immunoreactivity was observed in normal regions. This suggests that the accumulation of CMA in tissue proteins may contribute to the pathophysiologies associated with aging and age-related diseases.

Key words: N$^\omega$-(carboxymethyl)arginine; advanced glycation end products; monoclonal antibody; diabetes; atherosclerosis

Introduction

The long-term incubation of proteins with glucose leads, through formation of Schiff base and Amadori products, to the generation of advanced glycation end products (AGEs) of the Maillard reaction. AGE-modified proteins increase during the normal process of aging, but this is markedly accelerated in people with diabetes who have sustained hyperglycemia. Immunohistochemical studies have detected AGE modification in several pathological tissues. A recent study demonstrated that N$^\varepsilon$-(carboxymethyl)lysine (CML), one of the major AGE structures, accumulates in several tissue proteins including kidneys of patients with diabetic nephropathy and chronic renal failure.[1]

atherosclerotic lesions of arterial walls,[2] amyloid fibrils in hemodialysis-related amyloidosis,[3] and actinic elastosis of the photo-aged skin.[4] Furthermore, the AGE inhibitors aminoguanidine and pyridoxamine block CML formation and retard the development of early renal disease in the streptozotocin-diabetic rat.[5] These studies strongly suggest an association between CML and the development of diabetic complications.

N$^\omega$-(carboxymethyl)arginine (CMA) is an acid-labile AGE and was discovered in enzymatic hydrolysate of glycated collagen.[6] Subsequently, CMA was also detected in human serum by electrospray ionization/liquid chromatography/mass spectrometry, and its level was found to be higher in diabetic patients than people without diabetes.[7] Although CML is known to be generated not only from glucose but also from glyoxal and unsaturated fatty acids, little is known about the formation pathway of CMA (FIG. 1).[8] Furthermore, the histological localization of CMA and its pathophysiological significance remains poorly understood. In the present study, we raised a monoclonal

Address for correspondence: Ryoji Nagai, Ph.D., Department of Medical Biochemistry, Faculty of Medical and Pharmaceutical Sciences, Kumamoto University, Honjo 1-1-1, Kumamoto 860-8556, Japan. Fax: +81-96-364-6940.

nagai-883@umin.ac.jp

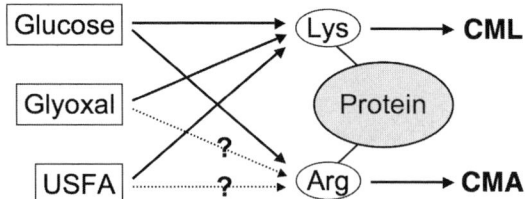

FIGURE 1. Possible formation pathways of N^ε-(carboxymethyl)lysine (CML) and N^ω-(carboxymethyl)arginine (CMA).

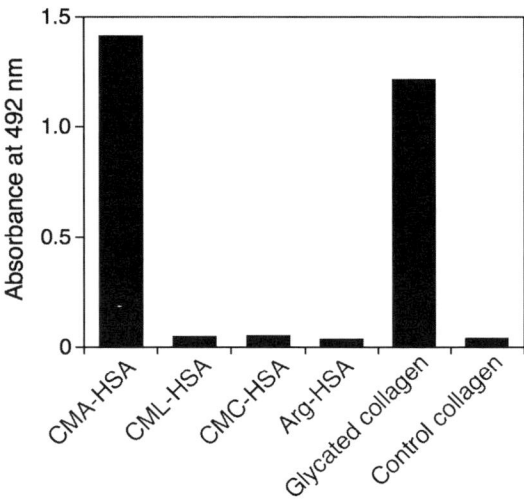

FIGURE 2. The immunoreactivity of monoclonal anti-CMA antibody. The samples (10 μg/mL) were coated on the ELISA plate and incubated for 1 h. The wells were washed and blocked with gelatin, followed by reaction for 1 h with anti-CMA antibody. The antibodies bound to wells were detected by horseradish peroxidase-conjugated anti-mouse immunoglobulin G antibody. Glycated collagen was prepared by incubation of type I collagen (1.5 mg/mL) with 100 mmol glucose in 100 mmol sodium phosphate buffer (pH 7.4) at 37°C for 4 weeks. Collagen incubated without glucose was used as a control.

antibody specific for CMA to detect the histological localization of CMA.

Materials and Methods

Preparation of a Monoclonal Antibody against CMA

CMA was conjugated to human serum albumin (HSA) or keyhole limpet hemocyanin (KLH) according to the method described by Jono et al.[9] Arginine-conjugated HSA (Arg-HSA), CML-conjugated HSA (CML-HSA), and S-(carboxymethyl) cysteine-conjugated HSA (CMC-HSA) were also similarly prepared. Splenic lymphocytes from a Balb/c mouse immunized with CMA-conjugated HSA (CMA-HSA) were fused to myeloma P3U1 cells. After successive screening, we obtained one cell line that was positive to CMA-conjugated KLH (CMA-KLH) but negative to CML-HSA and KLH. The cell line was used to prepare ascites fluid in mice, and monoclonal anti-CMA antibody was further purified by affinity chromatography on protein G-immobilized Sepharose gel to IgG. The immunoreactivity of the purified antibody was evaluated by enzyme-linked immunosorbent assay (ELISA) according to the method described by Mera et al.[10]

Immunohistochemistry

For immunohistochemical analysis of CMA accumulation in human atherosclerotic lesions, tissue samples of the aorta and internal carotid artery were obtained from three nondiabetic patients (two males, 68 and 74 years; one female, 78 years). The samples consisted of fibroatheroma with or without ulceration (histological classification type V or VI of atherosclerosis from the American Hearth Association). Serial frozen sections were made by cryostat and subjected to indirect immunohistochemical staining using the monoclonal anti-CMA antibody.

Results and Discussion

The Immunoreactivity of Monoclonal Anti-CMA Antibody

The immunoreactivity of monoclonal anti-CMA antibody was determined by noncompetitive ELISA. This antibody reacted with CMA-HSA but not with Arg-HSA, CML-HSA, or CMC-HSA (FIG. 2), suggesting that this antibody specifically recognizes the CMA structure. Since recent studies using the enzymatic hydrolysis method have shown that CMA was generated in glycated collagen,[6] we next measured the immunoreactivity of the monoclonal anti-CMA antibody to glycated collagen by noncompetitive ELISA. This antibody reacted with glycated collagen, whereas its reactivity with control collagen was negligible (FIG. 2), suggesting that this antibody is useful for detecting CMA in glycated proteins.

Detection of CMA in the Atherosclerotic Lesions of Human Aorta

A significant CMA immunoreactivity was found in atherosclerotic lesions, whereas no such immunoreactivity was observed in normal regions. Furthermore, the positive staining patterns of the anti-CMA antibody were significantly weakened when the antibody was pretreated with CMA-conjugated HSA. These results suggest that the accumulation of CMA in tissue proteins may contribute to the pathophysiology of aging and age-related diseases, including atherosclerosis.

Acknowledgments

We are grateful to Yasunori Iwao, Kazuhiro Takeo, Keita Motomura, and Mime Nagai of our laboratory for their collaborative endeavors. This work was supported in part by Grants-in-Aid for Scientific Research (No. 18790619 to R.N.) from the Ministry of Education, Science, Sports and Cultures of Japan. We thank The Naito Foundation for a travel scholarship to K.M. that enabled us to participate in the 9th International Symposium on the Maillard Reaction.

Conflict of Interest

The authors declare no conflicts of interest.

References

1. YAMADA, K. et al. 1994. Immunohistochemical study of human advanced glycosylation end-products (AGE) in chronic renal failure. Clin. Nephrol. **42:** 354–361.
2. KUME, S. et al. 1995. Immunohistochemical and ultrastructural detection of advanced glycation end products in atherosclerotic lesions of human aorta with a novel specific monoclonal antibody. Am. J. Pathol. **147:** 654–667.
3. MIYATA, T. et al. 1996. Identification of pentosidine as a native structure for advanced glycation end products in beta-2-microglobulin-containing amyloid fibrils in patients with dialysis-related amyloidosis. Proc. Natl. Acad. Sci. USA **93:** 2353–2358.
4. MIZUTARI, K. et al. 1997. Photo-enhanced modification of human skin elastin in actinic elastosis by N(epsilon)-(carboxymethyl)lysine, one of the glycoxidation products of the Maillard reaction. J. Invest. Dermatol. **108:** 797–802.
5. DEGENHARDT, T.P. et al. 2002. Pyridoxamine inhibits early renal disease and dyslipidemia in the streptozotocin-diabetic rat. Kidney Int. **61:** 939–950.
6. IIJIMA, K. et al. 2000. Identification of N(omega)-carboxymethylarginine as a novel acid-labile advanced glycation end product in collagen. Biochem. J. **347:** 23–27.
7. ODANI, H. et al. 2001. Identification of N(omega)-carboxymethylarginine, a new advanced glycation end-product in serum proteins of diabetic patients: possibility of a new marker of aging and diabetes. Biochem. Biophys. Res. Commun. **285:** 1232–1236.
8. FU, MX. et al. 1996. The advanced glycation end product, Nepsilon-(carboxymethyl)lysine, is a product of both lipid peroxidation and glycoxidation reactions. J. Biol. Chem. **271:** 9982–9986
9. JONO, T. et al. 2004. Nepsilon-(Carboxymethyl)lysine and 3-DG-imidazolone are major AGE structures in protein modification by 3-deoxyglucosone. J. Biochem. (Tokyo) **136:** 351–358.
10. MERA, K. et al. 2007. Hypochlorous acid generates N epsilon-(carboxymethyl)lysine from Amadori products. Free Radic. Res. **41:** 713–718.

Isolation and Partial Characterization of Four Fluorophores Formed by Nonenzymatic Browning of Methylglyoxal and Glutamine-derived Ammonia

CELINE NIQUET,[a] SERGE PILARD,[b] DAVID MATHIRON,[b] AND FREDERIC J. TESSIER[a]

[a]*Institut Polytechnique LaSalle Beauvais, Beauvais, France*
[b]*Plateforme Analytique, Faculté des Sciences, Amiens, France*

An aqueous solution of L-glutamine (50 mmol/L) and methylglyoxal (100 mmol/L) was incubated at 120°C for 3 h in a 200 mmol/L phosphate buffer (pH 7.4). Four major fluorophores were revealed on the HPLC chromatogram. The same four fluorophores were obtained from the heating of a mixture of ammonia and methylglyoxal. After purification and concentration, they were structurally characterized by electrospray ionization mass spectrometry (ESI-MS) using the high resolution and tandem mass spectrometry capabilities of a quadrupole time-of-flight MS. The accurate mass measurement of their $[M+H]^+$ ions, the MS fragment patterns, and the presence of one to two nitrogen indicate the formation of fluorophores with molecular formulas of $C_7H_7NO_3$, $C_8H_9NO_3$, $C_{12}H_{14}N_2O_4$, and $C_{12}H_{14}N_2O_5$. These results show that, in an aqueous solution, free glutamine undergoes a rapid degradation, leading to the formation of ammonia which reacts with methylglyoxal to form fluorescent heterocyclic Maillard products.

Key words: glutamine; ammonia; methylglyoxal; fluorophores; Maillard reaction

Introduction

The Maillard reaction is responsible for the formation of aroma taste-active compounds and potentially toxic molecules during thermal food processing. The N terminus amino groups and the N-containing side chains of amino acids on proteins are suspected to be the major precursors of the Maillard reaction products (MRPs) in food. However, the recent discovery of acrylamide in food products rich in free asparagine has raised major concerns regarding the implication of free amino acids in the formation of compounds in food cooked at high temperature.[1] Free glutamine, which is found at relatively high concentration in many raw foods[2–4] and which has a similar chemical structure as asparagines, may be a perfect target for the Maillard reaction. Methylglyoxal, a dicarbonyl also found in many foods, is known to be much more reactive than glucose with amino groups.[5] The purpose of this study was to identify the MRPs from methylglyoxal and glutamine.

Address for correspondence: Frederic J. Tessier, Institut Polytechnique LaSalle Beauvais, 19 rue Pierre Waguet, BP 30313, 60026 Beauvais Cedex, France. Voice: +33-3-44063851; fax: +33-3-44062526.
frederic.tessier@lasalle-beauvais.fr

Materials and Methods

Materials

L-glutamine and methylglyoxal 40% in aqueous solution were obtained commercially from Sigma (St. Louis, MO). Ammonia solution (35% HPLC grade), acetonitrile (HPLC grade), and formic acid (98/100%) were purchased from Fisher Bioblock Scientific (Ilkirch, France) and disodium hydrogen phosphate, sodium dihydrogen phosphate, and water for liquid chromatography (LC)–mass spectrometry (MS) analysis were purchased from VWR International (Fontenay-Sous-Bois, France). Water for model systems and HPLC was prepared in the laboratory using an Elga Purelab UHQ II system (Veolia Water STI, Le Plessis Robinson, France).

Incubation of Glutamine or Ammonia with Methylglyoxal

A mixture of 50 mmol/L glutamine (or ammonia) and methylglyoxal (100 mmol/L) was prepared in a 200 mmol/L phosphate buffer (pH 7.4) and heated in an oil bath for 3 d at 70°C in screw-capped Pyrex tubes. After incubation, the mixture was cooled at room temperature, filtered through a 0.45 μm nylon filter, and subjected to chromatographic separation.

Isolation of Four Fluorophores by Reversed-phase HPLC with Fluorescence Detection

Analytical HPLC was performed with a Thermo Separation Products HPLC instrument (Thermo Electron Corporation, Courtaboeuf, France) coupled to a fluorescence 2475 detector (Waters, Guyancourt, France). The chromatography was performed using a 250 × 4.6 mm, 4 μm particle size, Synergi MAX-RP column (Phenomenex, Le Pecq, France) eluted at a flow rate of 1 mL/min. After injection of the sample (10 μL), analysis was performed using a linear gradient of 2–5% acetonitrile from 0 to 10 min with 0.1% formic acid followed by an isocratic elution for 16 min. The eluent was monitored for fluorescence at the excitation and emission wavelengths of 330 and 450 nm, respectively. Each fluorophore of interest was collected using a Gilson FC203B fraction collector (Gilson SAS, Villiers Le Bel, France), concentrated in a Speed-Vac concentrator (Thermo Savant, Courtaboeuf, France), and analyzed by MS.

High-resolution Mass Spectrometry

High-resolution electrospray mass spectra (ESI-HRMS) in the positive-ion mode were obtained on a Q-TOF *Ultima Global* hybrid quadrupole time-of-flight instrument (Waters-Micromass, Manchester, UK) that was equipped with a pneumatically assisted electrospray (Z-spray) ionization source and an additional sprayer (Lock Spray) for the reference compound.

The purified compounds were dissolved in water, and the solutions were directly introduced (5 μL/min) through an integrated syringe pump into the electrospray source. The source and desolvation temperatures were 80 and 120°C, respectively. Nitrogen was used as the drying and nebulizing gas at flow rates of 350 and 50 L/h, respectively. The capillary voltage was 3 kV, the cone voltage 40 V, and the RF lens1 energy was optimized for each sample (10–50 V). For the tandem MS (MS/MS) experiments, argon was used as the collision gas at an indicated analyzer pressure of 5×10^{-5} Torr, and the collision energy was optimized for each specific precursor ion (10–25 V). Lock mass correction, using appropriate cluster ions of an orthophosphoric acid solution (0.1% in H_2O/CH_3CN 50/50, v/v), was applied for accurate mass measurements. The mass range was typically 50–1000 Da and spectra were recorded at 1 s/scan in the profile mode at a resolution of 10,000 FWMH. Data acquisition and processing were performed with MassLynx 4.0 software (Waters-Micromass).

Kinetic Studies on the Formation of the Four Compounds

Ammonia (50 mmol/L) was heated with methylglyoxal (100 mmol/L) at 120°C in 200 mmol/L phosphate buffer, pH 7.4. Samples were taken in triplicate at different heating times from 30 min to 24 h. Each sample was immediately cooled in ice water to stop any further reaction and filtered before analysis by HPLC–MS. The analyses were performed on a Surveyor system coupled to a Finnigan LTQ mass spectrometer (Thermo Electron Corporation) using a 150 × 2.0 mm, 5 μm particle size, Luna C18(2) column (Phenomenex). For elution, a gradient was used of 0–5% solvent B for 7 min and 5% solvent B from 7.01 to 17 min at a flow rate of 0.3 mL/min (solvent A, 0.1% formic acid in water; solvent B, 0.1% formic acid in acetonitrile). The four fluorophores were detected and quantified by MS/MS.

Results and Discussion

The instability of glutamine in aqueous model systems and in food has been known for several decades. Our recent data reveal that glutamine significantly influences the extent of fluorescence in simple model systems compared to other amino acids.[6] Despite extensive studies conducted to understand the Maillard reaction on free amino acids, the chemical structures of brown products and fluorophores formed from the reaction between glutamine and α-dicarbonyls have not yet been discovered.

The glutamine residues disappeared in 15 min in our model reaction, and four major fluorescent MRPs (A to D) separated by RP-HPLC fluorescence were formed in the presence of methylglyoxal (FIG. 1). The same unknown fluorophores were formed when glutamine was replaced by the same concentration of ammonia, despite a lower yield of formation for the latter (data not shown). The MRP-D probably has two isomers well separated in the HPLC chromatogram but that gives the same MS and MS/MS spectra. Taken together, these results clearly indicate that the glutamine is not the only precursor of the four MRPs, although it appears to be the major one.

After HPLC purification, the high resolution of the Q-TOF mass spectrometer allowed us to obtain unambiguous molecular formulas for each fluorophore (FIG. 2, TABLE 1). So, the MRP-A, -B, -C, and -D have the following formulas: $C_7H_7NO_3$, $C_8H_9NO_3$, $C_{12}H_{14}N_2O_4$, and $C_{12}H_{14}N_2O_5$. The odd or even number of nitrogen atoms per molecule was confirmed

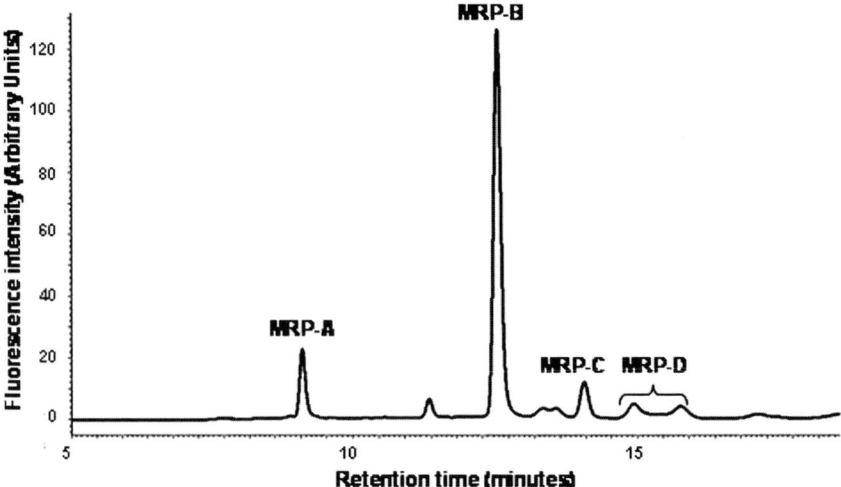

FIGURE 1. Reversed-phase HPLC profiles showing the presence of the four major fluorophores formed during the 72 h incubation mixture of glutamine and methylglyoxal at 70°C. The fluorescence was monitored at 330/450 nm.

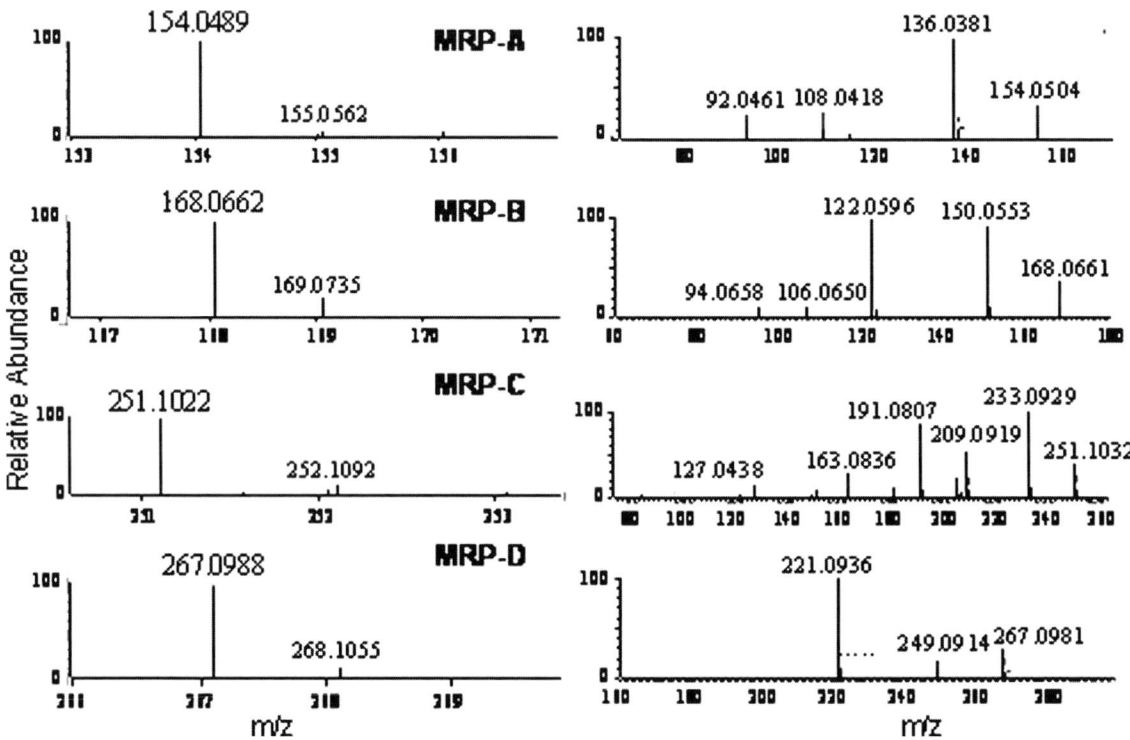

FIGURE 2. High-resolution electrospray ionization mass spectroscopy (*left*) and tandem mass spectroscopy (*right*) spectra of MRP-A, MRP-B, MRP-C, and MRP-D isolated from the reaction mixture of glutamine (or ammonia) and methylglyoxal heated at 70°C for 3 d.

by the even or odd mass of the corresponding molecular ion. The fragment patterns indicate that the MRP-A and -B are probably two substituted pyridine rings which differ only from one methyl group (FIG. 2). This was confirmed by the difference of 14 Da (CH_2: 14.0157) between the respective fragments of each molecule (150/136, 122/108, and 106/92). MRP-C and -D are two more complex heterocyclic

TABLE 1. Accurate mass measurements of the [M+H]+ ions of the Maillard reaction product compounds

Compound	Mass (Da)	Calc. mass (Da)	Mass difference (mDa)	Formula
MRP-A	154.0489	154.0504	−1.5	$C_7H_8NO_3$
MRP-B	168.0662	168.0661	0.1	$C_8H_{10}NO_3$
MRP-C	251.1022	251.1032	−1.0	$C_{12}H_{15}N_2O_4$
MRP-D	267.0988	267.0981	0.7	$C_{12}H_{15}N_2O_5$

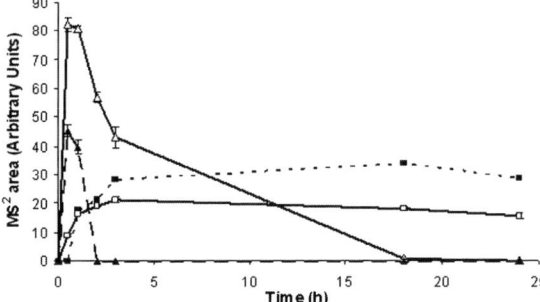

FIGURE 3. Kinetics of formation of the fluorophores MRP-A ■, MRP-B □, MRP-C △, and MRP-D ▲ obtained with the heat treatment (120°C, pH 7.4) of ammonia in the presence of methylglyoxal. Data are the mean values ± SD for triplicate incubations.

fluorophores which involve two molecules of ammonia and four of methylglyoxal. The difference of 16 Da observed for their $[M+H]^+$ ions at m/z 251, 267 and for their two fragments at m/z 233, 249 (-H_2O: 18.0106) was attributed to an additional oxygen atom (O: 15.9949). The fluorescence spectra of each isolated Maillard product, measured at pH 7.0, show specific excitation and emission maxima at 315 and 435, 320 and 440, 355 and 450, and 315 and 435 nm for the MRP-A, -B, -C, and -D, respectively.

When the kinetics of the formation of the four fluorophores at 120°C in a simple aqueous model system buffered at pH 7.4 are examined, significant differences in the kinetic profiles are observed (Fig. 3). The rate of formation of MRP-A and -B increased as a function of time and reached a plateau after 3 h. A slight degradation was observed after 18 h of incubation. On the other hand, the kinetics of the formation of MRP-C and -D revealed that these fluorophores are two unstable compounds. In these cases, the synthesis was followed, after the lapse of an hour of the heating treatment, by rapid degradation, which indicates that MRP-C and -D are two intermediates of the Maillard reaction.

In conclusion, our results clearly show that glutamine, through its degradation in ammonia, plays part in the formation of fluorescent MRPs. Their complete chemical characterization, their mechanisms of formation in the presence of methylglyoxal and their potential formation with amine other than glutamine and ammonia, and their presence in foodstuffs are all under investigation.

Conflict of Interest

The authors declare no conflicts of interest.

References

1. TAEYMANS, D. *et al.* 2004. A review of acrylamide: an industry perspective on research, analysis, formation, and control. Crit. Rev. Food Sci. Nut. **44:** 323–347.
2. ARISTOY, M.C. & F. TOLDRA. 1998. Concentration of free amino acids and dipeptides in porcine skeletal muscles with different oxidative patterns. Meat Sci. **50:** 327–332.
3. EPPENDORFER, W.H. & S.W. BILLE. 1996. Free and total amino acid composition of edible parts of beans, kale, spinach, cauliflower and potatoes as influenced by nitrogen fertilisation and phosphorus and potassium deficiency. J. Sci. Food Agric. **71:** 449–458.
4. PRATTA, G. *et al.* 2004. Glutamine and glutamate levels and related metabolizing enzymes in tomato fruits with different shelf-life. Sci. Hort. **100:** 341–347.
5. THORNALLEY, P.J. 2005. Dicarbonyl intermediates in the Maillard reaction. Ann. N.Y. Acad. Sci. **1043:** 111–117.
6. NIQUET, C. & F.J. TESSIER. 2007. Free glutamine as a major precursor of brown products and fluorophores in Maillard reaction systems. Amino Acids. **33:** 165–171.

Receptor for Advanced Glycation End Product Polymorphisms and Type 2 Diabetes

The CODAM Study

KATRIEN H.J. GAENS,[a,b] CARLA J.H. VAN DER KALLEN,[a,b] MARLEEN M.J. VAN GREEVENBROEK,[a,b] EDITH J. FESKENS,[c] COEN D.A. STEHOUWER,[a,b] AND CASPER G. SCHALKWIJK[a,b]

[a]*Department of Internal Medicine, Laboratory for Metabolism and Vascular Medicine, Maastricht University, Maastricht, the Netherlands*

[b]*Cardiovascular Research Institute Maastricht, Maastricht, the Netherlands*

[c]*Division of Human Nutrition, Section Nutrition and Epidemiology, Wageningen University, Wageningen, the Netherlands*

Genetic variation in the receptor for advanced glycation end products (RAGE) gene may alter the expression and function of RAGE and affect disease development and outcome. We investigated whether single nucleotide polymorphisms (SNPs) in RAGE were associated with diabetes and parameters of glucose homeostasis. In total, nine SNPs of RAGE were analyzed in individuals with and without type 2 diabetes in CODAM: a cohort study of diabetes and atherosclerosis, Maastricht. A significant difference in genotype frequency of SNP rs3134945 was observed between the nondiabetic control subjects, subjects with impaired glucose metabolism, and diabetic patients. The C allele of this polymorphism was significantly associated with higher fasting glucose concentrations, 2-h postload glucose concentrations, insulin levels, and homeostasis model assessment of insulin resistance. These results indicate that SNP rs3134945 or a locus in linkage disequilibrium with this polymorphism may be involved in the development of insulin resistance and diabetes. Because the functionality of this polymorphism is not known, the mechanism whereby this polymorphism contributes to the development of insulin resistance and diabetes has to be further elucidated.

Key words: RAGE; polymorphism; type 2 diabetes; insulin resistance

Introduction

The formation of advanced glycation end products (AGEs) and biological consequences have been shown to play an important role in the development of a variety of pathological processes, such as inflammation and diabetic complications.[1,2] AGEs exert their effects by a number of mechanisms including interaction with specific cellular receptors, which trigger signal transduction pathways resulting in the activation of proinflammatory cellular responses.[3]

The best characterized and most extensively studied AGE-binding protein is the receptor for AGEs (RAGE). RAGE is a member of the immunoglobulin (Ig) superfamily of cell surface molecules that can bind a heterogeneous group of ligands.[4] Binding of ligands to RAGE has been shown to activate multiple cellular signaling cascades, resulting in the expression of a number of genes responsible for cellular dysfunction and inflammation.[5] The expression of RAGE is low under normal circumstances, while pathogenic conditions, such as inflammation and diabetes, are associated with a sustained upregulation of RAGE.[6,7] The receptor protein is composed of three extracellular Ig-like regions, i.e., one V-type Ig domain and two C-type Ig domains. Molecular studies have shown that the ligand-binding site resides within the V-type Ig domain. These extracellular domains are followed by a single hydrophobic transmembrane domain and a short cytosolic tail that is essential for RAGE-mediated cellular effects upon engagement of the ligand.

Address for correspondence: Casper G. Schalkwijk, Ph.D., Department of Internal Medicine, Laboratory for Metabolism and Vascular Medicine, University Hospital Maastricht, Debeyelaan 25, P.O. Box 5800, 6202 AZ Maastricht, the Netherlands. Voice: +31-43-3882186; fax: +31-43-3875006.

C.Schalkwijk@intmed.unimaas.nl

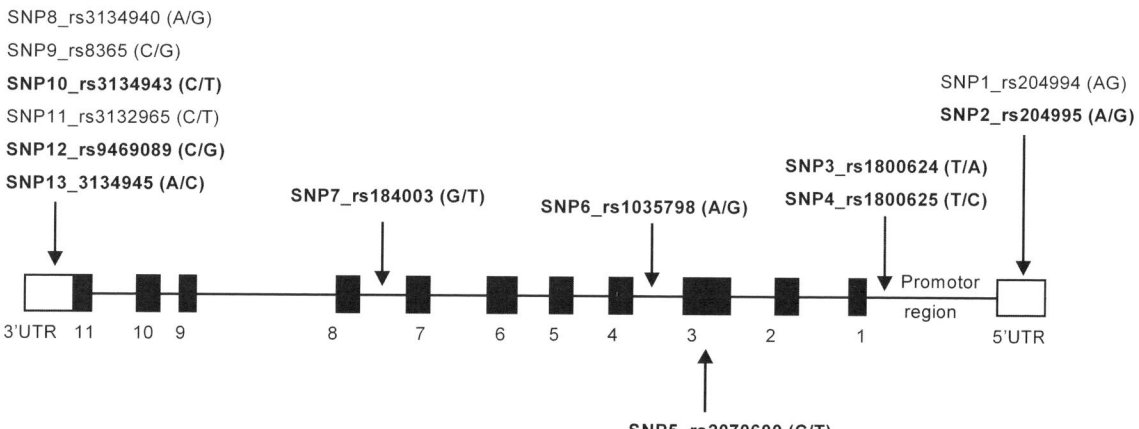

FIGURE 1. Genetic structure of the receptor for advanced glycation end product (RAGE). Exons are shown as *black boxes* (exon 1–11); promotor and intron sequences are drawn as *lines*. Open boxes represent 3′ and 5′ untranslated regions. Using the HapMap/Haploview database and literature, 13 SNPs are identified in the RAGE gene. Arrows indicate the locations of these 13 sequence changes. RAGE polymorphisms that are genotypes in our study are highlighted in *bold*.

Previous studies have demonstrated that the RAGE–AGE axis is an important contributing factor in the development of diabetes and diabetic complications.[1,2] Genetic variations in the gene coding for RAGE (AGER gene) (FIG. 1) may have an effect on the transcriptional activity of the RAGE gene, can alter AGE processing in tissues, and can change binding affinity of ligands to RAGE.[8,9] A relatively large number of single nucleotide polymorphisms (SNPs) in the RAGE gene have been identified. Most genetic studies have focused on the association between RAGE polymorphisms and diabetic complications, but studies investigating RAGE polymorphisms in relation to glucose tolerance status are limited. In this report, we studied the relationship between RAGE polymorphisms, glucose tolerance status, and parameters of glucose homeostasis in individuals with and without type 2 diabetes (T2DM).

Materials and Methods

Study Population

The genetic association between polymorphisms in RAGE, glucose tolerance status, and parameters of glucose homeostasis was investigated in a cohort study of diabetes and atherosclerosis, Maastricht (CODAM cohort). This cohort comprised 574 unrelated subjects that were phenotyped with respect to glucose metabolism with an oral glucose tolerance test according to the 1999 World Health Organization criteria. Subjects were classified as normal glucose tolerant (NGT) ($n = 301$); having an impaired glucose metabolism (IGM) ($n = 127$), i.e., impaired glucose tolerant or impaired fasting glucose; and T2DM ($n = 146$). Details of patient recruitments and inclusion and exclusion criteria are reviewed in Kruijshoop et al.[10]

SNP Selection

To include all potential SNPs of the RAGE gene and polymorphisms located at the promotor region or at the 5′ untranslated region or 3′ untranslated region of the gene, we enlarged the region of interest with 3000 bp upstream and downstream of the RAGE gene. Using the HapMap and Haploview database (http://www.hapmap.org/index.html.en), all SNPs that have a minor allele frequency (MAF) >1% were identified. Initially, 10 SNPs (SNP1, SNP2, SNP4, SNP6, SNP8, SNP9, SNP10, SNP11, SNP12, and SNP13) were identified for genotyping. Using Haploview, only six tag-SNPs, which cover the common variation of the RAGE gene including the region 3 kb upstream and downstream, were selected for genotyping using a linkage disequilibrium (LD) $r^2 > 0.8$. These tag-SNPs are SNP2 (in LD with SNP1), SNP4 (in LD with SNP8 and SNP9), SNP6, SNP10, SNP12, and SNP13 (in LD with SNP11). In addition to the selection of these tag-SNPs by HapMap and Haploview, three additional SNPs (SNP3, SNP5, and SNP7) were selected for their potential functional effects based on the literature. These SNPs were not identified by the HapMap database or had a MAF <1%. In total, nine polymorphisms were genotyped in the CODAM cohort. The gene structure of RAGE including the RAGE polymorphisms, their position, and sequence change are given in FIGURE 1.

TABLE 1. Characteristics of the NGT, IGM, and T2DM group

	NGT ($n = 301$)	IGM ($n = 127$)	T2DM ($n = 146$)
Age (years)	58.2 ± 7.3	59.4 ± 6.7	60.7 ± 6.2
Men (%)	60	60	66
Body mass index (kg/m^2)[a]	27.6 ± 3.9	28.9 ± 4.3*	30.3 ± 4.6*
Fasting glucose (mmol/L)[a]	5.3 ± 0.4	5.9 ± 0.5*	8.0 ± 1.8*
2-h postload glucose (mmol/L)[a]	5.6 ± 1.2	8.6 ± 1.7*	13.3 ± 3.3*
Fasting insulin (pmol/L)[a]	59.5 ± 28.2	77.7 ± 44.0*	101.6 ± 63.2*
HOMA-IR[a]	1.1 ± 0.5	1.5 ± 0.8	2.1 ± 1.3*

Data represented as mean ± SD or percentage. *$P < 0.01$ versus NGT. Determined by linear regression analysis with glucose tolerance status (NGT, IGM, and T2DM) as dummy variable and adjustment for age and gender. NGT = normal glucose tolerance; IGM = impaired glucose metabolism or impaired fasting glucose; T2DM = type 2 diabetes.

[a]Log transformed.

TABLE 2. Genotype frequency of SNP13 in the CODAM population

		NGT	IGM	T2DM	P value*
SNP13	Genotype A/A	9 (3.2%)	2 (1.8%)	0 (0.0%)	0.046
	Genotype A/C	109 (38.4%)	38 (34.5%)	46 (34.6%)	
	Genotype C/C	166 (58.5%)	70 (63.6%)	87 (65.4%)	

*Determined by a general linear model for adjustment of age and gender.

SNP Genotyping

DNA was isolated from peripheral blood leucocytes (obtained from EDTA-anticoagulated blood) by a standard DNA extraction method. SNP2, SNP5-7, SNP10, SNP12, and SNP13 were genotyped in the ABI PRISM 7900HT Sequence Detection System (Applied Biosystems, Foster City, CA) in a 5 μL reaction according to the manufacturer's instructions. Because of the presence of a 63 bp insertion–deletion close to SNP3 and SNP4, no assay could be developed for Taqman genotyping. Therefore, these promotor SNPs were determined together with the −407 to −345 63 bp deletion from a single PCR using the primers 5′-AAAAACATGAGAAACCCCAGAA-3′ and 5′-AGAGCCCCCGATCCTATTTA-3′. After amplification, the products (228 bp) were digested with the restriction endonucleases AluI and MfeI and were run on an ethidium–bromide-stained 4% agarose gel. Approximately 93% of the genotyping analyses were successful for all SNPs.

Statistical Analysis

Differences in genotype distributions from those expected for Hardy–Weinberg equilibrium were tested by a χ^2 test. The statistical significance of differences in genotype frequencies between the NGT group, IGM group, and T2DM group were tested by a general linear model adjusted for age and gender. A multivariate linear regression model was used to detect independent association of RAGE polymorphisms in the CODAM study population with glucose tolerance status, fasting glucose, 2-h postload glucose, insulin concentration, and homeostasis model assessment of insulin resistance (HOMA-IR) (an estimate of insulin resistance)[11] with adjustment for age and gender. RAGE polymorphisms were included in the model as dummy variables.

Results

Characteristics of the study cohort are presented in TABLE 1. In this study, we investigated whether the genetic variations in the RAGE gene were associated with glucose tolerance status. The genotype frequencies were in agreement with those predicted by the Hardy–Weinberg equilibrium for all polymorphisms in all groups.

In our study, no significant differences in genotype frequencies of the polymorphisms SNP2-7, SNP10, and SNP12 were observed between NGT subjects, IGM subjects, and diabetic patients (data not shown). Linear regression analysis with adjustment for age and gender demonstrated that these SNPs were not associated with parameters of glucose homeostasis and insulin resistance, such as fasting glucose, 2-h postload glucose, insulin concentration, and HOMA-IR (data not shown). In contrast, SNP13 located at the 3′ untranslated region of the RAGE gene showed a weak but significant association with glucose tolerance status ($P < 0.05$) (TABLE 2). Linear regression analysis

TABLE 3. Association between SNP13 and parameters of glucose homeostasis

	Allele A	Allele C	P value[*]
Fasting glucose concentration (mmol/L)[a]	6.0 ± 1.5	6.1 ± 1.5	0.050
2-h postload glucose concentration (mmol/L)[a]	7.3 ± 3.2	8.0 ± 3.8	0.035
Fasting insulin concentration (pmol/L)[a]	70.0 ± 47.6	76.8 ± 45.3	0.031
HOMA-IR[a]	1.4 ± 1.0	1.5 ± 0.9	0.039

HOMA-IR, homeostasis model assessment of insulin resistance.
Data represented as mean ± SD. [*]Determined by linear regression model adjusted for age and gender.
[a] Log transformed.

showed that the C allele of this polymorphism was associated with higher fasting glucose concentration, 2-h postload glucose concentration, insulin concentration, and HOMA-IR (TABLE 3).

Discussion

Genetic variation in the gene coding for RAGE may alter gene expression and function of the receptor and affect disease development and outcome.[12] In the present investigation, we studied the association between RAGE polymorphisms and glucose tolerance status and showed a significant difference in genotype frequencies of RAGE polymorphism SNP13 between the nondiabetic control subjects, the IGM subjects, and diabetic patients. We demonstrated that the C allele of this polymorphism was associated with higher fasting glucose concentration, 2-h postload glucose concentration, insulin concentration, and HOMA-IR. These results indicate that RAGE polymorphism SNP13 or a locus in LD with this SNP may contribute to the development of insulin resistance and diabetes. Our study is, to the best of our knowledge, the first study identifying this polymorphism and analyzing its association with glucose tolerance status and glucose parameters. This polymorphism is located in the 3′UTR of the RAGE gene, but the functionality of this polymorphism is not known. Further studies are needed to confirm our results and to find possible mechanisms explaining the association between SNP13 and diabetes.

Conflict of Interest

The authors declare no conflicts of interest.

References

1. BROWNLEE, M. 2005. The pathobiology of diabetic complications: a unifying mechanism. Diabetes **54:** 1615–1625.
2. YAN, S.F. *et al*. 2003. Glycation, inflammation, and RAGE: a scaffold for the macrovascular complications of diabetes and beyond. Circ Res. **93:** 1159–1169.
3. SCHMIDT, A.M. *et al*. 1994. Cellular receptors for advanced glycation end products. Implications for induction of oxidant stress and cellular dysfunction in the pathogenesis of vascular lesions. Arterioscler Thromb. **14:** 1521–1528.
4. NEEPER, M. *et al*. 1992. Cloning and expression of a cell surface receptor for advanced glycosylation end products of proteins. J. Biol. Chem. **267:** 14998–15004.
5. BIERHAUS, A. *et al*. 2001. Diabetes-associated sustained activation of the transcription factor nuclear factor-kappaB. Diabetes **50:** 2792–2808.
6. BRETT, J. *et al*. 1993. Survey of the distribution of a newly characterized receptor for advanced glycation end products in tissues. Am. J. Pathol. **143:** 1699–1712.
7. BUCCIARELLI, L.G. *et al*. 2002. RAGE is a multiligand receptor of the immunoglobulin superfamily: implications for homeostasis and chronic disease. Cell Mol Life Sci. **59:** 1117–1128.
8. HUDSON, B.I. *et al*. 2001. Effects of novel polymorphisms in the RAGE gene on transcriptional regulation and their association with diabetic retinopathy. Diabetes **50:** 1505–1511.
9. HOFMANN, M.A. *et al*. 2002. RAGE and arthritis: the G82S polymorphism amplifies the inflammatory response. Genes Immun. **3:** 123–135.
10. KRUIJSHOOP, M. *et al*. 2004. Validation of capillary glucose measurements to detect glucose intolerance or type 2 diabetes mellitus in the general population. Clin. Chim. Acta **341:** 33–40.
11. BONORA, E. *et al*. 2000. Homeostasis model assessment closely mirrors the glucose clamp technique in the assessment of insulin sensitivity: studies in subjects with various degrees of glucose tolerance and insulin sensitivity. Diabetes Care **23:** 57–63.
12. HUDSON, B.I. *et al*. 2005. Diabetic vascular disease: it's all the RAGE. Antioxid Redox Signal. **7:** 1588–15600.

Pentosidine Effects on Human Osteoblasts in Vitro

ROBERTA SANGUINETI,[a] DANIELA STORACE,[a] FIAMMETTA MONACELLI,[a] ALBERTO FEDERICI,[b] AND PATRIZIO ODETTI[a]

[a]*Division of Geriatrics, Deptartment of Internal Medicine and Medical Specialities, University of Genoa, Genoa, Italy*

[b]*Orthopaedics Surgery Unit, St. Anthony Hospital, Genoa, Italy*

Osteoporosis, a multifactorial and progressive skeletal metabolic disease, is characterized by low-mass density and structural deterioration of bone micro-architecture that leads to enhanced bone fragility and increased susceptibility to fractures. Recently, it has been proposed that age-related bone loss could be correlated with the glycoxidative process. The aim of the present study was to investigate the *in vitro* effects of pentosidine, a glycoxidative end product, on human osteoblasts (HOb). The mineralization rate, the specific bone markers (alkaline phosphatase [ALP], collagen Iα1 [COL Iα1], osteocalcin [BGP]), and the human receptor for advanced glycation end products (RAGE) gene expression have been evaluated. Pentosidine incubation of HOb caused a significant decrease in ALP, Col Iα1, and RAGE mRNA levels, but only the RAGE gene expression decreased with no dose dependency. Moreover, pentosidine incubation of osteoblasts hampered the formation of bone nodules. No effect was observed on BGP gene expression under all experimental conditions. Our data gives further support to a detrimental effect of AGEs on bone that leads to functional alterations of osteoblasts. This study addresses a crucial role of protein glycoxidation in the bone mineralization process. AGEs formation and accumulation in bone may be one of the first pathogenetic steps of bone remodeling in aging and in age-related diseases, leading to enhanced bone mass loss.

Key words: osteoporosis; nonenzymatic glycation; pentosidine; osteoblasts; bone nodule; mineralization

Introduction

Osteoporosis, a progressive age-related disease, is characterized by insufficient bone strength.[1] Bone matrix homeostasis results from the synergistic alteration of the neoapposition activity by osteoblasts and the resorptive activity accomplished by osteoclasts. The pathogenetic mechanisms influencing the onset of the disease are complex. Bone remodeling is heavily influenced by nutritional and hormonal factors, but the glycation pathway has been implicated as a strong contributor to this age-related disease. Recent literature data support the hypothesis that protein glycation may affect bone remodeling.[2,3]

Generation of advanced glycation end products (AGEs) is an inevitable process *in vivo*. The content of AGEs increases during aging in all tissues, including bone, and contributes to the structural and functional changes of bone proteins, leading to intramolecular or intermolecular cross-links,[4,5] partially explaining the deleterious effects of AGEs on bone biomechanical properties.[3] AGEs can regulate the proliferation and differentiation of osteoblasts, and AGEs-specific binding sites are present in cultured osteoblast-like cells.[6] For example, in the presence of AGEs, the receptor for AGE (RAGE) is able to elicit, in osteoblasts, the activation of nuclear factor kappa B, resulting in an increased expression of cytokines, growth factors, and adhesion molecules.[7] Therefore, AGEs accumulation in bone has been associated with a decrease of cortical and trabecular biomechanical properties and with functional impairment of bone cells.[8–10]

Pentosidine is a reliable and well-characterized marker of nonenzymatic glycation in collagen,[4,11,12] and its significant increase was first observed in diabetes mellitus, in end-stage renal failure, and during the aging process.[2,4] The aim of this study was to assess the *in vitro* pentosidine effects (0.25 μmol/L and

Address for correspondence: Patrizio Odetti, M.D., Department of Internal Medicine and Medical Specialties (Di.M.I.), Viale Benedetto XV, 6, 16132 Genova, Italy. Voice/fax: +39 010 353 7985.
odetti@unige.it

1 μmol/L) on a primary culture of human osteoblasts in order to evaluate the effects of pentosidine on bone matrix mineralization.

Materials and Methods

Chemicals

TRYPLE Express and TRIzol® reagents were provided by Invitrogen Life Technologies (Carlsbad, CA); specific reagents for reverse transcriptase (RT)-PCR were provided by Promega (Milan, Italy); tissue culture dishes were provided by BD Biosciences (Franklin Lakes, New Jersey); and all other reagents were provided by Sigma-Aldrich (Milan, Italy).

Technical Instruments

Images of calcium-rich deposits and mineralized nodules were collected using an optical Olympus IX71 microscope (Olympus Italia, Milan, Italy) and processed using the analySIS software (Soft Imaging System GmbH, Muenster, Germany).

Subjects

Trabecular bone samples were obtained from patients undergoing orthopedic surgery; all donors were free from bone-related diseases, diabetes, or autoimmune pathologies. Informed consent was given by all donors before surgery.

Isolation of Human Trabecular Bone-derived cells

Human osteoblast cells (HOb) were isolated from trabecular bone samples. Bone chips were washed with phosphate-buffered saline solution and then cut into $2\,mm^2$ pieces prior to digestion in collagenase P (1.2 mg/mL in Dulbecco's Modified Eagle Medium[DMEM]) for 2 h at 37 °C. The digested bone chips were seeded in $30\,cm^2$ tissue culture dishes with DMEM containing 10% fetal bovine serum, 2 mmol/L L-glutamine, and mixed antibiotics (Basal DMEM) at 37 °C in a 5% CO_2-humidified atmosphere. After several days, HOb outgrowing from the bone chips was observed. Medium was changed twice weekly, and bone chips were removed to prevent contamination. HOb became confluent within 3 weeks from the beginning of the culture period.

Bone Nodule Formation

HOb were grown until confluent, removed with TRYPLE Express, and plated at 3×10^5 in 10-cm^2 tissue culture dishes. Once confluence was achieved, nodule formation was induced using a Basal DMEM supplemented with 0.3 mmol/L ascorbic acid, 10 mmol/L sodium β-glycerophosphate, and 0.1 μmol/L dexamethasone (Osteogenic Medium). Medium was changed every 3–4 days. The experiment was carried out until the stage of maximal calcium influx (day 21).

Experimental Conditions

HOb were cultured for 21 days in Osteogenic Medium plus 0.25 and 1 μmol/L pentosidine (PENT). PENT was prepared as reported[13] and evaluated in accordance with the modified Odetti method.[14] The morphological characteristics and the proliferation rate of PENT-treated HOb were observed after 7, 14, and 21 days using an optical microscope (magnification 100×).

Cell Viability and Proliferation Assessment

The viability and proliferation rate were evaluated in control cells and PENT-treated HOb (0.25 μmol/L and 1 μmol/L) using the 3-(4,5-dimethylthiazolyl-2)-2,5-diphenyltetrazolium bromide (MTT) assay according to the manufacturer's instructions.

Detection and Quantification of Mineralization: Alizarin Red S Staining

Calcium-rich deposits and nodular patterns *in vitro* were determined by alizarin red S (ARS) staining.[15] Mineralized nodules were destained by cetylpyridinium chloride.[16] ARS extracts were read at 550 nm, and the results were reported as optical density (OD) compared to a standard curve.

mRNA Isolation and RT-PCR Analysis

Total RNA was extracted by means of a one-step phenol-guanidinium isothiocyanate method of Chomczynski and Sacchi,[17] using TRIzol reagent according to the manufacturer's recommendations. RT-PCR was carried out as described,[18] using specific primers designed according to sequences available in the data banks (TABLE 1). The mRNA content was normalized using the housekeeping gene (glyceraldehyde-3-phosphate dehydrogenase [GAPDH]). Digital images of PCR products were analyzed by means of a Gel-Doc system (Bio-Rad Laboratories, Milan, Italy) and the bands quantified by Quantity One software (Bio-Rad Laboratories). The results were expressed as percentage of test cells versus control.

Statistical Analyses

The results represent the mean number of at least three independent experiments performed under the same conditions. Analyses were carried out with the Graph Pad Prism 4.0 (Graph Pad Software, San Diego, CA). Results were expressed as mean ± SD. The statistical significance of experimental data was evaluated by one-way analysis of variance followed by Dunnett's post test; a P value < 0.05 was considered significant.

TABLE 1. Oligonucleotide primers used for conventional RT-PCR

Gene	Primer sequence (strand)	Product size (bp)	Temperature (°C)	PCR (cycles)
ALP	5'-3' ACGTGGCTAAGAATGTCATC	470	47	40
	3'-5' CTGGTAGGCGATGTCCTTA			
COL Iα1	5'-3' AAGAGGCGAGAGGGTTTCC	475	51	32
	3'-5' ATCACCAGGTTCACCTTTCG			
BGP	5'-3' GGCAGCGAGGTAGTGAAGAG	306	53	39
	3'-5' CTGGAGAGGAGCAGAACTGG			
RAGE	5'-3' GATCCCCGTCCCACCTTCTCCTGTAGC	556	68	40
	3'-5' CACGGCTCCTCCTCTTCCTCCTGGTTTTCTG			
GAPDH	5'-3' GCTCATGACCACAGTC	970	60	35
	3'-5' TGTAGGCCATGAGGTC			

Abbreviations: ALP, alkaline phosphatase; BGP, osteocalcin; COL Iα1, collagen Iα1; GAPDH, glyceraldehyde-3-phosphate dehydrogenase; RAGE, receptor for advanced glycation end products.

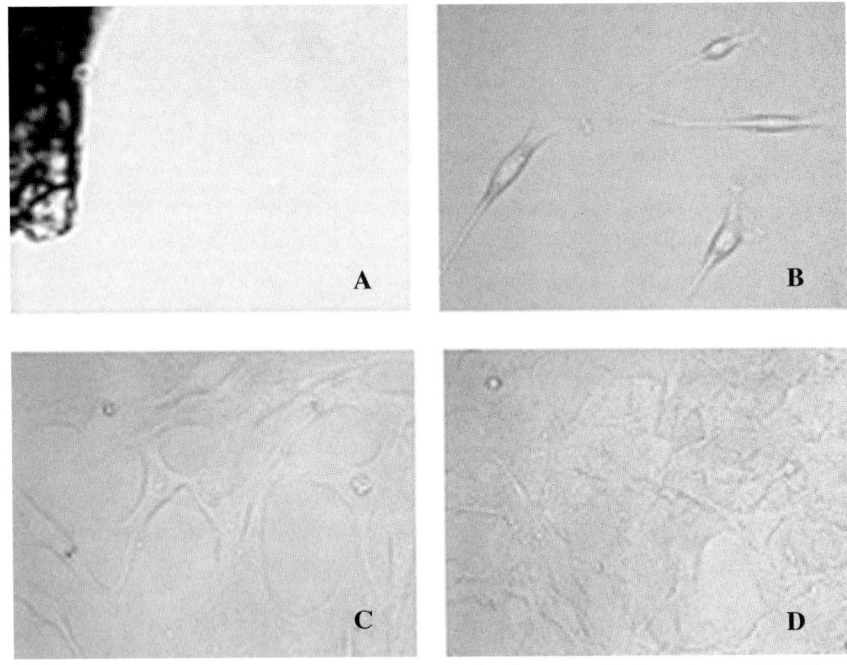

FIGURE 1. Isolation of human trabecular bone-derived cells. **(A)** 24 h after plating; **(B)** 4 days after plating; **(C)** 14 days after plating; **(D)** cells confluent after 21 days (magnification 400×).

Results

No cell growth was observed in any chip 24 h after plating (FIG. 1A). Round or polygonal HOb were observed migrating from the bone chips by day 4 after the start of the culture. By day 7 and day 14, cells began migrating and proliferating more rapidly (FIG. 1B and C), and after 3 weeks, the migrated cells exhibited a high proliferation rate (FIG. 1D).

Collagen Iα1 (COL Iα), alkaline phosphatase (ALP), and osteocalcin (BGP) genes were detected in HOb at day 21 in the Osteogenic Medium, confirming their proper differentiation toward the osteoblast phenotype (FIG. 2).

At day 21, the HOb cultured with 1 μmol/L PENT (experimental medium changed twice a week) did not reveal a cytotoxic effect as the HOb was approximately 90% viable compared to the control.

PENT exerted a dose-dependent (0.25 or 1 μmol/L) inhibition on HOb differentiation, as shown by RT-PCR (FIG. 3), on both ALP and Col Iα1 gene expression (-30% and -70% for 0.25 μmol/L and 1 μmol/L PENT, respectively, compared to the control; $n = 3$, $P < 0.01$) (FIG. 3A, B). No change was

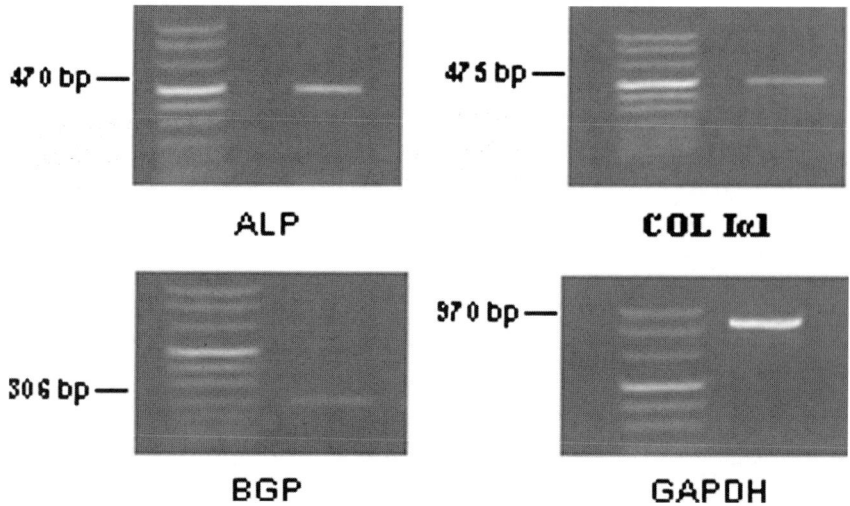

FIGURE 2. Osteoblast phenotype identification. Reverse transcriptase-PCR analysis of alkaline phosphatase (ALP), collagen Iα1 (COL Iα), and osteocalcin (BGP) quantified with a densitometer and Quantity One software.

observed on BGP mRNA levels under any of the experimental conditions (FIG. 3C). RAGE mRNA levels showed a significant PENT-mediated decrease with no dose dependency (FIG. 4).

Isolated bone cells, after a 3-week incubation period with Osteogenic Medium, exhibited osteogenic capacity after 7 days (FIG. 5A) that continued up to 21 days (FIG. 5B, C), producing calcified nodules detected by optical microscope.

PENT-challenged HOb (0.25 and 1 μmol/L) exerted a dose-dependent inhibition on bone matrix mineralization, resulting in a progressively impoverished matrix with the appearance of pale and disorganized nodules (FIG. 5D–I); this was confirmed by the ARS method (FIG. 6A–C).

Using a quantitative destaining procedure, our *in vitro* assessment of matrix mineralization detected a pentosidine dose-dependent inhibition on bone mineralization under all the experimental conditions compared to the control (−55% and −95% for 0.25 μmol/L and 1 μmol/L, respectively) (FIG. 6D).

Discussion

Bone formation may be impaired by AGEs. Bone collagen has a long lifetime, making it susceptible to glycation. The generation and accumulation of AGEs in bone tissue can contribute to the deterioration of bone quality. Katayama reported that AGEs-modified collagen is able to regulate osteoblast proliferation and differentiation,[19] inhibiting their phenotypic expression.[20] Thus, it seems that the glycation of bone proteins is able to affect osteoblast neoapposition of bone mineral matrix.

Our study shows that pentosidine exerts, in a human osteoblast primary culture, a dose-dependent detrimental role, inhibiting the gene expression of osteoblast-specific markers, such as ALP and COL Iα1. This result clearly addresses a pentosidine interference with osteoblast function. The formation of flawed nodules and matrix mineralization further support this finding.

In agreement with our data, Cortizo reported that, in murine cell lines, cellular functions could be influenced by AGEs–bovine serum albumin acute treatment, causing a RAGE gene overexpression through a positive-feedback mechanism, while a chronic AGEs exposition induced a decrease in RAGE gene expression.[21] Notwithstanding, pentosidine is a weak ligand of RAGE[22]; in our experimental conditions, we observed a decreased RAGE gene expression, suggesting, at any rate, a specific receptor involvement.

Furthermore, pentosidine did not exert any effect on BGP gene expression, which suggests a likely low-grade activated state of osteocalcin. When the Osteogenic Medium was supplemented with vitamin D+K (for a 3-week period), a significant increase of BGP gene expression compared to control was observed, while vitamins plus pentosidine incubation resulted in detection of BGP gene expression reduction (data not shown). Our data, demonstrating a pentosidine detrimental effect on human osteoblast functions and a dose-dependent inhibition on bone nodule formation,

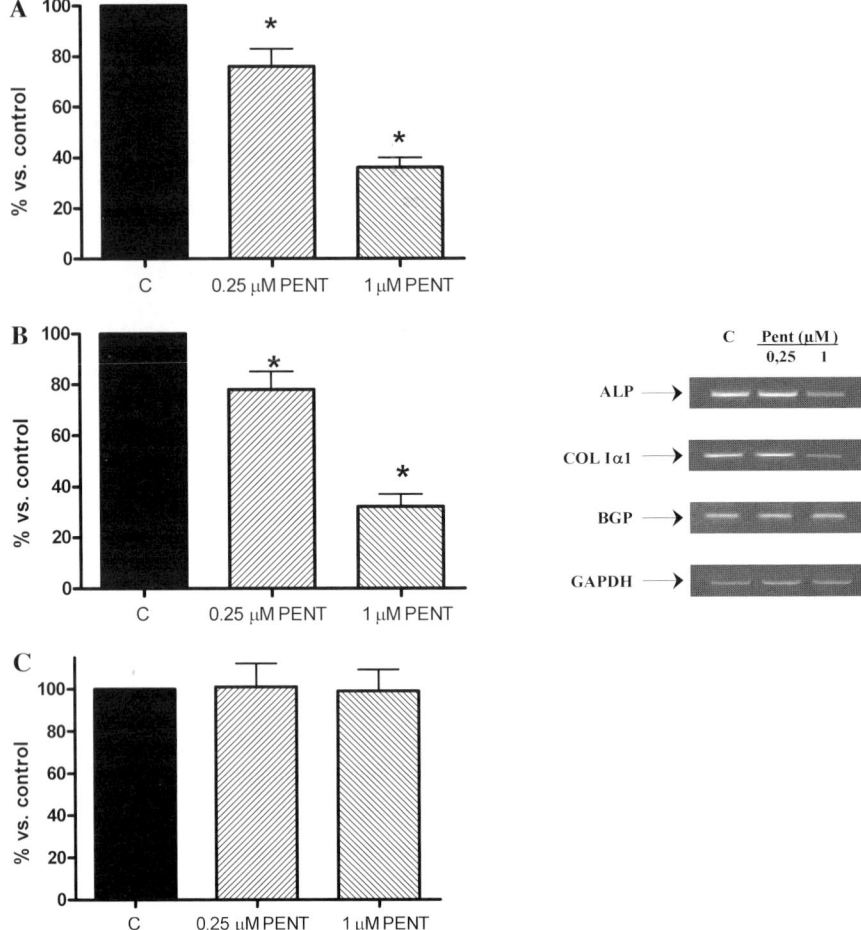

FIGURE 3. Pentosidine effects on gene expression of osteoblast markers. Gene expression of **(A)** ALP, **(B)** COL Iα1, **(C)** BGP. Solid columns are RT-PCR densitometric analysis of control cells, columns with left diagonal hatching are 0.25 μmol/L PENT-incubated cells, and columns with right diagonal hatching are 1 μmol/L PENT-incubated cells (mean of at least three experiments). Representative agarose gels are shown to the right (*$P < 0.01$ versus control).

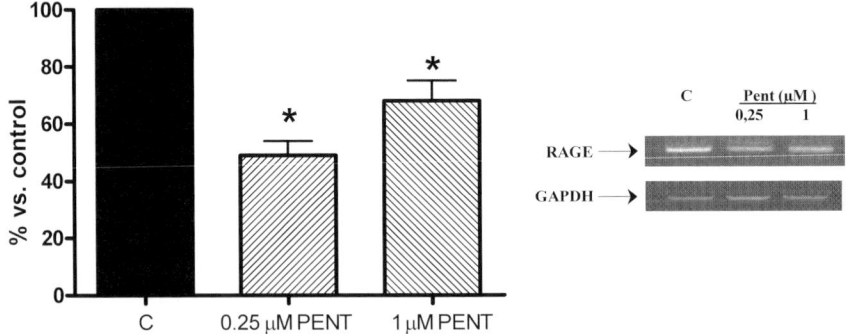

FIGURE 4. Pentosidine effects on RAGE gene expression. Solid column is RT-PCR densitometric analysis of control cells; columns with left diagonal hatching are 0.25 μmol/L PENT-incubated cells, and columns with right diagonal hatching are 1 μmol/L PENT-incubated cells (mean of at least three experiments). Representative agarose gels are shown to the right (*$P < 0.01$ versus control).

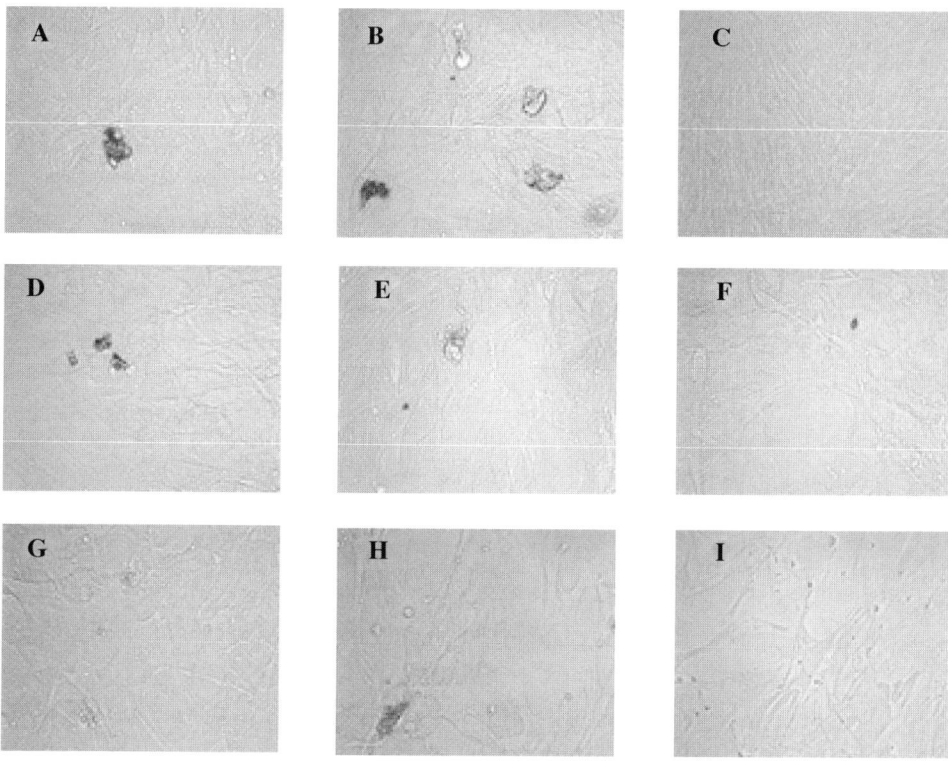

FIGURE 5. Evaluation of pentosidine effects on formation of mineralized nodules. Microphotographs of HOb cultured for 7, 14, 21 days in Osteogenic Medium without PENT (**A, B, C**), with 0.25 μmol/L PENT (**D, E, F**), and with 1 μmol/L PENT (**G, H, I**) (magnification 100×).

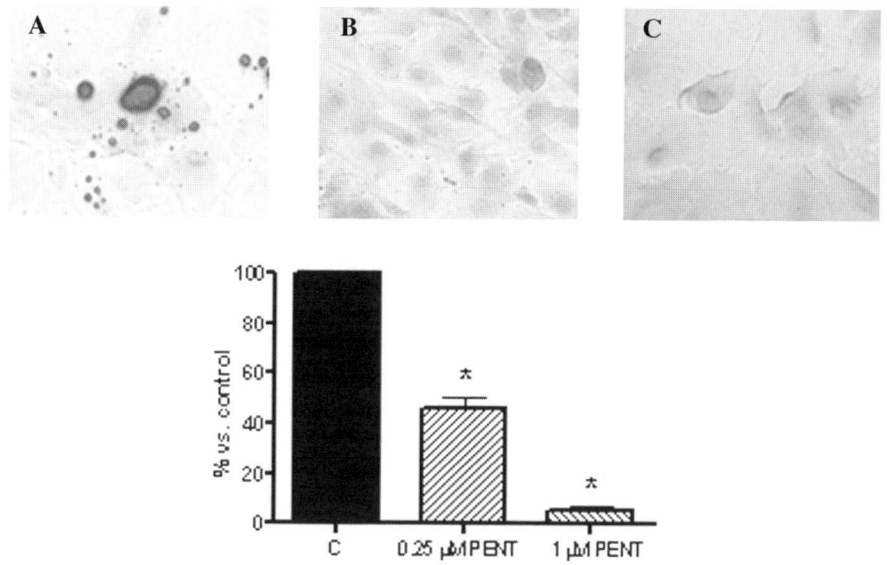

FIGURE 6. Alizarin red S staining of mineralized nodules. Microphotographs of HOb cultured for 21 days in Osteogenic Medium (**A**) without PENT, (**B**) with 0.25 μmol/L PENT, and (**C**) with 1 μmol/L PENT (magnification 100×). Optical density of matrix mineralization in control cells (solid column), in 0.25 μmol/L PENT-incubated cells (left diagonal hatch), and 1 μmol/L PENT-incubated cells (right diagonal hatch) (mean of at least three experiments) (*$P < 0.01$ versus control).

emphasize a conceivable pathogenetic relationship between pentosidine accumulation and age-related bone loss.

Acknowledgments

The authors thank Professor M. Pestarino and coworkers (Department of Biology, University of Genoa, Italy) and Professor U.M. Marinari, Professor A.M. Bassi, and co-workers (Department of Experimental Medicine, University of Genoa, Italy) for their technical support; and Dr M.R. Pantaleo (Alfa Wassermann S.p.A., Bologna, Italy) for providing cetylpyridinium chloride. This work has been partially supported by a grant from the University of Genoa and by a grant from Fondo per gli Investimenti della Ricerca di Base (FIRB) (RBAU01JBH8_004).

Conflict of Interest

The authors declare no conflicts of interest.

References

1. NIH Consensus development panel on osteoporosis prevention, diagnosis and therapy. 2000. Highlights of The Conference. South Med. J. **94:** 569–573.
2. ODETTI, P. *et al.* 2005. Advanced glycation end products and bone loss during aging. Ann. N.Y. Acad. Sci. **1043:** 710–717.
3. HEIN, G.E. 2006. Glycation endproducts in osteoporosis–is there a pathophysiologic importance? Clin. Chim. Acta **371:** 32–36.
4. SELL, D.R. *et al.* 1989. Structure elucidation of a senescence cross-link from human extracellular matrix. Implication of pentoses in the aging process. J. Biol. Chem. **36:** 21597–21602.
5. DUNN, J.A. *et al.* 1991. Age-dependent accumulation of N epsilon-(carboxymethyl)lysine and N-epsilon-(carboxymethylhydroxy)lysine in human skin collagen. Biochemistry **30:** 1205–1210.
6. HEIN, G.E. *et al.* 2006. Advanced glycation end product modification of bone proteins and bone remodelling: hypothesis and preliminary immunohistochemical findings. Ann. Rheum. Dis. **65:** 101–104.
7. THORNALLEY, P.J. 1998. Cell activation by glycated proteins. AGE receptors, receptor recognition factors and functional classification of AGEs. Cell Mol Biol (Noisy-le-grand) **44:** 1013–23.
8. BANSE, X. *et al.* 2002. Cross-link profile of bone collagen correlates with structural organization of trabeculae. Bone **31:** 70–76.
9. HERNANDEZ, J. *et al.* 2005. Trabecular microfracture and the influence of pyridinium and non-enzymatic glycation-mediated collagen cross-links. Bone **37:** 825–832.
10. WANG, X. *et al.* 2002. Age-related changes in the collagen network and toughness of bone. Bone **31:** 1–7.
11. DYER, D.G. *et al.* 1991. Formation of pentosidine during non-enzymatic browning of proteins by glucose. Identification of glucose and other carbohydrates as possible precursors of pentosidine in vivo. J. Biol. Chem. **266:** 11654–11660.
12. VERZIJL, N. *et al.* 2000. Age-related accumulation of Maillard reaction products in human articular cartilage collagen. Biochem. J. **350:** 381–387.
13. GRANDHEE, S.K. *et al.* 1991. Mechanism of formation of the Maillard protein cross-link pentosidine. J. Biol. Chem. **266:** 11649–11653.
14. ODETTI, P. *et al.* 1992. Chromathografic quantification of plasma and red blood cell pentosidine in diabetic and uremic subjects. Diabetes **41:** 153–159.
15. DAWSON, A.B. 1926. A note on the staining of the skeleton of cleared specimens with Alizarin Red S Stain. Technol **1:** 123–124.
16. GREGORY, C.A. *et al.* 2004. An Alizarin Red-based assay of mineralization by adherent cells in culture: comparison with cetylpyridinium chloride extraction. Anal. Biochem. **329:** 77–84.
17. CHOMCZYNSKI, P. *et al.* 1987. Single-step method of RNA isolation by acid guanidinium thiocyanate-phenol-chloroform extraction. Anal. Biochem. **162:** 156–159.
18. BASSI, A.M. *et al.* 2005. Antioxidant status in J774A.1 macrophage cell line during chronic exposure to glycated serum. Biochem. Cell. Biol. **83:** 176–187.
19. HEIN, G.E. 2006. Glycation endproducts in osteoporosis—Is there a pathophysiologic importance? Clin. Chim. Acta **371:** 32–36.
20. KATAYAMA, Y. *et al.* 1996. Role of nonenzymatic glycosylation of type I collagen in diabetic osteopenia. J. Bone Miner. Res. **11:** 931–937.
21. CORTIZO, A.M. *et al.* 2003. Advanced glycation end-products (AGEs) induce concerted changes in the osteoblastic expression of their receptor RAGE and in the activation of extracellular signal-regulated kinases (ERK). Mol. Cell. Biochem. **250:** 1–10.
22. VALENCIA, S.V. *et al.* 2004. Advanced glycation end product ligands for the receptor for advanced glycation end products: biochemicals characterization and formation kinetics. Anal. Biochem. **324:** 68–78.

Strategy for the Study of the Health Impact of Dietary Maillard Products in Clinical Studies

The Example of the ICARE Clinical Study on Healthy Adults

PHILIPPE POUILLART,[a] HÉLÈNE MAUPRIVEZ,[a] LAMIA AIT-AMEUR,[a] AMÉLIE CAYZEELE,[b] JEAN-MICHEL LECERF,[b] FRÉDÉRIC J. TESSIER,[a] AND I. BIRLOUEZ-ARAGON[a]

[a]*Institut Polytechnique LaSalle Beauvais, Beauvais, France*
[b]*Institut Pasteur de Lille, Paris, France*

The study of the health impact of dietary Maillard products (MPs) in realistic clinical studies requires the design of nutritionally equivalent diets with high and low levels of MPs. This difficult challenge may be achieved by setting the high-MP diet at the regular daily level, where the common use of grilling, frying, and roasting processes allows significant amounts of carboxymethyllysine, hydroxymethylfurfural and acrylamide to be formed. In such conditions, we show that major lipid degradation does not occur, nor does degradation of vitamin E or thiamine. Based on this finding, the low-MP diet; must be constructed accordingly, by replacing all high-temperature techniques with steam cooking or the absence of cooking. The cooking fat must be replaced with similar raw fat as seasoning in the low-MP diet, the high caloric density resulting from water loss in the high-MP diet must be compensated by higher food quantities offered in the low-MP diet, and the vitamin loss in fruit and vegetables resulting from high temperatures in the high-MP diet can be circumvented by increasing the corresponding portion size. In the ICARE study, equilibrated diets were proposed, fulfilling all nutritional needs, but with a 3- to 45-fold difference in MP concentrations. Individual quantification of nutritional and MP intakes will ensure the nutritional equivalence of the two diets and allow for quantification of the specific impact of ingested MPs.

Key words: clinical study; CML; acrylamide; HMF; adults

Introduction

The discovery that dietary Maillard products (MPs) may influence biological advanced glycation endproduct (AGE) levels and promote inflammatory reactions in people with diabetes[1] has opened an exciting new research topic. Quantifying MP oral exposure, bioaccessibility, and excretion is an important first issue to address. And further understanding the potential biological effects and health implications is a second major objective. Some studies have dealt with such questions, but the existing data are either controversial or not fully convincing. The reason is probably the different experimental and methodological strategies applied to answer these questions. Beneficial health impacts of MPs in experimental animals, especially with respect to gut metabolism, has often been reported when poorly characterized melanoidins, synthesized or extracted from bread crust or coffee, are administered.[2] Coffee, a melanoidin-rich product, has been confirmed to exert antioxidant[3] properties in humans. However, the results obtained with melanoidins in animals can hardly be extrapolated to humans, for whom such compounds are minimally represented within the total MP of the diet, and in whom biological effects are expected to be strongly modulated by interactions with other prooxidant and proinflammatory dietary MPs. In contrast, decreasing the MP concentration of animal chow through the application of mild processing, despite a decrease in the content of antioxidant melanoidins at the same time, resulted in beneficial effects in rodents, such as a decrease in diabetes-associated atherosclerotic lesions[4] and nephropathy.[5] But MPs in the standard chow were not characterized and are probably

Address for correspondence: Inès Birlouez-Aragon, Institut Polytechnique LaSalle Beauvais, Rue Pierre Waguet, BP 30313, F-60026 Beauvais, France. Voice: +33 03-44-06-38-76; fax: +33 03-44-06-25-26.
ines.birlouez@lasalle-beauvais.fr

very different from those in the human diet. A few clinical studies were carried out under more realistic conditions to examine the global impact of dietary MPs in humans. The common method used is to compare two diets, one composed of severely heat-treated foods, the other composed of foods prepared with low heat. These studies have all shown the deleterious impact of a high-MP diet compared with a low-MP diet.[6,7] However, some have criticized the experimental conditions, suggesting that they could limit the conclusions that can be drawn from such studies.[8] First, the severe heat treatment applied to produce high levels of MPs might induce many changes in the chemical composition of the diet, other than MP levels specifically. Numerous vitamins are heat-sensitive, specifically vitamins C, B1, B9, and B12. In the case of insufficient intake of fresh fruit and vegetables, some vitamin deficiency may be expected, with a possible decrease in antioxidative defense (vitamin C), an impact on glycolysis efficiency (B1) and an increase in some cardiovascular risk factors (B9 and B12). To compensate for such degradation, vitamin and mineral supplementation has been proposed.[1] Moreover, dietary polyunsaturated fatty acids (PUFA) may be severely damaged through peroxidation, cis-trans isomerization, or thermal degradation; all of these reactions are associated with cytotoxic activity, including inflammatory response. Consequently, the influence of these other factors in the high-MP diet could partly explain some of the biological changes observed, such as higher oxidative stress, inflammation, or vascular damage. Generally, no details are given on how the menus were composed or the food prepared, and of course this information is particularly difficult to obtain when, as is most often the case, the volunteers themselves prepare or choose their food in accordance with recommendations. Finally, no indication is given of individual nutritional intakes and MP ingestion levels, which may vary considerably among subjects.

However, despite the methodological difficulties of this type of experimental design, comparison of two realistic and similar diets with different MP levels seems the most pertinent way to explore the possible health impact of dietary MPs. As complex interactions are expected regarding bioaccessibility and intestinal absorption, or between the anti- and pro-oxidant activities of the different Maillard compounds, only a realistic approach may help to highlight the global impact of the MP mixture formed in current diets. In the ICARE clinical study, such an approach was proposed, with particular attention to the preparation of meals, the availability of vitamins, and the re-equilibration of the PUFA. The precise concentration of some MPs that are supposed to exert possible deleterious health effects, such as acrylamide, hydroxymethylfurfural (HMF) and carboxymethyllysine (CML), was assessed. A cross-over design was adopted, including 15 and 10 days of standardization and washout periods, respectively, where 62 healthy volunteers (18–24 years old) were randomly assigned to a high- or low-MP diet.

Design and Preparation of Diets with Different MP Levels and No Lipid Degradation

To control the preparation of meals and nutritional intakes strictly, the clinical study was designed to take place on a university campus, and the volunteers were selected from among the students currently eating all their meals in the self-service restaurant of the campus. For scientific and ethical reasons, the normal current diet was considered to be the high-MP diet. In fact, the problem of excessively increasing the MP concentration of a diet is that the severe cooking that must be applied induces potential adverse effects that may result in the formation of mutagenic and carcinogenic compounds derived from partial combustion of organic compounds (especially polycyclic aromatic hydrocarbons, cyclic monomers, and acroleine). Possible rejection of such food by the volunteers is also to be expected, with consequences on dietary intakes. Finally, the concentration of MPs often exponentially increases during heat treatment (especially for HMF, heterocyclic amines, and acrylamide), so that very high levels—which are no longer representative of current diets—can be reached. Accordingly, whereas the high-MP diet was the normal daily diet consumed on the campus, the low-MP diet was specifically prepared using raw and exclusively steam-cooked food to replace the conventional grilling, roasting, and frying techniques. A dietician suggested a complete menu for a week, prepared with the two alternative cooking processes and ensuring the volunteers' cooperation, which was later confirmed (94% acceptance for both diets). The menus were repeated for the 4 weeks of the experimental study. The resulting mean exposure to dietary MPs from the standard (high-MP) diet was similar to or lower than that estimated for the European population. In particular, 32 µg/day acrylamide was absorbed, similar to the 0.5 µg/kg/day estimated by the French Association of Food Safety (AFSSA). The standard diet provided 500 µg HMF per day, which is 100 times lower than the 0.94 mg/kg calculated to be present in a Spanish "white diet" composed of mildly processed food.[9] In the standard diet,

TABLE 1. Example of 1-day menu with alternative cooking conditions

	Steam diet	Standard diet
Breakfast		
Food	steam-cooked muesli (60 g)[a]	corn flakes (60 g)
	white bread (70 g)	bread with brown crust (70 g)
	butter (10 g), jam (30 g)	butter (10 g), jam (30 g), chocolate paste (20 g)
Drink	tea with white sugar	coffee with brown sugar
Lunch		
Entrée	mushroom salad (50 g)	grilled mushrooms (225 g)
	no cooking; oil and vinegar dressing (25 mL)	frying pan (8 min, 180 °C) with margarine (5 g)
Main Dish	hake slice with hollandaise sauce (120 g)	hake slice with fish sauce (120 g)
	steam oven (10 min, 130 °C) with margarine (5 g)	frying pan (10 min, 180 °C) with margarine (5 g)
	jacket potatoes (200 g)	French fries (260 g)
	steam oven (25 min, 100 °C) with butter (5 g)	frying (10 min, 180 °C; 7% fat absorption)
Dessert	rice pudding (100 g) steam oven	rice pudding steam oven with caramel sauce (100 g)
Bread	white bread (25 g)	bread with brown crust (25 g)
Drink	water	water, light soda
Snack		
	madeleine (120 g) commercial	biscuit with chocolate (120 g) commercial
Drink	orange or apple juice	orange or apple juice
Dinner		
Entrée	4 seasons wheat (25 g) salad	4 seasons mini pizza (110 g)
	boiling water (20 min, 120 °C) with raw vegetables and vinegar dressing (25 mL)	oven (15 min, 130 °C)
Main Dish	tandoori chicken (230 g)	tandoori chicken (230 g)
	steam oven (40 min, 100 °C)	oven (20 min, 140 °C; and 20 min, 190 °C)
	ratatouille (180 g)	ratatouille (310 g)
	steam oven (15 min, 100 °C) with margarine (10 g)	frying pan (15 min, 180 °C) with margarine (10 g)
Dessert	apple (150 g)	stewed apple (150 g)
	no cooking	pan (20 min, 180 °C) with butter (5 g)
Bread	white bread (25 g)	bread with brown crust (25 g)
Drink	water	water, light soda

[a] *Note:* Weights are given on a fresh basis.

the exposure to CML was found to reach 6 mg, similarly to what was found in a high MP Spanish diet based on fast food intake.[9] This MP was the most quantitatively important if we except fructosyllysine (12 mg/day), which is not considered to exert any adverse health effects. The formation of lipid thermodegradation products was confirmed to be as low as possible in the standard diet, resulting from limitation of fried foods (one portion of French fries per week). No measurable cyclic monomers or oxysterols could be found, and the total mean content of trans fatty acids was 1.5 g per day, which is much lower than the French mean exposure (2.5–3 g/day; AFSSA data). Despite the moderate MP exposure in the high-MP diet, a strong decrease in the mean MP level was achieved by applying minimal processing to food in the low-MP diet, allowing a decrease by a factor of 3 (CML), 5 (acrylamide), and 45 (HMF).

The low-MP, or steam, diet was prepared using the same raw food and ingredients used in the standard diet, but with modified cooking techniques. Regarding ready-to-eat food, similar products were used, and these were known to have been subjected to a milder process (steamed instead of grilled breakfast cereal, cookies, like madeleine, with higher moisture and less severely baked instead of dry cookies, white bread instead of bread with brown crust). TABLE 1 shows an example of one day of the study and describes the culinary procedures applied to prepare the food.

Re-equilibration of the Caloric and Nutritional Intakes

The higher temperatures used to cook the food in the standard diet, compared to the steam diet (TABLE 1), induced several changes; the major differences included higher water loss with a concomitant increase in the caloric density and a degradation of heat-sensitive micronutrients, such as vitamin C, folates, and thiamine. In addition, part of the fat used to fry foods in the standard diet was absorbed by the food, increasing the final fat and caloric content of the meal. To circumvent these nutritional differences between diets, a preliminary study was performed to measure such differences precisely and to propose corrections. On each day of the experimental week, the food offered at lunch and dinner was mixed and analyzed for water content, protein and lipid concentration, and for content of vitamins C, E, and B1. For example, the dry matter content was $43.77 \pm 4.31\%$ and $36.17 \pm 4.16\%$ for the standard and steam meals, respectively. Consequently, the caloric content of the 1500 g food offered in the standard diet was 28% higher than the same quantity in the steam diet, a difference that was insufficiently compensated for by spontaneous lower intakes in the former (a 10% difference in caloric content remained between diets composed of 1186 g and 1366 g for the standard and steam diets, respectively). This problem should be corrected by offering an inversely proportional amount of food in relation to the caloric density. By replacing the cooked fat in the standard diet with the same raw fat as seasoning in the steam diet, the energy profile of the two diets (calculated over the week from the chemical analysis on the complete daily diet, including breakfast, lunch, dinner, and snacking) was confirmed not to be significantly different and was in agreement with nutritional recommendations (53% carbohydrates, 15% proteins, and 32% fat). The vitamin C content in the standard diet, which was almost twice as low as in the other diet, was compensated for by doubling the amount of cooked vegetables consumed (TABLE 1). In addition, fresh fruits and juice were offered each day. Vitamin E and thiamine were verified not to differ significantly between the two diets (3.9 ± 1.5 and 3.4 ± 0.8 mg/g lipids of vitamin E in the high- and low-MP diets, respectively; 0.13 ± 0.06 and 0.018 ± 0.03 mg thiamin per 100 g meal, respectively), confirming again that the cooking process in the standard diet was not too severe. All vitamins and minerals were naturally present at the recommended levels.

Conclusion

This study shows that producing two diets that are similar from a nutritional point of view—despite being composed of foods processed by different heat treatments—is a challenge that can be met. This is achieved by adjusting fat intake and increasing the quantity of food consumed in the steam-cooked diet to compensate for the lower caloric density compared with the standard diet, and by increasing the cooked vegetable quantity in the latter to circumvent degradation of ascorbate and other antioxidants. Furthermore, we show that conventional moderate cooking, compared with steam cooking, is sufficient to induce a strong difference in MP content between the diets, without significantly degrading lipids, thiamine, or vitamin E. Finally, the individual nutritional intakes and MP exposure calculated from intake weights and direct chemical analysis of the meals (to be published elsewhere) will allow us to verify the compliance of the volunteers in eating the food in the diets and to correlate individual MP exposure levels with biological responses to assess the specific impact of dietary MPs on health indicators.

Acknowledgments

This study was carried out as part of the ICARE project, supported by an EC 6th FP grant No. COLL-CT-2005-516415 No. 516415.

Conflict of Interest

The authors declare no conflicts of interest.

References

1. VLASSARA, H. et al. 2002. Inflammatory mediators are induced by dietary glycotoxins, a major risk factor for diabetic angiopathy. Proc. Natl. Acad. Sci. USA **99:** 15596–15601.
2. SOMOZA, V. 2005. Five years of research on health risks and benefits of Maillard reaction products: an update. Mol. Nutr. Food Res. **49:** 663–672.
3. ESPOSITO, F. et al. 2003. Moderate coffee consumption increases plasma glutathione but not homocysteine in healthy subjects. Alim. Pharmacol. Therapeutics **17:** 595–601.
4. LIN, R.Y. et al. 2003. Dietary glycotoxins promote diabetic atherosclerosis in apolipoprotein E-deficient mice. Atherosclerosis **168:** 213–220.
5. ZHENG, F. et al. 2002. Prevention of diabetic nephropathy in mice by a diet low in glycoxidation products. Diabetes Metab. Res. Rev. **18:** 224–237.
6. STIRBAN, A. et al. 2006. Benfotiamine prevents macro- and microvascular endothelial dysfunction and oxidative stress following a meal rich in advanced glycation end products in individuals with type 2 diabetes. Diabetes Care **29:** 2064–2071
7. URIBARRI, J. et al. 2005. Diet-derived advanced glycation end products are major contributors to the body's AGE pool and induce inflammation in healthy subjects. Ann. N. Y. Acad. Sci. **1043:** 461–466.
8. BUETLER, T. 2007. Heated diets, AGEs and tissue dysfunction. IMARS Highlights **2:** 1–2.
9. SEIQUER, I. et al. 2006. Diets rich in Maillard reaction products affect protein digestibility in adolescent males aged 11–14 years. Am. J. Clin. Nutr. **83:** 1082–1088.

Plasma Concentration and Urinary Excretion of N^{ε}-(Carboxymethyl)lysine in Breast Milk– and Formula-fed Infants

KATARÍNA ŠEBEKOVÁ,[a] GISELLE SAAVEDRA,[b] CORNELIA ZUMPE,[b] VERONIKA SOMOZA,[c] KRISTÍNA KLENOVICSOVÁ,[a] AND INES BIRLOUEZ-ARAGON[b]

[a]*Slovak Medical University, Bratislava, Slovakia*
[b]*Polytechnical Institute LaSalle Beauvais, Beauvais, France*
[c]*German Research Center for Food Chemistry, Garching, Germany*

Industrial processing of infant formulas (IFs) induces the formation of Maillard products, namely N^{ε}-(carboxymethyl)lysine (CML). CML content is expected to be several times higher in IFs than in fresh human breast milk. To elucidate whether CML is absorbed from IFs into the bloodstream, CML concentration in the plasma and urine were analyzed in 6-month-old infants (34 breast fed and 25 fed exclusively with IFs) and in 56 samples of human breast milk and 16 commercial IFs. We found that IFs contain higher amounts of CML compared to mother's milk (median: 70-fold; range: 28- to 389-fold), and CML content was higher in hydrolyzed IFs than in nonhydrolyzed IFs ($P < 0.03$). Plasma CML levels were 46% higher ($P < 0.01$) and urinary excretion of CML was 60-fold higher ($P < 0.001$) in the formula-fed infants than in the breast-fed group. Infants fed with hydrolyzed IFs displayed significantly higher plasma CML levels than those on nonhydrolyzed formulations. We conclude that CML from IFs is absorbed into the circulatory system and is rapidly excreted in the urine.

Key words: N^{ε}-(carboxymethyl)lysine; mother's milk; infant formulas

Introduction

Pediatricians recommend that infants should be fed exclusively with breast milk or, when the mother's milk is not available, with infant formulas (IFs) during the first 6 months of age. Thereafter, the milk is gradually substituted by other sources of nutrition.

During the industrial production of IFs, many Maillard reaction products (MRPs), such as N^{ε}-(carboxymethyl)lysine (CML), a stable compound considered a good indicator of the Maillard reaction in food,[1] are formed in substantial amounts. CML in IFs is therefore several times higher than the CML content of human breast milk.[1–3] Studies have shown that that dietary CML in rodents is at least partially absorbed into the circulatory system.[4] CML represents one of the ligands of the receptor for advanced glycation end products (RAGE) and its interaction with RAGE may result in pro-oxidative, proinflammatory, and proatherogenic actions.[5] It is unknown whether CML from IFs is absorbed into the bloodstream. To address this, we compared plasma CML levels and urinary CML excretion in two groups of 6-month-old healthy infants fed since birth exclusively with breast milk or with IFs, representing low- and high-CML-containing diets, respectively.

Methods

The study was carried out in accordance with the Declaration of Helsinki and after approval by the Institutional Ethics Board (Slovak Medical University, Bratislava, Slovakia). Informed written consent was obtained from the legal representatives of the infants.

Infant Formulas

Sixteen powdered IFs from four different producers were purchased in pharmacies in Bratislava. The purchased IFs corresponded to the brands administered to the infants in the IF group, as indicated by pediatricians. Ten grams of each powder was sampled

Address for correspondence: MUDr. Katarína Šebeková, Dr.Sc., Slovak Medical University, Limbová 14, 83303 Bratislava, Slovakia. Voice: +421-2-59369-431; fax: +421-2-59369-170.
katarina.sebekova@szu.sk

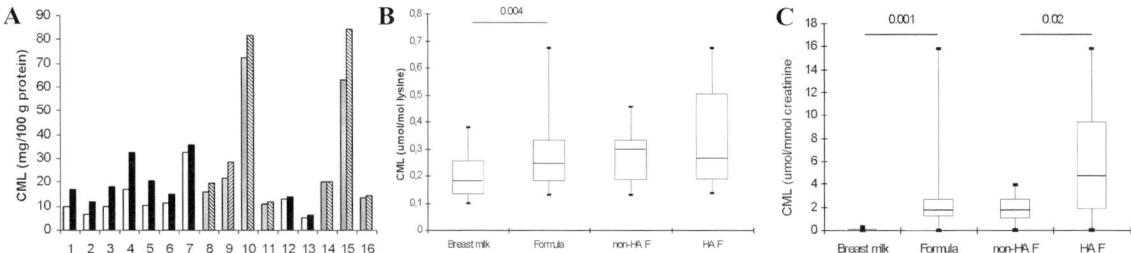

FIGURE 1. N^ε-(carboxymethyl)lysine (CML) content in 16 infant formulas and changes during the storage (**A**), total plasma CML concentration (**B**), and urinary excretion of CML (**C**) in 6-month-old infants fed exclusively with breast milk or infant formulas. **A**: Solid bars, nonhydrolyzed formulas ($n=9$); hatched bars, hydrolyzed formulas ($n=7$); first bar, sample taken immediately after purchase; second bar, sample taken at the end of the recommended storage period. Numbers on x-axis are 1–11, starting formulas; 12–16, follow-on formulas. **B** and **C**: HAF, hydrolyzed formula.

TABLE 1. The increase (x times) in N^ε-(carboxymethyl)lysine (CML) in powdered infant formulas (IFs) compared to human breast milk

	All IFs	HA IFs	Non-HA IFs	Starting IFs	Follow-on IFs
Mean	112	166	69	107	123
Median	70	110	57	61	71
Minimum	28	57	28	37	28
Maximum	389	389	175	389	338

HA, hydrolyzed IFs.

immediately after the purchase. A second sample was taken at the end of the manufacturer's recommended storage period (2–4 weeks).

Human Breast Milk

Fifty-six samples of human breast milk that were voluntarily donated by mothers of the studied breast-fed infants were investigated.

Infants: Only full-term infants were included in the study. Thirty-four infants (mean age, 5.9 ± 1.0 months) had been exclusively breast fed from birth until the time of our investigation, while 25 (mean age, 5.9 ± 1.0 months) had been fed exclusively with IFs (13 with hydrolyzed and 12 with nonhydrolyzed IFs). All infants were healthy, as indicated on the questionnaire forms completed by their pediatricians and confirmed by standard blood chemistry and blood count analyses (data not provided). The mean daily weight gain from birth up until the investigation was 23.6 ± 6.3 g in the breast-fed and 23.3 ± 5.0 g in the IF-fed infants. Blood (4.8 mL in total, used also for analyses not mentioned in this paper) and spot urine were sampled. Standard blood and urine chemistry (Vitros 250 analyzer, J&J, Rochester, NY) and blood count (Sysmex K-21, Kobe, Japan) analyses were performed to monitor the health of the children.

CML in plasma,[6] urine,[7] IFs,[7] and mother's milk[7] was determined by gas chromatography-tandem mass spectrometry (GS-MS/MS).

Sample Storage

Milk, IFs, plasma, and urine samples were stored frozen ($-80\,^\circ$C) and shipped on dry ice to cooperating laboratories (Beauvais, France; Garching, Germany) for analyses.

Statistics

Data are given as mean $\pm$ SD. Groups were compared using Student's t-test or Mann–Whitney U test, as appropriate; $P < 0.05$ was considered significant.

Results

CML in IFs versus Mother's Milk

IFs varied considerably in their CML content (FIG. 1A) and contained 28 to 389 times more CML than human breast milk ($P < 0.01$, TABLE 1). The hydrolyzed IFs generally contained more CML than the nonhydrolyzed IFs ($P < 0.03$), while no significant difference between starting formulas (assigned for infants up to 5 months old) and follow-on IFs was revealed. CML content rose during the storage period in the majority of formulas (FIG. 1A). Increase in CML content was higher in the starting IFs ($50\% \pm 36\%$; range, 10–99%) than in the follow-on IFs ($14\% \pm 13\%$; range, 0–33%; $P < 0.05$) and higher in the nonhydrolyzed ($55\% \pm 36\%$; range, 10–99%) than in the hydrolyzed IFs ($18\% \pm 13\%$; range, 0–34%; $P < 0.02$). No relationship between CML content of the breast milk and duration of lactation, mother's age, parturitions weight, body mass index, or weight gain during pregnancy was identified.

The theoretical daily burden of CML in the 6-month-old breast-fed infants was calculated as 0.01–0.03 mg/d, while that of the IF-fed infants was 4.6 ± 4.2 mg/d (range calculated from formulas for 6-month-old infants with the dosage specified by the manufacturer).

CML Concentration in Plasma

The breast-fed infants showed approximately 46% lower plasma CML levels compared to the IF-fed infants ($P < 0.004$). However, because of high interindividual variability, no significant difference was observed between the hydrolyzed- versus the nonhydrolyzed-IF groups (FIG. 1B).

CML Urinary Excretion

Urinary excretion of CML was considerably higher in the IF-fed group (60-fold) than in the breast milk–fed infants. Infants fed hydrolyzed IFs had approximately fourfold higher CML in their urine ($P < 0.02$) compared to infants fed nonhydrolyzed formulas. However, we failed to find any relationship if the consumed formula CML content was correlated to plasma CML level or CML urinary excretion (FIG. 1C).

Discussion

The most intriguing finding of our study was the substantially elevated plasma CML concentration and particularly the considerably increased urinary excretion in the healthy full-term IF-fed infants versus the breast-fed infants. The high CML content and variability in IFs versus mother's milk is well documented.[1,3] However, it was essential to determine the relevant CML concentration in the formulas or breast milk ingested by the infants to further explore CML bioavailability and urinary excretion. Our data suggest that CML from IFs is absorbed into the bloodstream and rapidly excreted via the urine. The absence of correlation between the CML content in the formula and the levels in biological material may be explained by the study design. On the one hand, we did not record the time interval between the last feeding and the collection of biological material, and on the other hand, we analyzed the CML content in IF that we purchased, not in the IF administered to each child. Thus, the variations in CML content in formulas as a result of storage and the potential batch-to-batch variations need to be considered.

Our data strongly suggest that dietary MRPs are significantly absorbed into the bloodstream.[4,8] However, recent experimental data showed that CML is not a substrate of intestinal amino acids and peptide transporters in human colon carcinoma Caco-2 cells and its low transepithelial flux was attributed to simple diffusion.[9] This finding does not exclude transport via different mechanisms, such as ion channels. Moreover, it is not clear to what extent the data obtained from Caco-2 cell lines can be extrapolated to the *in vivo* situation, particularly to the partially immature intestinal tract of infants. If we exclude, at least theoretically, any active absorption of CML from the intestine into the circulatory system, then the elevated CML levels in IF-fed infants should be attributed to the enhanced endogenous synthesis via unknown mechanisms. However, the design of our study (determination of total plasma CML, collection of spot urine) does not allow us to distinguish unequivocally whether markedly enhanced urinary CML excretion in IF-fed infants is derived from exogenous sources, from induced endogenous production, or both.

Taken together, infants consuming a high CML diet (IFs) show substantially elevated CML levels in plasma and urine, in comparison to breast-fed infants. CML from formulas is thought to be absorbed and rapidly excreted via urine. It is unknown whether these differences are only temporarily linked to very different dietary AGEs intake or will continue to persist when infants are on diversified diets. Moreover, diets high in MRPs were suggested to exert various negative health effects.[8] Whether the oral CML load at the levels found in IFs exerts potential biological effects in early childhood remains to be elucidated.

Acknowledgment

This study was supported by a European Commission 6th Framework Program Grant No. COLL-CT-2005-516415, entitled Impending neo-formed Contaminant Accumulation to Reduce their health Effects (ICARE).

Conflict of Interest

The authors declare no conflicts of interest.

References

1. BIRLOUEZ-ARAGON, I. *et al.* 2004. Loss of nutritional quality and chemical safety in heat-processed infant formulas. Food Chem. **87:** 253–259.

2. BIRLOUEZ-ARAGON, I. *et al.* 2005. Evaluation of the maillard reaction in infant formulas by means of front-face fluorescence. Ann. N.Y. Acad. Sci. **1043:** 308–318.
3. DITTRICH, R. *et al.* 2006. Concentrations of Nepsilon-carboxymethyllysine in human breast milk, infant formulas, and urine of infants. J. Agric. Food Chem. **54:** 6924–6928.
4. SOMOZA, V. *et al.* 2005. Influence of feeding malt, bread crust, and a pronylated protein on the activity of chemopreventive enzymes and antioxidative defense parameters in vivo. J. Agric. Food Chem. **53:** 8176–8182.
5. BIERHAUS, A. *et al.* 2005. Understanding RAGE, the receptor for advanced glycation end products. J. Mol. Med. **83:** 876–886.
6. SOMOZA, V. *et al.* 2006. Dose-dependent utilisation of casein-linked lysinoalanine, N(epsilon)-fructoselysine and N(epsilon)-carboxymethyllysine in rats. Mol. Nutr. Food Res. **50:** 833–841.
7. CHARISSOU, A. *et al.* 2007. Evaluation of a gas chromatography/mass spectrometry method for the quantification of carboxymethyllysine in food samples. J. Chromatogr. **A1140:** 189–194.
8. ŠEBEKOVÁ, K. & V. SOMOZA. 2007. Dietary advanced glycation endproducts (AGEs) and their health effects - PRO. Mol. Nutr. Food Res. **51:** 1079–1084.
9. HELLWIG, M. *et al.* 2007. Interaction of Maillard reaction products with intestinal transport systems. [abstract]. 9th International Symposium on the Maillard reaction. Munich, Germany, September 1–5, Abstracts, p. 164.

Maillard Reaction Products in the Escherichia coli–derived Therapeutic Protein Interferon Alfacon-1

ROUMYANA MIRONOVA,[a] ANGELINA SREDOVSKA,[a] IVAN IVANOV,[a] AND TOSHIMITSU NIWA[b]

[a]*Department of Gene Regulations, Institute of Molecular Biology, Bulgarian Academy of Sciences, Sofia, Bulgaria*

[b]*Department of Clinical Preventive Medicine, Nagoya University School of Medicine, Nagoya, Japan*

We have recently shown that recombinant human interferon-γ is affected by early stages of the Maillard reaction during its production in *Escherichia coli*. Over time, advanced glycation end products accumulated in the purified protein, accompanied with degradation, cross-linking, and a drop in the protein's biologic activity. Here, we provide further evidence for the presence of Maillard reaction products in another *E. coli*–derived therapeutic protein, interferon alfacon-1. These products might interfere with both treatment efficacy and patient safety.

Key words: Maillard reaction; glycation; advanced glycation end products; interferon alfacon-1; consensus interferon; antidrug antibodies; adverse events

Introduction

The development of antidrug antibodies in patients treated with human protein therapeutics appears to be a rule rather than the exception.[1] Immunoglobulin (Ig) E-mediated responses are sometimes associated with the onset of severe allergic reactions, including systemic anaphylaxis.[2] To date, the reasons for the immune response against human self-proteins remain poorly understood. Here, we provide evidence for the presence of Maillard reaction products (MRPs) in the protein therapeutic interferon (IFN) alfacon-1 that may, at least partially, contribute to immune-mediated adverse drug reactions.

The drug IFN alfacon-1, also referred to as consensus IFN, is used for the treatment of chronic hepatitis C. Unlike most IFN-α preparations, it does not contain human serum albumin as a stabilizer and is injected subcutaneously. The most frequent adverse events observed with IFN alfacon-1 are fever, arthralgia, myalgia, headache, nausea, vomiting, and rash. In addition, cases of severe focal glomerulosclerosis,[3] interstitial pneumonitis,[4] cutaneous mucinosis and skin necrosis,[5] immune thrombocytopenic purpura,[6] and destructive thyroiditis[7] have been reported as complications of consensus IFN therapy. These drug-induced toxicities are usually attributed to the pleiotropic biologic activities of IFN-α (e.g., immunomodulatory)[6] rather than to drug immunogenicity, although their pathogenesis remains, so far, largely unknown.

We have previously shown[8] that *Escherichia coli*–derived recombinant human IFN-γ (rhIFN–γ) is involved in a Maillard-type reaction during its production in the bacterial cells. The Maillard reaction, also known as nonenzymatic glycosylation (or glycation), introduces in proteins bulky moieties (reducing sugars) that are very unstable and ultimately involve proteins in a chain of chemical rearrangements lasting weeks and months. The process ceases with the formation of stable adducts called advanced glycation end products (AGEs), partial degradation, and cross-linking of target proteins. We have found that the accumulation of AGEs in rhIFN–γ results in structural and functional impairment of the protein.[8] IFN alfacon-1, like rhIFN–γ, is an *E. coli*–derived protein, and this fact prompted us to test it for glycation.

Address for correspondence: Roumyana Mironova, Ph.D., Department of Gene Regulations, Institute of Molecular Biology, Bulgarian Academy of Sciences, Acad. G. Bonchev str., bl. 21, 1113 Sofia, Bulgaria. Voice: +359-2-979-26-52; fax: +359-2-872-35-07.
rumym@bio21.bas.bg

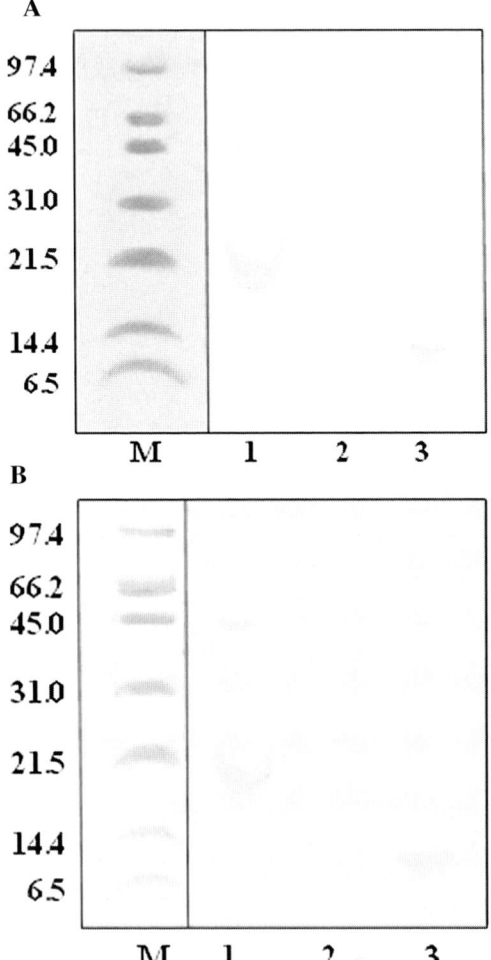

FIGURE 1. Western blots of interferon (IFN) alfacon-1 (lanes 1) stained with either an anti- N^ε-(carboxymethyl)lysine (CML) **(A)** or an anti-imidazolone **(B)** antibody. Normal (lanes 2) and amyloid (lanes 3) β_2-microglobulin were used as the negative or positive control, respectively. M: molecular weight standards in kDa.

Results and Discussion

The consensus IFN analyzed in this study was a liquid formulation containing 18 million IU consensus IFN per 0.6 mL of phosphate buffered saline. The presence of AGEs in the drug was tested by competitive enzyme-linked immunosorbent assay (ELISA) using two monoclonal antibodies specific for the advanced products of glycation, N^ε-(carboxymethyl)lysine (CML)[9] and 3-deoxyglucosone-derived imidazolone.[10] Bovine serum albumin (BSA) glycated in 0.5 mol aqueous glucose solution at 37°C for 3 months (AGE-BSA) was used as both coating (competitive) and standard antigen. The content of CML and imidazolone in the drug was expressed as AGE-BSA equivalents per vial (eq/vial), where one eq corresponds to the amount of CML and imidazolone present in 1 μg of the referent BSA. Both AGEs were detected in IFN alfacon-1. The content (mean ± SD, $n = 4$) of CML and imidazolone was 1.7 ± 0.2 AGE-BSA eq/vial and 2.3 ± 0.3 AGE-BSA eq/vial, respectively.

The presence of CML and imidazolone in IFN alfacon-1 was further confirmed by Western blotting. Consensus IFN is a monomer protein that contains four cysteine residues. In order to break down any covalent cross-links arising in the preparation as a result of the formation of intermolecular disulfide bridges, IFN alfacon-1 was reduced overnight at 37°C with 0.7 mol 2-mercaptoethanol. Western blotting was performed with the same anti-AGE antibodies used in the ELISA. The main immunoreactive band at approximately 20 kDa observed with both antibodies corresponds to the monomer IFN alfacon-1 (FIG. 1A and B). A second band reveals the presence of a higher molecular weight AGE-modified species. This heavy derivative, approaching the molecular weight of an IFN dimer, was reactive to an anti-IFN-α antibody (data not shown), excluding bacterial contamination. The blotting experiments thus demonstrated that IFN alfacon-1 had undergone noncysteine-mediated covalent dimerization, most probably caused by AGEs with cross-linking properties formed in IFN alfacon-1.

Silver-stained SDS-polyacrylamide gel (FIG. 2) confirmed the presence of heavy IFN alfacon-1 species. In addition, it demonstrated the existence of high mobility protein forms that could originate from either protein degradation or incomplete reduction. To distinguish between these two cases, IFN alfacon-1 was subjected to electrospray ionization mass spectrometry coupled to liquid chromatography. The m/z spectrum of IFN alfacon-1 (FIG. 3) was analyzed manually by matching the observed m/z values against the multiple m/z values expected for all possible combinations of (i) +Met[1]/-Met[1] protein species, (ii) full-length and truncated forms (1 to 20 amino acids missing from both the N and C terminus), and (iii) with one to four molecules of 2-mercaptoethanol bound per protein molecule. Detailed analysis of the mass spectrum confirmed the presence of partially reduced protein species, but it also revealed a site of C-terminal cleavage at the very end of the molecule, between Arg[164] and Lys[165]. It is noteworthy that arginine and lysine residues are the main targets for glycation and glycation-mediated proteolysis.[8]

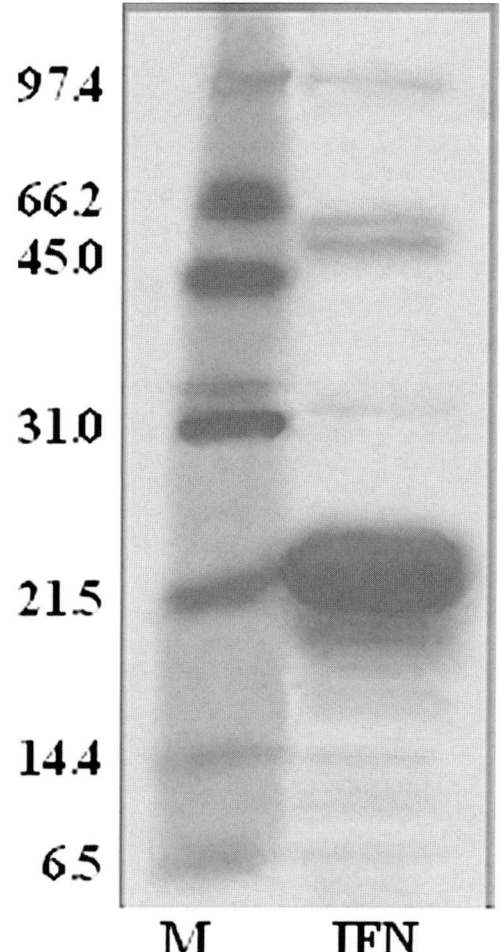

FIGURE 2. Silver-stained SDS-polyacrylamide gel of IFN alfacon-1. M: molecular weight standards in kDa.

The biologic activity of IFN alfacon-1 was measured a month prior to the drug expiration date and compared to that indicated on the drug vial. The antiviral activity of IFN alfacon-1 was calculated on the basis of three independent measurements and found to be $(7.4 \pm 2.0) \times 10^6$ IU per vial, more than two times lower than expected (1.8×10^7 IU). These data are in conformity with our previous finding showing that glycation-promoted degradation and covalent dimerization of rhIFN-γ lower the protein biologic activity upon storage.[8]

Consensus IFN is a genetically engineered protein which differs in amino acid sequence from all natural subtypes of IFN-α.[11] To this well-known fact our study adds the appearance of sugar-derived neoantigens (AGEs) as well as partial degradation and cross-linking of the protein. One could expect that all these structural alterations may render IFN alfacon-1 immunogenic. Special attention should be paid to the fact that, when injected, IFN alfacon-1 may trigger an autoimmune response by covalent binding to endogenous proteins via intermediate glycation adducts that are much more reactive than AGEs. Moreover, the cells of the innate immune system recognize as pathogens repeating arrays of carbohydrate moieties that are characteristic of microbial surfaces but are not found on host cells. These structures, known as pathogen-associated molecular patterns, are recognized by pattern recognition (PR) receptors. Protein drugs containing AGEs could interact with PR receptors on macrophages, NK cells, mast cells, and $\gamma\delta$ T cells, thus provoking an immune response. Assays for antibody formation are not routine to the ongoing pilot studies and clinical trails with IFN alfacon-1. Among 12 studies, only one observed patients for the formation of anti-IFN alfacon-1 antibodies and found that 13.0% (12/92) of patients developed such antibodies.[12] In this report, the Ig class of the antibodies, as well as their nature, whether binding or neutralizing, has not been specified.

Importantly, AGEs accumulate endogenously in the healthy and diseased human body and are consumed with food as well. A direct association between dietary AGE intake and some markers of systemic inflammation has been reported.[13] AGEs have also been found to confer increased allergenicity to the main peanut allergens Ara h1 and Ara h2.[14] In the last decade, it has been convincingly demonstrated that AGEs are recognized by receptors for advanced glycation end products (RAGEs) belonging to the immunoglobulin superfamily that are expressed on a number of cells including inflammatory ones (e.g., monocytes). The advanced product of glycation, CML, which we detected in IFN alfacon-1, has been reported to interact with RAGEs and to activate proinflammatory pathways.[15] Taken together, the formation of novel epitopes in IFN alfacon-1 through the Maillard reaction as well as AGE-mediated inflammation and autoimmunity may cause at least some of the adverse events observed in patients on medication with IFN alfacon-1.

In conclusion, to the best of our knowledge, this study provides, for the first time, evidence for the presence of non-native antigens of a well-defined origin (AGEs) in a human protein therapeutic. Further clinical and laboratory investigations are required to evaluate the impact of AGEs on the drug-induced toxicities in patients treated with IFN alfacon-1. In addition, we recommend close monitoring of patients on therapy with this and other protein drugs for the formation of antidrug antibodies and autoantibodies as these may

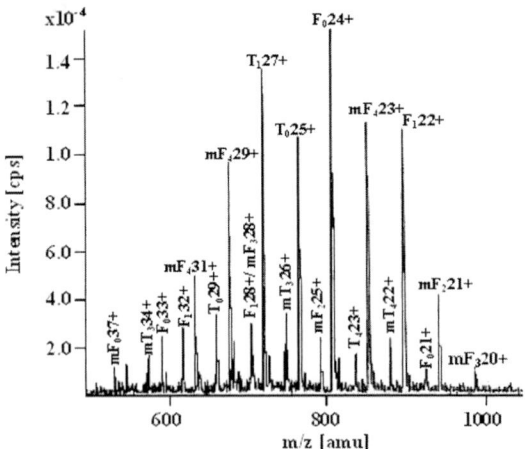

FIGURE 3. Mass to charge (m/z) spectrum of IFN alfacon-1. **F**, -Met^1 full-size species; **mF**, +Met^1 full-size species; **T**, -Met^1 species cleaved at Arg^{164}/Lys^{165}; **mT**, +Met^1 species cleaved at Arg^{164}/Lys^{165}. Positive numbers correspond to the charge state (z) and subscript numbers indicate the number of 2-mercaptoethanol molecules bound per protein molecule.

seriously compromise treatment efficacy and patient safety.

Acknowledgments

This work was supported by Contract 02/008 from the International Center for Genetic Engineering and Biotechnology (Trieste, Italy) and Contract B-1501/05 from the Bulgarian Ministry of Education and Science. We thank the operator of the mass spectrometer, Yoshinori Yamakawa, for his assistance in the mass spectral experiments and Dr. Georgi Milchev for critically reading the manuscript.

Conflict of Interest

The authors declare no conflicts of interest.

References

1. SCHELLEKENS, H. 2003. Immunogenicity of therapeutic proteins. Nephrol. Dial. Transplant. **18:** 1257–1259.
2. DIOUN, A.F. et al. 1998. IgE-mediated allergy and desensitization to factor IX in hemophilia B. J. Allergy Clin. Immunol. **102:** 113–117.
3. STEIN, D.F. et al. 2001. Collapsing focal segmental glomerulosclerosis with recovery of renal function: an uncommon complication of interferon therapy for hepatitis C. Dig. Dis. Sci. **46:** 530–535.
4. HILLIER, A.E. et al. 2006. Consensus interferon induced interstitial pneumonitis in a patient with HCV. Am. J. Gastroenterol. **101:** 200–202.
5. SIEWERT, E. et al. 2005. Cutaneous mucinosis and skin necrosis complicates interferon alfacon-1 (consensus interferon) treatment of chronic hepatitis C. Eur. J. Med. Res. **10:** 63–67.
6. DIMITROULOPOULOS, D. et al. 2004. Immune thrombocytopenic purpura in a patient treated with interferon alfacon-1. J. Viral. Hepat. **11:** 477–478.
7. MAZZIOTTI, G. et al. 2002. Temporal relationship between the appearance of thyroid autoantibodies and development of destructive thyroiditis in patients undergoing treatment with two different type-1 interferons for HCV-related chronic hepatitis: a prospective study. J. Endocrinol. Invest. 25: 624–630.
8. MIRONOVA, R. et al. 2003. Glycation and post-translational processing of human interferon-gamma expressed in *Escherichia coli*. J. Biol. Chem. **278:** 51068–51074.
9. NIWA, T. et al. 1996. Amyloid beta 2-microglobulin is modified with N epsilon-(carboxymethyl)lysine in dialysis-related amyloidosis. Kidney Int. **50:** 1303–1309.
10. NIWA, T. et al. 1997. Immunohistochemical detection of imidazolone, a novel advanced glycation end product, in kidneys and aortas of diabetic patients. Clin. Invest. **99:** 1272–1280.
11. BLATT, L. et al. 1996. The biologic activity and molecular characterization of a novel synthetic interferon-alpha species, consensus interferon. J. Interferon Cytokine Res. **16:** 489–499.
12. SUZUKI, H. & TANGO, T. 2002. A multicenter, randomized, controlled clinical trial of interferon alfacon-1 in comparison with lymphoblastoid interferon-alpha in patients with high-titer chronic hepatitis C virus infection. Hepatol. Res. **22:** 1–12.
13. URIBARRI, J. et al. 2005. Diet-derived advanced glycation end products are major contributors to the body's AGE pool and induce inflammation in healthy subjects. Ann. N.Y. Acad. Sci. **1043:** 461–466.
14. MALEKI, S.J. et al. 2000. The effects of roasting on the allergenic properties of peanut proteins. J. Allergy Clin. Immunol. **106:** 763–768.
15. BIERHAUS, A. et al. 2001. Diabetes-associated sustained activation of the transcription factor nuclear factor-kappaB. Diabetes **50:** 2792–2808.

Acrolein Induces Inflammatory Response Underlying Endothelial Dysfunction

A Risk Factor for Atherosclerosis

YONG SEEK PARK[a] AND NAOYUKI TANIGUCHI[b]

[a]*Department of Microbiology and MRC for Bioreaction to ROS, Kyung Hee University School of Medicine, Seoul, Korea*

[b]*Department of Disease Glycomics, Research Institute for Microbial Diseases, Osaka University, Osaka, Japan*

Endothelial dysfunction by proinflammatory stimuli represents an important link between risk factors and the pathologic mechanisms underlying atherosclerosis. Thus, control of the inflammatory status of endothelial cells is crucial to limiting the disease. Tobacco smoking induces inflammatory reactions and promotes atherosclerosis; however, the mechanism that links cigarette smoking to an increased incidence of atherosclerosis is poorly understood. Our study demonstrates that acrolein, a known toxin in tobacco smoke, elevates oxidative stress via inactivation of thioredoxin reductase and stimulates expression of cyclooxygenase-2 through activation of the protein kinase C, p38 mitogen-activated protein kinase, and cAMP response element-binding protein pathway in endothelial cells. Our finding suggests that acrolein may play a role in the progression of atherosclerosis.

Key words: acrolein; COX-2; atherosclerosis; endothelial cells; oxidative stress; TR

Introduction

Acrolein ($CH_2=CH\text{-}CHO$), the most reactive α,β-unsaturated aldehyde, is found widely in the environment and is hazardous to human health.[1] Acrolein is produced by a wide variety of both natural and synthetic processes, including the incomplete combustion of organic materials and the degradation of polyamines.[2] Acrolein also has been found to be formed from threonine by neutrophil myeloperoxidase at sites of inflammation[3] and has been identified as both a product and initiator of lipid peroxidation.[4] Recent studies have shown that acrolein levels are increased in many diseases, such as atherosclerosis, Alzheimer's disease, and diabetes; acrolein may be related to pathogenesis in these conditions.[5–7] We and others have reported that acrolein elevates intracellular reactive oxygen species (ROS) levels, which leads to cell dysfunction.[6,8] ROS-mediated cell damage is an important etiologic factor in the pathogenesis of atherosclerosis.[9] ROS has been reported to induce the production of various atherogenic factors, including inflammatory proteins.[10] In this paper, we review a possible role of acrolein in the progression of atherosclerosis on the basis of our studies.

Inactivation of Thioredoxin Reductase Activity by Acrolein

Many processes can increase intracellular oxidative stress. It appears that one such process is the inactivation of antioxidant enzymes, as previously reported.[11,12] Thioredoxin reductase (TR), a central antioxidant enzyme, has the potential to detoxify hydrogen peroxide and lipid hydroperoxides, regenerate ascobate, induce superoxide dismutase, reduce S-nitrosoglutathione, maintain the cell redox state, and influence the activity of the transcription factors of a number of important genes.[13] Our data show that acrolein specifically inhibits TR activity both *in vitro* and in human umbilical vein endothelial cells (HUVEC) (FIG. 1). These data suggest that the inactivation of TR by acrolein is responsible for the observed

Address for correspondence: Naoyuki Taniguchi MD, PhD, Department of Disease Glycomics, Research Institute for Microbial Diseases, Osaka University, 2-2 Yamadaoka, Suita, Osaka 565-0871, Japan. Voice: 81-6-6879-4137; fax: 81-6-6879-4137
tani52@wd5.so-net.ne.jp

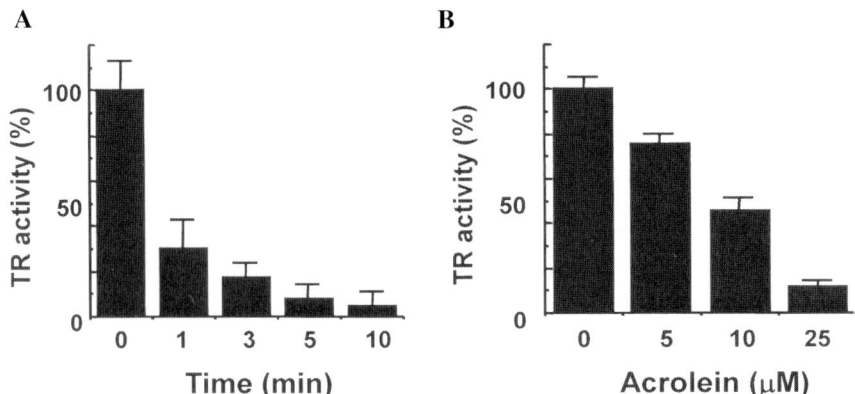

FIGURE 1. Effect of acrolein on TR activity. (**A**) TR activities of purified human TR incubated with 5 μM acrolein for various times. 25 nM TR was used in the experiments. (**B**) TR activity in HUVEC incubated with various concentrations of acrolein for 30 min. TR activity was determined as thioredoxin-dependent insulin reduction activity as described in Ref. 8. Activities are given as the percentage of the control value. Data are presented as the means ± SD of triplicate experiments. Modified from Ref. 8.

increase in intracellular oxidative stress and subsequent cellular damage.

Acrolein Induces Cyclooxygenase-2 Expression and Prostaglandin E$_2$ Production

Cyclooxygenase (COX) catalyzes the oxygenation of arachidonic acid to prostaglandin endoperoxides, which are converted enzymatically into prostaglandins (PGs) and thromboxane A2, both of which play physiologic and pathologic roles in vascular function. Two distinct isoforms of COX have been identified in mammalian cells. COX-1 is constitutively expressed in a variety of cells, such as vascular cells, fibroblasts, platelets, and epithelia, whereas COX-2 is absent from most normal tissues but is expressed in response to proliferative and inflammatory stimuli.[14] COX-2 is expressed in atherosclerotic lesions and is increased after vascular injury. Because chronic inflammation plays an important role in atherosclerosis, COX-2 may participate in the genesis of atherosclerosis.

To test the effects of acrolein on HUVEC, we incubated cells in medium containing acrolein at different concentrations and analyzed COX-2 levels by Western blot. COX-2 was strongly induced in HUVEC (FIG. 2A). We tested whether this COX-2 enzyme was responsible for PGE$_2$ production in the culture media of cells stimulated with acrolein because COX-2 catalyzes biosynthesis of PGs. As shown in FIGURE 2B, acrolein increased PGE$_2$ secretion by eight-fold at 16 h. The results indicate that acrolein can lead to COX-2 protein expression and subsequently PGE$_2$ biosynthesis in HUVEC.

Signaling Pathway Involved in the Induction of COX-2 by Acrolein in Endothelial Cells

It is well known that COX-2 expression has been linked with activation of mitogen-activated protein kinase (MAPK) pathways and that the particular signaling pathway involved is dependent on the type of stimuli. Our data demonstrate that acrolein induces COX-2 expression by activating p38 MAPK (FIG. 2C). In contrast, extracellular signal-regulated kinase (ERK) and c-Jun N-terminal kinase (JNK) did not contribute to acrolein-mediated COX-2 induction. The activation of protein kinase C (PKC) induces COX-2 expression in many cell types, such as astrocytic and endothelial cells.[15] Activation of PKC may be a key event in the signaling pathway leading to COX-2 expression. Our data show that COX-2 expression is reduced by a PKC inhibitor (FIG. 2D), and PKC activation leads to p38 MAPK activation and COX-2 expression.[16] In addition, we found involvement of cAMP response element-binding protein (CREB) activation in acrolein-induced COX-2 (FIG. 2E).

Conclusion

Atherosclerosis is a chronic endothelial inflammatory process. Endothelial cells are primary targets of

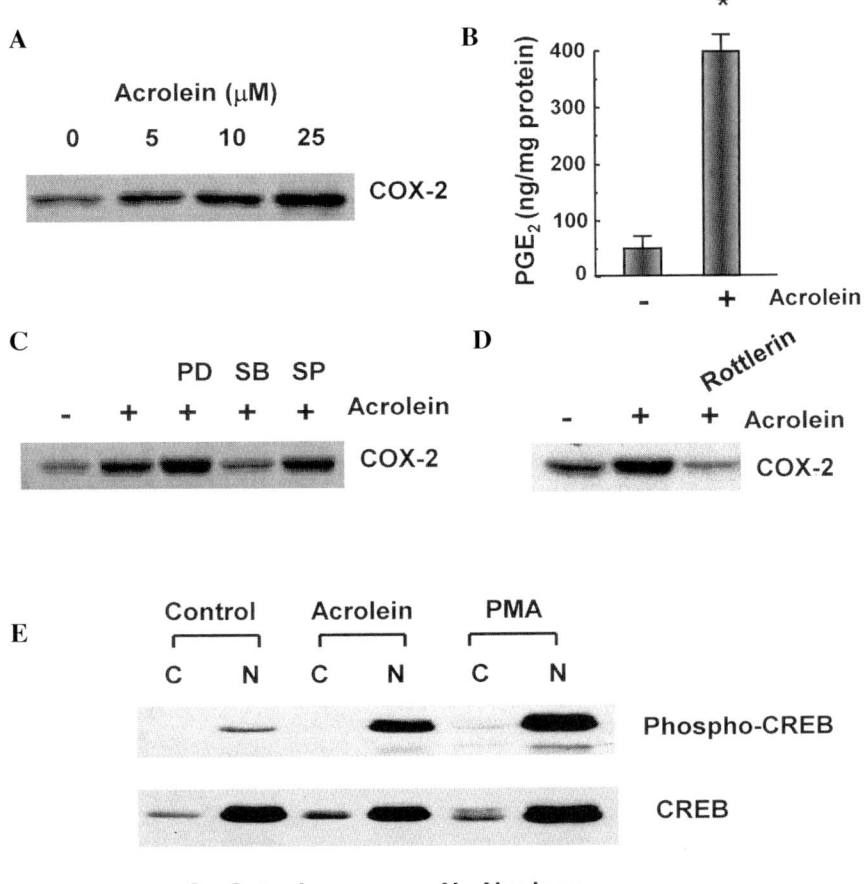

FIGURE 2. Increase of COX-2 expression and PGE$_2$ production, and involvement of p38 MAPK, PKCδ, and CREB activation by acrolein in HUVEC. After treatment of HUVEC with various concentrations of acrolein for 16 h (**A**), the cells were washed twice with phosphate-buffered saline. The cell lysates were prepared and 20-μg samples of proteins were subjected to Western blotting using anti-COX-2. (**B**), Cells were treated with 10 μM acrolein, and the release of PGE$_2$ was then measured from supernatants as described in Ref 16. The values shown for PGE$_2$ production are the mean ± SD of three independent experiments. (**C**), HUVEC were pre-incubated with the ERK inhibitor, PD (PD98059, 40 μM); the p38 MAPK inhibitor, SB (SB203580, 10 μM); and the JNK inhibitor, SP (SP600125, 20 μM) for 30 min. The cells were then treated with 10 μM acrolein for 16 h and 20-μg samples of whole-cell lysates were subjected to Western blotting using anti-COX-2 antibody. (**D**), HUVEC were pre-incubated with the PKCδ inhibitor, rottlerin (5 μM), for 30 min. The cells were treated with 10 μM acrolein for 16 h and subjected to Western blotting for COX-2. (**E**), CREB activation. HUVEC were treated with 10 μM acrolein and 0.1 μM phorbol 12-myristate 13-acetate (PMA), as a positive control for CREB activation, for 30 min and 30-μg samples of lysates were subjected to Western blotting using anti-phospho-CREB antibody and anti-CREB antibody.

oxidative stress.[17] Upon exposure to ROS, endothelial cells display a variety of adverse biological effects, including the production of inflammatory mediators, expression of adhesion molecules, and increased cell permeability. ROS-induced endothelium damage or dysfunction is considered to be a primary pathogenetic mechanism in various cardiovascular diseases. Acrolein levels are increased in patients with atherosclerosis and in cigarette smokers[6]; this increase is strongly associated with an increased risk of vascular disease. Thus, increased acrolein levels might be involved in pathogenesis of atherosclerosis. On the other hand, augmented COX-2 expression or prostaglandin overproduction in atherosclerotic lesions has also been reported. We have shown that acrolein, a known toxin in tobacco smoke,[18] may contribute to the

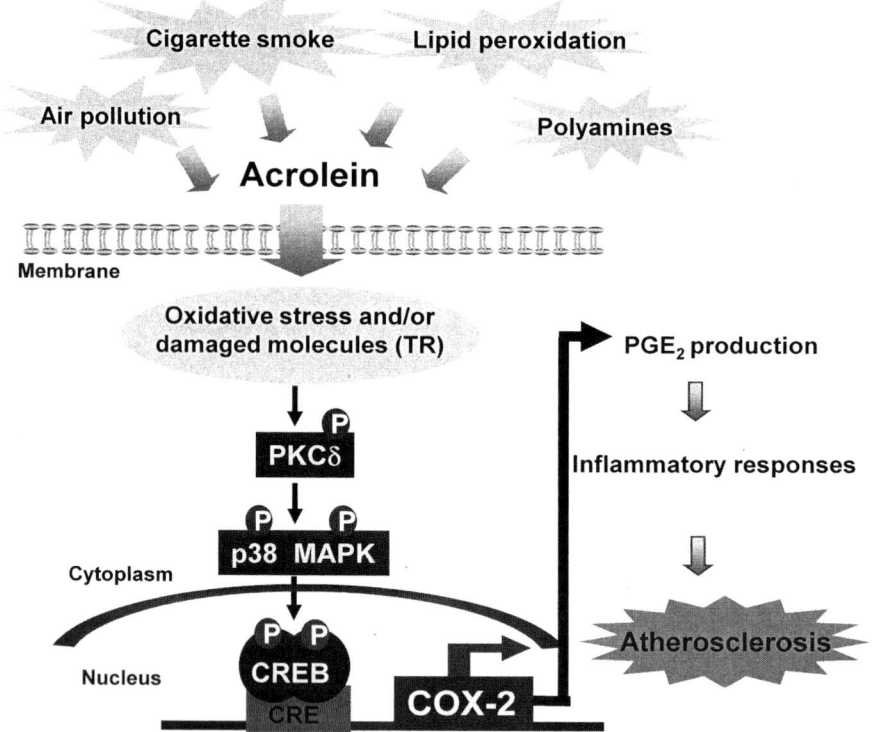

FIGURE 3. The main signaling pathway for induction of COX-2 by acrolein. Modified from Ref. 16.

development of vascular disease. Our data demonstrate that acrolein elevates oxidative stress via inactivation of TR, stimulates the expression of COX-2, and enhances prostaglandin synthesis in HUVEC through the activation of the PKC, p38 MAPK, and CREB pathway (FIG. 3). Our findings suggest that acrolein may play an important role in the progression of atherosclerosis via an inflammatory response involving COX-2 expression.

Acknowledgements

This study was supported, in part, by Grants-in Aid for 21 century Center of Excellence from the Ministry of Education, Science, Sports and Culture of Japan and by Brain Korea 21 of Korea.

Conflict of Interest

The authors declare no conflicts of interest.

References

1. KEHER, J.P. & S.S. BISWAL. 2000. The molecular effect of acrolein. Toxicol. Sci. **57:** 6–15
2. SHARMIN, S. *et al*. 2001. Polyamine cytotoxicity in the presence of bovine serum amine oxidase. Biochem. Biophys. Res. Commun. **282:** 228–235.
3. ANDERSON, M.M. *et al*. 1997. Human neutrophils employ the myeloperoxidase-hydrogen peroxide-chloride system to convert hydroxy-amino acids into glycolaldehyde, 2-hydroxypropanal, and acrolein. A mechanism for the generation of highly reactive alpha-hydroxy and alpha,beta-unsaturated aldehydes by phagocytes at sites of inflammation. J. Clin. Invest. **99:** 424–432.
4. UCHIDA, K. *et al*. 1998. Acrolein is a product of lipid peroxidation reaction. Formation of free acrolein and its conjugate with lysine residues in oxidized low density lipoproteins. J. Biol. Chem. **273:** 16058–16066.
5. LOVELL, M.A., C. XIE & W.R. MARKESBERY. 2001. Acrolein is increased in Alzheimer's disease brain and is toxic to primary hippocampal cultures. Neurobiol. Aging **22:** 187–194.
6. UCHIDA, K. *et al*. 1998. Protein-bound acrolein: potential markers for oxidative stress. Proc. Natl. Acad. Sci. USA **95:** 4882–4887.
7. DAIMON, M. *et al*. 2003. Increased urinary levels of pentosidine, pyrraline and acrolein adduct in type 2 diabetes. Endocr. J. **50:** 61–67.
8. PARK, Y.S. *et al*. 2005. Induction of thioredoxin reductase as an adaptive response to acrolein in human umbilical vein endothelial cells. Biochem. Biophys. Res. Commun. **327:** 1058–1065.

9. LEOPOLD, J.A. & J. LOSCALZO. 2005. Oxidative enzymopathies and vascular disease. Arterioscler. Thromb. Vasc. Biol. **25:** 1332–1340.
10. NAPOLI, C. et al. 2001. Multiple role of reactive oxygen species in the arterial wall. J. Cell Biochem. **82:** 674–682.
11. PARK, Y.S. et al. 2003. Identification of the binding site of methylglyoxal on glutathione peroxidase: methylglyoxal inhibits glutathione peroxidase activity via binding to glutathione binding sites Arg 184 and 185. Free Radic. Res. **37:** 205–211.
12. MIYAMOTO, Y. et al. 2003. Oxidative stress caused by inactivation of glutathione peroxidase and adaptive responses. Biol. Chem. **384:** 567–574.
13. ARNER, E.S. & A. HOLMGREN. 2000. Physiological functions of thioredoxin and thioredoxin reductase. Eur. J. Biochem. **267:** 6102–6109.
14. DUBOIS, R.N. et al. 1998. Cyclooxygenase in biology and disease. FASEB J. **12:** 1063–1073.
15. COSENTINO, F. et al. 2003. High glucose causes upregulation of cyclooxygenase-2 and alters prostanoid profile in human endothelial cells: role of protein kinase C and reactive oxygen species. Circulation **107:** 1017–1023
16. PARK, Y.S. et al. 2007. Acrolein induces cyclooxygenase-2 and prostaglandin production in human umbilical vein endothelial cells: roles of p38 MAP kinase. Arterioscler. Thromb. Vas. Biol. **27:** 1319–1325
17. HALLIWELL, B. & J.M.C. GUTTERIDGE. 1999. Free Radicals in Biology and Medicine. Oxford University Press. Oxford.
18. FUJIOKA, K. & T. SHIBAMOTO. 2006. Determination of toxic carbonyl compounds in cigarette smoke. Environ. Toxicol. **21:** 47–54.

Oxidative Stress and Advanced Glycation in Diabetic Nephropathy

MELINDA T. COUGHLAN, AMY L. MIBUS, AND JOSEPHINE M. FORBES

JDRF Albert Einstein Centre for Diabetes Complications, Diabetes and Metabolism Division, Baker Heart Research Institute, Melbourne, Victoria, Australia

Nephropathy remains a significant cause of morbidity and mortality in the diabetic population and is the leading cause of end-stage renal failure in the Western World. As a result of the diabetic milieu, increased generation of reactive oxygen species (ROS) is thought to play a key role in the progression of diabetic nephropathy. Recent experimental studies have suggested that the receptor for advanced glycation end products (RAGE), which is central to the advanced glycation pathway, may mediate renal structural and functional damage via oxidative stress. This review focuses on how RAGE and subsequent ROS generation play a deleterious role in the diabetic kidney, promoting cross-talk among signaling pathways, ultimately leading to renal dysfunction.

Key words: advanced glycation end products (AGEs); diabetic nephropathy; reactive oxygen species (ROS); receptor for advanced glycation end products (RAGE)

Introduction

Diabetes mellitus is currently at epidemic proportions in Western countries and is an emerging health problem in developing nations. Both macrovascular and microvascular end-organ injury as a result of diabetes constitutes a significant proportion of the mortality attributed to this condition. Microvascular pathology from diabetes occurs at various sites, particularly in the retina, the renal glomerulus, and the peripheral nerve, and, as a consequence, diabetes is a leading cause of blindness, end-stage renal disease, and neuropathy. Indeed, diabetic nephropathy is now the leading cause of renal failure in adults. It has been established that hyperglycemia is the likely initiating factor in tissue damage observed clinically in diabetes.[1,2] Specifically, it has been hypothesized that hyperglycemia-induced damage in diabetic complications is primarily from mitochondrial superoxide overproduction, which then activates major biochemical pathways of dysfunction.[3]

Oxidative Stress and Diabetes Complications

In the "unifying hypothesis" described by Brownlee, it is postulated that the generation of mitochondrial reactive oxygen species (ROS) is the primary initiating event that subsequently activates a number of these other pathways implicated in the development of the complications of diabetes.[3] Nevertheless, debate continues as to whether oxidative stress is an important early link between hyperglycemia and complications or just a byproduct of various pathogenic mechanisms.[4] ROS can directly inflict macromolecular damage to DNA, protein, and lipids.[3,5] In addition, ROS can function as signaling molecules and induce a number of stress-sensitive pathways that cause cellular damage.[5] There are a number of sources which could potentially generate excess ROS in diabetes, such as autooxidation of glucose, transition metal-catalyzed Fenton reactions, mitochondrial respiratory chain deficiencies, xanthine oxidase activity, and NAD(P)H oxidase.[4,6]

Mitochondrial Production of ROS

ROS are generated within the mitochondria as a result of electron flux through the electron transport chain. During oxidative phosphorylation (OXPHOS), a low proportion of molecular oxygen is converted to superoxide and subsequently to

Address for correspondence: Dr. Melinda T. Coughlan, Albert Einstein Centre for Diabetes Complications, Baker Heart Research Institute, PO Box 6492, St. Kilda Road Central, Melbourne, 8008, Australia. Voice: +61 3 8532 1278; fax: +61 3 8532 1100.
melinda.coughlan@baker.edu.au

hydrogen peroxide and the hydroxyl radical, which, under normal conditions, are scavenged by antioxidant enzymes, including mitochondrial manganese superoxide dismutase (MnSOD, SOD2) and glutathione peroxidase (GPx). Damaged or dysfunctional mitochondria, however, overgenerate superoxide radicals, creating a state of redox imbalance.[7] Excess mitochondrial ROS production is often mediated by disruption of the activity of OXPHOS enzymes via electron leakage at complex I (NADH:ubiquinone oxidoreductase) or complex III (ubiquinol:cytochrome c oxidoreductase).[7–9] A recent study has found that renal cortical mitochondria from rats with diabetes exhibited a diminution of OXPHOS via decreased complex III activity and increased superoxide formation.[10] Complex III activity negatively correlated with increasing methylglyoxal modification of mitochondrial proteins. Our own group has demonstrated overproduction of superoxide from the renal mitochondria in rats with diabetic nephropathy in parallel with increased intramitochondrial carboxymethyllysine accumulation and deficiencies in MnSOD activity.[11] We have also shown evidence of a deficiency in complex I of the mitochondrial respiratory chain in renal cortex from rats with diabetes.[12] Intramitochondrial superoxide production initiates a range of damaging reactions through the production of hydrogen peroxide, ferrous iron, hydroxyl radical, and peroxynitrite, which can damage lipids, proteins, and nucleic acids. Mitochondria are particularly susceptible to ROS-mediated damage, resulting in altered ATP synthesis, cellular calcium dysregulation, and induction of mitochondrial permeability transition, all of which predispose the cell to necrosis or apoptosis.[13]

NAD(P)H Oxidase and ROS Production

NAD(P)H oxidase is a cytosolic enzyme complex that was initially discovered in neutrophils where it plays a vital role in nonspecific host defense by producing millimolar quantities of superoxide.[14] In addition to residing in phagocytic cells, NAD(P)H oxidase is present in nonphagocytic cell types, such as renal mesangial, proximal tubular, vascular smooth muscle cells, and podocytes.[15–18] In these cell types, small amounts of ROS generated by NAD(P)H oxidase may participate in second-messenger redox signaling, and, when upregulated, the ensuing increased ROS production contributes to oxidative imbalances.[15] The expression of NAD(P)H oxidase components is increased in microvascular and macrovascular tissues of diabetic animals, including the kidney. Studies from our own laboratory have demonstrated increases in AGEs and nitrotyrosine accumulation in association with the enhanced expression of RAGE and the NAD(P)H oxidase subunit gp91^{phox} as well as nuclear factor kappa-B (NF-κB) activation in the kidneys of rats with diabetes.[19] Pharmacological inhibition of NAD(P)H oxidase with apocynin prevents renal NAD(P)H oxidase subunit p47^{phox} and gp91^{phox} upregulation and retards mesangial matrix expansion in diabetic rats.[20] In addition, we have observed that diabetes induces increases in NAD(P)H oxidase-driven ROS formation in rat renal cortex.[11] Furthermore, a recent study has demonstrated that in mesangial cells *in vitro*, NAD(P)H oxidase was responsible for high glucose- and AGE-induced superoxide production.[21] These observations highlight the possibility that cytosolic sources of ROS, such as NAD(P)H oxidase, may also be equally important in the progression of diabetic nephropathy in addition to those ROS originating from the mitochondria.

Advanced Glycation End Products

The advanced glycation pathway is considered to be a key process in mediating tissue damage in diabetic microvascular disease. Clinical studies in patients with type 1 diabetes demonstrate a strong correlation between AGE accumulation and the severity of microvascular and macrovascular complications.[22–24] In particular, serum concentrations of AGEs are significantly increased with the progression to microalbuminuria and subsequently to overt nephropathy.[22] In addition to the circulation, AGEs have been found in increased concentrations at various sites of injury in diabetes, such as renal glomeruli[25] and tubules.[19] Pharmacological inhibition of AGE accumulation ameliorates both structural and functional features of experimental diabetic nephropathy.[26]

Within diabetic tissues, AGEs are thought to contribute to end-organ injury via a number of processes, such as by inducing cross-linking of proteins and through ligand-receptor binding. Further to this, elevated intracellular glucose degradation products resulting from glycolysis and the tricarboxylic acid cycle initiate the glycation of proteins far more rapidly than glucose itself inside the cell.[27] The generation of intracellular AGEs can disturb redox homeostasis by modifying protein and enzyme structure and function. For example, glycation of antioxidants, such as copper and zinc containing SOD (CuZn-SOD or SOD1) contribute to the decline in activity of antioxidant enzymes characteristic of diabetic microvascular disease.[28] While oxidative stress can augment the formation of AGEs through glycoxidation, AGEs can also lead to enhanced formation of free radicals directly through catalytic sites in their molecular structure[29] and via stimulation of membrane-bound NAD(P)H oxidase.[30]

AGEs can also mediate their effects via interactions with specific receptors and binding proteins, inducing inflammatory, oxidative, metabolic, and hemodynamic pathways. These include the RAGE, AGE-R1 (p60), AGE-R2 (80k-H, protein kinase C substrate), AGE-R3 (galectin-3), lysozyme,[31] as well as the macrophage scavenger receptors ScR-II and CD-36[32] and the more recently identified members of the ezrin–radixin–moesin family.[33] There is an increasing body of evidence to support the concept that interactions between AGEs and their receptors, in particular RAGE, are involved in the pathogenesis of diabetic nephropathy. Studies in RAGE transgenic mice reveal that these rodents have increased glomerulosclerosis following the induction of diabetes.[34] Conversely, RAGE knockout mice have decreased renal injury in response to diabetes,[35] and long-term administration of a RAGE-neutralizing antibody to $db/db^{(+/+)}$ mice confers renoprotection.[36] In the glomeruli of patients with diabetic nephropathy, RAGE expression is upregulated and positively correlates with AGE accumulation.[37]

Studies indicate that ROS generation with subsequent increased oxidant stress is a potent factor initiating signal transduction and altered gene expression as a result of the AGE–RAGE interaction.[38,39] RAGE can activate NAD(P)H oxidase[30] potentially through a protein kinase C–dependent process.[40] Induction of intracellular ROS can promote cellular dysfunction particularly targeting the mitochondria, such as the induction of mitochondrial permeability transition, leading to apoptosis.[41] In renal cells, hydrogen peroxide triggers a deleterious cycle of NF-κB activation and consequent induction of NF-κB–dependent inflammatory signals.[42] Finally, antioxidant treatment of cultured cells prevents AGE-induced activation of NF-κB, transforming growth factor-β1, and cell death.[43] These studies emphasize the significance of AGE-receptor-mediated ROS production in the pathogenesis of diabetic nephropathy.

Conclusion

It is clear that both RAGE and oxidative stress play an important role in the progression of diabetic complications, including nephropathy. The exact source of the increased ROS remains to be fully determined, and the relative importance of altered antioxidant defense in the kidney has not been adequately examined. Nevertheless, increasing elucidation of oxidative stress pathways and how they interact with other mediators of disease, such as AGEs, should lead to new therapeutic targets for this major complication of diabetes.

Conflict of Interest

The authors declare no conflicts of interest.

References

1. THE DIABETES CONTROL AND COMPLICATIONS TRIAL RESEARCH GROUP. 1993. The effect of intensive treatment of diabetes on the development and progression of long-term complications in insulin-dependent diabetes mellitus. N. Engl. J. Med. **329:** 977–986.
2. UKPDS GROUP. 1998. Intensive blood-glucose control with sulphonylureas or insulin compared with conventional treatment and risk of complications in patients with type 2 diabetes (UKPDS 33). UK Prospective Diabetes Study. Lancet **352:** 837–853.
3. BROWNLEE, M. 2001. Biochemistry and molecular cell biology of diabetic complications. Nature **414:** 813–820.
4. BAYNES, J.W. & S.R. THORPE. 1999. Role of oxidative stress in diabetic complications: a new perspective on an old paradigm. Diabetes **48:** 1–9.
5. ROSEN, P. et al. 2001. The role of oxidative stress in the onset and progression of diabetes and its complications: a summary of a Congress Series sponsored by UNESCO-MCBN, the American Diabetes Association and the German Diabetes Society. Diabetes Metab. Res. Rev. **17:** 189–212.
6. CAMERON, N.E. & M.A. COTTER. 1999. Effects of antioxidants on nerve and vascular dysfunction in experimental diabetes. Diabetes Res. Clin. Pract. **45:** 137–146.
7. PITKANEN, S. & B.H. ROBINSON. 1996. Mitochondrial complex I deficiency leads to increased production of superoxide radicals and induction of superoxide dismutase. J. Clin. Invest. **98:** 345–351.
8. BEYER, R.E. 1992. An analysis of the role of coenzyme Q in free radical generation and as an antioxidant. Biochem. Cell Biol. **70:** 390–403.
9. KOOPMAN, W.J. et al. 2005. Inhibition of complex I of the electron transport chain causes O^{2-} mediated mitochondrial outgrowth. Am. J. Physiol. Cell Physiol. **288:** C1440–C1150.
10. ROSCA, M.G. et al. 2005. Glycation of mitochondrial proteins from diabetic rat kidney is associated with excess superoxide formation. Am. J. Physiol. Renal Physiol. **289:** F420–F430.
11. COUGHLAN, M.T. et al. 2007. Combination therapy with the advanced glycation end product cross-link breaker, alagebrium, and angiotensin converting enzyme inhibitors in diabetes: synergy or redundancy? Endocrinology **148:** 886–895.
12. COUGHLAN, M.T. et al. 2005. Renal intra-mitochondrial glycation drives deficiencies in the activity of manganese superoxide dismutase and complex I of the mitochondrial

13. JAMES, A.M. & M.P. MURPHY. 2002. How mitochondrial damage affects cell function. J. Biomed. Sci. **9:** 475–487.
14. BABIOR, B.M., J.D. LAMBETH & W. NAUSEEF. 2002. The neutrophil NADPH oxidase. Arch. Biochem. Biophys. **397:** 342–344.
15. GRIENDLING, K.K., D. SORESCU & M. USHIO-FUKAI. 2000. NAD(P)H oxidase: role in cardiovascular biology and disease. Circ. Res. **86:** 494–501.
16. JONES, S.A. et al. 1995. The expression of NADPH oxidase components in human glomerular mesangial cells: detection of protein and mRNA for p47phox, p67phox, and p22phox. J. Am. Soc. Nephrol. **5:** 1483–1491.
17. RADEKE, H.H. et al. 1991. Functional expression of NADPH oxidase components (alpha- and beta-subunits of cytochrome b558 and 45-kDa flavoprotein) by intrinsic human glomerular mesangial cells. J. Biol. Chem. **266:** 21025–21029.
18. SHIOSE, A. et al. 2001. A novel superoxide-producing NAD(P)H oxidase in kidney. J. Biol. Chem. **276:** 1417–1423.
19. FORBES, J.M. et al. 2002. Reduction of the accumulation of advanced glycation end products by ACE inhibition in experimental diabetic nephropathy. Diabetes **51:** 3274–3282.
20. ASABA, K. et al. 2005. Effects of NADPH oxidase inhibitor in diabetic nephropathy. Kidney Int. **67:** 1890–1898.
21. LIN, C.L. et al. 2006. Ras modulation of superoxide activates ERK-dependent fibronectin expression in diabetes-induced renal injuries. Kidney Int. **69:** 1593–1600.
22. MIURA, J. et al. 2003. Serum levels of non-carboxymethyllysine advanced glycation endproducts are correlated to severity of microvascular complications in patients with Type 1 diabetes. J. Diabetes Complications **17:** 16–21.
23. MONNIER, V.M. et al. 1999. Skin collagen glycation, glycoxidation, and crosslinking are lower in subjects with long-term intensive versus conventional therapy of type 1 diabetes: relevance of glycated collagen products versus HbA1c as markers of diabetic complications. DCCT Skin Collagen Ancillary Study Group. Diabetes Control and Complications Trial. Diabetes **48:** 870–880.
24. WENDT, T. et al. 2002. Receptor for advanced glycation endproducts (RAGE) and vascular inflammation: insights into the pathogenesis of macrovascular complications in diabetes. Curr. Atheroscler. Rep. **4:** 228–237.
25. HORIE, K. et al. 1997. Immunohistochemical colocalization of glycoxidation products and lipid peroxidation products in diabetic renal glomerular lesions. Implication for glycoxidative stress in the pathogenesis of diabetic nephropathy. J. Clin. Invest. **100:** 2995–3004.
26. SOULIS-LIPAROTA, T. et al. 1991. Retardation by aminoguanidine of development of albuminuria, mesangial expansion, and tissue fluorescence in streptozocin-induced diabetic rat. Diabetes **40:** 1328–1334.
27. HAMADA, Y. et al. 1996. Rapid formation of advanced glycation end products by intermediate metabolites of glycolytic pathway and polyol pathway. Biochem. Biophys. Res. Commun. **228:** 539–543.
28. FUJII, J. et al. 1996. Oxidative stress caused by glycation of Cu,Zn-superoxide dismutase and its effects on intracellular components. Nephrol. Dial. Transplant. 11(Suppl 5): 34–40.
29. YAGIHASHI, S. 1997. Pathogenetic mechanisms of diabetic neuropathy: lessons from animal models. J. Peripher. Nerv. Syst. **2:** 113–132.
30. WAUTIER, M.P. et al. 2001. Activation of NADPH oxidase by AGE links oxidant stress to altered gene expression via RAGE. Am. J. Physiol. Endocrinol. Metab. **280:** E685–E694.
31. VLASSARA, H. 2001. The AGE-receptor in the pathogenesis of diabetic complications. Diabetes Metab. Res. Rev. **17:** 436–443.
32. SMEDSROD, B. et al. 1997. Advanced glycation end products are eliminated by scavenger-receptor-mediated endocytosis in hepatic sinusoidal Kupffer and endothelial cells. Biochem. J. **322**(Pt 2): 567–573.
33. MCROBERT, E.A. et al. 2003. The amino-terminal domains of the ezrin, radixin, and moesin (ERM) proteins bind advanced glycation end products, an interaction that may play a role in the development of diabetic complications. J. Biol. Chem. **278:** 25783–25789.
34. YAMAMOTO, Y. et al. 2001. Development and prevention of advanced diabetic nephropathy in RAGE-overexpressing mice. J. Clin. Invest. **108:** 261–268.
35. WENDT, T. et al. 2003. Glucose, glycation, and RAGE: implications for amplification of cellular dysfunction in diabetic nephropathy. J. Am. Soc. Nephrol. **14:** 1383–1395.
36. FLYVBJERG, A. et al. 2004. Long-term renal effects of a neutralizing RAGE antibody in obese type 2 diabetic mice. Diabetes **53:** 166–172.
37. TANJI, N. et al. 2000. Expression of advanced glycation end products and their cellular receptor RAGE in diabetic nephropathy and nondiabetic renal disease. J. Am. Soc. Nephrol. **11:** 1656–1666.
38. MIYATA, T. et al. 1996. The receptor for advanced glycation end products (RAGE) is a central mediator of the interaction of AGE-beta2microglobulin with human mononuclear phagocytes via an oxidant-sensitive pathway. Implications for the pathogenesis of dialysis-related amyloidosis. J. Clin. Invest. **98:** 1088–1094.
39. WAUTIER, J.L. et al. 1994. Advanced glycation end products (AGEs) on the surface of diabetic erythrocytes bind to the vessel wall via a specific receptor inducing oxidant stress in the vasculature: a link between surface-associated AGEs and diabetic complications. Proc. Natl. Acad. Sci. USA **91:** 7742–7746.
40. NITTI, M. et al. 2007. PKC delta and NADPH oxidase in AGE-induced neuronal death. Neurosci. Lett. **416:** 261–265.
41. TAKEYAMA, N. et al. 2002. Role of the mitochondrial permeability transition and cytochrome C release in hydrogen peroxide-induced apoptosis. Exp. Cell Res. **274:** 16–24.
42. MORIGI, M. et al. 2002. Protein overload-induced NF-kappaB activation in proximal tubular cells requires H(2)O(2) through a PKC-dependent pathway. J. Am. Soc. Nephrol. **13:** 1179–1189.
43. LAL, M.A. et al. 2002. Role of oxidative stress in advanced glycation end product-induced mesangial cell activation. Kidney Int. **61:** 2006–2014.

Vitamin C–mediated Maillard Reaction in the Lens Probed in a Transgenic-mouse Model

XINGJUN FAN[a] AND VINCENT M. MONNIER[a,b]

Departments of [a]Pathology and [b]Biochemistry, Case Western Reserve University, Cleveland, Ohio, USA

Aging human lens crystallins are progressively modified by yellow glycation, oxidation, and cross-linked carbonyl compounds that have deleterious properties on protein structure and stability. In order to test the hypothesis that some of these compounds originate from oxidized vitamin C, we have overexpressed the human vitamin C transporter 2 (hSCVT2) in the mouse lens. We find that levels of ascorbic and dehydroascorbic acid are highly elevated compared to the wild type and that the lenses have accumulated yellow color and advanced Maillard reaction products identical with those of the human lens. Treatment of the mice with nucleophilic inhibitors can slow down the process, opening new avenues for the pharmacological prevention of senile cataractogenesis.

Key words: glycation; ascorbic acid; crystallin; cross-linking; aging

Introduction

The aging human lens, by virtue of having little to no protein turnover, is susceptible to accumulation of damage to its proteins, the crystallins. During early development, the crystallins undergo a set of postsynthetic modifications that include phosphorylation, deamidation, and enzyme-mediated protein truncation.[1–3] While, however, the phosphorylation status appears to be little changed once fiber-like cells are differentiated, deamidation is an ongoing process that participates in chemical modification of side-chain residues resulting from oxidation, cleavage of the peptide backbone through reactive oxygen species, and covalent modification of amino acid residues.[4] This is associated with enhanced risk of crystalline aggregation.[5]

In contrast, senescence-related changes typically relate to stochastic forms of damage, i.e., post-translations by low-molecular weight compounds. All processes occur throughout life. While they are unlikely to occur rapidly enough to trigger cataract formation per se, they are thought to predispose lens crystallins toward aggregation and formation of light scattering high-molecular-weight aggregates. In that regard, unequivocal evidence for age-related accumulation of multimeric protein cross-links in the aging human lens has been reported.[6]

The Maillard reaction is the reaction that is commonly observed during cooking or baking foods. It involves reactive carbonyl compounds from reducing sugars, methyglyoxal, oxidized lipids, or ascorbic acid (ASA) oxidation products, which, upon reaction with proteins, lead to the formation of yellow-colored protein adducts and cross-links. The Maillard reaction proceeds throughout aging in the human lens and other tissues.[7] Because of the high reactivity of ASA degradation products with proteins and its presence in high concentrations in the human lens, we and others have postulated that it is, in part, responsible for the progressive damage to the aging human lens crystallins.[8–10] However, it has not been possible to unequivocally implicate vitamin C in this aging process because the products formed are not specific for this vitamin.

To confirm the role of vitamin C in lens ag, we took advantage of the fact that the mouse lens has very low levels of the vitamin and engineered a mouse that selectively overexpresses the human vitamin C transporter 2 (hSVCT2)[11] in both the epithelium and fiber-like lens cells. Consequently, lenticular levels of vitamin C and its oxidation products dramatically increased, resulting in accelerated formation of several advanced glycation end products (AGEs) identical with those of the aging human lens. With the use of this humanized mouse model, we conducted a first intervention study with potential inhibitors to test the feasibility of preventing ascorbylation in aging lenses.

Address for correspondence: Vincent M. Monnier, CWRU, 2103 Cornell Rd., Wolstein Research Building, Cleveland, OH 44106. Voice: +1-216-368-6613; fax: +1-216-368-1357.

vmm3@cwru.edu

Study Design

Generation of Transgenic Mice

Mice were housed under diurnal lighting conditions and allowed free access to food and water. All animals were used in accordance with the Association for Research in Vision and Ophthalmology Statement for the Use of Animals in Ophthalmology and Vision Research. The hSVCT2 driven by αA crystallin promoter with chick δ1-crystallin enhancer was microinjected into fertilized eggs of B6SJLF1 mice, and the eggs were then transferred into the oviducts of pseudopregnant female mice (Transgenic Animal Model Core, University of Michigan, Ann Arbor, MI). hSVCT2 transgenic mice were identified via PCR screening from genomic DNA extracted from tails. Transgenic mice were crossbred to C57BL/6 gene background by at least eight generations.

Aging Study Design

Transgenic and age-matched control mice, 10 mice per group, were maintained on a standard mouse diet (Prolab 5P75 Isopro 3000; LabDiet, Richmond, IN). Mice were killed at 6, 9, and 12 months. Eyes were removed from mice and decapsulated to release lenses.

Intervention Study Design

Pyridoxamine (PM) dihydrochloride, DL-penicillamine (PA), aminoguanidine (AG), and nucleophilic compounds NC-I and NC-II were all purchased from Sigma Company (St. Louis, MI). The mouse diet with 0.1% w/w potential inhibitors listed above was produced by Bio-Serv using standard diet Isopro 3000. Transgenic and age-matched control mice, 10 mice per group, were maintained on a standard mouse diet or special medical diet started at 2 months and continued until 9 months of age. The body weight and food intake were monitored monthly. Nine-month-old mice were sacrificed, and eyes were removed and decapsulated to release the lenses.

Measurement of Ascorbic Acid and Dehydroascorbic Acid

ASA and dehydroascorbic acid (DHA) concentrations were determined based on dimethyl-o-phenylene-diamine derivatization as previously described,[12] with few modifications.[36] In brief, after mice were killed, eyes were removed and decapsulated immediately. Mouse lenses were homogenized in 200 μL of 10% cold trichloroacetic acid and kept on ice for 10 min.

After oxidation with 10 μL of 0.01 M iodine in 2.7% potassium iodide, 20 μL of 0.01 M thiosulfate was added to reduce excess iodine. Samples were then derivatized by adding 50 μL of sodium phosphate buffer (pH 5.4) and 20 μL of dimethyl-o-phenylene-diamine (1 mg/mL in 0.1 M HCl). The derivatized samples were injected into an HPLC and separated by C18 reverse phase. The derivative was detected by using a fluorescent detector with excitation at 360 nm and emission at 440 nm. For quantitation of DHA, the lens extract was directly derivatized under the same conditions but without the oxidation step.

Measurement of Advanced Glycation End Products

Several AGEs were analyzed in both aging and intervention studies. Pentosidine, carboxymethyllysine (CML), carboxyethyllysine (CEL), furosine, and vesperlysine A were analyzed after hydrolysis of lens tissue with 6 M HCL. CML, CEL, and furosine were determined with a gas chromatography/mass spectrometry method as reported by a previous study.[13] Vesperlysine A was determined by two-step HPLC as reported previously.[14] K2P and fluorescence measurements were completed after enzymatic digestion with a series protease following a previous report.[15] Two types of fluorescence were recorded at $\lambda_{335/385}$ and $\lambda_{370/440}$.

Statistical Analysis

All values were expressed as mean ± SE. Statistical significance of difference in mean values was assessed by repeated measures ANOVA or Student's t-test. Only P values of <0.05 were considered statistically significant.

Results

Expression of hSVCT2 in lens epithelium and fiber cells leads to increased lenticular levels of ASA and DHA. Evidence of lens-specific expression of the transporter was obtained in several ways. First, immunohistochemical staining with an hSVCT2 antibody (Alpha Diagnostic International, San Antonio, TX) revealed staining throughout the epithelium and the fiber-like cells of the transgenic-mouse lens, whereas only the lens epithelium of the wild type was positive (FIG. 1A).

Finally, lenticular levels of ASA and DHA measured by HPLC in trichloroacetic acid extract were five-fold to 15-fold elevated. ASA increased from approximately 0.2 mM to 0.5–2.5 mM, and DHA increased from 10 μM to 40–250 μM in the transgenic lenses (FIG. 1B and C). ASA levels are shown for comparison in human lenses obtained at autopsy and stored frozen (FIG. 1B). Because of auto-oxidation during storage, these ascorbate levels are likely to be at the lower end. Most importantly, in contrast to previous lens-specific

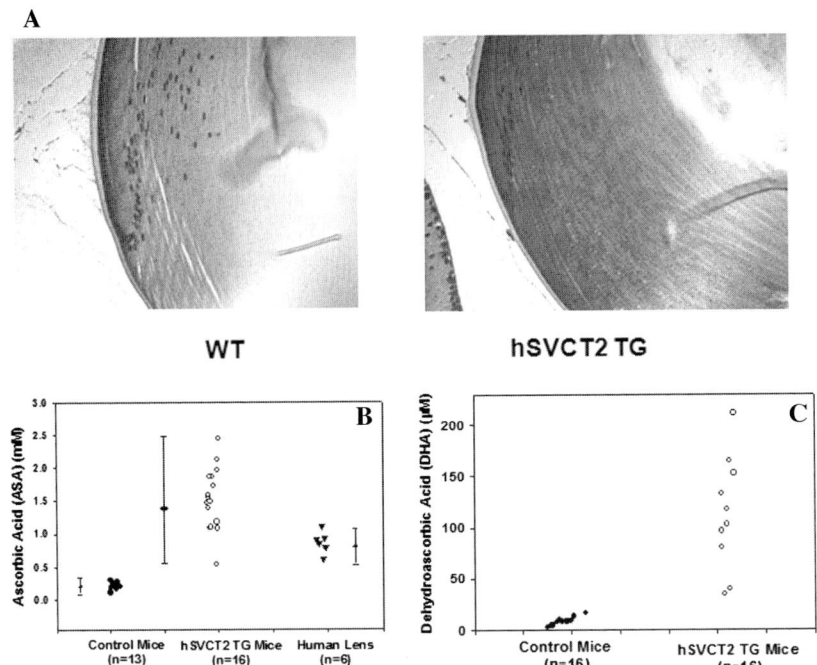

FIGURE 1. Effects of hSVCT2 overexpression in mouse lens epithelial and fiber cells on uptake and lenticular levels of vitamin C and dehydroascorbic acid (DHA). (**A**) Immunohistochemical staining of a typical control (*left*) and transgenic (*right*) lens with an antibody against the C-terminal portion of hSVCT2 reveals a diffuse presence of hSVCT2 along the cell membrane in cortical layers of the lens. (**B**) ASA concentration in 16 lenses from hSVCT2 mice, 13 controls from the same litter, and six human lenses. (**C**) DHA concentration in 10 lenses from hSVCT2 and control mice. WT = wild type; TG = transgenic.

transgenes,[16] sustained age-related expression of the hSVCT2 transgene persisted throughout the 12-month testing period (data not shown).

Expression of hSVCT2 in Mouse Lens Leads to Increased Levels of Ascorbic Acid–derived Advanced Ascorbylation End Products

The incubation of ASA and other reducing sugars with proteins under oxidative conditions is known to result in the formation of various protein-bound fluorophores and cross-links, such as pentosidine and vesperlysine A, as well as the colorless, glycoxidation, lipoxidation product CML, all of which have been documented in the aging human lens.[14,17] In addition, a lysine–lysine cross-link named K2P was recently identified in the human lens as a major fluorophore and UVA-active protein modification.[18] K2P could be synthesized from the reaction of ASA with human lens proteins.[19] All these compounds, including protein-bound fluorescence, were assayed at 6, 9, and 12 months of age either in the enzymatic digest (K2P and protein-bound fluorescence) or acid hydrolyzate of the mouse lens crystallins (pentosidine, vesperlysine A, and CML) following established procedures. The glucose-derived Amadori product fructose–lysine (assayed as furosine) and the methylglyoxal (MGO)-derived CEL were also measured. Protein-bound fluorescence at $\lambda_{335/385\,nm}$ and $\lambda_{370/440\,nm}$ was significantly increased in the transgenic versus the wild-type lenses and increased with age ($P = 0.0001$) (TABLE 1). The fluorescent cross-links K2P, pentosidine, and vesperlysine A were all significantly increased in the transgenic lenses at all time points ($P = 0.001$ to $P = 0.0001$) (TABLE 1). CML was also increased, although to a lesser degree ($P = 0.001$ or less) (TABLE 1). CEL was barely increased (TABLE 1), suggesting that MGO is likely only mildly elevated, although other markers of MGO levels will be needed to confirm this conclusion. Similarly, furosine was not increased (TABLE 1), excluding the presence of high lenticular glucose concentrations as a cause for the elevated AGEs in the transgenic lenses.

Finally, we noticed in the course of this work that the 12-month-old transgenic lenses had acquired a yellow

TABLE 1. Lens protein bound fluorescence and advanced glycation end products (AGEs) in transgenic-mouse model and wild-type mouse at age 6, 9, and 12 months

Fluorescence and AGEs	Wild-type mouse lens				Transgenic-mouse lens				Human lens	
	6 mos	9 mos	12 mos	P value with age	6 mos	9 mos	12 mos	P value with age	Reported Human lens results with age	Ref.
λ 335/385(unit/unit protein)	6.57 ± 2.02	9.11 ± 0.97	13.91 ± 1.07	0.05	15.04 ± 2.53	18.97 ± 3.36	45.50 ± 11.17	0.0001	Significant	35
λ 370/440(unit/unit protein)	5.20 ± 1.76	5.88 ± 1.02	6.95 ± 1.53	N.S	8.16 ± 3.41	10.89 ± 3.37	17.59 ± 4.20	0.0001	Significant	36
K2P (pmol/μmol leu equi)	0.00	0.10 ± 0.00	0.11 ± 0.0	N.S	3.08 ± 1.33	3.50 ± 1.51	19.94 ± 4.27	0.0001	Significant	18
CML (μmol/mol lysine)	159.73 ± 13.05	228.82 ± 30.47	253.03 ± 35.81	0.01	367.35 ± 76.96	444.41 ± 101.14	583.33 ± 105.37	0.005	Significant	14
CEL (μmol/mol lysine)	205.37 ± 6.71	48.10 ± 5.11	53.30 ± 9.75	N.S	86.58 ± 14.32	65.04 ± 18.80	77.62 ± 28.72	N.S	N.S	37
Furosine (μmol/mol lysine)	1035.95 ± 175.88	1383.21 ± 267.32	1250.16 ± 263.30	N.S	1323.13 ± 319.68	1391.42 ± 272.52	1138.92 ± 215.27	N.S	Significant	14
Vesperlysine A (pmol/μmol leu equi)	0	0	0	N.S	2.93 ± 1.35	10.35 ± 1.59	14.17 ± 2.29	0.008	Significant	14
Pentosidine (pmol/μmol leu equi)	0.0006 ± 0.0	0.056 ± 0.053	0.111 ± 0.071	N.S	0.44 ± 0.27	1.13 ± 0.48	1.94 ± 0.79	0.0001	Significant	9
Lens discoloration	no	no	no		no	mild	yellow		Yellow around 55 yrs	20

Abbreviations: CEL, carboxyethyllysine; CML, carboxymethyllysine; K2P ;N.S., not significant.

color similar to that observed in older human lenses (FIG. 2).

Effect of Inhibitors on Ascorbic Acid-derived Advanced Glycation End Products

The incubation of ASA and other reducing sugars with proteins under oxidative conditions is known to result in the formation of various protein cross-links. This has been unequivocally demonstrated in our transgenic mice. This humanized mouse model condensed into 12 months the age-related lens discoloration process that usually develops over several decades in the human; thus, we expect it to be a useful model for testing several inhibitors to prevent the ascorbylation process *in vivo*.

Toward this end, we have tested five candidate inhibitors in a 7-month-long intervention study. NC-I was able to significantly decrease the lysine–arginine cross-link pentosidine by more than 50% (TABLE 2) ($P < 0.04$) and suppressed by 60% the increase in the lysine-modified AGE product CML (TABLE 2) ($P < 0.007$). NC-II-treated mice showed significantly decreased pentosidine level (TABLE 2) ($P < 0.003$) but no change in CML. NC-I but not NC-II tended to decrease the lysine-linked modification CEL. It should be noted that the latter was only mildly elevated, as previously observed,[15] suggesting that MGO formation is not an important pathway in the oxidative degradation of ASA. Animals treated with AG, PM, NC-I, and NC-II showed improvements of lysine–lysine cross-link compound K2P. Because of large standard deviations, none of them reached significance. Animals treated with AG, PM, and pentosidine showed no significant improvement in any of the other measured AGEs.

Discussion

The above data strongly implicate vitamin C in at least one form of the chemical processes that affect the aging human lens crystallins. Furthermore, the hSVCT2–δenαA mouse is an animal model capable of reproducing, within a very short time, the yellow discoloration and crystallin modifications that increase over several decades in the human lens. These findings suggest that a substantial part of the yellowing process to aging human lens crystallins likely arises from vitamin C. These two statements and their significance for the aging human lens and cataractogenesis need to be examined. At the outset, it should be noted that

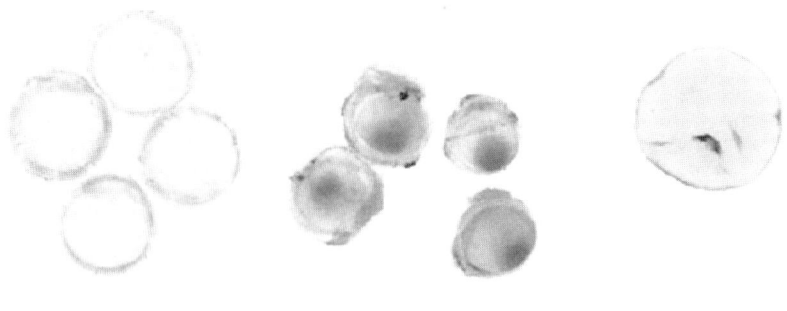

FIGURE 2. Macroscopic appearance of the hSVCT2 transgenic-mouse lenses at 12 months. The transgenic lenses are colored yellow. All mouse lenses were photographed in the same dish to allow accurate comparison between wild-type and transgenic lenses. A fresh 67-year-old human lens obtained at autopsy is shown for comparison. (In color in *Annals* online.)

cataracts, i.e., the formation of light-scattering protein aggregates, can occur at any age as a result of mutations in any of the lens proteins that are critically involved in lens transparency. Thus, chemical modification of lens crystallins is not needed for opacification to occur, and, in the absence of overt cataractogenic conditions and risk factors, the healthy lens can remain transparent for decades while being progressively discolored. Concerning this latter process, however, several epidemiological and clinical studies have revealed a strong association among lens color, lens fluorescence, and the nuclear sclerosis that accompany age-onset cataract formation.[20,21] It is, therefore, reasonable to postulate that the accumulation of protein modifications by vitamin C oxidation and other Maillard reaction products may predispose lens crystallins toward destabilization and aggregation, as supported by several *in vitro* studies.[22–24]

PM and AG are two AGE inhibitors that have been extensively studied both *in vitro* and *in vivo*.[25–27] Studies revealed that PM and AG can dramatically inhibit AGE formation in diabetic animals. PA, a thiol compound, is another potential AGE inhibitor and antioxidant that has been shown to inhibit AGEs formation and nitric oxide synthase activity in cultured rabbit proximal tubular epithelial cells[28] as well as AGE formation in bovine eyes incubated with glucose or glucose-6-phosphate.[29] Both NC-I and NC-II have a guanidino group in the structure that can trap dicarbonyl compounds and block AGE formation. In addition, guanidino compounds can also serve as free radical scavengers.[30] Increased guanidino compounds were found present in brain from hyperargininemia patients.[31,32] However, their role in this condition as well as in normal humans is still unclear.[32]

We decided to choose these five prototypic inhibitors for our first pharmacological intervention in the δenαA-hSVCT2 transgenic mouse. NC-I and NC-II stood out showing significant inhibition of AGE formation by ascorbylation. NC-I and NC-II also showed significant improvement of protein-bound fluorescence at both bands of λex/em 335/385 nm and 370/440 nm. Other inhibitors, PM, AG, and PA had no significant effect. All five inhibitors showed potential reduction of the lysine–lysine cross-link K2P with, however, large standard deviations such that none of them reached significance. A previous study on K2P levels in the human lens showed that levels remain quite low until middle age and significantly increase at late age.[33] This was also confirmed in our previous δenαA-hSVCT2 transgenic-mouse aging study.[15] K2P was maintained at a fairly low level from 6 to 9 months and dramatically increased at 12 months. For budgetary considerations, we chose to stop the intervention at 9 months of age. This may explain the high standard deviations in some of the assays, and it is, therefore, possible that other inhibitors besides NC-I and NC-II may have significant effects beyond 9 months. Similar to a previous study, CEL level was only mildly elevated and not significantly affected by treatment, except for a tendency of decreased NC-I. If MGO is indeed a CEL precursor, it is not a significant ASA oxidation product. Finally, the glucose indicator furosine also was not significantly increased in the transgenic lenses, confirming that is an unlikely source of the AGEs measured in this study. PM and AG are two widely studied AGE inhibitors. Surprisingly, they were unable to inhibit formation of most AGEs in spite of data showing they can definitely block the *in vitro* ascorbylation.[34] To our knowledge, there are no

TABLE 2. Inhibitors effect on ascorbylation in transgenic-mouse compared to wild-type mouse lens

Fluorescence and AGEs	WT	TG	PD	AG	PA	NC-I	NC-II	P value compare to TG
$\lambda_{335/385}$ (unit/unit protein)	9.10 ± 0.97	17.54 ± 2.28	16.87 ± 5.52	17.36 ± 3.26	15.51 ± 3.95	*13.41 ± 2.41	**13.24 ± 2.19	*0.04
$\lambda_{370/440}$ (unit/unit protein)	5.88 ± 1.02	8.76 ± 1.26	8.66 ± 1.17	8.51 ± 1.70	7.97 ± 2.25	*6.63 ± 1.19	**6.53 ± 1.00	**0.004 / *0.05
K2P (pmol/μmol leu equi)	50.10 ± 0.00	2.11 ± 1.21	0.98 ± 0.48	0.91 ± 0.50	2.02 ± 1.43	1.10 ± 0.36	1.33 ± 0.51	*0.02 / N.S
CML (μmol/mol lysine)	156.00 ± 13.05	224.68 ± 30.21	223.21 ± 21.17	226.34 ± 56.46	226.66 ± 74.91	**183.70 ± 22.63	210.83 ± 36.08	**0.007
CEL (μmol/mol lysine)	48.09 ± 5.11	62.02 ± 5.98	61.11 ± 14.45	62.02 ± 8.24	61.12 ± 5.94	52.61 ± 14.39	59.70 ± 11.74	N.S
Pentosidine (pmol/μmol leu equi)	0.055 ± 0.053	0.795 ± 0.366	0.725 ± 0.341	0.532 ± 0.203	0.637 ± 0.143	*0.401 ± 0.120	*0.380 ± 0.149	*0.04
Furosine (μmol/mol lysine)	1383.21 ± 267.32	1331.91 ± 351.85	1235.71 ± 426.73	1597.49 ± 376.74	1566.04 ± 274.86	1168.56 ± 356.43	1408.13 ± 533.32	**0.003 / N.S
GSH	2.46 ± 0.12	2.10 ± 0.09	2.03 ± 0.07	2.15 ± 0.13	2.15 ± 0.11	2.16 ± 0.16	2.01 ± 0.07	N.S

Abbreviations: AG, aminoguanidine; GSH, glutathione; NC-1, nucleophilic compound I; NC-2, nucleophilic compound II; PA, penicillamine; PD, pyridoxamine; TG, transgenic; WT, wild type.

in vitro inhibition data on ascorbylation by PA. However, based on its ability to inhibit fluorescence in experiments involving glucose,[29] PA may still be able to block ascorbylation *in vitro*.

The discrepant effects of NC-I and NC-II versus PM, AG, or PA may be, in part, linked to their ability to reach and be taken up by the lens. In fact, two effective inhibitors support our hypothesis. NC-I and NC-II have been shown elsewhere penetrating through the blood barrier. NC-I and NC-II may be able to achieve millimolar concentrations in the lens and thus better trap dicarbonyl compounds. The glutathione level (GSH) was unchanged with inhibitor treatment (data not shown), suggesting that two inhibitors did not change the redox balance, and the GSH was efficient at maintaining ASA in a reduced form.

Acknowledgments

We thank the National Eye Institute (NEY-07099) and the Visual Sciences Research Center Grant (NE 1 EY-11373) for support of these studies.

Conflicts of Interest

The authors declare no conflicts of interest.

References

1. SCHEY, K.L., J.G. FOWLER, J.C. SCHWARTZ, M., *et al.* 1997. Complete map and identification of the phosphorylation site of bovine lens major intrinsic protein. Invest. Ophthalmol. Vis. Sci. **38:** 2508–2515.
2. TAKEMOTO, L. & D. BOYLE. 2000. Increased deamidation of asparagine during human senile cataractogenesis. Mol. Vis. **6:** 164–168.
3. SRIVASTAVA, O.P., K. SRIVASTAVA & V. HARRINGTON. 1999. Age-related degradation of betaA3/A1-crystallin in human lenses. Biochem. Biophys. Res. Commun. **258:** 632–638.
4. TAKEMOTO, L. 2001. Deamidation of Asn-143 of gamma S crystallin from protein aggregates of the human lens. Curr. Eye. Res. **22:** 148–153.
5. GUPTA, R. & O.P. SRIVASTAVA. 2004. Effect of deamidation of asparagine 146 on functional and structural properties of human lens alphaB-crystallin. Invest. Ophthalmol. Vis. Sci. **45:** 206–214.
6. SRIVASTAVA, O.P., M.C. KIRK & K. SRIVASTAVA. 2004. Characterization of covalent multimers of crystallins in aging human lenses. J. Biol. Chem. **279:** 10901–10909.
7. BAYNES, J.W. 2000. From life to death–the struggle between chemistry and biology during aging: the Maillard reaction as an amplifier of genomic damage. Biogerontology **1:** 235–246.
8. BENSCH, K.G. J.E. FLEMING, & W. LOHMANN. 1985. The role of ascorbic acid in senile cataract. Proc. Natl. Acad. Sci. USA **82:** 7193–7196.

9. NAGARAJ, R.H., D.R. SELL, M. PRABHAKARAM, *et al.* 1991. High correlation between pentosidine protein crosslinks and pigmentation implicates ascorbate oxidation in human lens senescence and cataractogenesis. Proc. Natl. Acad. Sci. USA **88:** 10257–10261.
10. CHENG, R., B. LIN, K.W. LEE & B.J. ORTWERTH. 2001. Similarity of the yellow chromophores isolated from human cataracts with those from ascorbic acid–modified calf lens proteins: evidence for ascorbic acid glycation during cataract formation. Biochim. Biophys. Acta **1537:** 14–26.
11. HEDIGER, M.A.. 2002. New view at C. Nat. Med. **8:** 445–446.
12. TESSIER, F., I. BIRLOUEZ-ARAGON, C. TJANI & J.C. GUILLAND. 1996. Validation of a micromethod for determining oxidized and reduced vitamin C in plasma by HPLC-fluorescence. Int. J. Vitam. Nutr. Res. **66:** 166–170.
13. DUNN, J.A., D.R. MCCANCE, S.R. THORPE, *et al.* 1991. Age-dependent accumulation of N epsilon-(carboxymethyl)lysine and N epsilon-(carboxymethyl)hydroxylysine in human skin collagen. Biochemistry **30:** 1205–1210.
14. TESSIER, F., M. OBRENOVICH & V.M. MONNIER. 1999. Structure and mechanism of formation of human lens fluorophore LM-1. Relationship to vesperlysine A and the advanced Maillard reaction in aging, diabetes, and cataractogenesis. J. Biol. Chem. **274:** 20796–20804.
15. FAN, X., L.W. RENEKER, M.E. OBRENOVICH, *et al.* 2006. Vitamin C mediates chemical aging of lens crystallins by the Maillard reaction in a humanized mouse model. Proc. Natl. Acad. Sci. USA **103:** 16912–16917.
16. CAMMARATA, P.R., C. ZHOU, G. CHEN, *et al.* 1999. A transgenic animal model of osmotic cataract. Part 1: overexpression of bovine Na+/myo-inositol cotransporter in lens fibers. Invest. Ophthalmol. Vis. Sci. **40:** 1727–1737.
17. DUNN, J.A., M.U. AHMED, M.H. MURTIASHAW, *et al.* 1990. Reaction of ascorbate with lysine and protein under autoxidizing conditions: formation of N epsilon-(carboxymethyl)lysine by reaction between lysine and products of autoxidation of ascorbate. Biochemistry **29:** 10964–10970.
18. CHENG, R., Q. FENG, O.K. ARGIROV & B.J. ORTWERTH. 2004. Structure elucidation of a novel yellow chromophore from human lens protein. J. Biol. Chem. **279:** 45441–45449.
19. CHENG, R., Q. FENG & B.J. ORTWERTH. 2006. LC-MS display of the total modified amino acids in cataract lens proteins and in lens proteins glycated by ascorbic acid in vitro. Biochim. Biophys. Acta **1762:** 533–543.
20. CHYLACK, L.T., JR., J.K. WOLFE, J. FRIEND, *et al.* 1993. Quantitating cataract and nuclear brunescence, the Harvard and LOCS systems. Optom. Vis. Sci. **70:** 886–895.
21. SIIK, S., L.T. CHYLACK, JR., J. FRIEND, *et al.* 1999. Lens autofluorescence and light scatter in relation to the lens opacities classification system, LOCS III. Acta Ophthalmol. Scand. **77:** 509–514.
22. NAGARAJ, R.H., T. OYA-ITO, P.S. PADAYATTI, *et al.* 2003. Enhancement of chaperone function of alpha-crystallin by methylglyoxal modification. Biochemistry **42:** 10746–10755.
23. DICKERSON, J.E., JR., M.F. LOU & R.W. GRACY. 1995. Ascorbic acid mediated alteration of alpha-crystallin secondary structure. Curr. Eye. Res. **14:** 163–166.
24. PRABHAKARAM, M. & B.J. ORTWERTH. 1992. The glycation and cross-linking of isolated lens crystallins by ascorbic acid. Exp. Eye. Res. **55:** 451–459.
25. STITT, A., T.A. GARDINER, N.L. ALDERSON, *et al.* 2002. The AGE inhibitor pyridoxamine inhibits development of retinopathy in experimental diabetes. Diabetes **51:** 2826–2832.
26. TANIMOTO, M., T. GOHDA, S. KANEKO, *et al.* 2007. Effect of pyridoxamine (K-163), an inhibitor of advanced glycation end products, on type 2 diabetic nephropathy in KK-A(y)/Ta mice. Metabolism **56:** 160–167.
27. LIN, Y.T., Y.Z. TSENG & K.C. CHANG. 2004. Aminoguanidine prevents fructose-induced arterial stiffening in Wistar rats: aortic impedance analysis. Exp. Biol. Med. (Maywood) **229:** 1038–1045.
28. VERBEKE, P., M. PERICHON, B. FRIGUET & H. BAKALA. 2000. Inhibition of nitric oxide synthase activity by early and advanced glycation end products in cultured rabbit proximal tubular epithelial cells. Biochim. Biophys. Acta **1502:** 481–494.
29. STEVENS, A.. 1995. The effectiveness of putative anti-cataract agents in the prevention of protein glycation. J. Am. Optom. Assoc. **66:** 744–749.
30. NAKAI, K., M.B. KADIISKA, J.J. JIANG, *et al.* 2006. Free radical production requires both inducible nitric oxide synthase and xanthine oxidase in LPS-treated skin. Proc. Natl. Acad. Sci. USA **103:** 4616–4621.
31. DE DEYN, P.P., B. MARESCAU, R. D'HOOGE, *et al.* 1995. Guanidino compound levels in brain regions of non-dialyzed uremic patients. Neurochem. Int. **27:** 227–237.
32. MIZUTANI, N., C. HAYAKAWA, Y. OHYA, *et al.* 1987. Guanidino compounds in hyperargininemia. Tohoku J. Exp. Med. **153:** 197–205.
33. CHENG, R., Q. FENG, O.K. ARGIROV & B.J. ORTWERTH. 2005. K2P–a novel cross-link from human lens protein. Ann. N.Y. Acad. Sci. **1043:** 184–194.
34. ARGIROVA, M. & O. ARGIROV. 2003. Inhibition of ascorbic acid–induced modifications in lens proteins by peptides. J. Pept. Sci. **9:** 170–176.
35. NAGARAJ, R.H., M. PRABHAKARAM, B.J. ORTWERTH & V.M. MONNIER. 1994. Suppression of pentosidine formation in galactosemic rat lens by an inhibitor of aldose reductase. Diabetes **43:** 580–586.
36. MONNIER, V.M., R.R. KOHN & A. CERAMI. 1984. Accelerated age-related browning of human collagen in diabetes mellitus. Proc. Natl. Acad. Sci. USA **81:** 583–587.
37. AHMED, M.U., E. BRINKMANN FRYE, T.P. DEGENHARDT, *et al.* 1997. N-epsilon-(carboxyethyl)lysine, a product of the chemical modification of proteins by methylglyoxal, increases with age in human lens proteins. Biochem. J. **324**(Pt 2): 565–570.

N^ε-(Carboxymethyl)lysine during the Early Development of Hypertension

MARCUS BAUMANN,[a,b] COEN STEHOUWER,[c] JEAN SCHEIJEN,[c] UWE HEEMANN,[b] HARRY STRUIJKER BOUDIER,[a] AND CASPER SCHALKWIJK[c]

[a]*Department of Pharmacology and Toxicology, University Maastricht, Maastricht, the Netherlands*

[b]*Department of Nephrology, Klinikum rechts der Isar, Technical University Munich, Munich, Germany*

[c]*Department of Internal Medicine, Division of General Internal Medicine, University Hospital Maastricht, Maastricht, the Netherlands*

Advanced glycation end products (AGEs) are associated with hypertension. Whether N^ε-(carboxymethyl)lysine (CML) contributes to the development of hypertension in young spontaneously hypertensive rats (SHR) remains to be established compared to WKY. We determined blood pressure, renal function, marker for oxidative stress (OS), and CML in young WKY rats and SHR. We found blood pressure was increased in SHR with no difference in renal function and OS compared to WKY. CML was elevated in plasma (2.3 ± 0.3 vs. 1.3 ± 0.2 μmol/L) and kidney (1.0 ± 0.1 vs. 0.5 ± 0.1 μmol/L) compared to WKY. Early CML accumulation may contribute to the development of hypertension potentially by inducing early renal inflammation independent of glomerular dysfunction or oxidative stress.

Key words: N^ε-(carboxymethyl)lysine; hypertension; spontaneously hypertensive rat; renal function; oxidative stress

Introduction

N^ε-(carboxymethyl)lysine (CML) is a noncrosslinking advanced glycation end product (AGE) formed in the Maillard reaction either from glucose or lipids. CML has been recognized as an indicator of carbonyl overload *in vivo*.[1] AGEs have been linked to the development of nephropathy in diabetes, although the exact role in this process remains unclear.[2] Recent reports additionally demonstrated an association of AGEs with hypertension. This increase was associated with an increase of aldehyde conjugate levels in kidney but not in heart or liver.[3] This is of particular interest as the kidneys plays a major role in the development and maintenance of hypertension as evidenced by cross-transplantation studies in rats and humans.[4,5]

One aspect discussed in this context is the role of inflammation as a crucial factor in the development and maintenance of hypertension. Beside oxidative stress, AGEs are known to act as proinflammatory agents by activating NFκB, in particular in the proximal tubular cells.[2] Therefore, AGEs could participate in the development of hypertension by inducing inflammation in the kidney.

In spontaneously hypertensive rat (SHR), the development of hypertension is most prominent between weeks 4 to 8 of age.[6] This clear time frame allows the analysis of the role of AGEs in the context of blood pressure, oxidative stress, and renal function. The aim of the present study was to investigate whether CML is associated with the development of hypertension in SHR and whether this can be attributed to renal dysfunction and oxidative stress.

Methods

Four- and 8-week-old WKY rats and SHR ($n = 9$ per time point and strain) were housed under

Address for correspondence: Dr. Marcus Baumann, Department of Nephrology, Klinikum rechts der Isar, Technical University Munich, Ismaninger Str. 22, 81675 Munich, Germany. Voice: +0049-89-4140-6704; fax: +0049-89-4140-4878.

Marcus.baumann@lrz.tum.de

TABLE 1. General characteristics and renal function

	BW (g)	MAP (mm Hg)	HR (bpm)	Creatinine clearance (mL/min/100 g)	Urinary albumin (mg/kg/24 h)
WKY (4 weeks)	108 ± 16	106 ± 6	419 ± 37	0.93 ± 0.15	0.45 ± 0.07
SHR (4 weeks)	90 ± 7	131 ± 4*	437 ± 21	0.88 ± 0.11	0.47 ± 0.16
WKY (8 weeks)	252 ± 16$	118 ± 9$	379 ± 39	1.11 ± 0.25	0.45 ± 0.09
SHR (8 weeks)	229 ± 6*$	156 ± 11*$	369 ± 38	1.05 ± 0.41	0.51 ± 0.15

BW, body weight; HR, heart rate; MAP, mean arterial pressure.
*$P < 0.05$, WKY versus age-matched SHR; $P < 0.05$, 4-weeks versus 8-weeks-old WKY or SHR, respectively.

controlled conditions of temperature (21°C) and light (12-h light:dark cycle 0700–1900 h) and were maintained on normal rat chow and water *ad libitum*. All experiments were approved by the animal ethics committee of Maastricht University and performed in accordance with institutional guidelines. Intra-arterial mean arterial pressure (MAP) was measured in the conscious unrestrained state through the left femoral artery at 4 and 8 weeks of age.[7] Heparinized blood samples were drawn via the arterial catheter. Rats were kept in metabolic cages for 24 h at 4 and 8 weeks of age. Plasma and urine were collected and kept at −20°C until further processing. Albumin concentrations were measured with the rat albumin enzyme immunoassay obtained from SPI-BIO (Montigny le Bretonneux, France).[8] Creatinine was determined nephelometrically in plasma and urine, and creatinine clearance was calculated.[9]

For immunohistochemistry of CML and ED-1, paraffin-embedded tissue sections (4 μm) were used and stained as previously described.[10] The intensity of staining was semiquantitatively scored for positive renal tubular tissue: 1 <25%, 2 <50%, 3 <75%, 4 >75%.

Serum and renal CML and N^ε-(carboxyethyl)lysine (CEL) were detected by liquid chromatography (LC)–mass spectrometry (MS)/MS as previously described.[11] Briefly, proteins were precipitated, hydrolyzed, and dissolved in 5 mmol nonafluoropentanoic acid. CML and CEL were resolved by reversed-phase high-performance liquid chromatography (HPLC) and analyzed on an API 3000 mass spectrometer (PE Sciex, Mississauga, Ontario, Canada).

In homogenized renal tissue, malondialdehyde (MDA) was determined by HPLC.[12] Briefly, reagents were added (0.12 mol *tert*-butyl alcohol, 0.32 mol *o*-phosphoric acid, 0.68 mmol butylated hydroxytoluene, and 0.01% [mass/vol] EDTA) and incubated for 1 h at 100°C. MDA products were extracted and injected on a fluorescent HPLC system (Agilent, Palo Alto, CA) (excitation: 532 nm, emission: 553 nm) using malonaldehyde bis(diethylacetal) as the standard. Renal supernatant was used for assay of the glutathione disulphide (GSSG)-reductase 5′,5′-dithio-bis (2-nitrobenzoic acid) recycling procedure to measure the intracellular glutathione (GSH) and GSSG contents, followed by HPLC analysis as reported previously.[13] Renal superoxide dismutase (SOD) activity was calculated based on the ability of the plasma to decrease the rate of nitroblue tetrazolium reduction induced by xanthine and xanthine oxidase. Human SOD1 activity in phosphate buffer was used for calibration.[14] For statistical analysis the student's *t*-test was used.

TABLE 2. Advanced glycation end products and oxidative stress in 8-week-old rats

	WKY	SHR
CML plasma (μmol/L)	1.3 ± 0.2	2.3 ± 0.3*
CEL plasma	2.5 ± 0.1	3.4 ± 0.2*
CML renal homogenate	0.5 ± 0.1	1.0 ± 0.1*
CEL renal homogenate	0.4 ± 0.1	0.4 ± 0.1
Renal GSH/GSSG ratio	3.15 ± 0.71	2.63 ± 0.86
SOD activity (U/mg kidney)	65.9 ± 21.4	55.3 ± 18.0
MDA (μmol/g kidney)	1.2 ± 0.1	1.3 ± 0.4

CML, N^ε-(carboxymethyl)lysine; CEL, N^ε-(carboxyethyl) lysine; GSH, glutathione; GSSG, glutathione disulphide-reductase; SOD, superoxide dismutase; MDA, malondialdehude.
*$P < 0.05$ versus age-matched WKY.

Results

Four-week-old SHR demonstrated higher intra-arterial mean arterial pressure than age-matched WKY, while heart rate did not differ between the strains (TABLE 1). The reduction in body weight in SHR was not significant ($P < 0.06$) compared to WKY. Between week 4 and 8, rats gained a significant amount of weight and had increased blood pressure, whereas heart rate decreased significantly in both strains. At 8 weeks of age, MAP remained significantly higher in SHR compared to age-matched WKY. Moreover,

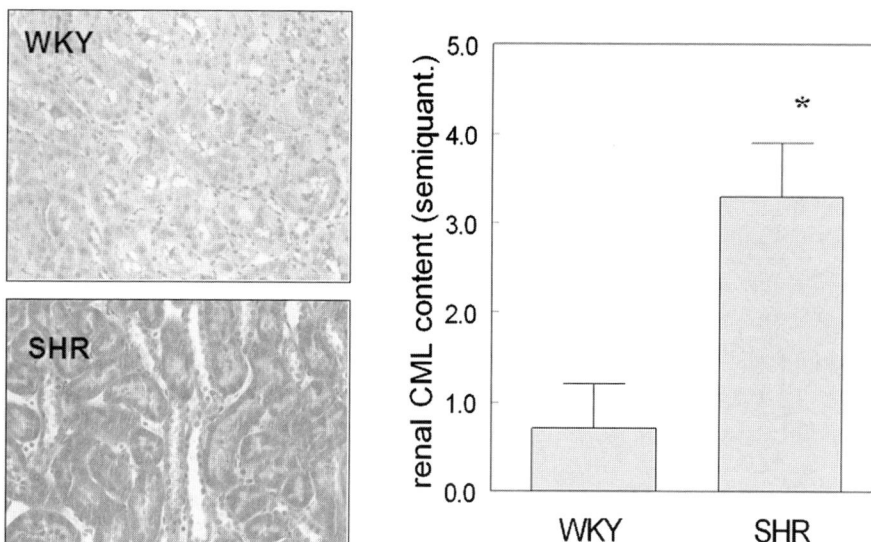

FIGURE 1. Renal N^ϵ-(carboxymethyl)lysine (CML) content in 8-week old WKY and SHR. Staining of the renal cortical CML content in 8-week-old WKY rats (*upper left*) and SHR (*lower left*) is depicted. *$P < 0.05$.

blood pressure differences further increased at 8 weeks compared to 4 weeks of age. Creatinine clearance and albuminuria did not differ between the strains at both time points (TABLE 1).

Protein-bound CML determined at 8 weeks of age in plasma and renal homogenate showed significantly elevated values in SHR compared to age-matched WKY (plasma: 2.3 ± 0.3 vs. $1.3 \pm 0.2\,\mu$mol/L; kidney: 1.0 ± 0.1 vs. $0.5 \pm 0.1\,\mu$mol/L; TABLE 2). In contrast, CEL was significantly elevated in plasma but not in the renal tissue of SHR compared to age-matched WKY strain (plasma: 3.4 ± 0.2 vs. $2.5 \pm 0.1\,\mu$mol/L; kidney: 0.4 ± 0.1 vs. $0.4 \pm 0.1\,\mu$mol/L). Immunohistochemistry of the renal section confirmed the elevated tandem MS values in SHR and showed that the vast majority of CML staining was present in the proximal tubuli (0.7 ± 0.6 vs. 3.3 ± 0.8; FIG. 1).

Oxidative stress determined at 8 weeks of age did not show significant differences between the strains for renal GSH/GSSG ratio, SOD, and MDA (TABLE 2). Numbers of ED-1 positive cells in the kidney, determined by immunohistochemistry in 8-week-old rats, were low in both strains. However, SHR showed significantly more ED-1 positive cells than age-matched WKY (2.8 ± 2.0 vs. 1.4 ± 0.8 per 40X view; $P < 0.05$).

Discussion

The main finding of this study was that CML had already accumulated during the development of hypertension of SHR, independent of renal function and before oxidative stress was apparent.

Wang et al. demonstrated that CML is increased in plasma and renal tubules of adult SHR,[15] an association which seems to be clinically relevance.[16] The investigated period represents the strongest blood pressure increase in SHR.[7] Thus, the increased CML levels of young SHR imply an increased systemic source of AGEs already present during blood pressure development in young SHR.

The kidney plays a crucial role in the development of hypertension.[4,5] CML was particularly localized in proximal tubuli, confirming previous findings in SHR.[17] This is of interest as proximal tubules regulate the clearance of AGE-bound proteins,[2,18] differ in function in young SHR,[6] and represent a major source of proinflammatory signaling during the development of hypertension.[19]

Renal dysfunction has a large effect on the accumulation of CML.[20] However, renal dysfunction can be excluded as cause for CML accumulation since renal function was similar between both strains at these young ages.[21] Oxidative stress participates in increasing blood pressure and stimulates AGE production in SHR.[22,23] However, neither renal SOD activation[22] nor the cellular GSH/GSSG ratio were impaired in young SHR. Therefore, the data is in line with previous studies demonstrating that significant oxidative stress starts to appear between weeks 10–13 in SHR,[15,17,24] after hypertension has been established.[7]

The number of macrophages was significantly increased in young SHR, suggesting a stronger proinflammatory signaling in young SHR compared to WKY.[25] As no differences in oxidative stress were apparent, increased renal CML deposition may induce renal inflammation and thereby participate in the increase in blood pressure.[12,26]

In summary, this study provides evidence that CML accumulates in renal proximal tubuli during the development of hypertension. This process is independent of renal dysfunction or oxidative stress. Thus, CML accumulation may participate in the development of hypertension potentially by inducing renal inflammation.

Conflict of Interest

The authors declare no conflicts of interest.

References

1. SINGH, R. et al. 2001. Advanced glycation end-products: a review. Diabetologia **44:** 129–146.
2. MORCOS, M. et al. 2002. Activation of tubular epithelial cells in diabetic nephropathy. Diabetes **51:** 3532–3544.
3. VASDEV, S. et al. 1998. Role of aldehydes in fructose induced hypertension. Mol. Cell Biochem. **181:** 1–9.
4. BIANCHI, G. et al. 1974. Blood pressure changes produced by kidney cross-transplantation between spontaneously hypertensive rats and normotensive rats. Clin. Sci. Mol. Med. **47:** 435–448.
5. GUIDI, E. et al. 1996. Hypertension may be transplanted with the kidney in humans: a long-term historical prospective follow-up of recipients grafted with kidneys coming from donors with or without hypertension in their families. J. Am. Soc. Nephrol. **7:** 1131–1138.
6. BAUMANN, M., E.H. VAN, J.J.R. HERMANS, et al. 2004. Functional and Structural Postglomeruar Aterations in the Kidney of Prehypertensive Spontaneously Hypertensive Rats. Clin. Exp. Hypertens **26:** 663–672.
7. BAUMANN, M. et al. 2007. Transient AT1 receptor-inhibition in prehypertensive spontaneously hypertensive rats results in maintained cardiac protection until advanced age. J. Hypertens **25:** 207–215.
8. SANDERS, M.W. et al. 2005. High sodium intake increases blood pressure and alters renal function in intrauterine growth-retarded rats. Hypertension **46:** 71–75.
9. LUTZ, J. et al. 2006. Angiotensin type 1 and type 2 receptor blockade in chronic allograft nephropathy. Kidney Int. **70:** 1080–1088.
10. SCHALKWIJK, C.G. et al. 2004. Increased accumulation of the glycoxidation product Nepsilon-(carboxymethyl)lysine in hearts of diabetic patients: generation and characterisation of a monoclonal anti-CML antibody. Biochim. Biophys. Acta **1636:** 82–89.
11. SCHRAM, M.T. et al. 2005. Advanced glycation end products are associated with pulse pressure in type 1 diabetes: the EURODIAB Prospective Complications Study. Hypertension **46:** 232–237.
12. RODRIGUEZ-ITURBE, B. et al. 2002. Reduction of renal immune cell infiltration results in blood pressure control in genetically hypertensive rats. Am. J. Physiol. Renal Physiol. **282:** F191–F201.
13. JONES, D.P. 2002. Redox potential of GSH/GSSG couple: assay and biological significance. Methods Enzymol. **348:** 93–112.
14. DEN HARTOG, G.J. et al. 2003. Superoxide dismutase: the balance between prevention and induction of oxidative damage. Chem. Biol. Interact. **145:** 33–39.
15. WANG, X. et al. 2004. Increased methylglyoxal and advanced glycation end products in kidney from spontaneously hypertensive rats. Kidney Int. **66:** 2315–2321.
16. WANG, X. et al. 2007. Attenuation of hypertension development by aminoguanidine in spontaneously hypertensive rats: role of methylglyoxal. Am. J. Hypertens **20:** 629–636.
17. WANG, X. et al. 2005. Vascular methylglyoxal metabolism and the development of hypertension. J. Hypertens **23:** 1565–1573.
18. WIHLER, C. et al. 2005. Renal accumulation and clearance of advanced glycation end-products in type 2 diabetic nephropathy: effect of angiotensin-converting enzyme and vasopeptidase inhibition. Diabetologia **48:** 1645–1653.
19. STEWART, T. et al. 2005. Kidney immune cell infiltration and oxidative stress contribute to prenatally programmed hypertension. Kidney Int. **68:** 2180–2188.
20. WAGNER, Z. et al. 2001. N(epsilon)-(carboxymethyl)lysine levels in patients with type 2 diabetes: role of renal function. Am. J. Kidney Dis. **38:** 785–791.
21. OFSTAD, J. & B.M. IVERSEN. 2005. Glomerular and tubular damage in normotensive and hypertensive rats. Am. J. Physiol. Renal Physiol. **288:** F665–F672.
22. NAKAZONO, K. et al. 1991. Does superoxide underlie the pathogenesis of hypertension? Proc. Natl. Acad. Sci. USA **88:** 10045–10048.
23. WILCOX, C.S. 2005. Oxidative stress and nitric oxide deficiency in the kidney: a critical link to hypertension? Am. J. Physiol. Regul. Integr. Comp. Physiol. **289:** R913–R935.
24. TOUYZ, R.M. & E.L. SCHIFFRIN. 2001. Increased generation of superoxide by angiotensin II in smooth muscle cells from resistance arteries of hypertensive patients: role of phospholipase D-dependent NAD(P)H oxidase-sensitive pathways. J. Hypertens **19:** 1245–1254.
25. STRUIJKER BOUDIER, H.A. et al. 2003. The heart, macrocirculation and microcirculation in hypertension: a unifying hypothesis. J. Hypertens Suppl. **21:** S19–S23.
26. RODRIGUEZ-ITURBE, B. et al. 2004. Oxidative stress, renal infiltration of immune cells, and salt-sensitive hypertension: all for one and one for all. Am. J. Physiol. Renal Physiol. **286:** F606–F616.

Aging, Diabetes, and Renal Failure Catalyze the Oxidation of Lysyl Residues to 2-Aminoadipic Acid in Human Skin Collagen

Evidence for Metal-catalyzed Oxidation Mediated by α-Dicarbonyls

DAVID R. SELL,[a] CHRISTOPHER M. STRAUCH,[a] WEI SHEN,[a] AND VINCENT M. MONNIER[a,b]

Departments of [a]Pathology and [b]Biochemistry, Case Western Reserve University, Cleveland, Ohio, USA

The ε-amino group of lysyl residues oxidatively deaminates in the presence of α-dicarbonyl sugars and redox-active metals forming α-aminoadipic acid-δ-semialdehyde (allysine; Suyama's hypothesis), which can further oxidize into 2-aminoadipic acid. Here we show that 2-aminoadipic acid is significantly ($P < 0.05$) correlated with 6-hydroxynorleucine, carboxyethyllysine (CEL), and carboxymethyllysine (CML) in human skin collagen. Since CEL and CML can originate from carbohydrate and lipid by oxidative decomposition and α-dicarbonyl formation, these results provide support for Suyama's hypothesis. Allysine, in turn, is readily converted by oxidation into 2-aminoadipic acid, which accumulates to high levels in skin (i.e., > 2 nmol/mg collagen).

Key words: Maillard reaction; glycation; lysine; redox-active metals; oxidation; dicarbonyls; methylgloxal; glyoxal; sugars; lipids

Introduction

Physiological cross-linking of collagen in growth and maturation is initiated by the enzyme lysyl oxidase (LOX) which, in the presence of Cu^{2+} and the quinone cofactors, oxidizes the ε-amino group of a lysine residue to an aldehyde. This forms the collagen cross-linking precursor α-aminoadipic acid-δ-semialdehyde, also known as allysine.[1,2] However, evidence over the last 15+ years has shown that allysine can form by other mechanisms as well, all involving oxidation but unrelated to LOX, and are ultimately related to aging and age-related disease processes, such as diabetes.[3] In this, and most importantly for the present investigation, allysine can form from metal-catalyzed oxidation (MCO) as originally reported by Stadtman[4] and more recently by Suyama and associates[5] who have proposed a mechanism for allysine formation based upon MCO reactions of α-dicarbonyl sugars with lysyl residues. Of further significance is that various investigators over the past 30 years or more have speculated that allysine itself probably undergoes a further oxidative reaction leading to the formation of 2-aminoadipic acid *in vivo*.[6] However, no systematic studies on this topic were reported.

We recently reported levels of both allysine, measured as 6-hydroxynorleucine, and 2-aminoadipic acid in insoluble human skin collagen as a function of age, diabetes, and renal failure. Surprisingly, 2-aminoadipic acid, but not 6-hydroxynorleucine, significantly and progressively increased with age to very large levels (i.e., >2 nmol/mg collagen) and was significantly elevated by diabetes and renal failure.[3] A second finding was that septicemia catalyzed 2-aminoadipic acid formation in nondiabetic patients, many of whom were diagnosed with acute renal failure. Sepsis was found to significantly ($P < 0.05$) correlate with levels of the methyglyoxal-derived advanced glycation end products (AGEs), carboxyethyllysine (CEL), argpyrimidine, and methylgloxal-derived imidazolium crosslink (MODIC) in nondiabetic patients, which supports the Suyama pathway of allysine formation as mediated by α-dicarbonyls (methylglyoxal).[3] The purpose

Address for correspondence: David R. Sell, Ph.D., Department of Pathology, Case Western Reserve University, Wolstein Research Bldg.-Room 5144, 2103 Cornell Road, Cleveland, OH 44106-7288. Voice: +1-216-368-2930; fax: +1-216-368-1357.

drs7@po.cwru.edu

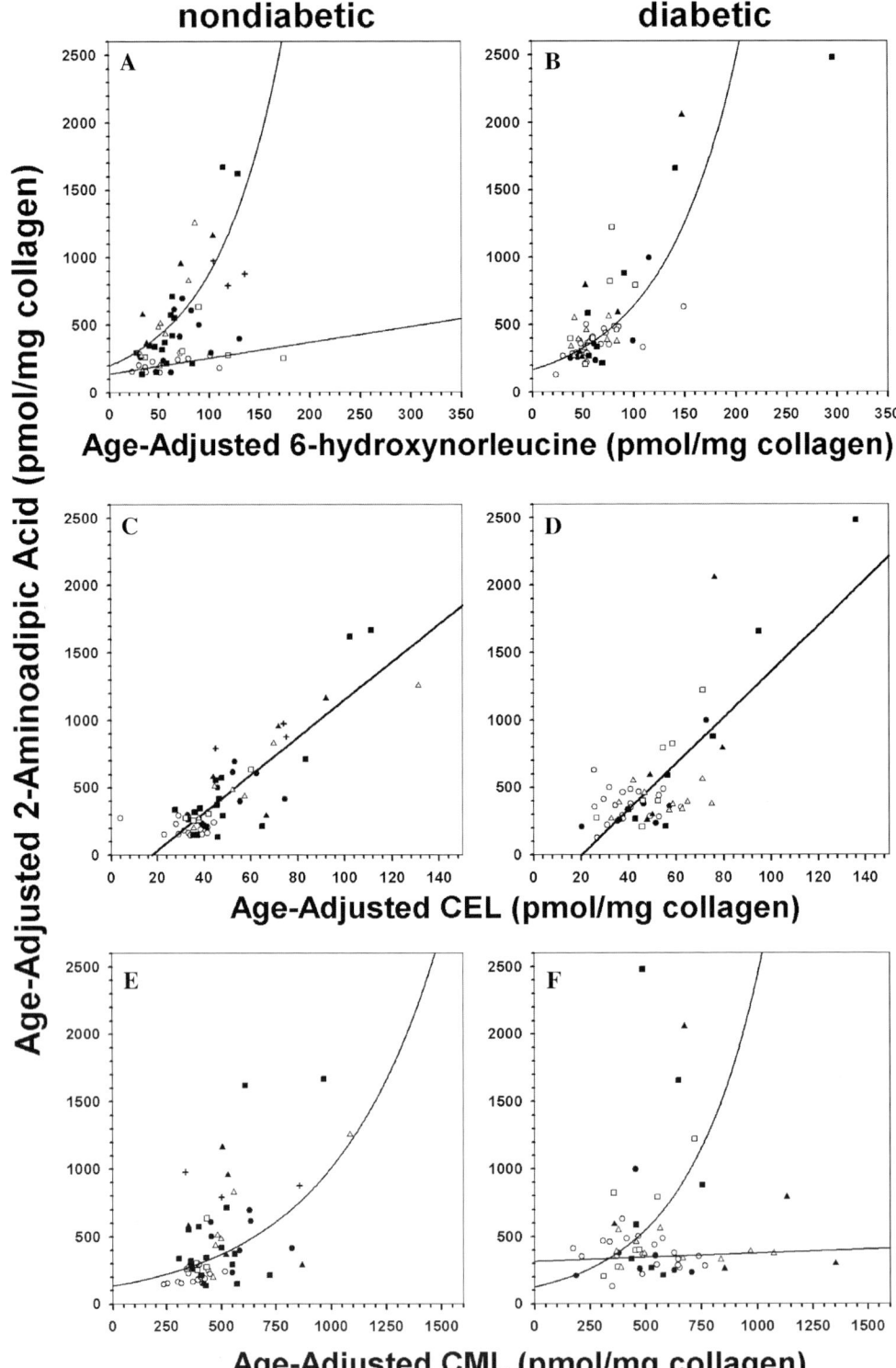

FIGURE 1. Correlation plots for 6-hydroxynorleucine, carboxyethyllysine (CEL), and carboxymethyllysine (CML) versus 2-aminoadipic acid in insoluble human skin collagen after age adjustment of levels to 50 years in nondiabetic (plots **A, C, E**) and diabetic individuals (plots **B, D, F**). For each plot, the regression line(s) represents the best-fit response to the overall

TABLE 1. Partial correlations of advanced glycation end product markers with 2-aminoadipic acid after controlling for age and sepsis

Marker	Nondiabetic $n = 61$		Diabetic $n = 56$	
	Corr. (r)	Prob. (P)	Corr. (r)	Prob. (P)
	No renal failure			
	$n = 29$		$n = 27$	
6-hydroxynorleucine	0.63	<0.0001*	0.75	<0.0001*
CEL	0.71	<0.0001*	0.42	0.036*
CML	0.54	0.004*	−0.33	0.11
	Acute renal failure			
	$n = 20$		$n = 14$	
6-hydroxynorleucine	0.85	<0.0001*	0.87	<0.0001*
CEL	0.89	<0.0001*	0.93	<0.0001*
CML	0.55	0.018*	0.46	0.13
	Chronic renal failure			
	$n = 12$		$n = 15$	
6-hydroxynorleucine	0.94	<0.0001*	0.84	<0.0001*
CEL	0.91	<0.0001*	0.44	0.13
CML	0.52	0.12	−0.22	0.48

*The correlation is statistically significant ($P < 0.05$).
Abbreviations: CEL, carboxyethyllysine; CML, carboxymethyllysine; Corr., correlation; Prob., probability.

of the present study was to investigate the relationship of 2-aminoadipic acid formation with levels of 6-hydroxynorleucine and AGE markers in insoluble human skin collagen.

Methods

Procedures, including tissue donor information, the preparation of insoluble collagen from human skin, borohydride reduction, and acid hydrolysis, as well as analytical assays are described in detail elsewhere.[3] Each collagen sample was reacted with sodium borohydride before acid hydrolysis, which converted allysine into the more stable 6-hydroxynorleucine. After acid hydrolysis, levels of CEL, carboxymethyllysine (CML), 6-hydroxynorleucine, and 2-aminoadipic acid were measured in individual samples by gas chromatography–mass spectrometry (GC/MS), as previously described.[3] In these assays, analytes were derivatized as trifluoroacetyl methyl esters and mass ion fragments were quantitated as previously described.[3] Assays for AGEs used in this study have been published elsewhere as follows: pentosidine by HPLC using a fluorescence detector[7]; furosine by GC/MS[7]; levels of argpyrimidine, ornithine, glucosepane, deoxyglucosone-derived imidazolium crosslink (DODIC), oxidized DODIC (DODIC-OX), and MODIC by liquid chromatography–mass spectrometry (LC/MS).[8,9] Statistical analyses used in this study are detailed elsewhere[7] and for the most part were done by SPSS (SPSS, Inc., Chicago, IL).

Results

In an attempt to investigate the molecular origin and mechanism of 2-aminoadipic acid formation *in vivo*, levels were correlated with allysine, measured as 6-hydroxynorleucine, and AGE markers previously measured by this laboratory as listed in the Methods. Their correlations with 2-aminoadipic acid were evaluated in nondiabetic and diabetic cohorts with and without renal failure initially by partial correlation analysis controlling for age and sepsis, which previously were found to significantly intercorrelate among these markers. In short, these results showed that 2-aminoadipic acid correlated most significantly ($P < 0.05$) with its precursor allysine (6-hydroxynorleucine), CEL, and CML in both nondiabetic and diabetic cohorts with and without acute and chronic renal failure (TABLE 1).

In further analysis, individual levels of markers listed in TABLE 1 were age-adjusted to 50 years and correlation plots were made between 6-hydroxynorleucine,

---

relationship for individuals with and without renal failure and/or septicemia. Individuals diagnosed with septicemia are shown by closed symbols. Three nondiabetic individuals with 2-aminoadipic acid levels previously classified as outliers[3] but with suspected septicemia are depicted by crosshair symbols. Regression line equations for model data: **(A)** $y = 184e^{0.016x}$, $n = 35$, $r = 0.84$, $P < 0.0001$; $y = 141 + 1.13x$, $n = 26$, $r = 0.65$, $P < 0.0001$. For all data in plot A, the relationship was significantly affected by sepsis ($P = 0.004$) and renal failure ($P < 0.0001$). **(B)** $y = 167e^{0.013x}$, $n = 55$, $r = 0.71$, $P < 0.0001$ (one outlier). The effect of diabetes is significant ($P = 0.025$) versus plot A. **(C)** $y = 14x − 248$, $n = 60$, $P < 0.0001$ (one outlier). **(D)** $y = 17x − 347$, $n = 56$, $r = 0.77$, $P < 0.0001$. For plots C and D, there were no significant ($P > 0.05$) effects from sepsis and renal failure on the relationship. **(E)** $y = 135e^{0.002x}$, $n = 61$, $r = 0.51$, $P < 0.0001$. The relationship in plot E was significantly affected by renal failure ($P = 0.014$), while sepsis approached statistical significance ($P = 0.06$). **(F)** $y = 125e^{0.003x}$, $n = 27$, $r = 0.55$, $P = 0.003$; $y = 316 + 0.06x$, $n = 29$, $r = 0.13$, $P = 0.42$ (nonsignificant). Symbol key: ○, ● no renal failure; □, ■ acute renal failure; △, ▲ chronic renal failure.

CEL, or CML level versus 2-aminoadipic acid (FIG. 1). Nondiabetic and diabetic individuals were separately plotted for each of these relationships and were further characterized for the presence or absence of acute and chronic renal failure as well as septicemia by the use of different typographic symbols. The results showed significant relationships existed between 2-aminoadipic acid with 6-hydroxynorleucine ($P < 0.0001$), CEL ($P < 0.0001$), and CML ($P < 0.003$) for both nondiabetic (FIG. 1A, C, E) and diabetic individuals (FIG. 1B, D, F). As shown for 6-hydroxynorleucine in FIGURE 1A, two divergent groups of nondiabetic individuals were identified as modeled by a linear and exponential function (FIG. 1A). One group showed a gradual but significant ($P < 0.0001$) rate increase in 2-aminoadipic acid formation relative to that of 6-hydroxynorleucine as depicted by a linear regression line in FIGURE 1A. This group consisted of all nondiabetic patients in the study diagnosed without septicemia and renal failure along with some patients positively identified with the same pathologies but, albeit, in less severe progression (FIG. 1A). In contrast, a second group of individuals showed a highly significant ($P < 0.0001$) and greatly accelerated rate increase in 2-aminoadipic acid formation versus 6-hydroxynorleucine. This group, as depicted by an exponential regression line in FIGURE 1A, consisted of nondiabetic individuals diagnosed with severe cases of septicemia and/or renal failure. In comparison, for the diabetic cohort shown in FIGURE 1B, the same type of accelerated relationship existed between 2-aminoadipic acid and 6-hydroxynorleucine for all diabetic patients with and without renal failure and septicemia, which was modeled by a single exponential function ($P < 0.0001$).

The relationship between 2-aminoadipic acid and CEL was highly significant ($P < 0.0001$) for both nondiabetic (FIG. 1C) and diabetic cohorts (FIG. 1D), which, in both cases, were modeled by a linear function (FIG. 1C, D). Conversely, for CML, the relationship with 2-aminoadipic acid was much more variable for individuals of both nondiabetic (FIG. 1E) and diabetic (FIG. 1F) cohorts. For nondiabetics, an exponential function was used to model this relationship which still proved to be highly significant ($P < 0.0001$) despite the noted variability (FIG. 1E). For diabetics, as similar to FIGURE 1A, two divergent groups were identified. First, one group, consisting mainly of patients with renal failure diagnosed with and without septicemia, showed a highly accelerated rate increase in 2-aminoadipic acid formation relative to that of CML, as modeled by an exponential function ($P = 0.003$, FIG. 1F). Likewise, a second group, consisting of all diabetic patients diagnosed without septicemia and renal failure plus some patients positively diagnosed with these pathologies, showed no relationship with CML levels as modeled by a nonsignificant ($P > 0.05$) linear function (FIG. 1F).

Overall, the analyses showed that almost all the correlation between 6-hydroxynorleucine, CEL, and CML with 2-aminoadipic acid was essentially from the acceleration in marker formation because of renal failure, septicemia, and diabetes. In this regard, when nondiabetic and diabetic patient data consisting of those without renal failure and septicemia were considered (FIG. 1), the only correlation plots that showed significance were 6-hydroxynorleucine versus 2-aminoadipic acid in diabetic patients ($P < 0.0001$, linear relationship, FIG. 1B) and CML versus 2-aminoadipic in nondiabetic patients ($P = 0.032$, exponential relationship, FIG. 1E) (regression lines not shown).

Discussion

The finding that 6-hydroxynorleucine was the foremost marker ($P < 0.0001$) related to 2-aminoadipic acid levels in this study (TABLE 1) provides further proof that allysine is a precursor for 2-aminoadipic acid in human skin collagen. Second, since both CEL ($P < 0.0001$) and CML ($P = 0.005$) were also significantly correlated with 2-aminoadipic acid levels, these results suggest that the oxidative mechanism for 2-aminoadipic acid formation is related to formation of these products *in vivo*. Previously it has been shown that CEL and CML originate from auto-oxidation and decomposition of both carbohydrate and lipids.[10] Undoubtedly, methylglyoxal and glyoxal are reactive with lysyl residues *in vivo* and form CEL and CML. Thus, these results support Suyama's hypothesis[5] for allysine formation based upon oxidative deamination of lysyl residues mediated by α-dicarbonyls compounds. Allysine, in turn, undergoes a further oxidative reaction spontaneously forming 2-aminoadipic acid, which accumulates to high levels in skin collagen as shown in FIGURE 1.

Acknowledgments

This research was supported by Grants NIA AG18629 from the National Institutes of Health, Grant JDRF 2004-128 from the Juvenile Diabetes Research Foundation International, as well as a Mentorship Grant from the American Diabetes Association.

Conflict of Interest

The authors declare no conflicts of interest.

References

1. ASGHAR, A. & R.L. HENRICKSON. 1982. Chemical, biochemical, functional, and nutritional characteristics of collagen in food systems. Adv. Food Res. **28:** 231–372.
2. LUCERO, H.A. & H.M. KAGAN. 2006. Lysyl oxidase: an oxidative enzyme and effector of cell function. Cell. Mol. Life Sci. **63:** 2304–2316.
3. SELL, D.R., C.M. STRAUCH, W. SHEN, et al. 2007. 2-Aminoadipic acid is a marker of protein carbonyl oxidation in the aging human skin: effects of diabetes, renal failure and sepsis. Biochem. J. **404:** 269–277.
4. STADTMAN, E.R. 1992. Protein oxidation and aging. Science **257:** 1220–1224.
5. AKAGAWA, M., T. SASAKI & K. SUYAMA. 2002. Oxidative deamination of lysine residue in plasma protein of diabetic rats. Eur. J. Biochem. **269:** 5451–5458.
6. STAHMANN, M.A. 1977. Cross-linking of protein by peroxidase. Adv. Exp. Med. Biol. **86B:** 285–298.
7. SELL, D.R., N.R. KLEINMAN & V.M. MONNIER. 2000. Longitudinal determination of skin collagen glycation and glycoxidation rates predicts early death in C57BL/6NNIA mice. FASEB J. **14:** 145–156.
8. SELL, D.R. & V.M. MONNIER. 2004. Conversion of arginine into ornithine by advanced glycation in senescent human collagen and lens crystallins. J. Biol. Chem. **279:** 54173–54184.
9. SELL, D.R., K.M. BIEMEL, O. REIHL, et al. 2005. Glucosepane is a major protein cross-link of the senescent human extracellular matrix. Relationship with diabetes. J. Biol. Chem. **280:** 12310–12315.
10. JANUSZEWSKI, A.S., N.L. ALDERSON, T.O. METZ, et al. 2003. Role of lipids in chemical modification of proteins and development of complications in diabetes. Biochem. Soc. Trans. **31:** 1413–1416.

α-Dicarbonyl Compounds—Key Intermediates for the Formation of Carbohydrate-based Melanoidins

LOTHAR W. KROH,[a] THORSTEN FIEDLER,[b] AND JANINE WAGNER[a]

[a]*Berlin University of Technology, Department of Food Chemistry and Food Analysis, Berlin, Germany*

[b]*Pfeifer & Langen KG, Development/Technology/Analysis, Elsdorf, Germany*

The Maillard reaction of carbohydrates and amino acids is the chemical basis for flavor and color formation in many processed foods. Dicarbonyl compounds, such as 1-, 3-deoxyosones and 1,4-dideoxyosones, as well as short-chain dicarbonyls, such as methylgyoxal or glyoxal, are key compounds of the Maillard browning reaction. The α-dicarbonyls are also starting materials for polymerization reactions which lead to formation of carbohydrate-based melanoidins. With regard to the dicarbonyl compound, different possible chemical structures of melanoidins will be discussed. The analysis by size-exclusion chromatography revealed that those colored compounds differ in their molecular size and are directly associated with reactions having specific α-dicarbonyl compounds.

Key words: dicarbonyl compounds; melanoidin; formation

Introduction

The consumer often judges the quality of a processed food by its aroma, color, and/or shelf life. The compounds that are formed during the Maillard reaction have an important influence on these properties. The chemistry behind the formation of flavor and aroma compounds is well researched. However, not much is known about those partly dark-brown colored reaction products, the melanoidins. To date, neither the mechanism of their formation nor the structure of those important organic compounds has been fully clarified. In the early 1950s, Hodge discovered that α-dicarbonyl compounds play a central role in the course of the Maillard reaction as flavor and especially color (melanoidins) are formed by the reaction of those key intermediates.[1] Ledl investigated the role of the dicarbonyl compounds and systematized their role in a fundamental paper about the chemistry of the Maillard reaction.[2]

The formation of melanoidins can be divided into two reaction pathways. The formation of protein-based melanoidins (so-called *melanoproteins*) from deoxysones and other sugar degradation products was investigated by Hofmann.[3] Hofmann demonstrated that this reaction leads to the formation of colored compounds of low-molecular weight with furan and pyrrolidin structures as well as brown compounds of high-molecular weight whose molecular weight and molecular size is determined by the reacting protein. During another reaction, which is responsible for the caramelization of sugars as well as for the Maillard reaction, the previously formed dicarbonyl compounds react directly to high-molecular-weight structures. This leads to the formation of carbohydrate-based melanoidins, a reaction that can also be catalyzed by amino compounds.[4] The investigation of those high-molecular-weight compounds poses many difficulties as melanoidins are a complex mixture of different partly polymeric compounds, especially when various sugars react with amino acids or proteins.

In order to gain a better understanding about the role of α-dicarbonyl compounds in color formation, we investigated melanoidins formed in a model reaction of sucrose and γ-aminobutyric acid by size-exclusion chromatography (SEC). This method allowed us to investigate the molecular-size distribution of the formed sucrose–melanoidins.[5]

Materials and Methods

D-glucose, D-fructose, sucrose, γ-aminobutyric acid (GABA), acetic anhydride, pyridine, kieselguhr,

Address for correspondence: Lothar W. Kroh, Berlin University of Technology, Department of Food Chemistry and Food Analysis, Gustav-Meyer-Allee 25, D-13355 Berlin, Germany. Voice: +49-30-31472584; fax: +49-30-31472585.
lothar.kroh@tu-berlin.de

glyoxal, quinoxaline, and 1-butanol were obtained from Merck (Darmstadt, Germany); methylquinoxaline and *o*-phenylenediamine were obtained from Fluka (Buchs, Switzerland); toluene Pestanal® was obtained from Riedel-de Haën (Seelze, Germany); methanol (HPLC grade), 2,3-diphenylquinoxaline, and methylglyoxal were obtained from Sigma Aldrich (Steinheim, Germany); and dimethylquinoxaline was obtained from Lancaster (Morecambe, England).

Carbohydrate Model Solutions

The sucrose–Maillard reaction solutions contained 65% (w/w) sucrose, 0.1% invert sugar (1:1), and 0.1% GABA; this corresponds to a thick juice of the last stage of evaporation during the sugar processing. The aqueous α-dicarbonyl solutions had a concentration of 500 mg/kg. All models were adjusted to pH 8.0. The pH value was not regulated during the reaction.

Thermal Treatment and Derivatization

The solutions were heated in sealed ampoules for up to 300 min at 130°C or 80°C ± 1°C by means of a thermoblock (Behr Labortechnik, behrotest ET 2, Düsseldorf, Germany). The original samples were used for color measurement and for determining the molecular size by SEC–diode-array detector (DAD)/refractive index (RI). For quantification of the α-dicarbonyls after a defined reaction time, the samples were stirred with 0.05 mol/L *o*-phenylenediamine to convert α-dicarbonyls into quinoxalines (postderivatization), which were analyzed after filtration by HPLC-DAD and by gas chromatography–mass spectrometry (GC/MS) after acetylation.

HPLC-DAD

Instrumentation and criteria used in the study were as follows: Degasser, Degasys DG-13000 (Knauer, Berlin, Germany); pump, Shimadzu (Duisburg, Germany) LC-10 AT; thermostat, 30°C, Shimadzu CT0-6A; guard column, Nucleosil 120-5 C18 Macherey-Nagel (Düren, Germany); column, Nucleosil 5 C18 (250 × 4.6 mm); detector, DAD Gynkotek (Munich, Germany) UVD 340S; flow, 1.0 mL/min; injection volume, 40 μL; eluent, methanol–water gradient as follows: 0–5 min, 30% methanol; 5–12 min, 30%–50%; 12–20 min, 50%–100%; 20–30 min, 100% methanol.

GC/MS

Before quantification, the reaction mixture was extracted with 1-butanol. The solvent was evaporated, the residue was dissolved in a mixture of toluene and pyridine (30:1), and acetic anhydride was added afterward. The following instrumentation and methods

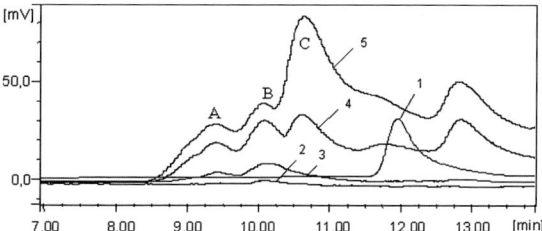

FIGURE 1. Size-exclusion chromatograms of model sucrose melanoidins obtained after 30 min (2), 90 min (3), 180 min (4), and 300 min (5) of thermal treatment at 130°C (pH 8.0). Sucrose (1) is used as a retention-time marker. A, B, and C: molecular-size domains.

were used: gas chromatograph, Finnigan (San Jose, CA) GCQTM; capillary column, BPX-5 (SGE; 30 m, 0.25 mm internal diameter, 0.5 μm film thickness); carrier gas, helium 4.6; detector, Finnigan Ion Trap Mass Analyzer GCQTM; injection temperature, 270°C. The temperature program was as follows: initial temperature 95°C, hold 1 min; 95–200°C, 15°C/min; 200°C, 1 min; 200–280°C, 3°C/min; 280°C, 5 min; 280–300°C, 5°C/min; 300°C, 5 min. Column effluents were analyzed by selected ion monitoring. Quinoxalines prepared by Hollnagel[6] were used as standards for quantification.

SEC–DAD/RI

The following instrumentation and criteria were used: Degasser, Degasys DG-1310 (Knauer); pump, Gynkotek model 480; column, PLaquagel-OH mixed and PLaquagel-OH 30 each 8 μm, 300 × 7.5 mm; detector, DAD Kontron 440 and RI Shimadzu RID-6A; flow, 1.5 mL/min; injection volume, 20 μL; eluent, water. All chromatograms show the melanoidins that were detected by DAD at wavelengths between 260 nm–285 nm. The sucrose standards were measured with the RI detector.

Color Measurement

The color was measured by a modified method based on the International Commission for Uniform Methods of Sugar Analysis (ICUMSA) at an absorption wavelength of 420 nm and is referred to as ICUMSA-Units (IU) (Abbe-Refractometer; Carl-Zeiss Jena, Jena, Germany; photometer; Novaspec II Pharmacia LKB, Freiburg, Germany).

Results and Discussion

Our sucrose reaction model is adapted to the technology of sugar production and consists of a mixture of sucrose, γ-amino butyric acid, and 0.1% invert sugar,

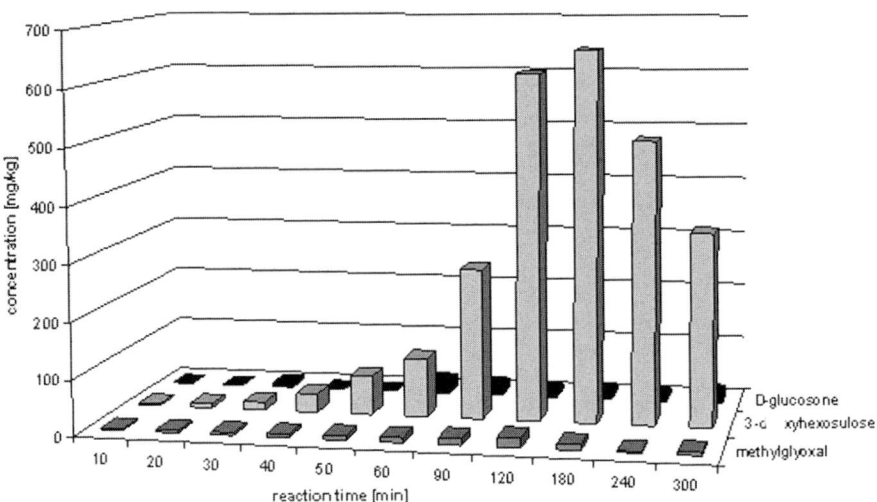

FIGURE 2. Time-dependent HPLC determination of the α-dicarbonyl compounds D-glucosone, 3-deoxyhexosulose, and methylglyoxal as quinoxalines [mg/kg] in thick juice samples.

which was heated at a temperature of 130°C and at a pH of 8.0.[7] The formed melanoidins were then separated by SEC and characterized by their molecular-weight distribution. FIGURE 1 shows the molecular-size separation of melanoidins of a thermally treated sucrose solution where the unheated sucrose is used as a retention-time marker. The chromatogram shows the molecular-size distribution after 30 and 90 min and the development of two small domains, A and B. The intensity of the domains A and B increases after 180 min up to 300 min, and a further chromatographical peak C of lower molecular size and showing the strongest increase is formed in the advanced reaction. It can be seen that the sucrose melanoidins can be separated by SEC into three very discrete molecular-size domains, A, B, and C. It is now of interest to understand the mechanism behind this phenomenon and to be able to explain the distribution. Of special interest is the strong increase of domain C in the advanced reaction, which also correlates to the strong color formation that was measured at the same time. We believe that α-dicarbonyl compounds are the most important key intermediates, leading to the formation of melanoidins in concentrated carbohydrate solutions. For the quantification by HPLC and GC/MS, a trapping reaction with o-phenylendiamine was used and the α-dicarbonyl compounds were analyzed as quinoxalines.[6] It was observed that 3-deoxyosone, methylglyoxal, and glucosone were the three most important intermediates that were formed during the browning reaction at 130°C (FIG. 2). The strong increase of 3-deoxyosone to 650 mg/kg in the advanced-reaction phase correlates well with the formation of domain C (FIG. 1) as well as with the color formation. This suggests a direct relationship between the dicarbonyl formation, the increase of the discrete melanoidin domains, and the color formation.

The attempt to clarify the relationship between the formation of α-dicarbonyl compounds and melanoidins involved the preparation of melanoidins based on several α-dicarbonyl compounds as well as analyzing them by SEC. Two important dicarbonyls, 3-deoxyosone and glucosone, were synthesized and thermally treated at 130°C. The result of the size-exclusion separation of the melanoidins formed during the reaction of those α-dicarbonyls is shown in FIGURE 3. It was very surprising to see that such very small, sharp, molecular-size domains based on α-dicarbonyl compounds were formed. By comparing the melanoidins formed by methylglyoxal (FIG. 3) to the melanoidins formed by 3-deoxyosone and glucosone, it can clearly be seen that the latter are of lower molecular size. In order to explain the different molecular-size distribution of the melanoidins formed by methylglyoxal and 3-deoxyosone, two different melanoidin skeletons based on aldol condensation were postulated. It could be possible that side-chain cross-links of the methylglyoxal skeleton are the reason for the higher molecular size of those melanoidins compared to the 3-deoxyosone melanoidins.[8] An interesting result was obtained by mixing the two α-dicarbonyls and treating them thermally at 130°C. The result of the size-exclusion separation of the reaction of both dicarbonyls together is presented in FIGURE 4. We observed that, in spite of having a mixture of dicarbonyls, two

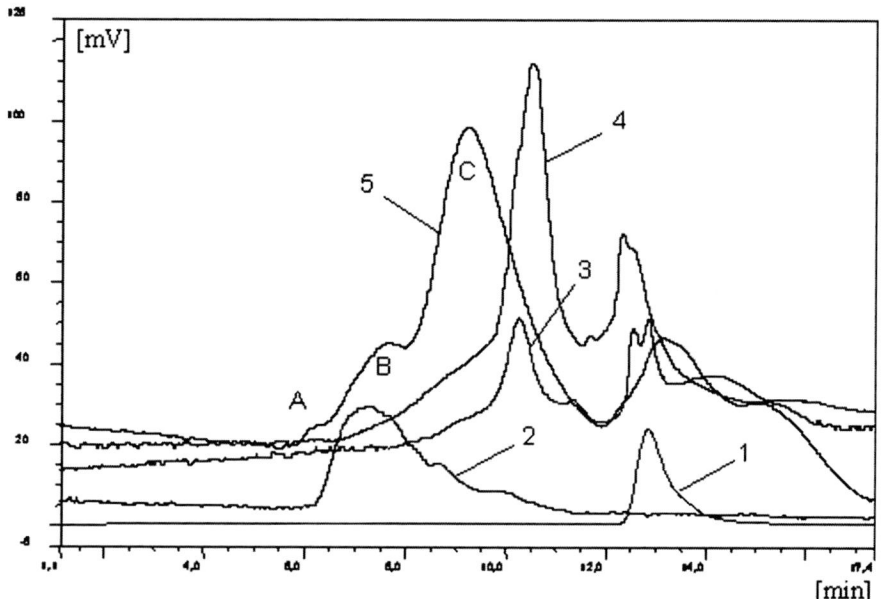

FIGURE 3. Size-exclusion chromatograms of D-glucosone melanoidin (3), 3-deoxyhexosulose melanoidin (4), and methylglyoxal melanoidins (2) in comparison to the sucrose melanoidin (5) (300 min, 130°C, pH 8.0). Sucrose is used as a retention-time marker (1). A, B, and C: see Figure 1.

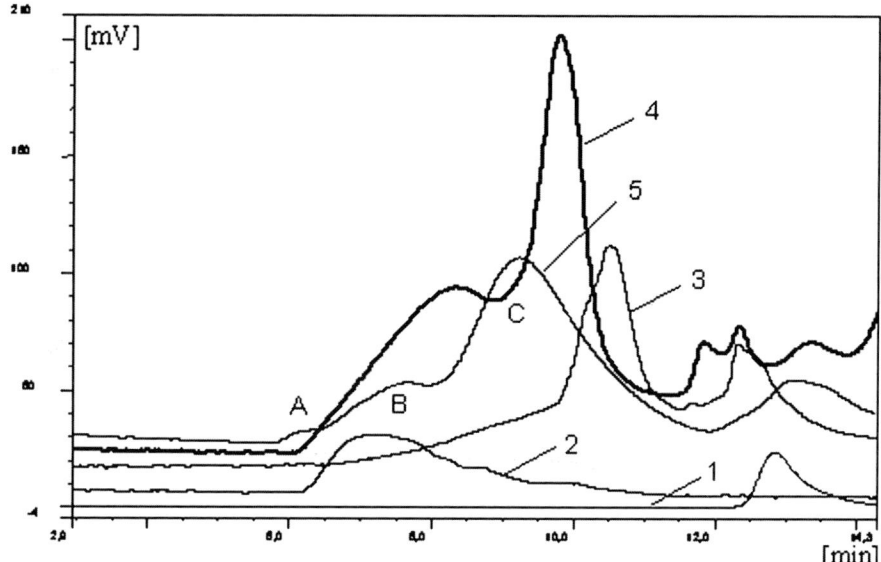

FIGURE 4. Size-exclusion chromatograms of 3-deoxyhexosulose melanoidin (3), methylglyoxal melanoidins (2), and sucrose melanoidin (5) compared to a mixture (4) of both (300 min, 130°C, pH 8.0). Sucrose is used as a retention-time marker (1). A, B, and C: see Figure 1.

separate domains are detectable. The domains were shifted in their retention time and therefore consist of melanoidins of different molecular size compared to the solutions containing only one α-dicarbonyl compound. This phenomenon can be explained by partial cross-link reactions between the two α-dicarbonyl compounds.[8]

Finally, we wanted to show that the aldol reaction is truly a way of forming melanoidins. Therefore, the melanoidins formed in a heated methylglyoxal solution and a heated glyoxal solution were analyzed by SEC, which is shown in FIGURE 5. No peak containing high-molecular compounds was detected by UV measurement for the glyoxal solution. However, the

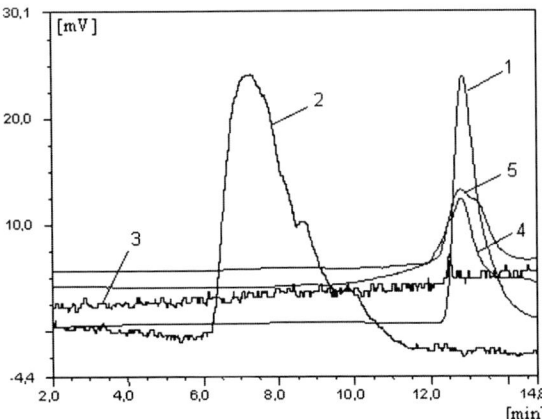

FIGURE 5. Size-exclusion chromatograms of the formation of melanoidins from methylglyoxal (2), UV signal 30 min; glyoxal (3), UV signal 30 min; glyoxal (4), refractive index (RI) signal 30 min; and glyoxal (5), RI signal 90 min. Sucrose is used as a retention-time marker (1).

index signal (RI) showed a peak near the retention time of the retention-time marker sucrose, suggesting that these are the unreacted glyoxal molecules. Moreover, no change of the RI signal was observed after 90 min (FIG. 5). The conclusions are that, based on glyoxal, no melanoidins were formed and, in correlation to this, no color formation was measured. This behavior is in contrast to what was observed when methylglyoxal was used (FIG. 5). The explanation for this difference in browning behavior is that at alkaline pH methylglyoxal forms a carbanion on the third carbon atom in the first step and proceeds to form melanoidins by aldol condensation. Glyoxal cannot form a carbanion, and, therefore, neither melanoidins nor color can be formed.

Conclusions

The presented findings support the hypothesis that the reactions of α-dicarbonyl compounds are the basis for the formation of carbohydrate-based melanoidins and, as a result, the melanoidin structure proposed in FIGURE 6 and published in Ref. 4 gains more importance. Discrete molecular-size domains of carbohydrate-based melanoidins were formed from the reaction of several α-dicarbonyl compounds with high-molecular-size melanoidin domains having a more intensive color than low-molecular-size melanoidin domains. This study confirmed that the melanoidin formation in carbohydrate solutions is based on the aldol condensation of α-dicarbonyl

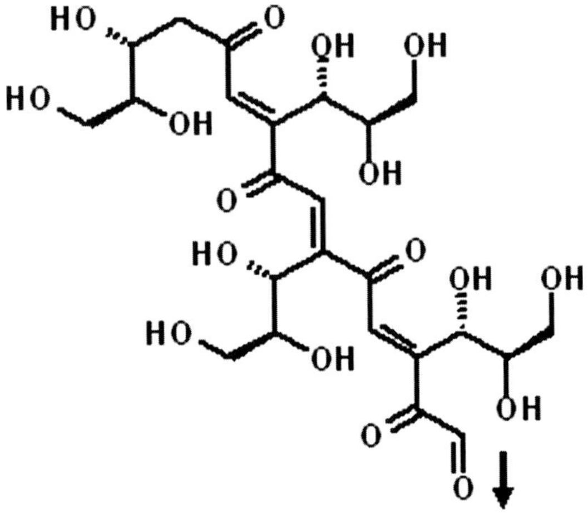

FIGURE 6. Proposal of a carbohydrate-based melanoidin structure.[4]

compounds. The formation of the carbohydrate-based melanoidins was dependent on temperature and pH.

Acknowledgments

This research project was supported by the FEI (Forschungskreis der Ernährungsindustrie e.V., Bonn), the AIF (Arbeitskreis industrielle Forschung), and the Ministry of Economics and Labour. AIF-Project No. AIF-FV 13584 supported by the Pfeifer & Langen KG and also by the Nordzucker AG.

Conflict of Interest

The authors declare no conflicts of interest.

References

1. HODGE, J. 1953. Chemistry of browning reaction in model systems. J. Agric. Food Chem. **1:** 928–943.
2. LEDL, F. & E. SCHLEICHER. 1990. Die Maillard-Reaktion in Lebensmitteln und im menschlichen Körper – neue Ergebnisse zu Chemie, Biochemie und Medizin. Angew. Chem. **102:** 597–734.
3. HOFMANN, T. 1998. 4-Alkylidene-2-imino-5-4-alkylidene-5-oxo-1,3-imidazol-2-inyl]aza-methylidene-1,3-imidazolidine – A novel colored substructure in melanoidins formed by maillard reactions of bound arginine with glyoxal and furan-2-carboxaldehyde. J. Agric. Food Chem. **46:** 3896–3901.
4. CÄMMERER, B., W. JALYSCHKO & L.W. KROH. 2002. Intact carbohydrate structure as part of the melanoidin skeleton. J. Agric. Food Chem. **50:** 2083–2087.
5. FIEDLER, T. & L.W. KROH. 2007. Formation of discrete molecular size domains of melanoidins depending on the

involvment of several a-dicarbonyl compounds: Part 2. Eur. Food Res. Technol. **225:** 473–481.

6. HOLLNAGEL, A. & L.W. KROH 2000. Degradation of oligosaccharides in nonenzymatic browning by formation of a-dicarbonyl compounds via a "Peeling-Off" mechanism. J. Agric. Food Chem. **48:** 6219–6226.

7. KROH, L.W., T. FIEDLER & J. WAGNER. 2007. The formation of coloured compounds in technical sucrose sollutions – new knowledge about their molecular basics and molecular size distribution. Sugar Industry **124:** 25–46.

8. FIEDLER, T. 2006. Ph. D. Thesis. Berlin University of Technology.

Approaches to Wine Aroma

C_1 Transfer during the Reaction between Diacetyl and Cysteine

JOHN ALMY AND GILLES DE REVEL

UMR 1219 œnologie, ISVV, Université Victor Segalen Bordeaux 2, Talence, France

The model reaction of cysteine and diacetyl under winemaking conditions resulted in the identification of over 40 products. Of these, 12 were also identified in wine samples. Several products of the model system contained unexpected structures that do not fit the four-carbon skeletal pattern of diacetyl. Possible synthetic routes to one of these products, trimethylpyrazine, are presented.

Key words: diacetyl; cysteine; winemaking conditions; trimethylpyrazine

Introduction

Of ongoing interest to wine producers are reactions between α-dicarbonyl compounds (byproducts of alcoholic and malolactic fermentations) and cysteine. The products are frequently heterocycles having characteristic odors that significantly influence wine quality and character. Prior work in this area surveyed glyoxal, methylglyoxal, diacetyl, and pentane-2,3-dione with several amino acids[1] and then the same dicarbonyls with cysteine.[2] These studies were carried out under conditions used in winemaking: 12% ethanol in water, pH 3.5 (tartrate). The results of similar studies at higher temperature or pH[3–5] provided valuable information for identification of products as well as insights on their formations.

An up-to-date study using a specific reaction pair, diacetyl and cysteine, was needed to provide a more comprehensive set of identified products; this would facilitate a more detailed organization of the reaction paths taken during this critical phase of odorant production. Diacetyl, whose average levels rise during malolactic fermentation of wine from 0.9 to 2.8 mg/L, was chosen as its symmetry reduces the number of products and its methyl groups aid in product identification. Cysteine, found at 0.5 to 4.6 mg/L in newly fermented wine,[2] can react with diacetyl to form sulfur compounds that influence wine aroma.

The present paper describes the identification of 12 products from the diacetyl–cysteine model system that were subsequently identified in wine samples. In addition, possible reaction pathways leading to one of the products, trimethylpyrazine, is discussed in detail.

Material and Methods

Reagents and solvents were used as purchased. To 100 mL of a stock solution of 27 mmol/L tartaric acid in 12% ethanol in water was added cysteine and diacetyl, both at 20 mmol/L. The mixture was stored at room temperature for 30 d in a sealed, clear, glass bottle. No precautions were taken to avoid light or oxygen. Parallel experiments in which cysteine was replaced by various reagents (see below) were carried out under the same conditions for 12 d.

A 5 mL sample of the mixture was then extracted with dichloromethane; the aqueous phase was made basic with sodium bicarbonate and a second extract was taken. These extracts were dried, concentrated, treated with 1.6 μg of pentamethylene sulfide internal standard, and then analyzed by gas chromatography/mass spectrometry (GC-MS) (details are reported elsewhere[2]). Product identifications were based on MS and linear retention time data from authentic samples and the literature.

For the identification of compounds in wine, 400 mL samples from three wines having discernable nutty aromas were saturated with ammonium sulfate and then extracted three times with 0.5 mL of dichloromethane. The combined extracts were rinsed with 1 mL of water,

---

Address for correspondence: John Almy, Department of Chemistry, California State University, Turlock, CA 95380. Voice: +209-667-3468; fax: +209-667-3845.

almy@chem.csustan.edu

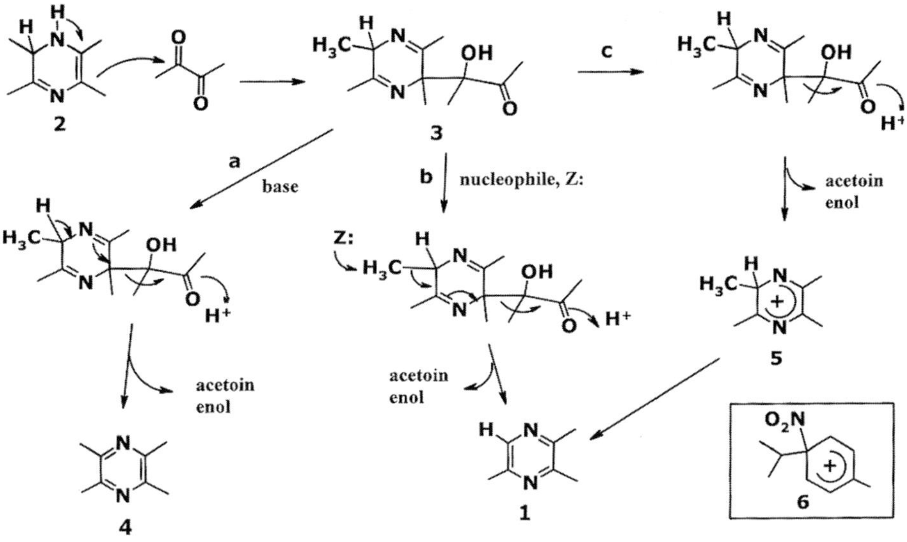

FIGURE 1. Formation of trimethylpyrazine, **1**, from dihydrotetramethylpyrazine, **2**, via adduct **3** and a nucleophile, **Z:**. Path **a**, the most favorable, leads to tetramethylpyrazine, **4**. In path **b**, adduct 3 undergoes nucleophilic substitution of the methyl group shown with concomitant loss of acetoin enol to form 1. This one-step conversion is split into two steps as shown in path **c**. Path **c** leads from adduct 3 to 1 via intermediate, **5**, similar to one **6**, that also undergoes dealkylation.[12]

separated, dried, and concentrated for injection on GC-MS. The select ion monitoring mode was programmed for two or more masses for each compound. Masses monitored were absent from the spectra of coelutants having large concentrations, such as fusel oils. Positive identification was based on matching data with known samples: retention times, relative ion abundances, and peak shapes of the ions monitored.

Results and Discussion

The Model Reaction and Wine Analysis

Over 40 products were identified from the model reaction: five furans; eight oxazoles and oxazolines; three pyrroles; one pyridine; four pyrazines; four thiophenes; 16 thiazoles, thiazolines, and thiazolidines; and four open-chain compounds, including 3-mercapto-2-butanone.

Three wine samples were examined for the presence of the products identified in the model reaction. The 12 compounds identified (and their intrinsic aromas, if known) were: 5-acetyldihydro-2-hydroxy-2,5-dimethylfuran-3(2H)-one, 2,5-dimethyl-3(2H)-furanone (sugary, fruity); 2-acetyl-4,5-dimethylfuran, 2-acetyl-4,5-dimethylpyrrole, 2-acetyl-4,5-dimethylthiophene, 2-acetylthiophene (roasty, meaty); 2,5,6-trimethyl-3-pyridinol, trimethylpyrazine (roasty, nutty); tetramethylpyrazine, 2-ethyl-3,5,6-trimethylpyrazine, 2-acetylthiazole (roasty, nutty); and 2-acetyl-2-thiazoline (roasty, popcorn).

Pathways Leading to the Formation of Trimethylpyrazine

Most of the products of the model reaction retain the four-carbon remnants of diacetyl. Examples are (1-hydroxyethyl)-2,4,5-trimethyl-3-thiazoline, 2,5,6-trimethyl-3-pyridinol, 2-acetyl-2-methylthiazolidine, 5-acetyldihydro-2-hydroxy-2,5-dimethylfuran-3(2H)-one, 2-acetyl-4,5-dimethylpyrrole, and tetramethylpyrazine.

Trimethylpyrazine (**1**, Fig. 1) is an exception. Its structure suggests that it was formed after the loss of a carbon atom from diacetyl's four-carbon skeleton. Over 200 times as much tetramethylpyrazine (formed from two molecules of diacetyl) than trimethylpyrazine was produced. However, the amount of trimethylpyrazine was higher than could be accounted for by traces of methylglyoxal impurity in the diacetyl reagent. Trimethylpyrazine has been reported in the reactions of diacetyl with amino acids other than cysteine[6] and with cystine.[3]

In our study, parallel reactions in which cysteine was replaced by either cysteamine, ammonium chloride, ammonium sulfide, or glycine gave less than 1% of the tetramethylpyrazine produced with cysteine. Trimethylpyrazine yields were proportionately lower, at or below the detection limit. Main paths to these pyrazines, therefore, require at least two of cysteine's functionalities: nucleophilic character, the ability to undergo the Strecker degradation to reduce α-dicarbonyl

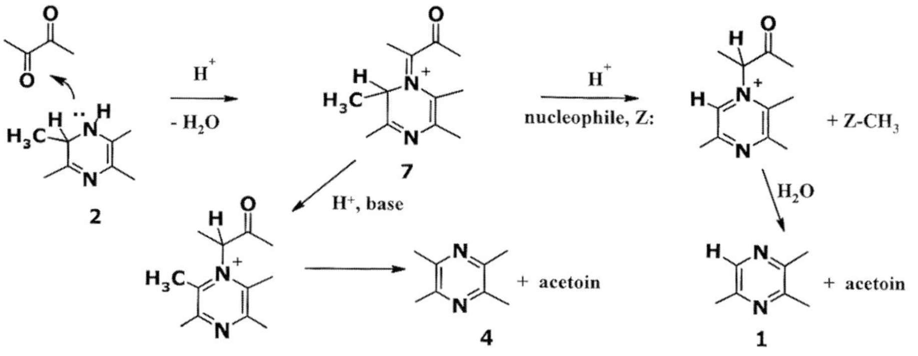

FIGURE 2. Formation of trimethylpyrazine from dihydrotetramethylpyrazine via an immonium intermediate, **7**.

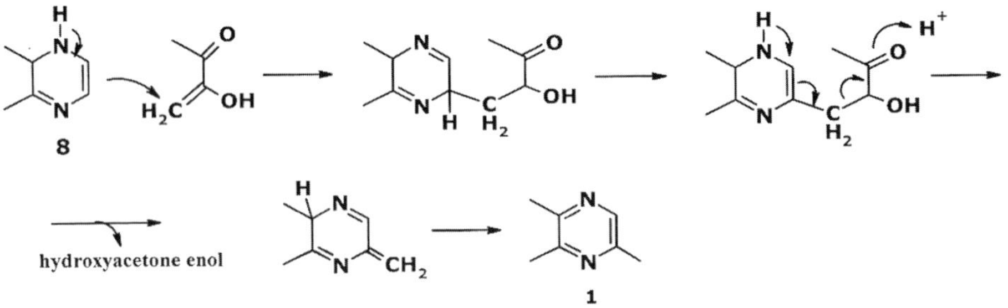

FIGURE 3. Formation of trimethylpyrazine from dihydro-2,3-dimethylpyrazine, **8**, and diacetyl enol.

compounds, and the ability to form reusable products, such as H_2S, NH_3, and acetaldehyde.

One possibility is that cysteine and 3-amino-2-butanone (the Strecker amine of diacetyl) condense to form 3,5,6-trimethyl-2(1H)-pyrazinone[7] following the loss of H_2S and tautomerization. This compound could reduce to form trimethylpyrazine, but it is already aromatic and this step is unlikely. A search for 3,5,6-trimethyl-2(1H)-pyrazinone in the product mix was unsuccessful.

Pathways leading to trimethylpyrazine from a precursor of tetramethylpyrazine, dihydrotetramethylpyrazine, **2**, are proposed in FIGURE 1. Dihydropyrazines are very slow to oxidize in the presence of air,[8,9] leaving diacetyl as the oxidant.[10] Adduct **3** is formed between **2** and a molecule of diacetyl (a similar addition of a dihydropyrazine was shown[11] to add to formaldehyde). This adduct undergoes simple elimination to form tetramethylpyrazine, **4** (Fig. 1, path a). The less-favored path b involves a nucleophilic substitution of methyl, a relatively high-energy step (compensated by the formation of an aromatic ring) that leads to **1** (Fig. 1, path b). Likely nucleophiles, Z, (Fig. 1), are water, H_2S, or the thiol group of cysteine. This transition proposed in path b is a single step. An alternative two-step process is shown in path c: an initial loss of a molecule of acetoin enol forms a delocalized intermediate **5** that loses methyl to form **1** (Fig. 1, path c). A similar intermediate **6** (Fig. 1) undergoes carbon–carbon scission to form an aromatic product during the nitration of p-isopropyltoluene; this yields p-nitrotoluene.[12]

A variation of the methyl loss from dihydrotetramethylpyrazine is shown in FIGURE 2. A nitrogen atom of **2** adds directly to diacetyl to form an imminoum ion, **7**, that undergoes either elimination to form **4** or loss of methyl to form **1**. A similar dealkylation involving loss of a secondary cation in 70–90% yield from certain dihydropyridines has been reported.[13]

A third path, outlined in FIGURE 3, begins with dihydrodimethylpyrazine (**8**, Fig. 3), a precursor of 2,3-dimethylpyrazine, also found in the product mixture. Addition of a molecule of formaldehyde to **8** would have led to trimethylpyrazine.[11] However, formaldehyde was not introduced to the model reaction, and any formaldehyde present would be almost quantitatively converted by cysteine to thiazolidine-4-carboxylic acid.[14] In our proposed pathway, the role of formaldehyde is taken by the enol of diacetyl, 3-hydroxy-3-butene-2-one.

Acknowledgments

This research was supported by a Grant from a Fulbright Aquitaine Research Award through the Franco American Commission.

During discussion, it was pointed out that methylglyoxal could be generated in the model system by an aldol condensation of diacetyl with acetaldehyde (a byproduct of cysteine during Strecker degradation). The route would proceed via 5-hydroxy-2,3-hexanedione followed by enolization and tautomerization to 3-hydroxy-2,5-hexanedione. A retroaldol step would then produce acetone enol and methylglyoxal, the C_3 homologue of diacetyl. Methylglyoxal could react with diacetyl to form trimethylpyrazine. We thank Dr. C. Cerny for this suggestion.

Conflict of Interest

The authors declare no conflicts of interest.

References

1. PRIPIS-NICOLAU, L., G. DE REVEL, A. BERTRAND & A. MAUJEAN. 2000. Formation of flavor compounds by the reaction of amino acid and carbonyl compounds in mild conditions. J. Agric. Food Chem. **48:** 3761–3766.
2. MARCHAND, S., G. DE REVEL & A. BERTRAND. 2000. Approaches to wine aroma: release of aroma compounds from reactions between cysteine and carbonyl compounds in wine. J. Agric. Food Chem. **48:** 4890–4895.
3. HARTMAN, G.J. & C.T. HO. 1984. Volatile products of the reaction of sulfur-containing amino acids with 2,3-butanedione. Lebensm. -Wiss. u. -Technol. **17:** 171–174.
4. HO, C.T. & G.J. HARTMAN. 1982. Formation of oxazolines and oxazoles in Strecker degradation of DL-alanine and L-cysteine with 2,3-butanedione. J. Agric. Food Chem. **30:** 793–794.
5. GRIFFITH, R. & G. HAMMOND. 1988. Generation of Swiss cheese flavor components by the reaction of amino acids with carbonyl compounds. J. Dairy Sci. **72:** 604–613.
6. PILOTY, M. & W. BALTES. 1979. Investigations on the reaction of amino acids with α-dicarbonyl compounds II. Volatile products of the reaction of aminoacids with diacetyl (2,3-butanedione). Z. Lebensm. Unters. Forsch. **168:** 374–380
7. SHU, C.K. & B.M. LAWRENCE. 1995. 3-Methyl-2(1H)-pyrazinones, the asparagine-specific Maillard products formed from asparagine and monosaccharides. J. Agric. Food Chem. **43:** 779–781.
8. RIZZI, G.P. 1988. A mechanistic study of alkylpyrazine formation in a model system. J. Agric. Food Chem. **36:** 349–352.
9. SHIBAMOTO, T., T. AKIYAMA, M. SAKAGUCHI, *et al*. 1979. A study of pyrazine formation. J. Agric. Food Chem. **27:** 1027–1031.
10. YAYLAYAN, V.A, L. HAFFENDEN, F.L CHU & A. WNOROWSKI. 2005. Oxidative pyrolysis and postpyrolytic derivatization techniques for the total analysis of Maillard model systems: investigation of control parameters of Maillard reaction pathways. Ann. N.Y. Acad. Sci. **1043:** 41–54.
11. AMRANI-HEMAIMI, M., C. CERNY, & L.B. FAY. 1995. Mechanisms of formation of alkylpyrazines in the Maillard reaction. J Agric. Food Chem. **43:** 2818–2882.
12. OLAH, G.A. & S.J. KUHN. 1964. Aromatic substitution. XX. Intact and dealkylating nitration of propylated and butylated alkylbenzenes with nitronium tetrafluoroborate. J. Am. Chem. Soc. **86:** 1067–1070.
13. LOEV, B. & K.M. SNADER. 1965. The Hantzsch reaction. I. Oxidative dealkylation of certain dihydropyridines. J. Org. Chem. **30:** 1914–1916.
14. KALLEN, R.G. 1971. Equilibria for the reaction of cysteine and derivatives with formaldehyde and protons. J. Am. Chem. Soc. **93:** 6227–6235.

Antioxidant Activity and Chemical Properties of Crude and Fractionated Maillard Reaction Products Derived from Four Sugar–Amino Acid Maillard Reaction Model Systems

XIU-MIN CHEN AND DAVID D. KITTS

University of British Columbia, Food, Nutrition and Health, Vancouver, Canada

Antioxidant activity of Maillard reaction products (MRPs) derived from four sugar–amino acid Maillard reaction model systems (glucose [Glc] or ribose [Rib] reacted with glycine [Gly] and L-lysine [Lys]) were examined in terms of chemical properties and molecular weight fractionation of reaction products. Rib–amino acid model systems produced MRPs with higher antioxidant activity than Glc–amino acid model MRPs ($P < 0.05$, Rib–Lys > Rib–Gly > Glc–Lys > Glc–Gly). In the same sugar or same amino acid model systems, antioxidant activity of MRPs was negatively related to the final pH, fluorescent intensity, and the content of dicarbonyl compounds. Antioxidant activity positively related to the production of late-stage browning MRPs. Fraction I from the Glc–Lys model system separated by gel filtration chromatography had the highest oxygen radical absorbance capacity (ORAC) value (1736 μmol Trolox/g MRP). Fraction IV from the Rib–Lys model system had a higher ORAC value compared to Fraction III. This result indicated that high molecular weight MRPs do not necessarily have higher antioxidant activity compared to low molecular weight MRPs.

Key words: Maillard reaction; antioxidant activity; oxygen radical absorbance capacity; gel filtration chromatography

Introduction

It is well known that Maillard reaction products (MRPs) produced in both heat-treated food systems and in sugar–amino acid model systems have antioxidant activity.[1–3] However, to date, it is not clear what Maillard reaction (MR) compounds contribute to the antioxidant activity of MRPs and how this activity is derived over time. Some studies have indicated that the antioxidant capacity is a result of intermediate and low molecular weight (LMW) MRPs,[4,5] but other studies have suggested that high molecular weight (HMW) MRPs have higher antioxidant activity than LMW MRPs.[1,6] The production of MRPs is widely affected by temperature, pH, reaction time, and reactants,[7,8] which in turn result in different chemical composition and variable antioxidant activity of MRPs. The relationship between the chemical composition of MRPs and antioxidant activity remains unclear. The purpose of this study was to investigate the possible relationship between different compositions of MRPs and the antioxidant activity of both fractionated and crude MRPs derived from different sugar–amino acid substrate model systems using the oxygen radical absorbance capacity (ORAC) assay. The effect of sugar and amino acid type on the chemical composition and antioxidant activity of MRPs was evaluated.

Materials and Methods

Preparation of Sugar–Amino Acid Maillard Reaction Products

For this study, 0.8 mol/L glucose (Glc) or ribose (Rib) was mixed as a 1:1 molar ratio with glycine (Gly) or L-lysine (Lys) (Sigma, St. Louis, MO) in 100 mL ddH_2O and adjusted to pH 7.0. The mixtures were heated at 121°C for 60 min in an autoclave (Barnstead, Boston, MA). The heated solutions were rapidly cooled on ice and lyophilized.

Final pH and Spectroscopic Measurement

The final pH of the heated mixture was recorded. The absorbance of the heated mixtures at 290 and 420 nm was measured after appropriate dilution with

---

Address for correspondence: David D. Kitts, University of British Columbia, Food, Nutrition and Health, 2205 East Mall, Vancouver, BC, CAN V6T 1Z4. Voice: +1-604-822-5560.
ddkitts@interchange.ubc.ca

TABLE 1. Final pH, UV-vis absorbance at 290 and 420 nm, and oxygen radical absorbance capacity (ORAC) value of the crude sugar–amino acid Maillard reaction products (MRPs)

	Glc–Gly	Glc–Lys	Rib–Gly	Rib–Lys
Final pH	4.23 ± 0.02^a	3.45 ± 0.02^b	3.59 ± 0.03^c	2.88 ± 0.02^d
UV290	0.72 ± 0.04^a	1.85 ± 0.03^b	3.46 ± 0.05^c	4.98 ± 0.02^d
UV420	0.13 ± 0.01^a	0.37 ± 0.01^b	0.68 ± 0.02^c	1.40 ± 0.01^d
ORAC (μmol Trolox/ MRP)	205.4 ± 10.8^a	292.9 ± 26.8^b	348.5 ± 10.5^c	562.2 ± 43.5^d

^{abcd} letters signify significant difference from each other ($P \leq 0.5$).

phosphate-buffered saline solution (5×, pH 7.4). The fluorescence spectra of the heated mixtures were measured with an excitation wavelength of 400 nm and emission from 350 to 550 nm.

Quantification of Dicarbonyl Compounds

Dicarbonyl compounds were quantified by reversed-phase-HPLC after derivation with 2,3-diaminonaphthaline (DAN) according to the method of Odani et al.,[9] with some modification. Briefly, MRPs in 10 mmol/L phosphate buffer (pH 7.4) were incubated with DAN in the presence of 3,4-hexanedione (internal standard) overnight at 4°C. The reaction mixture was extracted by ethyl acetate and evaporated until dry under nitrogen gas. The extract was reconstituted in methanol and injected into a Sphereclone ODS2 column (Phenomenex, Torrance, CA) eluted with gradient acetonitrile (ACN) and 0.2% formic acid: 0–13 min, 28–45% ACN; 13–25 min, 45–85% ACN; 25–28 min, 85% ACN, with a flow rate of 0.8 mL/min. The quinoxaline derivatives were detected by diode array detector (265 nm) and fluorescent detectors (excitation at 267 nm and emission at 503 nm).

Fractionation of Maillard Reaction Products

Lyophilized MRPs were dissolved in ddH$_2$O and fractionated by a gel filtration chromatography column (2.5 × 50 cm, packaged with fine Biogel P10; Bio-Rad Laboratories Inc., Hercules, CA) and eluted with ddH$_2$O at a flow rate of 0.2 mL/min. The eluent was collected every 40 min using a Bio-Rad Econo system and monitored at 290 nm. The fractions were pooled and lyophilized.

Oxygen Radical Absorbance Capacity Assay

Antioxidant activity was determined by the ORAC method used by Kitts and Hu.[10] Trolox (6-Hydroxy-2,5,7,8-tetramethylchroman-2-carboxylic acid (Sigma) was used to quantify ORAC value.

Statistical Analysis

Two batches of heated mixture were used. All data are expressed as means ± SD. Means were compared by one-way analysis of variance, followed by Tukey's test. The level of confidence required for a significant difference was selected at $P \leq 0.05$.

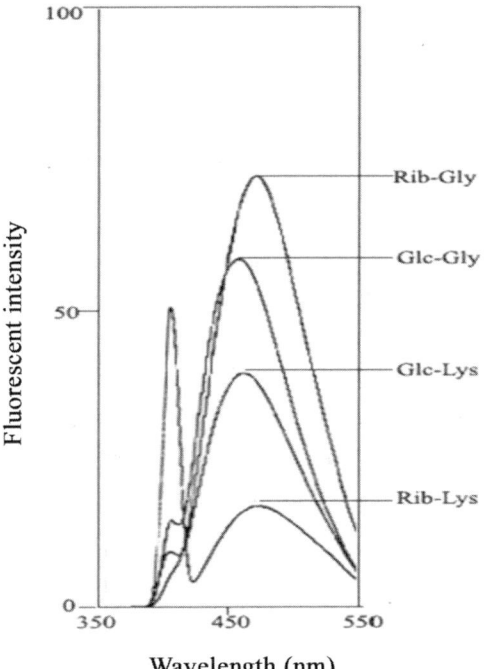

FIGURE 1. Fluorescence emission spectra of heated sugar–amino acid model systems.

Results and Discussion

The final pH of four sugar–amino acid model systems decreased dramatically after heating for 60 min based on a relative rate of change where Rib–Lys < Glc–Lys < Rib–Gly < Glc–Gly ($P < 0.05$) (TABLE 1). The final pH of the MR mixture is often used as an indicator of the MR rate and can be attributed to the presence of derived acetic acid.[9] In general, LMW reactants are more reactive than HMW reactants because of lesser steric hindrance, explaining why Rib–Gly and Rib–Lys MRPs had lower pH values than Glc–Gly and Glc–Lys, respectively. Similarly, the lower pH values in Glc–Lys and Rib–Lys are attributed to the presence of the ϵ-amino group in lysine.

The fluorescent emission spectra of Glc–Gly and Glc–Lys (FIG. 1), obtained at an excitation wavelength

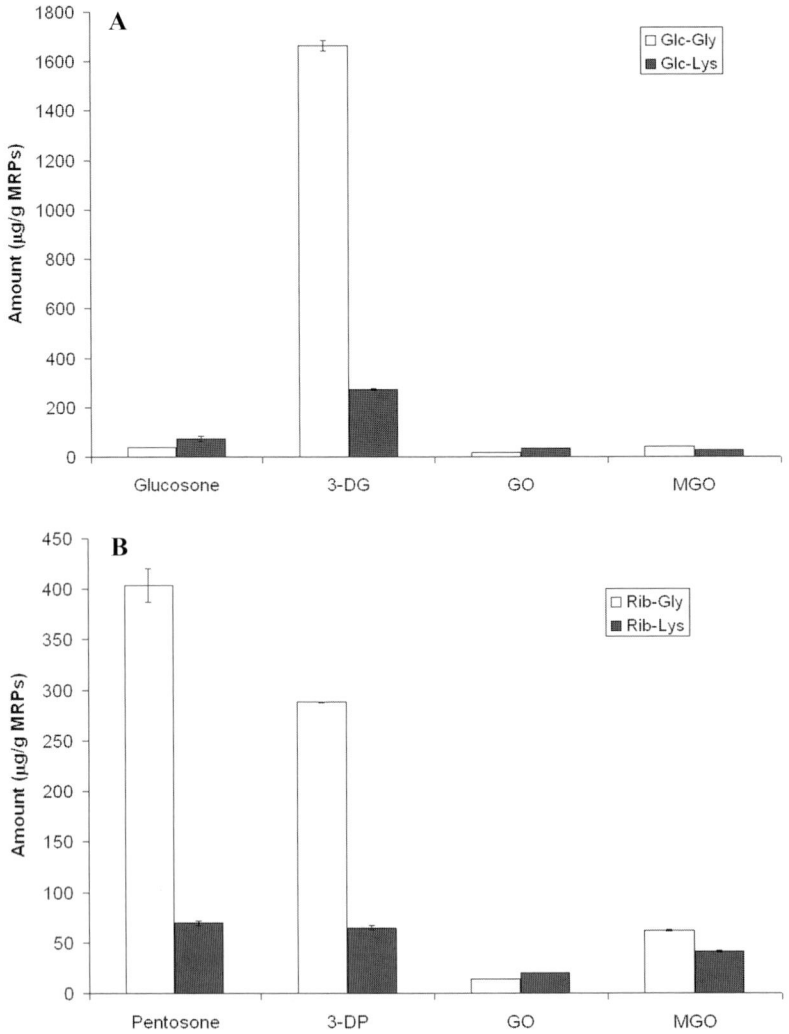

FIGURE 2. Dicarbonyl content produced in sugar–amino acid model systems. **(A)** Glc–amino acid; **(B)** Rib–amino acid.

of 400 nm, showed emission wavelength peaks at 460 nm, whereas Rib–Gly and Rib–Lys both give an emission peak at about 475 nm. The fluorescence intensity of sugar–Gly was higher than that of sugar–Lys MRPs, thus indicating a slower transition of reactant compounds to higher HMW MRPs.

Four dicarbonyl compounds, including glucosone, 3-deoxyglucosone (3-DG), glyoxal (GO), and methylglyoxal (MGO) were identified in Glc–Gly and Glc–Lys model systems. Pentosone, 3-deoxypentosone (3-DP), GO, and MGO were identified in Rib–Gly and Rib–Lys model systems (FIG. 2). In Glc–Gly and Glc–Lys model systems, 3-DG was the major dicarbonyl compound and its concentration was higher in the Glc–Gly model system. The concentration of pentosone and 3-DP was higher than GO and MGO in both Rib–Gly and Rib–Lys model systems. The amount of pentosone and 3-DP produced in the Rib–Gly model system was greater than in the Rib–Lys model system. The variable amount of dicarbonyl compounds produced in the MR was greatly influenced by the type of sugars and amino acids. The influence of the sugar type was greatest on the type of dicarbonyl compounds produced, whereas the amino acid had a relatively greater effect on the amount of dicarbonyl compounds produced.

The antioxidant activity of crude MRPs differs significantly ($P < 0.5$) in all four model systems (TABLE 1). In general, ORAC values of Rib–amino acid model systems were higher than that of the Glc–amino

TABLE 2. Molecular weight distribution and recovery rate of MRPs

MRPs*	Fraction (mg)				Total amount	Recovery (%)
	I	II	III	IV		
Glc–Gly	371.8 ± 10.5				371.8 ± 10.5	74.4 ± 2.1
Glc–Lys	30.6 ± 2.3	219.3 ± 14.5	190.3 ± 13.4		440.2 ± 13.5	88.0 ± 5.7
Rib–Glc	382.1 ± 17.5	79.0 ± 4.2			462.4 ± 13.7	92.5 ± 4.8
Rib–Lys	33.4 ± 2.0	50.2 ± 4.6	162.9 ± 10.9	201.8 ± 10.0	448.3 ± 8.8	89.7 ± 5.8

*The loading amount each time was 500 mg.

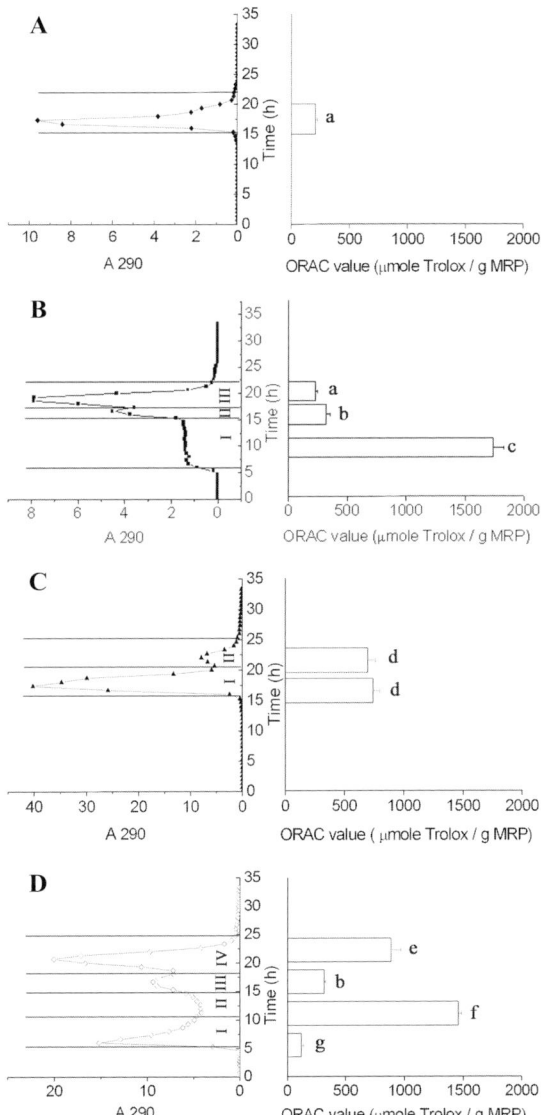

FIGURE 3. Elution curve and oxygen radical absorbance capacity (ORAC) value of fractionated Maillard reaction products (MRPs) from (**A**) Glc–Gly; (**B**) Glc–Lys; (**C**) Rib–Gly; (**D**) Rib–Lys model systems. Data is expressed as mean ± SD ($n = 6$). Bars with different letters are significantly different from each other ($P \leq 0.05$).

acid model systems ($P < 0.05$). This result indicated that the antioxidant activity in these four different model systems was influenced more by the type of sugar than by the type of amino acid. In these four model systems, higher UV-VIS absorbance at 290 and 420 nm corresponded to higher antioxidant activity. UV290 is an indicator of the production of low or intermediate molecular weight MRPs, whereas absorbance at UV 420 nm reflects the presence of HMW browning MRPs.[1] Glc–amino and Rib–amino acid model systems resulted in a lower pH, reduced fluorescent intensity, and lower amount of dicarbonyl compounds, which correspond to higher antioxidant activity. A similar result was achieved when sugar–Gly and sugar–Lys model systems were compared.

The distribution and the recovery rate of MRPs are shown in TABLE 2. The number of MRP fractions obtained from these four different model systems was found to be specific to the individual reactants used to generate crude MRPs (FIG. 3). Fraction I collected from the Glc–Lys model system had the highest ORAC values (1736 μmol Trolox /g MRP). The ORAC value of the fractions derived from the Glc–Gly model system increased as the molecular weight of MR constituents increased and was found to be significantly different ($P < 0.05$). Fractions I from both the Glc–Gly and Rib–Gly model system were recovered almost at the same time; however, greater antioxidant activity was obtained from the Rib–Gly model system compared to the Glc–Gly model system. This result suggests that although the average molecular weight of MRPs is similar, the antioxidant activity is different because of the nature of the reactants. Moreover, fraction IV collected from the Rib–Lys model system had a relatively higher antioxidant activity than fraction III, demonstrating that antioxidant activity of MRPs is not always attributed to the HMW components.

Acknowledgments

This work was supported by the Natural Sciences and Engineering Research Council-discovery grant to D.D.K. We thank Dr. Chun Hu for his kindly help.

Conflict of Interest

The authors declare no conflicts of interest.

References

1. JING, H. & D.D. KITTS. 2004. Antioxidant activity of sugar-lysine Maillard reaction products in cell free and cell culture systems. Arch. Biochem. Biophys. **429:** 154–163.
2. YEN, W.J. et al. 2005. Antioxidant properties of roasted coffee residues. J. Agric. Food Chem. **53:** 2658–2663.
3. WIJEWICKREME, A.N., Z. KREJPCIO & D.D. KITTS. 1999. Hydroxyl scavenging activity of glucose, fructose, and ribose-lysine model Maillard products. J. Food Sci. **64:** 457–461.
4. MORALES, F.J. & M.B. BABBEL. 2002. Melanoidins exert a weak antiradical activity in watery fluids. J. Agric. Food Chem. **50:** 4657–4661.
5. HOFMANN, T. et al. 2001. Determination of the molecular weight distribution of non-enzymatic browning products formed by roasting of glucose and glycine and studies on their effects on NADPH-cytochrome c-reductase and glutathione-S-transferase in Caco-2 cells. Nahrung-Food **45:** 189–194.
6. MONTI, S. M. et al. 1999. LC/MS analysis and antioxidative efficiency of Maillard reaction products from a lactose-lysine model system. J. Agric. Food Chem. **47:** 1506–1513.
7. WIJEWICKREME, A.N., D.D. KITTS & T.D. DURANCE. 1997. Reaction conditions influence the elementary composition and metal chelating affinity of nondialyzable model Maillard reaction products. J. Agric. Food Chem. **45:** 4577–4583.
8. JING, H. & D.D. KITTS. 2003. Chemistry of Maillard reaction products. Recent Res. Devel. Mol. Cell. Biochem. 77–95.
9. ODANI, H. et al. 1999. Increase in three alpha,beta-dicarbonyl compound levels in human uremic plasma: Specific in vivo determination of intermediates in advanced Maillard reaction. Biochem. Biophys. Res. Commun. **256:** 89–93.
10. KITTS, D.D. & C. HU. 2005. Biological and chemical assessment of antioxidant activity of sugar-lysine model Maillard reaction products. Ann. N.Y. Acad. Sci. **1043:** 501–512.

Advanced Glycation Endproducts in Chronic Heart Failure

ANDRIES J. SMIT,[a] JASPER W. L. HARTOG,[b] ADRIAAN A. VOORS,[b] AND DIRK J. VAN VELDHUISEN[b]

Departments of [a]Medicine and [b]Cardiology, University Medical Center Groningen and University of Groningen, Groningen, the Netherlands

Advanced glycation endproducts (AGEs) have been proposed as factors involved in the development and progression of chronic heart failure (CHF). Cross-linking by AGEs results in vascular and myocardial stiffening, which are hallmarks in the pathogenesis of CHF. Additionally, stimulation of receptors by AGEs may affect endothelial function and myocardial calcium uptake and may perpetuate coronary sclerosis in CHF. CHF is common in conditions with AGE accumulation, such as diabetes and renal failure. This review describes in detail the interrelation of plasma AGEs, renal function, and the severity and prognosis in clinical CHF patients with mild to moderate loss of renal function. This association is compared with the relation between tissue AGE accumulation (marked by skin autofluorescence) and diastolic dysfunction in renal failure. The evidence reviewed here provides support for the assumed role of AGEs in determining the severity and prognosis of CHF, but also highlights the differences in this relation between plasma and tissue AGEs and between patients with and without advanced renal failure. Ongoing clinical intervention trials to reduce AGE accumulation in patients with CHF may elucidate the causal role of AGEs in the development and course of CHF.

Key words: heart failure; advanced glycation endproducts; renal function

Introduction

Chronic heart failure (CHF) is a common condition that continues to have a high mortality rate. In contrast to the drop in mortality in acute coronary syndromes, roughly 50% of CHF patients still die within 5 years of diagnosis.[1] CHF is characterized by symptoms of dyspnea at exercise and fatigue, related to an inability of the heart to pump sufficient blood to fulfill the requirements of the body, or to its inability to do so only at elevated filling pressures. Etiologically, 50% of CHF cases are due to ischemic causes, which include those related to preceding acute coronary syndromes or other manifestations of predominantly coronary atherosclerosis. The main cause in the nonischemic group is idiopathic dilatory cardiomyopathy. The increase in the prevalence of CHF is related to the increase in the prevalence of some of the risk factors for CHF: increasing age, diabetes mellitus, and renal failure are important among these and are all conditions well known to be associated with the accumulation of Maillard reaction products (MRP) and advanced glycation endproducts (AGEs). Although this link was recognized long ago, and AGEs were suggested to be important causes of the development of diabetic cardiomyopathy, it has received surprisingly little attention. One reason may have been that CHF is not recognized as a characteristic chronic complication of diabetes mellitus (or of renal failure for that matter).

The severity of HF is clinically graded according to the New York Heart Association (NYHA) functional classification, but can also be quantified using markers such as ejection fraction (EF) or maximal total body oxygen consumption (VO_2max), or biochemically by levels of NT-pro-BNP. NYHA class and EF are also prominent prognostic factors, but the same holds for renal function, which may reflect the degree of forward failure. However, renal failure may also have an important causative or perpetuator role in the prognosis of CHF, including the mechanism of accelerated AGE accumulation.[2] Recent data on the prognostic value of AGE in CHF in patients with renal failure will be reviewed here.

CHF is preceded by diastolic dysfunction (DD), which is characterized by reduced relaxation and compliance of the left ventricle (LV). Because cross-linking of the extracellular matrix by AGEs is a classic

Address for correspondence: A. J. Smit, MD, PhD, Department of Medicine, University Medical Center Groningen, Hanzeplein 1, P.O. Box 30001, 9700 RB Groningen, the Netherlands. Voice: +31 0 50 3614854; fax: +31 0 50 3619069.

a.j.smit@int.umcg.nl

explanation for the deleterious effects of AGEs, increased myocardial AGE accumulation has been related to DD; supportive evidence of this link is discussed below.[3,4] Interventions with AGE formation inhibitors or breakers in CHF are also discussed.

Reasons for a Link Between AGE Accumulation and CHF

-*Epidemiologic*: *High prevalence of HF and DD in conditions associated with AGE accumulation, diabetes, and renal failure. Diabetic cardiomyopathy.*

The risk of HF is higher in patients with diabetes, especially in females.[5] This is partly because of the well-known increase in macrovascular complications, including coronary heart disease (CHD). However, in the absence of CHD, diabetic cardiomyopathy is often invoked as another precursor to CHF. Diabetic cardiomyopathy is common in diabetes mellitus.[6,7] Originally it was defined by specific pathological abnormalities. Diabetic cardiomyopathy is first manifested by asymptomatic DD that later progresses to systolic dysfunction.[7,8] Eventually diabetic cardiomyopathy can deteriorate into CHF and sudden cardiac death. Although AGE accumulation has been recognized as one of the possible contributing factors to diabetic cardiomyopathy, this issue has so far received limited attention.[9,10]

-*Pathogenetic*: *Cross-linking, secondary to AGE accumulation, in myocardium and vessels, gives compliance loss, a cornerstone in the development of HF. Role of AGE–receptor interaction.*

Accumulation of AGEs results in stiffening of interstitial and vascular tissue, clinically manifested by reduced arterial compliance and increased pulse pressure. AGEs form cross-links among protein fibers, such as matrix collagen, elastin, and laminin, resulting in conformational changes that alter both the structural and functional characteristics of these cross-linked proteins.[11] They affect physiological properties of matrix proteins, such as charge, hydrophobicity, turnover, and elasticity.

These pathophysiological changes do not only result from indirect effects of AGEs on endothelial dysfunction, increased thrombogenicity and accelerated atherosclerosis at the (coronary) vascular level, but direct myocardial changes due to cross-linking of proteins in the myocardial matrix also occur. The systemic vascular effects also result in increased afterload for the heart.[9]

Stimulation by AGE of receptors such as the RAGE receptor, leads to prolonged cellular activation and release of inflammatory cytokines, which may also result in the development and progression of cardiac dysfunction and subsequent CHF.

One receptor-mediated effect of AGEs is the induction of fibrosis via the upregulation of transforming growth factor-β.[12] AGE receptor activation also influences calcium metabolism in cardiac myocytes, causing a significant delay in calcium reuptake and a consequent increase in the duration of the repolarization phase.[13]

-*Contribution of renal hypoperfusion and function loss as perpetuator in HF through decreased clearance of circulating and renal MRP and AGE.*

A hallmark of clinical CHF is renal hypoperfusion and loss in renal function. Considered a functional consequence of forward failure and thereby systolic HF, it is also an important marker of the severity and the prognosis of CHF.[14,15] The fall in glomerular filtration rate (GFR) and renal blood flow is associated with a reduced filtration and tubular handling of MRP, such as AGEs and AGE peptides, resulting in accelerated accumulation of AGE. Therefore, reduced clearance and accelerated accumulation of AGE resulting from renal hypoperfusion in CHF may perpetuate the dismal course of CHF.

-*Interventional*: *CHF chosen as model in AGE-reducing pharmaceutical interventions.*

Preliminary evidence for AGEs in the development of cardiac dysfunction originates from two trials with the AGE cross-link breaker Alagebrium (ALT-711) in patients with HF. In the DIAMOND trial, Little et al.[16] treated 23 patients with stable diastolic HF open-label with ALT-711. After 16 weeks, left ventricular mass was reduced and diastolic function improved. The PEDESTAL trial is an open-label study investigating the effects of ALT-711 on diastolic function and left ventricular mass in patients with systolic HF and DD. Preliminary results confirm those of the DIAMOND trial.[17]

Clinical Evidence for Link Between AGE and Diastolic Function or Systolic HF?

The role of AGEs in the development of DD has been investigated extensively in animals.[4,18–20] In such studies, the effect of various AGE-lowering strategies confirms an active role for AGEs in the development of DD. Only limited human studies have been published.[16,17,20]

Berg et al.[21] analyzed serum AGE levels and LV diastolic function in type 1 diabetes. They found correlations between serum AGEs and both isovolumetric relaxation time and LV end-diastolic diameter, but did not find correlations between AGEs and other parameters for diastolic function. Notably, tissue velocity imaging was not used to determine diastolic function.

Plasma AGE levels were previously evaluated in CHF by Heidland et al.[22] in a small group of patients with severe CHF, heart transplant recipients, and normal controls. Paradoxically, a decrease was found in carboxymethyllysine (CML) and AGE fluorescence in patients with CHF. Heart transplant recipients did, however, show an increase in measured AGE data. Unfortunately, data on NYHA functional class, NT-pro-BNP levels, and prognosis were not provided. The authors suggested that the results may have been biased by hypervolemia, lowered plasma protein concentrations, and decreased dietary intake of AGEs in patients with CHF.

Steine et al.[23] reported that serum AGEs and duration of diabetes were predictors of systolic strain, assessed by Doppler tissue imaging, in type 1 diabetes (duration 32 ± 5 years). However, no relation existed between serum AGE and other systolic or echocardiographic indices.

Koyama et al.[24] evaluated the prognostic value of serum AGEs as risk factors in CHF for the first time. Serum pentosidine levels predicted cardiac death and rehospitalization, independent of other risk factors in CHF.

As discussed above, indirect evidence for a role of AGEs in cardiac dysfunction also originates from the DIAMOND and PEDESTAL trials in CHF patients.[16,17]

Recently, a study in our center on patients with CHF and mild to moderate impairment of renal function assessed whether plasma CML and carboxyethyllysine (CEL), assayed by liquid chromatography–tandem mass spectrometry, are related to the severity and prognosis of CHF; this study focused on the role of renal function and perfusion in this relation.[9,25] One hundred and two CHF patients with an average LVEF of $28 \pm 9\%$ were followed for 1.4 ($1.1–1.9$) years. NYHA functional class and NT-pro-BNP were used as estimates of the severity of CHF. GFR and effective renal plasma flow (ERPF) were measured using radiolabeled iothalamate and hippurate infusion. Survival analysis was performed for the combined endpoint of the first occurrence of either death, heart transplantation, or hospitalization for CHF. CML levels were associated with NYHA class and NT-pro-BNP. Furthermore, CML predicted outcome, even after correction for age and gender (hazard ratio [HR] 3.9). In a multivariate analysis, GFR was the only remaining determinant of CML. Regarding the prognostic value of CML, model correction for GFR, ERPF, and microalbuminuria resulted in an HR for CML of 1 after correction for GFR, and showed that replacing GFR by ERPF in the model led again to a nonsignificant HR. After correction for urinary albumin excretion, the HR was 4.4 (2.2–9). Similar analysis revealed that CEL was again primarily determined by GFR (B = -0.28, P = 0.004), but was not significantly associated with CHF severity or prognosis.

For comparison, provisional results are discussed from a cross-sectional study in dialysis patients—both hemodialysis and peritoneal dialysis—in which diastolic and systolic function were evaluated by echocardiography (unpublished results). DD was assessed with early diastolic velocity (Em) using tissue velocity imaging. Skin autofluorescence (AF), as a marker of tissue AGE accumulation, was assessed using an AGE reader as described previously.[26] Among 43 patients, aged 58 ± 15 years and with a median duration of dialysis of 2.8 ($1.3–5.3$) years, systolic dysfunction (LVEF $\leq 45\%$) was present in 9 patients (24%), and DD (Em < 8 cm/s) was present in 35 patients (81%). No correlations were found between plasma AGE levels and any of the diastolic function parameters. Skin AF revealed high levels (mean 3.7 ± 0.7 a.u.), and was correlated with diastolic filling, including Em, E/A ratio, and E/E' (p all < 0.02). No correlations were found between LVEF, LV and LA diameters, and skin AF. In multivariate regression analysis 45% of the variance of average Em was determined by age and skin AF.

Discussion

Despite reasons summarized in this review for a link between AGE and CHF, and despite experimental support, clinical evidence based on circulating AGE levels remains limited and has provided no convincing proof. This review stresses that the relation between AGE and CHF differs for plasma and tissue AGEs and for patients with and without advanced renal failure. A more complex interaction seems to exist between renal function and the behavior of AGE in circulation and at the tissue level, not only in CHF, but also in other conditions. Plasma AGEs are related to severity and prognosis of CHF. Levels of circulating AGEs are determined to a major extent by metabolism of plasma proteins, which is orders of magnitude faster (days to weeks) than in tissues with long-lived proteins like the skin or the myocardial interstitial tissue (often many years). Because tissue levels of AGE directly reflect the cross-linking of interstitial and intracellular proteins and the receptor interactions, it is not surprising that tissue function is better represented by tissue AGE markers, such as skin autofluorescence than by circulating AGE levels. Moreover, as for the renal failure group in our center's study, plasma or serum AGE levels are known to be quite variable depending on the dialysis procedures and sampling times in hemodialysis patients.[27] Skin AF is independent of hemodialysis sessions (unpublished results) and has been reported to have a day-to-day coefficient of variation of less than 6 percent.[25]

The prognostic value of AGEs has been studied in populations other than patients with CHF. In patients with renal failure, results vary widely. Although Schwedler et al.[28] and Busch et al.[29] reported that circulating AGE levels were not related to prognosis in renal failure, Wagner et al.[30] and Roberts et al.[31] did find prognostic value in circulating AGEs. Perhaps these discrepancies in the predictive value of circulating AGE levels illustrate again that circulating AGEs are less reliable as prognostic markers, especially in renal failure, because of the reasons given above. Our group demonstrated that AGE accumulation, measured by skin AF, was a strong and independent determinant of total and cardiovascular mortality both in patients on dialysis and in diabetic patients.[32,33] Kilhovd et al.[34] showed that high levels of circulating AGEs predicted cardiovascular mortality in nondiabetic women. However, they presented their data uncorrected for renal function; their results may, therefore, be biased. The relation of plasma CML to severity and prognosis in CHF in our patients seems to be restricted to mild and moderate renal failure, and may actually be a mere reflection of the effects of renal function loss, considering the predominant role of GFR on plasma CML in our multivariate analysis. On the other hand, tissue AGE accumulation remains related to DD in advanced renal failure, whereas plasma AGEs are not. Unfortunately, no skin AF or other indices of tissue AGE accumulation were measured in our CHF study.

Because the prognostic value of plasma CML in our CHF group disappeared after correction for GFR or for ERPF, but not with microalbuminuria/proteinuria as a structural renal damage marker, this suggests that the direct functional effects of HF on GFR and ERPF account for the prognostic effects of circulating CML in CHF. Although this may seem to further diminish the clinical value of circulating AGEs as prognostic markers in CHF, it also serves to explain why renal function is such a determining factor for the course and prognosis of CHF in the longer term, independent of actual filling state. Accumulation of AGE and other MRP resulting from impaired renal clearance in CHF with forward failure may serve as a pivotal perpetuator of myocardial and vascular damage. Thus, AGE accumulation may be a possible explanation for the fact that the prognosis of patients with a history of cardiovascular disease becomes worse as soon as CHF develops, independent of the degree of local cardiac damage.

This mechanism underlying the bleak prognosis supports interventions specifically directed at the reduction of AGE accumulation in CHF patients. Experimental, and still limited clinical, evidence suggests that angiotensin receptor blockers and angiotensin converting enzyme inhibitors may reduce AGE formation, but no studies have addressed this in patients with CHF.[35] In conditions other than CHF, attempts to reduce AGE formation, and thereby ameliorate clinical endpoints, with interventions with benfothiamine—and, earlier, pyridoxamine—may serve as examples for another approach. As a follow-up to the DIAMOND and PEDESTAL studies, the BENEFICIAL study, a single-center, double-blind, placebo-controlled, randomized study for 9 months of Alagebrium in addition to conventional medication in patients with CHF with systolic dysfunction (EF < 40%), has recently started (Hartog, personal communication), with the effect on aerobic capacity (VO_2max) and safety as primary endpoints. Secondary endpoints include diastolic/systolic function with echocardiography, NT-pro-BNP, NYHA HF class, AGEs in tissue as assessed by skin AF with an AGE reader, and in blood, and questionnaire scores. Results of such trials to reduce AGE accumulation are needed to elucidate the causal and/or perpetuating role of AGEs in the development and course of HF.

Funding Sources and Disclosures

J. Hartog is supported by, and D. J. van Veldhuisen and A. Voors (2006T37) are Clinical Established Investigators of, the Netherlands Heart Foundation.

Conflict of Interest

The authors declare no conflicts of interest.

References

1. LEVY, D., S. KENCHAIAH, M.G. LARSON, et al. 2002. Long-term trends in the incidence of and survival with heart failure. N. Engl. J. Med. **347:** 1397–1402.
2. VIRGA, G., B. STOMACI, A. MUNARO, et al. 2006. Systolic and diastolic function in renal replacement therapy: a cross-sectional study. J. Nephrol. **19:** 155–160.

3. SCHAFER, S., J. HUBER, C. WIHLER, et al. 2006. Impaired left ventricular relaxation in type 2 diabetic rats is related to myocardial accumulation of N(epsilon)-(carboxymethyl) lysine. Eur. J. Heart Fail. **8:** 2–6.

4. ARONSON, D. 2003. Cross-linking of glycated collagen in the pathogenesis of arterial and myocardial stiffening of aging and diabetes. J. Hypertens **21:** 3–12.

5. ARONOW, W.S. & C. AHN. 1999. Incidence of heart failure in 2,737 older persons with and without diabetes mellitus. Chest **115:** 867–868

6. GALDERISI, M., K.M. ANDERSON, P.W. WILSON & D. LEVY. 1991. Echocardiographic evidence for the existence of a distinct diabetic cardiomyopathy. Am. J. Cardiol. **68:** 85–89.

7. GALDERISI, M. 2006. Diastolic dysfunction and diabetic cardiomyopathy: evaluation by Doppler echocardiography. J. Am. Coll. Cardiol. **48:** 1548–1551.

8. BOUDINA, S. & D. ABEL. 2007. Diabetic cardiomyopathy revisited. Circulation **115:** 3213–3223.

9. HARTOG, J.W.L., A.A. VOORS, S.J.L. BAKKER, et al. 2007. Advanced Glycation end-products (AGEs) and heart failure: pathophysiology and clinical implications. Eur. J. Heart. Fail. **9:** 1146–1155.

10. VAITKEVICIUS, P.V., M. LANE & H. SPURGEON. 2001. A crosslink breaker has sustained effects on arterial and ventricular properties in older rhesus monkeys. Proc. Natl. Acad. Sci. USA **98:** 1171–1175.

11. SMIT, A.J. & H.L. LUTGERS. 2004. The clinical relevance of advanced glycation endproducts and recent developments in pharmaceutics to reduce AGE accumulation. Curr. Med. Chem. **11:** 2767–2784.

12. STRIKER, L.J. & G.E. STRIKER. 1996. Administration of AGEs *in vivo* induces extracellular matrix gene expression. Nephrol. Dial. Transpl. **11:** S62–S65.

13. PETROVA, R., Y. YAMAMOTO, K. MURAKI, et al. 2002. Advanced glycation endproduct-induced calcium handling impairment in mouse cardiac myocytes. J. Mol. Cell Card. **34:** 1425–1431.

14. SMITH, G.L., J.H. LICHTMAN, M.B. BRACKEN, et al. 2006. Renal impairment and outcomes in heart failure: systematic review and meta-analysis. J. Am. Coll. Cardiol. **47:** 1987–1996.

15. SMILDE, T.D., D.J. VAN VELDHUISEN, G. NAVIS, et al. 2006. Drawbacks and prognostic value of formulas estimating renal function in patients with chronic heart failure and systolic dysfunction. Circulation **114:** 1572–1580.

16. LITTLE, W.C., M.R. ZILE, D.W. KITZMAN, et al. 2005. The effect of alagebrium chloride (ALT-711), a novel glucose cross-link breaker, in the treatment of elderly patients with diastolic heart failure. J. Card Fail. **11:** 191–195.

17. THOHAN, V., M.M. KOERNER, C.M. PRATT & G.A. TORRE. 2005. Improvements in diastolic function among patients with advanced systolic heart failure utilizing alagebrium (an oral advanced glycation endproduct cross-link breaker). Circulation **112**(Suppl 2): 2647.

18. NORTON, G.R., G. CANDY & A.J. WOODIWISS. 1996. Aminoguanidine prevents the decreased myocardial compliance produced by streptozotocin-induced diabetes mellitus in rats. Circulation **93:** 1905–1912.

19. AVENDANO, G.F., R.K. AGARWAL, R.I. BASHEY, et al. 1999. Effects of glucose intolerance on myocardial function and collagen-linked glycation. Diabetes **48:** 1443–1447.

20. ASIF, M., J. EGAN, S. VASAN, et al. 2000. An advanced glycation endproduct cross-link breaker can reverse age-related increases in myocardial stiffness. Proc. Natl. Acad. Sci. USA **97:** 2809–2813.

21. BERG, T.J., O. SNORGAARD, J. FABER, et al. 1999. Serum levels of advanced glycation end products are associated with left ventricular diastolic function in patients with type 1 diabetes. Diabetes Care **22:** 1186–1190.

22. HEIDLAND, A., K. SEBEKOVA, A. FRANGIOSA, et al. 2004. Paradox of circulating AGE concentrations in patients with CHF and after heart transplantation. Heart **90:** 1269–1274.

23. STEINE, K., J. LARSEN, M. STUGAARD, et al. 2007. LV systolic impairment in patients with asymptomatic coronary heart disease and type 1 diabetes is related to coronary atherosclerosis, glycaemic control and advanced glycation endproducts. Eur. J. Heart Failure **9:** 1044–1050.

24. KOYAMA, Y., Y. TAKEISHI & T. ARIMOTO. High serum level of pentosidine, an advanced glycation end product, is a risk factor of patients with heart failure. J. Card. Fail. **13:** 199–206.

25. HARTOG, J.W.L., A.A. VOORS, C.G. SCHALKWIJK, et al. 2007. Clinical and prognostic value of advanced glycation endproducts in chronic heart failure. Eur. Heart J. **28:** 2879–2885.

26. MEERWALDT, R., R. GRAAFF, P. OOMEN, et al. 2004. Simple non-invasive assessment of advanced glycation endproduct accumulation. Diabetologia **47:** 1324–1330.

27. JADOUL, M., Y. UEDA, Y. YASUDA, et al. 1999. Influence of hemodialysis membrane type on pentosidine plasma level, a marker of "carbonyl stress". Kidney Int. **55:** 2487–2492.

28. SCHWEDLER, S.B., T. METZGER, R. SCHINZEL & C. WANNER. 2002. Advanced glycation end products and mortality in hemodialysis patients. Kidney Int. **62:** 301–310.

29. BUSCH, M., S. FRANKE, G. WOLF, et al. 2006. The advanced glycation end product N(epsilon)-carboxymethyllysine is not a predictor of cardiovascular events and renal outcomes in patients with type 2 diabetic kidney disease and hypertension. Am. J. Kidney Dis. **48:** 571–579.

30. WAGNER, Z., M. MOLNAR, G.A. MOLNAR, et al. 2006. Serum carboxymethyllysine predicts mortality in hemodialysis patients. Am. J. Kidney Dis. **47:** 294–300.

31. ROBERTS, M.A., M.C. THOMAS, D. FERNANDO, et al. 2006. Low molecular weight advanced glycation end products predict mortality in asymptomatic patients receiving chronic haemodialysis. Nephrol. Dial. Transplant. **21:** 1611–1617.

32. MEERWALDT, R., J. HARTOG, R. GRAAFF, et al. 2005. Skin autofluorescence, a measure of cumulative metabolic stress and advanced glycation end products, predicts mortality in hemodialysis patients. J. Am. Soc. Nephrol. **16:** 3687–3693.

33. MEERWALDT, R., H.L. LUTGERS, T.P. LINKS, et al. 2007. Skin autofluorescence is a strong predictor of cardiac mortality in diabetes. Diabetes Care **30:** 107–112.

34. KILHOVD, B., A. JUUTILAINEN, S. LEHTO, *et al.* 2005. High serum levels of advanced glycation end products predict increased coronary heart disease mortality in nondiabetic women but not in nondiabetic men: a population-based 18-year follow-up study. Art. Thromb. Vasc. Biol. **25:** 815–820.

35. MIYATA, T. & S. VAN YPERSELE. 2003. Angiotensin II receptor blockers and angiotensin converting enzyme inhibitors: implication of radical scavenging and transition metal chelation in inhibition of advanced glycation end product formation. Arch. Biochem. Biophys. **419:** 50–54.

Methylglyoxal and Methylglyoxal-arginine Adducts Do Not Directly Inhibit Endothelial Nitric Oxide Synthase

OLAF BROUWERS,[a] TOM TEERLINK,[b] JAN VAN BEZU,[b] ROB BARTO,[b] COEN D.A. STEHOUWER,[a] AND CASPER G. SCHALKWIJK[a]

[a]*Department of Internal Medicine, Division of General Internal Medicine, Laboratory for Metabolism and Vascular Medicine, Maastricht University, Maastricht, the Netherlands*

[b]*Department of Clinical Chemistry, VU University Medical Center, Amsterdam, the Netherlands*

Increased formation of the reactive dicarbonyl compound methylglyoxal (MGO) and MGO-derived advanced glycation end products (AGEs) seems to be implicated in endothelial dysfunction and the development of diabetic vascular complications. MGO reacts with arginine residues in proteins to generate the major glycated adducts 5-hydro-5-methylimidazolone (MG-H1) and argpyrimidine (AP). We investigated whether the free forms of these adducts contribute to vascular cell dysfunction by inhibition of endothelial nitric oxide synthase (eNOS). MG-H1 and AP were synthesized and purified by reversed-phase chromatography, and the conversion of labeled L-arginine to L-citrulline was used to monitor eNOS activity. In contrast to the endogenous eNOS inhibitor asymmetric dimethylarginine (half maximal inhibitory concentration, approximately 5 μmol/L), pathophysiological concentrations of MGO and MG-H1 and AP did not inhibit eNOS activity. Although MGO-derived AGEs are implicated in the development of diabetic vascular complications, this study indicates that this is not mediated via direct inhibition of eNOS activity.

Key words: AGEs; methylglyoxal; glycation; asymmetric dimethylarginine; eNOS activity

Introduction

Vascular disease is the foremost complication of diabetes mellitus, accounting for much excess morbidity and mortality. Endothelial dysfunction is an early event in the pathogenesis of vascular disease, but the mechanism whereby diabetes mellitus leads to endothelial dysfunction is incompletely understood. The accumulation of advanced glycation end products (AGEs) and impaired nitric oxide (NO) metabolism are proposed mechanisms to explain the link between hyperglycemia and the development of diabetic vascular complications.[1] The increase in AGEs and the decrease in NO are closely related to the development and progression of diabetic complications.[2]

AGEs are formed from the reaction of reducing sugars with amino acid groups of proteins. AGEs can be formed extracellularly, but it is now apparent that intracellular sugars also participate in AGE formation. In endothelial cells, the highly reactive dicarbonyl compound methylglyoxal (MGO) has been identified as a major precursor in the formation of AGEs.[3] MGO is formed nonenzymatically by dephosphorylation of triose phosphates and is efficiently catabolized to D-lactate by the glyoxalase pathway. We recently demonstrated in human endothelial cells that hyperglycemia produced higher levels of MGO.[4] MGO reacts primarily with arginine residues to form argpyrimidine (AP) and 5-hydro-5-methylimidazolone (MG-H1), the latter representing a major AGE in mammalian cells.[5]

Disturbances in NO metabolism are probably of critical importance in the development of vascular complications in diabetes. The NO synthase reaction involves oxidation of a guanidino nitrogen of L-arginine to NO.[6] The synthesis of NO is selectively inhibited by guanidino-substituted analogues of arginine, including endogenously produced asymmetric dimethylarginine (ADMA).[7] Because of the structural homology between NO inhibitors and MGO–arginine adducts (see FIG. 1), we hypothesized that these AGEs are endogenous inhibitors of endothelial nitric oxide synthase (eNOS) activity.

Address for correspondence: Casper G. Schalkwijk, Department of Internal Medicine, University Hospital Maastricht, P. Debyelaan 25, P.O. Box 5800, 6202 AZ Maastricht, the Netherlands. Voice: +31 43 3882186; fax: +31 43 3875006.

C.Schalkwijk@INTMED.unimaas.nl

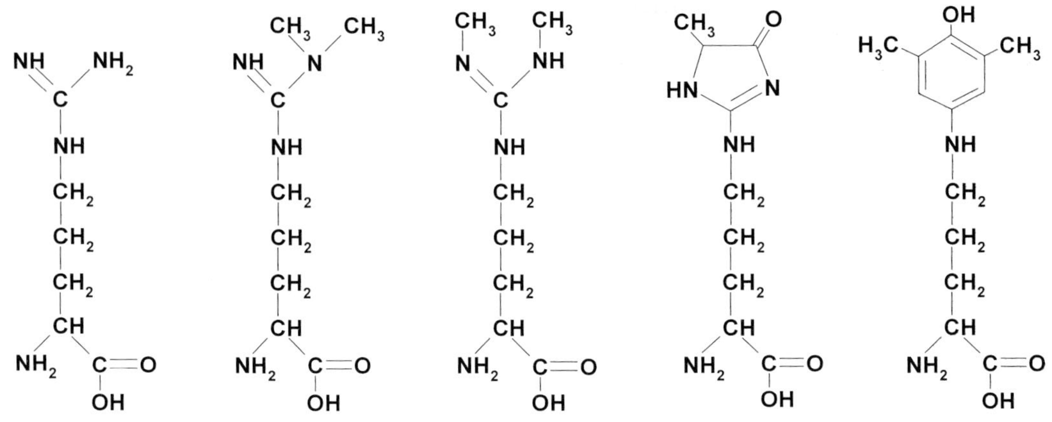

FIGURE 1. Structure of L-Arginine, asymmetric dimethylarginine (ADMA), symmetric dimethylarginine (SDMA), 5-hydro-5-methylimidazolone (MG-H1), and argpyrimidine (AP). ADMA is known as a nitric oxide synthase inhibitor.

Methods

Transduction of Human Umbilical Vein Endothelial Cells with eNOS

Human umbilical vein endothelial cells (HUVECs) were isolated and characterized as described previously.[4] For infection with eNOS, 5×10^5 cells, between passages 2–3, were seeded in six-well plates. Twenty-four hours later, cells were washed twice and subsequently infected for 2 h with increasing amounts of recombinant adenovirus. E1-deleted, replication-deficient, adenoviral vector (Ad5.eNOS) containing 1×10^{12} virus particles per mL were generated on PER.C6 cells as described earlier.[8]

Measurement of eNOS Activity

The kinetics of NO production were assessed in total cell homogenates from Ad5.eNOS transfected cells by measuring the conversion of L-[^{14}C]-arginine to L-[^{14}C]-citrulline in the presence of the necessary cofactors as previously reported.[8] Briefly, cell homogenates of transfected cells were incubated for 60 min at 37°C in 0.1 mL 25 mM Tris-HCl, pH 7.4, supplemented with 3 μmol/L tetrahydrobiopterine, 1 μmol/L flavin adenine dinucleotide, 100 nM calmoduline, 1 μmol/L flavin mono-nucleotide, 1 mM NADPH, and 0.1 μCi/ml L-[^{14}C]-arginine and non-labeled L-arginine. The final concentration of labeled and nonlabeled L-arginine was 10 μmol/L. The reaction was started by the addition of 75 μmol/L calcium chloride.

Preparation of Methylglyoxal Adducts

AP and MG-H1 were synthesized as described earlier.[5,9]

Results

Under our experimental conditions, the citrulline assay was not sensitive enough for NO measurements in homogenates of nontransfected endothelial cells. We therefore transfected cells with eNOS. Western blot analysis demonstrated a virus-concentration-dependent increase in expression of eNOS protein in Ad5.eNOS-transduced HUVEC that correlated well with activity and reached a plateau value of ≈40 pmol citrulline min^{-1} mg protein^{-1} after exposure to a virus concentration of 2×10^8 pfu/mL (data not shown). In Ad5.empty-infected cells, NOS activity remained below the detection limit of the assay. The NO production of Ad5.eNOS-transduced cells was calcium-dependent and was inhibited by omission of the cofactor β-NADPH and by L-N^G-nitroarginine N^G-nitro-L-arginine. These results demonstrate that transduced eNOS is functionally active and can be used to study the effects of inhibitors on eNOS activity.

The ability of free MGO and MGO adducts to inhibit eNOS was investigated using the citrulline formation assay under similar conditions. As shown in FIGURE 2, incubation of eNOS with increasing concentrations of MG-H1 led to a slight decrease in eNOS activity at the highest concentration. Also, AP had a mild effect on eNOS activity, but this effect was not dose dependent. Preincubations of eNOS with MGO or MGO adducts did not change the activity of eNOS. Under the same experimental conditions, ADMA, but not symmetric dimethylarginine (SDMA), exhibited a strong concentration-dependent inhibition of eNOS with a half maximal inhibitory concentration (IC$_{50}$) of approximately 5 μmol/L.

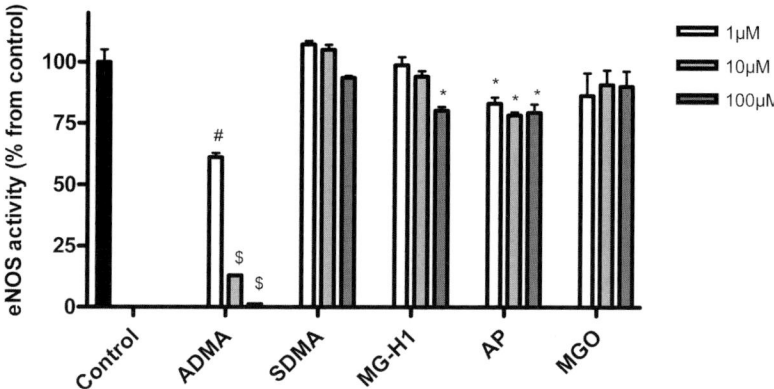

FIGURE 2. The effects of 1, 10, and 100 μmol/L of ADMA, SDMA, MG-H1, AP, and free methylglyoxal (MGO) on endothelial nitric oxide synthase (eNOS) activity. The symbols *, #, and $ indicate $P < 0.05$, $P < 0.005$, and $P < 0.0005$, respectively. Data are expressed as mean ± standard error of the mean ($n = 3$).

Discussion

MGO, a reactive dicarbonyl metabolite of glucose, has received considerable attention as the most reactive AGE precursor in endothelial cells.[3] Although MGO is involved in a variety of detrimental processes, we found that pathophysiological concentrations of MGO or MGO-arginine adducts do not directly inhibit eNOS activity.

The plasma concentrations of free MGO adducts in the healthy population are around 0.1 μmol/L for MG-H1 and 0.05 μmol/L for AP. In patients with renal failure these adducts can accumulate to 5 μmol/L and 0.1 μmol/L respectively.[10] In our in vitro system we tested 1–100 μmol/L of MG-H1 and AP, but we only saw a mild inhibition of eNOS activity with a 20-fold higher concentration (100 μmol/L MG-H1 and >1 μmol/L AP) than pathophysiological concentrations.

Plasma MGO levels are estimated to be around 0.5 μmol/L in healthy individuals and can increase to approximately 1 μmol/L in diabetic individuals.[11] It should be taken into account that intracellular levels of MGO are probably higher but are difficult to measure. Nevertheless, in our experiments, even the highest concentration of MGO (100 μmol/L) did not show any effect on eNOS activity.

The physiological range of ADMA concentrations in plasma has been determined to be 0.39–0.63 μmol/L and in several pathophysiological situations the ADMA levels can increase two- to sevenfold.[12,13] In our study, the IC_{50} of ADMA was approximately 5 μmol/L, which is in the same range as previously reported.[14]

MGO and AGEs are also a source of oxidative stress, which can lead to ADMA accumulation by inhibition of dimethylaminohydrolases (DDAH), enzymes that selectively degrade ADMA. It is known that hyperglycemia leads to lower DDAH activity and thereby to ADMA accumulation.[15] Indeed, ADMA was shown to be elevated in streptozotocin-induced diabetic rats and could be inhibited by glycemic control.[16] Furthermore, Yin et al. reported that glycated albumin is able to inhibit DDAH activity, thereby impairing NO synthesis. This impairment could be restored by pravastatin and the antioxidant PDTC.[17] In line with this, we cannot exclude that the MGO adducts indirectly cause inhibition of eNOS.

We demonstrated that pathophysiological levels of free MGO and the MGO adduct MG-H1 and AP do not directly inhibit eNOS activity like ADMA does. We conclude that MGO and MGO-derived AGEs are implicated in the development of diabetic vascular complications, but that this is not mediated via direct inhibition of eNOS activity.

Conflict of Interest

The authors declare no conflicts of interest.

References

1. BUCALA, R., K.J. TRACEY & A. CERAMI. 1991. Advanced glycosylation products quench nitric oxide and mediate defective endothelium-dependent vasodilatation in experimental diabetes. J. Clin. Invest. **87:** 432–438.
2. TAN, K.C. et al. 2002. Advanced glycation end products and endothelial dysfunction in type 2 diabetes. Diabetes Care **25:** 1055–1059.
3. SHINOHARA, M. et al. 1998. Overexpression of glyoxalase-I in bovine endothelial cells inhibits intracellular advanced

glycation endproduct formation and prevents hyperglycemia-induced increases in macromolecular endocytosis. J. Clin. Invest. **101:** 1142–1147.
4. SCHALKWIJK, C.G. *et al.* 2006. Heat-shock protein 27 is a major methylglyoxal-modified protein in endothelial cells. FEBS Lett. **580:** 1565-1570.
5. AHMED, N. *et al.* 2002. Assay of advanced glycation endproducts (AGEs): surveying AGEs by chromatographic assay with derivatization by 6-aminoquinolyl-N-hydroxysuccinimidyl-carbamate and application to Nepsilon-carboxymethyl-lysine- and Nepsilon-(1-carboxyethyl)lysine-modified albumin. Biochem. J. **364:** 1–14.
6. HUANG, P.L. *et al.* 1995. Hypertension in mice lacking the gene for endothelial nitric oxide synthase. Nature **377:** 239–242.
7. VALLANCE, P. *et al.* 1992. Accumulation of an endogenous inhibitor of nitric oxide synthesis in chronic renal failure. Lancet **339:** 572–575.
8. HAVENGA, M.J. *et al.* 2001. Simultaneous detection of NOS-3 protein expression and nitric oxide production using a flow cytometer. Anal. Biochem. **290:** 283–291.
9. OYA, T. *et al.* 1999. Methylglyoxal modification of protein. Chemical and immunochemical characterization of methylglyoxal-arginine adducts. J. Biol. Chem. **274:** 18492–18502.
10. THORNALLEY, P.J. *et al.* 2003. Quantitative screening of advanced glycation endproducts in cellular and extracellular proteins by tandem mass spectrometry. Biochem. J. **375:** 581–592.
11. LAPOLLA, A. *et al.* 2003. Glyoxal and methylglyoxal levels in diabetic patients: quantitative determination by a new GC/MS method. Clin. Chem. Lab. Med. **41:** 1166–1173.
12. TEERLINK, T. 2007. HPLC analysis of ADMA and other methylated L-arginine analogs in biological fluids. J. Chromatogr. B Analyt. Technol. Biomed. Life Sci. **851:** 21–29.
13. HOROWITZ, J.D. & T. HERESZTYN. 2007. An overview of plasma concentrations of asymmetric dimethylarginine (ADMA) in health and disease and in clinical studies: methodological considerations. J. Chromatogr. B Analyt. Technol. Biomed Life Sci. **851:** 42–50.
14. TSIKAS, D. *et al.* 2000. Endogenous nitric oxide synthase inhibitors are responsible for the L-arginine paradox. FEBS Lett. **478:** 1–3.
15. NOYMAN, I. *et al.* 2002. Hyperglycemia reduces nitric oxide synthase and glycogen synthase activity in endothelial cells. Nitric Oxide **7:** 187–193.
16. XIONG, Y. *et al.* 2003. Elevated levels of the serum endogenous inhibitor of nitric oxide synthase and metabolic control in rats with streptozotocin-induced diabetes. J. Cardiovasc. Pharmacol. **42:** 191–196.
17. YIN, Q.F. & Y. XIONG. 2005. Pravastatin restores DDAH activity and endothelium-dependent relaxation of rat aorta after exposure to glycated protein. J. Cardiovasc. Pharmacol. **45:** 525–532.

Kinetic Study of the Reaction of Glycolaldehyde with Two Glycation Target Models

MIQUEL ADROVER, BARTOLOMÉ VILANOVA, FRANCISCO MUÑOZ, AND JOSEFA DONOSO

Institut Universitari d'Investigació en Ciències de la Salut, Departament de Química, Universitat de les Illes Balears, Palma de Mallorca, Spain

We have studied the reactivity of glycolaldehyde (GLA) with N-acetyl-Cys and N-acetyl-Phe-Lys at physiological conditions of pH and temperature. The reaction between the N-Ac-Phe-Lys and GLA was studied in the presence of $NaCNBH_3$ and then by using high-performance liquid chromatography (HPLC)-UV/Vis. The reaction between N-Ac-Cys and GLA was followed by stopped-flow spectroscopy with UV/Vis detection. Both the reduced Schiff base and thiohemiacetal were identified by ^1H-NMR and HPLC-mass spectrometry detection. The kinetic rate constant for the thiohemiacetal formation is four orders of magnitude higher than that for the Schiff base formation. This result suggests that the thiol group represents the most important target in protein glycation.

Key words: kinetics; protein glycation; glycolaldehyde; side-chain protein models

Introduction

Protein glycation is a modification of proteins that occurs *in vivo* by a chemical reaction between sugars and amino groups of protein side chains. After the initial formation of a Schiff base (SB) between the carbonyl and the amine moiety, the aldimine rearranges to a more stable ketoamine structure, the Amadori compound, which represents a crucial key in the formation of advanced glycation end products (AGEs).[1] It has been shown that the thiol group in Cys residues can also represent a target to the protein glycation front of small carbonyl compounds, especially in intracellular space.[2] The chemical addition of the thiol group to a carbonyl compound forms a thiohemiacetal (TH) adduct, which also evolves to AGEs formation. The emphasis of research in recent years has been on AGEs,[2] but early glycation products can have a damaging effect and may be more important in damaging nonstructural proteins.[3]

Glycolaldehyde (GLA) is a potent glycation agent[4–8] and is formed from SB degradation (Namiki pathway)[4] and also from L-serine by the myeloperoxidase-hydrogen peroxidase-chloride system.[7] GLA is responsible for protein modification by N^ε-(carboxymethyl)lysine (CML) formation[4,5] and protein cross-linking.[6] Recently, Thorpe and Baynes have proposed that GLA contributes to protein glycation by Cys residues modification, leading a new class of intracellular AGEs (e.g., S-(carboxymethyl)cysteine [CMC]).[2]

In this study we examined the SB formation between GLA and an amino group dipeptide model by HPLC. Additionally, we investigated the TH formation between GLA and N-Ac-Cys by stopped-flow spectroscopy. The reaction products were identified by HPLC-mass spectrometry detection (MSD) and NMR spectroscopy, and the kinetic rate constants for both reactions were determined.

Materials and Methods

Materials

GLA was purchased from Sigma-Aldrich (Madrid, Spain); Ac-Phe-Lys was purchased from Bachem Inc. (Weilam Rhein, Germany); sodium cyanoborohydride and Ac-Cys were purchased from Acros Organics (Geel, Belgium). All reagents were used as received. The buffering material was reagent grade, and mili-Q water was used throughout.

Reaction Mixture Analysis

The reaction between GLA and Ac-Phe-Lys was monitored on an HPLC-Shimadzu-LC 10AT equipped with a Shimadzu SPD-M20A UV/Vis photodiode array detector (Shimadzu Deutschland GmBH, Duisburg, Germany) reading at 230 nm. The column was a Tracer Excel 120 ODSB model (Teknokroma, Barcelona, Spain) (25 × 0.46 cm, 5 μm).

Address for correspondence: Bartolomé Vilanova, Institut Universitari d'Investigació en Ciències de la Salut, Departament de Química, Universitat de les Illes Balears, Cra. Valldemossa km 7.5, E-07122 Palma de Mallorca, Spain. Voice: +34-971-173005; fax +34-971-173426.
bartomeu.vilanova@uib.es

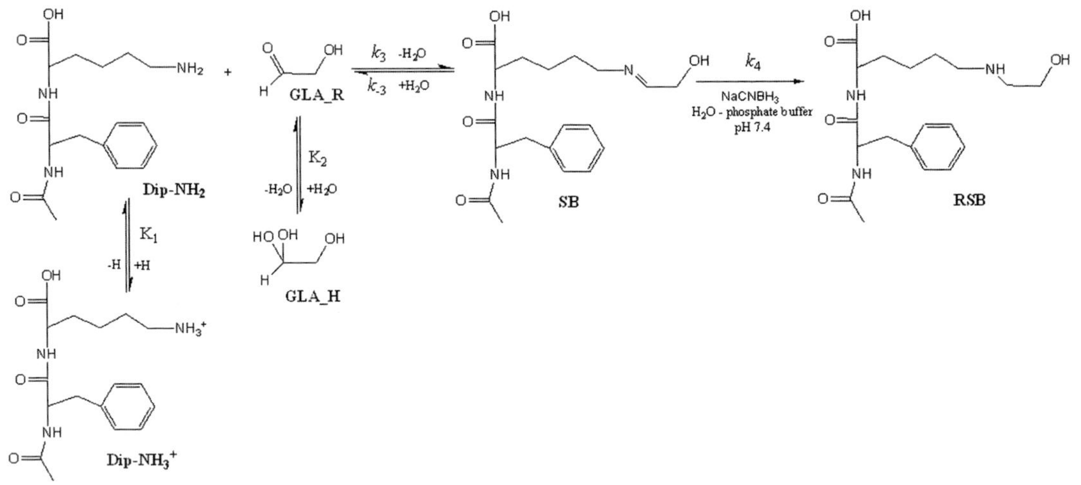

SCHEME 1. Mechanism of the secondary amine formation by reduction of the Schiff Base formed between glycolaldehyde (GLA) and Ac-Phe-Lys under physiological conditions in the presence of NaCNBH$_3$.

An isocratic method (1 mL/min) of water-50 mM potassium phosphate (pH 6.0)/MeCN (93:7) was used to separate the compounds. The reaction mixture (0.1 mM in Ac-Phe-Lys and 3 mM in GLA) was prepared in phosphate buffer (0.5 M) at pH 7.4 and kept at 37°C. NaCNBH$_3$ (1 mM) was added to the reaction mixture. This reagent selectively reduces imine groups at physiological pH, leaving intact carbonyl groups.[9] This methodology allows the scavenging of SBs and determines their formation rate constant.[10]

Reaction kinetics of interaction between Ac-Cys and GLA was conducted with a Bio-Logic SFM-20 mixer (Bio-Logic SA, Claix, France) equipped with a TC-100/10T quartz cell of 1 cm path length and coupled to a J&M Tidas16 256 diode array detector (Bio-Logic SA) for multi-wavelength data collection at 37°C. A 75-W Xe lamp and a coupled monochromator were used to irradiate samples with a selected wavelength in order to avoid photo-induced processes on GLA. Solutions of 8 mM Ac-Cys and 100 mM GLA were prepared in phosphate buffer (0.5 M) at pH 7.4 and kept at 37°C. These solutions were mixed at different rates in the stopped-flow spectrometer (V$_{Ac-Cys}$:V$_{GLA}$, 1:1, 1:2, and 1:2.5), and their absorbance increase at 286 nm were analyzed.

Reaction Products Characterization

NMR spectra were recorded on a Bruker AMX-300 spectrometer (Bruker Biospin Corporation, Billerica, MA). The solutions in heavy water (D$_2$O) were stabilized at pD 7.4 (pD = −log [D$^+$]) by using a 0.5 M phosphate buffer.

Mass analyses were performed on an Agilent 1110 Series HPLC-MSD instrument (Agilent Technologies, Palo Alto, CA). The reaction mixture of Ac-Phe-Lys and GLA was injected through a Tracer Excel 120 ODSB column using the same chromatographic conditions previously described, whereas for Ac-Cys and GLA, a columnless flow injection analysis was used. Mass spectral detection of the compounds was done by using an electrospray ionization interface and a quadrupole mass analyzer. The mobile phase was 5 mM ammonium acetate at pH 6.0 (NH$_4$OAc initiates the ionization for mass spectrometry [MS] detection). The mobile phase was nebulized into an electrospray mass analyzer by using gaseous nitrogen at 350°C at a flow rate of 10 mL/min. The detector was used to count positive ions in the scan mode over the m/z range 100–800, except in the Ac-Cys study where the detector was used to count negative ions in the same range. A nebulization pressure of 415.6 kPa, a fragmentor voltage of 70 V, and a capillary voltage of 3000 V were used.

Kinetic Analysis for the Reaction between Ac-Phe-Lys and GLA

SCHEME 1 shows the kinetic mechanism for the formation of the SB and subsequent reduction by sodium cyanoborohydride. The rate of disappearance of Ac-Phe-Lys from the reaction medium conformed to the following law:

$$\frac{-d[\text{Dip-NH}_2]}{dt} = k_3 [\text{GLA\_R}] \times [\text{Dip-NH}_2] - k_{-3}[\text{SB}] \quad (1)$$

Application of the steady-state approximation to the SB yields:

$$\frac{-d\,[\text{SB}]}{dt} = k_3[\text{GLA\_R}]\,[\text{Dip-NH}_2]$$
$$-k_{-3}\,[\text{SB}] - k_4\,[\text{SB}] = 0 \qquad (2)$$

$$[\text{SB}] = \frac{k_3[\text{GLA\_R}]\,[\text{Dip-NH}_2]}{(k_{-3} + k_4)} \qquad (3)$$

Substituting EQUATION 3 into EQUATION 1 yields:

$$\frac{-d\,[\text{Dip-NH}_2]}{dt} = \left(\frac{k_3 k_4}{k_{-3} + k_4}\right)$$
$$\times [\text{GLA\_R}][\text{Dip-NH}_2] \qquad (4)$$

Taking into account the approximation that $k_4 \ggg k_{-3}$, then,

$$\frac{-d\,[\text{Dip-NH}_2]}{dt} = k_3[\text{GLA\_R}]\,[\text{Dip-NH}_2] \qquad (5)$$

GLA in solution occurs as two different forms with the carbonyl group in free (GLA_R) and hydrated (GLA_H) form. The dehydration equilibrium constant is given by

$$K_2 = \frac{[\text{GLA\_R}]}{[\text{GLA\_H}]} \qquad (6)$$

Based on the applicable mass balance to GLA,

$$[\text{GLA}]_T = [\text{GLA\_R}] + [\text{GLA\_H}] \qquad (7)$$

it follows that

$$[\text{GLA\_R}] = \frac{K_2\,[\text{GLA}]_T}{1 + K_2} \qquad (8)$$

At physiological pH, the ε-amino group in the Ac-Phe-Lys dipeptide shows ionization equilibrium for such a group, being its equilibrium constant:

$$K_1 = \frac{[\text{Dip-NH}_2][\text{H}_3\text{O}^+]}{[\text{Dip-NH}_3^+]} \qquad (9)$$

Taking into account the mass balance,

$$[\text{Ac-Phe-Lys}]_T = [\text{Dip-NH}_2]$$
$$+ [\text{Dip-NH}_3^+] \qquad (10)$$

EQUATION 9 can be rewritten as:

$$[\text{Dip-NH}_2] = \frac{K_1[\text{Ac-Phe-Lys}]_T}{[\text{H}_3\text{O}^+] + K_1} \qquad (11)$$

Substitution of EQUATIONS 8 and 11 into EQUATION 5 yields:

$$\frac{-d\,[\text{Dip-NH}_2]}{dt} = \frac{-d\,[\text{Ac-Phe} - \text{Lys}]_T}{dt}$$
$$= k_3 \frac{K_1 K_2 [\text{GLA}]_T [\text{Ac-Phe-Lys}]_T}{[\text{H}_3\text{O}^+] + K_1 + [\text{H}_3\text{O}^+]K_2 + K_1 K_2} \qquad (12)$$

By collecting the constant terms in EQUATION 12, one can define a new constant such that,

$$k_{\text{obs}} = k_3 \frac{K_1 K_2 [\text{GLA}]_T}{[\text{H}_3\text{O}^+] + K_1 + [\text{H}_3\text{O}^+]K_2 + K_1 K_2} \qquad (13)$$

Therefore, EQUATION 12 can be rewritten as follows:

$$\frac{-d\,[\text{Ac-Phe-Lys}]_T}{dt}$$
$$= k_{\text{obs}}\,[\text{Ac-Phe-Lys}]_T \qquad (14)$$

which can be integrated to:

$$\ln \frac{[\text{Ac-Phe-Lys}]_T}{[\text{Ac-Phe-Lys}]_{T0}} = -k_{\text{obs}} t \qquad (15)$$

The k_{obs} value can be calculated by fitting the experimental data to EQUATION 15, and the knowledge of GLA concentration, pH value, and equilibrium rate constants for the reagents allowed us to determine the microscopic kinetic rate constant for the SB formation (k_3).

Kinetic Analysis for the Reaction between Ac-Cys and GLA

The microscopic kinetic rate constant for the reaction between the reactive tautomers of GLA and Ac-Cys (k_6) was determined by fitting the temporal variation of the absorbance at 286 nm to SCHEME 2. The fitting process was performed by using the DynaFit software package (BioKin, Pullman, WA). The molar absorption coefficient for each tautomer at 286 nm was determined and included in the input data ($\varepsilon_{\text{Ac-Cys-SH}}^{286nm} = 3.8$ M^{-1}·cm^{-1}; $\varepsilon_{\text{Ac-Cys-S}^-}^{286nm} = 9.8$ M^{-1}·cm^{-1}; $\varepsilon_{\text{GLA\_H}}^{286nm} = 1.4$ M^{-1}·cm^{-1}). The absorbance of the GLA_R at 286 nm was negligible because of its low concentration at physiological

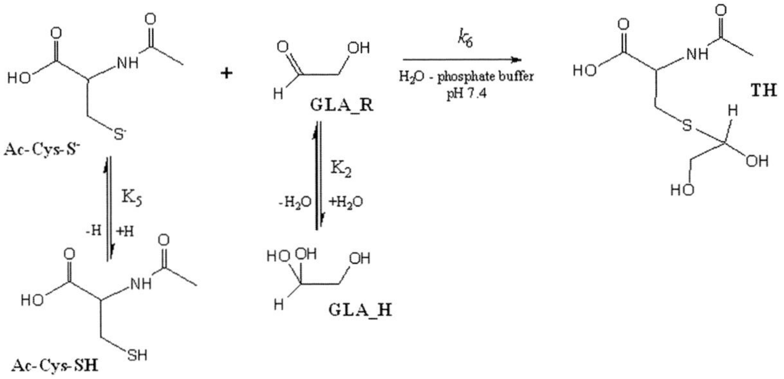

SCHEME 2. Mechanism of thiohemiacetal formation between GLA and Ac-Cys at physiological conditions.

conditions.[11] Also, k_{obs} can be obtained by mathematical treatment from the kinetic mechanism. The kinetic equation for the TH formation can be written as:

$$\frac{-d\left[\text{Ac-Cys}-S^-\right]}{dt} = k_6\left[\text{GLA\_R}\right]\left[\text{Ac-Cys}-S^-\right] \quad (16)$$

Taking into account the equilibrium constants and the material balance for the tautomers of the reagents, EQUATION 16 is transformed into EQUATION 17:

$$\frac{-d\left[\text{Ac-Cys}-S^-\right]}{dt} = \frac{k_6 K_2 K_5 [\text{GLA}]_T [\text{Ac-Cys}]_T}{[\text{H}_3\text{O}^+] + K_5 + [\text{H}_3\text{O}^+]K_2 + K_2 K_5} \quad (17)$$

where the constant terms can be grouped as follows:

$$k_{obs} = \frac{k_6 K_2 K_5 [\text{GLA}]_T}{[\text{H}_3\text{O}^+] + K_5 + [\text{H}_3\text{O}^+]K_2 + K_2 K_5} \quad (18)$$

In order to compare the magnitude of the k_{obs} with the previously k_{obs} values obtained for the reaction between Ac-Phe-Lys and GLA, we have defined k_{obs}' because the concentration of GLA used in the reaction mixtures was different:

$$k_{obs}' = \frac{k_{obs}}{[\text{GLA}]_T} \quad (19)$$

Results and Discussion

FIGURE 1 shows the temporal variation of chromatograms corresponding to the reaction between Ac-Phe-Lys and GLA in the presence of NaCNBH$_3$ at physiological conditions of pH and temperature. The initial chromatogram exhibited a single signal corresponding to Ac-Phe-Lys (t_R 8 min) that decreased with time as a new major signal at t_R 7.3 min appeared (m/z 380.4 according to the signal [M+H]$^+$ of the reduced SB). The value of m/z for the reduced SB and the ^{1}H-NMR experiments (data not shown: a new triplet signal appeared for the reaction mixture in D$_2$O corresponding the -CH$_2$- of the reduced SB) confirm that the amino group of the dipeptide reacts with the carbonyl group in GLA to form an SB that is subsequently reduced.

The rate constants k_{obs} (0.33 ± 0.03 h^{-1}) and k_{obs}' (110 ± 5 M^{-1} h^{-1}) were determined by EQUATION 15 and EQUATION 19, respectively. By using EQUATION 13, $k_3 = (7.2 \pm 0.8) \cdot 10^5$ M^{-1} h^{-1} was determined. The equilibrium constant for Ac-Phe-Lys was obtained by titration of the dipeptide solution (K$_1$ = [7.9 ± 1]·10^{-11} mol), and K$_2$ = 0.086 was obtained from the bibliography.[11]

FIGURE 2 shows the temporal increase in the absorbance for the reaction between GLA and Ac-Cys. This increase was assigned to the consequent TH formation, according to previous work reported by other authors for a similar reaction.[12] This hypothesis was confirmed by ^{1}H-NMR and HPLC-MSD experiments. The mass spectrum obtained for the reaction mixture showed a major signal of m/z 222.1 according to the signal [M-H]$^-$ for the TH. In addition, these experiments showed that the reaction was completely shifted to TH and allowed us to dismiss the TH degradation to reagents formation in SCHEME 2. The fitting process of the temporal variation of the absorbance to SCHEME 2 yields the kinetic rate constant for the TH formation $k_6 = (4.6 \pm 0.3) \cdot 10^9$ M^{-1} h^{-1} and its molar absorption coefficient ($\varepsilon_{TH}^{286\,nm} = 20 \pm$

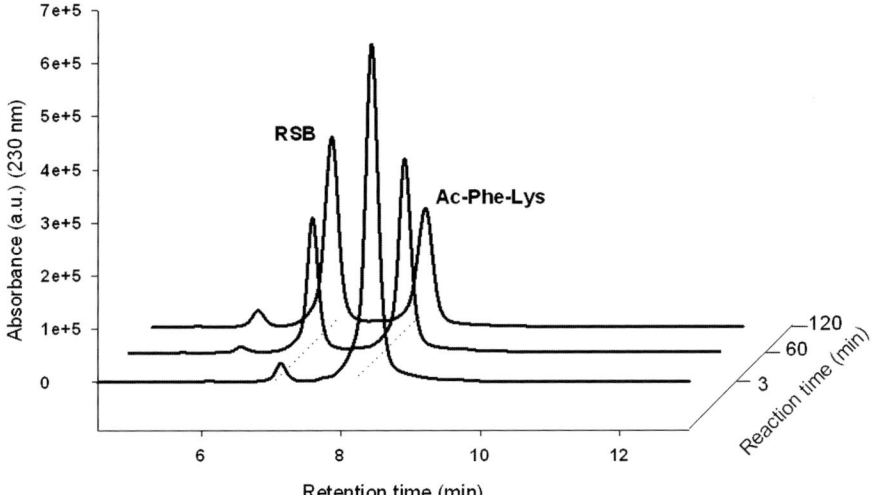

FIGURE 1. Time-dependent high-performance liquid chromatography chromatograms for the reaction between 0.1 mM Ac-Phe-Lys and 3 mM GLA in the presence of NaCNBH$_3$ in a phosphate-buffered medium at pH 7.4 and 37°C; UV/Vis detection at 230 nm.

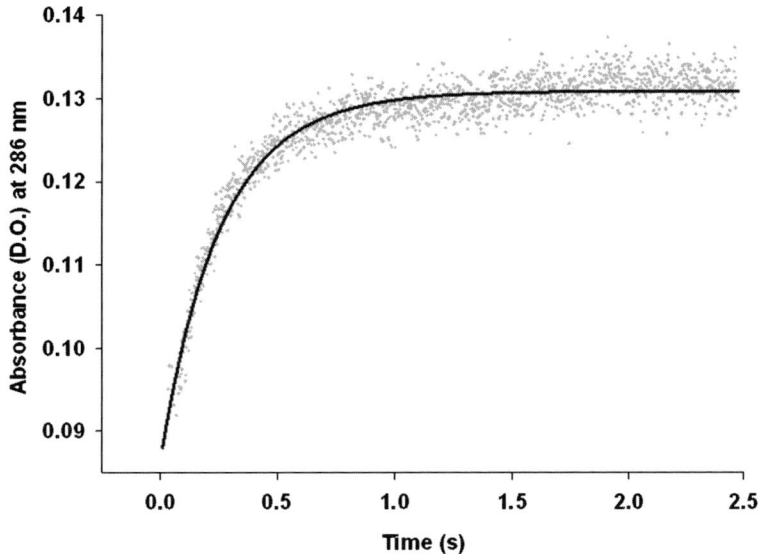

FIGURE 2. Time course of the absorbance variation in the reaction mixture at 286 nm for the reaction between 4 mM Ac-Cys and 50 mM GLA in a phosphate-buffered medium at pH 7.4 and 37°C.

3 M$^{-1} \cdot$cm^{-1}). By using EQUATION 18 and EQUATION 19 $k_{\text{obs}}' = (6.7 \pm 1.1) \cdot 10^6$ M^{-1} h^{-1} was obtained. The equilibrium constant for the Ac-Cys was obtained from titration of the amino acid solution (K$_5 = (8.5 \pm 1) \cdot 10^{-10}$ mol).

The microscopic rate constant obtained for the condensation between the thiolate group of Ac-Cys and the carbonyl group of GLA (k_6) is four orders of magnitude higher than the microscopic kinetic rate constant obtained when the unprotonated amino group of the dipeptide acts as a nucleophilic agent (k_3). This difference is essentially a result of the negative charge on the sulfur atom in thiolate in contrast to the neutral status of the amino group of the dipeptide. The second-order kinetic rate constant (k_{obs}') for the Ac-Cys is still higher than that determined for the dipeptide. This difference is essentially because of two factors: the most nucleophilic character of the thiolate group in front to the amino group of the dipeptide, and the highest percentage of the reactive tautomer of Ac-Cys in front to the Ac-Phe-Lys at physiological pH.

Conclusions

GLA, a carbonyl compound precursor of AGEs, especially in hyperglycemic people, reacts with both amino groups and thiol groups from side-chain protein models. The reaction between GLA and the amino residue of Ac-Phe-Lys yields an SB, whereas the thiol group of Ac-Cys reacts with GLA to form a TH. The kinetic rate constant for the TH formation is four orders of magnitude higher than for the SB formation. This result supports the previous hypothesis[2,13] on the strong role of Cys residues as targets in nonenzymatic protein glycation.

Acknowledgments

This work was made possible by a grant from the Spanish Government (DGICYT CTQ 2005-00250) and from the Balear Government (PROGECIB-28A).

Conflict of Interest

The authors declare no conflicts of interest.

References

1. ULRICH, P. & A. CERAMI. 2001. Protein glycation, diabetes, and aging. Recent. Prog. Horm. Res. **56:** 1–22.
2. THORPE, S.R. & J.W. BAYNES. 2003. Maillard reaction products in tissue proteins: new products and new perspectives. Amino Acids **25:** 275–281.
3. COHEN, M.P. & F.N. ZIYADEH. 1996. Role of Amadori-modified nonenzymatically glycated serum proteins in the pathogenesis of diabetic nephropathy. J. Am. Soc. Nephrol. **7:** 183–190.
4. GLOMB, M.A. & V.M. MONNIER. 1995. Mechanism of protein modification by glyoxal and glycolaldehyde, reactive intermediates of the Maillard reaction. J. Biol. Chem. **270:** 10017–10026.
5. NAGAI, R. et al. 2000. Glycolaldehyde, a reactive intermediate for advanced glycation end products, plays an important role in the generation of an active ligand for the macrophage scavenger receptor. Diabetes **49:** 1714–1723.
6. Acharya, A.S. & J.M. MANNING. 1983. Reaction of glycolaldehyde with proteins: Latent crosslinking potential of α-hydroxyaldehydes. Proc. Natl. Acad. Sci. USA **80:** 3590–3594.
7. ANDERSON, M.M. et al. 1999. The myeloperoxidase system of human phagocytes generates N^ε-(carboxymethyl)lysine on protein: a mechanism for producing advanced glycation end products at sites of inflammation. J. Clin. Invest. **104:** 103–113.
8. UNNO, Y. et al. 2005. Glycolaldehyde-modified bovine serum albumin downregulates leptin expression in mouse adipocytes via CD36-mediated pathway. Ann. N.Y. Acad. Sci. **1043:** 696–701.
9. BORCH, R.F. et al. 1971. The cyanohydridoborate anion as a selective reducing agent. J. Am. Chem. Soc. **93:** 2897–2904.
10. BUNN, H.F. & P.J. HIGGINS. 1981. Reaction of monosaccharides with proteins: possible evolutionary significance. Science **213:** 222–224.
11. BEEBY, A. et al. 1987. Photochemistry and photophysics of glycolaldehyde in solution. J. Am. Chem. Soc. **109:** 857–861.
12. Lo, T.W.C. et al. 1994. Binding and modification of proteins by methylglyoxal under physiological conditions. J. Biol. Chem. **269:** 32299–32305.
13. ZENG, J. & M.J. DAVIES. 2006. Protein and low molecular mass thiols as targets and inhibitors of glycation reactions. Chem. Res. Toxicol. **19:** 1658–1676.

Origin and Yields of Acetic Acid in Pentose-based Maillard Reaction Systems

TOMAS DAVIDEK,[a] ELISABETH GOUÉZEC,[a] STÉPHANIE DEVAUD,[b] AND IMRE BLANK[a]

[a]*Nestlé Product Technology Centre Orbe, Nestec LTD., Orbe, Switzerland*
[b]*Nestlé Research Center, Lausanne, Switzerland*

The formation of acetic acid from pentoses was studied in aqueous buffered systems (90–120°C, pH 6.0–8.0) containing equimolar concentrations of ^{13}C-labeled xylose and glycine. Acetic acid was quantified by gas chromatography–mass spectroscopy using an isotope dilution assay. Acetic acid was mainly formed from the C-1/C-2 carbon atoms of xylose (77–87%), while small amounts were also formed from the C-4/C-5 atoms of the pentose sugar (9–15%). Temperature and pH had only a small effect on the relative contribution of the sugar carbon atoms to acetic acid. These results support β-dicarbonyl cleavage of 1-deoxypento-2,4-diulose as a major pathway leading to acetic acid in pentose-based Maillard reaction systems under food processing conditions. Acetic acid was confirmed as a major degradation product of pentoses at the early stage of the Maillard reaction, yielding 16 mol% and 28 mol% at pH 6.0 and pH 8.0, respectively.

Key words: Maillard reaction; acetic acid; D-[^{13}C]-xyloses; mechanism; β-dicarbonyl cleavage

Introduction

Acetic acid is a well-known thermal degradation product of saccharides. Its formation from hexoses under alkaline conditions was first reported in 1926.[1] Recent investigations have shown that acetic acid is mainly formed at the early stage of the Maillard reaction cascade and that it is the major reaction product under neutral and alkaline conditions.[2–5] It has been demonstrated that in hexose-based systems, acetic acid is almost exclusively formed via hydrolytic β-dicarbonyl cleavage.[5] The key intermediate, 1-deoxy-2,3-hexodiulose, undergoes successive β-dicarbonyl cleavages forming acetic acid from all six carbon atoms. The α-hydroxycarbonyl compounds that are formed as counterpart (e.g., tetroses, 2-hydroxy-3-oxobutanal or glycolaldehyde) are highly reactive and rapidly enter the reaction cascade.

Contrary to hexoses, the origin and the yield of acetic acid from pentoses as well as the importance of hydrolytic β-dicarbonyl cleavage for the formation of organic acid and α-hydroxycarbonyl compounds is not well understood. The β-dicarbonyl cleavage of pentoses was first proposed by Hayami to explain the formation of acetol and glycolic acid.[6] However, the formation of acetic acid was not studied. The aim of this present work is to clarify the origin of acetic acid in pentose-based Maillard reaction systems.

Materials and Methods

Glycine and D-(+)xylose were obtained from Fluka (Buchs, Switzerland); monosodium dihydrogenphosphate and disodium monohydrogenphosphate were obtained from Merck (Darmstadt, Germany); D-[1-^{13}C]-xylose (99%), D-[2-^{13}C]-xylose (99%) and D-[5-^{13}C]-xylose (99%) were obtained from Cambridge Isotope Laboratories (Andover, MA). [^{2}H$_3$]Acetic acid was obtained from Isotec (Miamisburg, OH). All the chemicals were of analytical grade.

Model Reactions

A solution of glycine (0.35 mmol) and D-xylose (0.35 mmol) in phosphate buffer (3.5 mL, 0.2 mol/L) was dispatched (0.4 mL/vial) into 1.5 mL screw-cap vials (Infochroma, Zug, Switzerland) and thermally treated in a silicone bath (90°C, 15 h or 120°C, 4 h). The experiments were performed at pH 6.0 and pH 8.0. Similar experiments were carried out using [1-^{13}C]-xylose, [2-^{13}C]-xylose, and [5-^{13}C]-xylose instead of unlabeled xylose.

Analysis of Acetic Acid

Acetic acid was quantified by solid-phase microextraction in combination with gas

Address for correspondence: Dr. Tomas Davidek, Nestle Product Technology Centre Orbe, CH-1350, Orbe, Switzerland. Voice: +41-24-4427342; fax: +41-24-4427444.
tomas.davidek@rdor.nestle.com

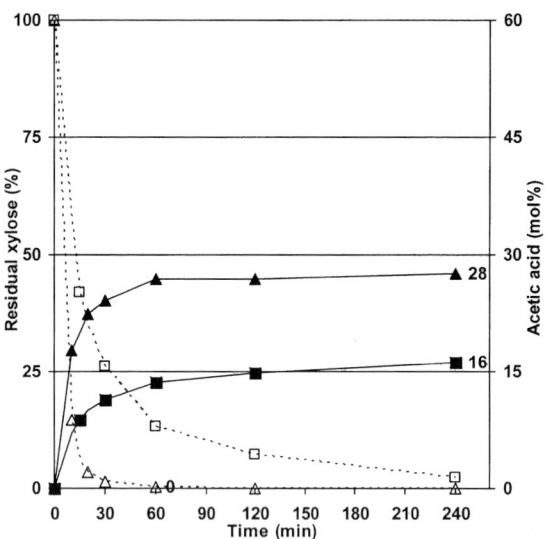

FIGURE 1. Formation of acetic acid (solid lines) and residual xylose level (dotted lines) in xylose–glycine reaction systems in phosphate buffer at 120 °C at pH 6.0 (squares) and pH 8.0 (triangles).

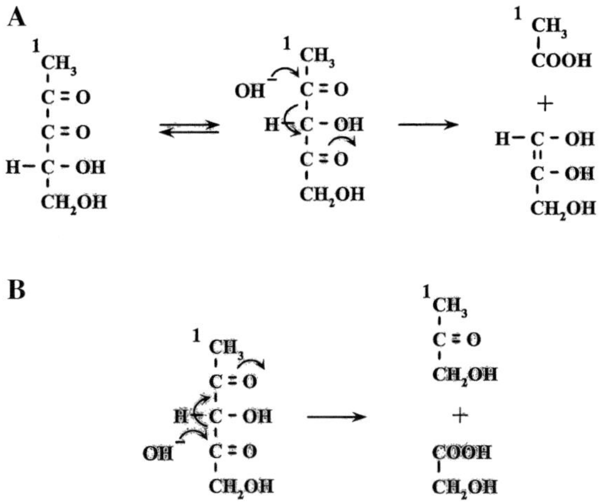

chromatography mass spectrometry (SPME–GC/MS) using [^{2}H$_3$]acetic acid as an internal standard as already described.[5]

Results and Discussion

The effect of pH, temperature, and reaction time on the origin and yields of acetic acid from xylose was studied in aqueous-buffered systems containing equimolar concentrations of glycine and unlabeled or ^{13}C-labeled xylose (0.1 mol/L). The initial experiments with unlabeled xylose at 120°C indicated that the level of acetic acid increased with increasing pH and reaction time (FIG. 1). Almost twice as much acetic acid was formed at pH 8.0 (28 mol%) compared to pH 6.0 (16 mol%) after 4 h of reaction. Acetic acid was mainly formed at the early stages of the reaction cascade as its formation almost stopped when no more xylose was available. These results are similar to those obtained in the glucose–glycine system and indicate that, as in pentose-based systems, acetic acid is most likely formed via the 2,3-enolization pathway by hydrolytic β-dicarbonyl cleavage of the corresponding 1-deoxy-2,4-diulose (FIG. 2A).[5]

To confirm this hypothesis, labeling experiments were performed using different ^{13}C-labeled xylose isotopomers reacted with unlabeled glycine (90°C and 120°C) at pH 6.0 and 8.0. The acetic acid was quantified by SPME-GC/MS using [^{2}H$_3$]acetic acid as the internal standard. As shown in FIGURE 3, singly labeled

FIGURE 2. Hydrolytic β-dicabonyl cleavage of 1-deoxypento-2,3-diulose yielding acetic acid (**A**) and glycolic acid (**B**).

acetic acid was formed from all three ^{13}C-labeled xylose isotopomers at 120°C at pH 6.0. However, the proportion of labeled acetic acid was much higher when [1-^{13}C]-xylose or [2-^{13}C]-xylose (85–89%) was used compared to [5-^{13}C]-xylose (8–11%). High and almost identical proportions of labeled acetic acid formed from [1-^{13}C]-xylose and [2-^{13}C]-xylose indicating that acetic acid is mainly formed by fragmentation between the C-2 and C-3 carbon atoms of the pentose sugar. These findings are in line with the mechanism shown in FIGURE 2A. In parallel to the formation of acetic acid, glycolic acid must be formed by the attack of the nucleophile HO$^-$ at the carbonyl in position C-4 instead of C-2, as proposed by Hayami (FIG. 2B).[6] In addition to this major pathway, small amounts of acetic acid are formed by fragmentation between the C-3 and C-4 carbon atoms of pentoses as indicated by the labeled acetic acid formed from [5-^{13}C]-xylose. The most probable mechanism should involve fragmentation of 2,3,4-pentanetrione formed by dehydratation of 1-deoxypento-2,3-diulose. The former compound was shown to form acetic acid after hydration of the middle carbonyl group.[7] Experiments confirming this hypothesis will be reported elsewhere. The contribution of the minor reaction pathway to acetic acid slightly increased with increasing reaction time. This effect of reaction time was more pronounced at pH 8.0 (contribution of C-4/C-5 atoms changed from 12% at the beginning of the reaction to 17% at the end).

The pH had relatively little effect on the origin of acetic acid from xylose (FIG. 4.). The relative contribution of C-1/C-2 carbon atoms of xylose was slightly decreased in favor of C-4/C-5 carbon atoms when pH

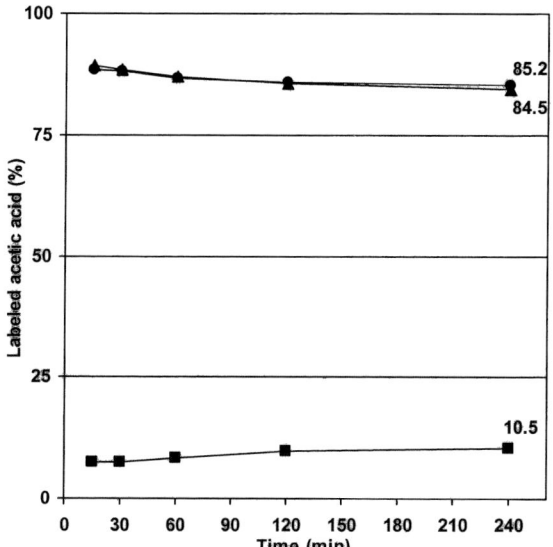

FIGURE 3. Percentage of singly labeled acetic acid arising from [1-^{13}C]-xylose (circles), [2-^{13}C]-xylose (triangles), and [5-^{13}C]-xylose (squares) in phosphate buffer at 120°C and pH 6.0.

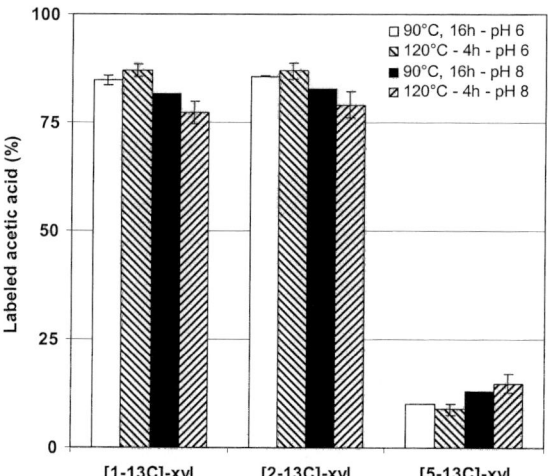

FIGURE 4. Percentage of singly labeled acetic acid arising from ^{13}C-labeled xylose isotopomers under different reaction conditions.

was increased from pH 6.0 to pH 8.0. On the other hand, the pH had a strong effect on the yield of acetic acid, which increased with increasing pH (FIG. 1). The temperature had only a negligible effect on the origin of acetic acid from xylose. The data shown in FIGURE 4 demonstrate that under cooking conditions the Maillard reaction systems containing pentoses similar to those containing hexoses yield acetic acid mainly from the C-1/C-2 carbon atoms of sugars. However, the relative contribution of C-1/C-2 carbon atoms of pentoses (77–85%) is more important than that of hexoses (67–71%).[5]

The results reported in this paper support the β-dicarbonyl cleavage of 1-deoxypento-2,4-diulose as a major pathway leading to acetic acid in pentose-based Maillard reaction systems under food processing conditions. Due to high amounts of acetic acid generated in both hexose- and pentose-based model systems, the hydrolytic β-dicarbonyl cleavage can be considered as a major Maillard reaction pathway under neutral and alkaline conditions.

Conflict of Interest

The authors declare no conflicts of interest.

References

1. EVANS, W.L. *et al.* 1926. Mechanism of carbohydrate oxidation. IV. The action of potassium hydroxide on D-glucose and D-galactose. J. Am. Chem. Soc. **48:** 2665–2677.
2. BRANDS, C.M.J. & M.A.J.S. VAN BOEKEL. 2001. Reaction of monosaccharides during heating of sugar-casein system: building of reaction network model. J. Agric. Food Chem. **49:** 4667–4675.
3. DAVIDEK, T. *et al.* 2002. Degradation of the Amadori compound N-(1-deoxy-D-fructos-1-yl)glycine in aqueous model systems. J. Agric. Food Chem. **50:** 5472–5479.
4. MARTINS, S.I.F.S. *et al.* 2003. Kinetics modeling of Amadori N-(1-deoxy-D-fructos-1-yl)-glycine degradation pathways. Part I – Reaction mechanism. Carbohydr. Res. **338:** 1651–1663.
5. DAVIDEK, T. *et al.* 2006. Sugar fragmentation in the Maillard reaction cascade: isotope labeling studies on the formation of acetic acid by hydrolytic β-dicarbonyl cleavage mechanism. J. Agric. Food Chem. **54:** 6667–6676.
6. HAYAMI, J. 1961. Studies on the chemical decomposition of simple sugars. XII. Mechanism of acetol formation. Bull. Chem. Soc. Japan **34:** 927–932.
7. DAO, L.H. *et al.* 1970. Alkalishe Spaltung und Benzillsäureumlagerung von 1,2,3-Triketonen. Helv. Chim. Acta **57:** 2215–2223.

The Peptide-catalyzed Maillard Reaction

Characterization of ^{13}C Reductones

LEIF ALEXANDER GARBE, ALEXANDER WÜRTZ, CHRISTIAN T. PIECHOTTA, AND ROLAND TRESSL

Technische Universität Berlin, Molecular Analysis, Berlin, Germany

The reaction pathways of amino acids and reducing sugars are now fully understood. The focus in the last few years, however, has turned to the reaction of peptides and proteins with reducing sugars. We have investigated the reaction of γ-aminobutanoic acid, the heptapeptide Nα-Acetyl-Lys-Lys-β-Ala-Lys-β-Ala-Lys-Gly, and the model protein β-casein in Maillard reactions with 1-^{13}C arabinose. Characterization of ^{13}C-labeled acetic acid and norfuraneol by gas chromatography–mass spectrometry and nuclear magnetic resonance revealed new formation pathways. The results demonstrate significant differences in the labeling pattern of the products depending on the amine used, indicating different formation pathways of acetic acid and norfuraneol.

Key words: norfuraneol; ^{13}C labeling; pathway; casein; peptide

Introduction

The basic Maillard reaction pathways within degradation of carbohydrates and amines were established many years ago.[1] Ever since Hodge[2] presented his fundamental scheme in 1953, only minor improvements in the understanding of the Maillard reaction product formation have been developed. When Ledl and Schleicher[3] published their comprehensive review on the Maillard reaction, the general pathways still remained the same. Most experiments to elucidate the formation pathways of aroma-active compounds used small molecules, e.g., amino acids or γ-aminobutanoic acid (GABA), as amine sources. As the focus in Maillard reaction research emerges into biological- and medicinal-relevant reaction sites, the specific reactivity of proteins and peptides is gaining more interest.[4,5] The development of "biocompatible" mass spectrometry ionization techniques (matrix-assisted laser desorption/ionization [MALDI] and electrospray ionization [ESI]) in the late 1990s as a tool to characterize peptides and proteins supported the bio-related and polymer research in the Maillard reaction.[6]

The presented work describes the formation pathways of norfuraneol and acetic acid from 1-^{13}C arabinose and GABA, a heptapeptide, and β-casein.

Experimental

Three different types of amines and 1-^{13}C arabinose were used for incubation experiments: 1) The model substance GABA was used to simulate the reactivity of the ε-amino group of lysine. 2) A heptapeptide with a sequence Nα-Acetyl-Lys-Lys-β-Ala-Lys-β-Ala-Lys-Gly was synthesized and used for incubation experiments. Instead of α-alanine, β-alanine was used as a spacer to separate the reactive lysine residues. This amino acid sequence was used to gain insight into the different influences of neighboring amine groups on the reactivity of the amine residues in the peptide. 3) As a model protein, β-casein was used. β-Casein is an important milk ingredient and its importance for the Maillard reaction in milk products is undisputable. β-Casein contains 11 lysine residues and five phosphorylated sites and therefore exhibits many Maillard reactive sites that should show a significant reactivity. 1-^{13}C arabinose was incubated with the three different amine compounds to elucidate the formation pathways of ^{13}C-labeled norfuraneol and acetic acid by gas chromatography–mass spectrometry (GC–MS).

Reaction Conditions

The substrates were dissolved in phosphate buffer, pH 7.4. The incubation temperatures were 120°C (GABA and heptapeptide) and 140°C (β-casein), and the reaction times were 60 min and 80 min, respectively.

Address for correspondence: Leif Alexander Garbe, TU Berlin, Molecular Analysis, Seestrasse 13, Berlin, Berlin, DE 13353.
leif-a.garbe@tu-berlin.de

TABLE 1. ^{13}C-labeling pattern of acetic acid analyzed from the Maillard reaction of 1-^{13}C arabinose with different amine sources

Intermediate Nitrogen Source	![NHR, OH, =O, =O structure] ^{12}C	![=O, =O, OH, OH structure] ^{13}C
GABA, γ-aminobutanoic acid.	18	82
Heptapeptide	74	26
β-Casein	95	5

Analysis

The organic compounds of the reaction mixture were extracted with diethyl ether and analyzed by GC–MS (HP 5890 Series 2 coupled to a MAT 8200 double-focusing mass spectrometer; Mascom, Bremen, Germany); data were recorded and interpreted by Maspec II^{32} GC-MS Software (Mascom).

The analysis of the ^{13}C labeling of acetic acid and norfuraneol was performed by electron impact (EI)–MS (cf. TABLES 1 and 2), and norfuraneol was also characterized by C-13 nuclear magnetic resonance (NMR) spectroscopy (JEOL 400 Lambda NMR in CD$_3$OD [Jeol, Tokyo, Japan], data not shown).

The analysis of the glycation site of the heptapeptide was also performed by proton and ^{1}H- and 13C-NMR (in D$_2$O) study and additionally by ESI-MS (Finnigan LCQ Ion Trap mass spectrometer; Thermo Scientific, Waltham, MA). These glycation site results will be published elsewhere.

Results

The results of the 1-^{13}C arabinose incubation with GABA, heptapeptide N$_\alpha$-Acetyl-Lys-Lys-β-Ala-Lys-β-Ala-Lys-Gly, and β-casein are shown in TABLES 1 and 2 and FIGURE 1. Labeled and unlabeled acetic acid can be distinguished by their EI–MS molecular ion signal at m/z 60 (unlabeled) and m/z 61 (^{13}C labeled). The fragmentation pattern of norfuraneol is more complex. The ^{13}C norfuraneol molecular ion M$^{+\circ}$ (m/z 115), representing the intact carbon chain, shows a rearrangement and an internal fragmentation resulting in m/z 43 (^{13}CH$_2$) and m/z 44 (^{13}CH$_3$) as shown in FIGURE 2 and TABLE 2. Thereby, the ^{13}C-labeling site of norfuraneol was characterized. The dependence of the amine source on the labeling pattern of norfuraneol and acetic acid could be clearly demonstrated.

Discussion

The results evidently show a shift in the position of the ^{13}C label in norfuraneol using the heptapeptide or β-casein as the amine compound compared to GABA (TABLES 1 and 2). This can be explained by a longer reaction time of the initially formed Amadori reaction product (ARP) and a further transformation of ARP. Three possible formation pathways of norfuraneol are evident and summarized in FIGURE 1: A) Cyclization of the amine catalyzed and released 1-^{13}C-1-deoxyosone into ^{13}CH$_3$ norfuraneol. B) Rearrangement of the ARP with migration of the carbonyl

TABLE 2. ^{13}C-labeling pattern of norfuraneol analyzed from the Maillard reaction of 1-^{13}C arabinose with different amine sources

	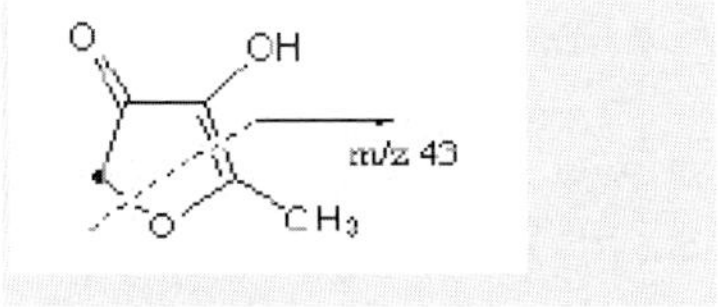 m/z 43	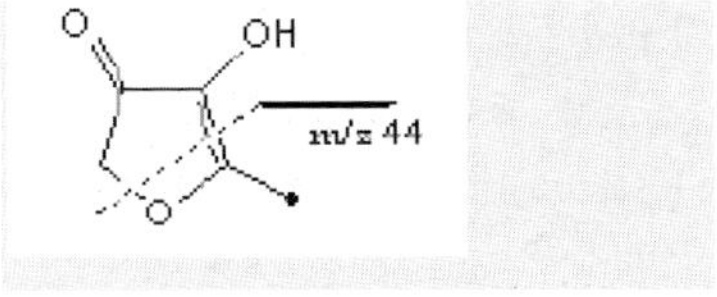 m/z 44
GABA	20 ^{13}CH$_2$	80 ^{13}CH$_3$
Heptapeptide	22 ^{13}CH$_2$	78 ^{13}CH$_3$
β-Casein	40 ^{13}CH$_2$	60 ^{13}CH$_3$

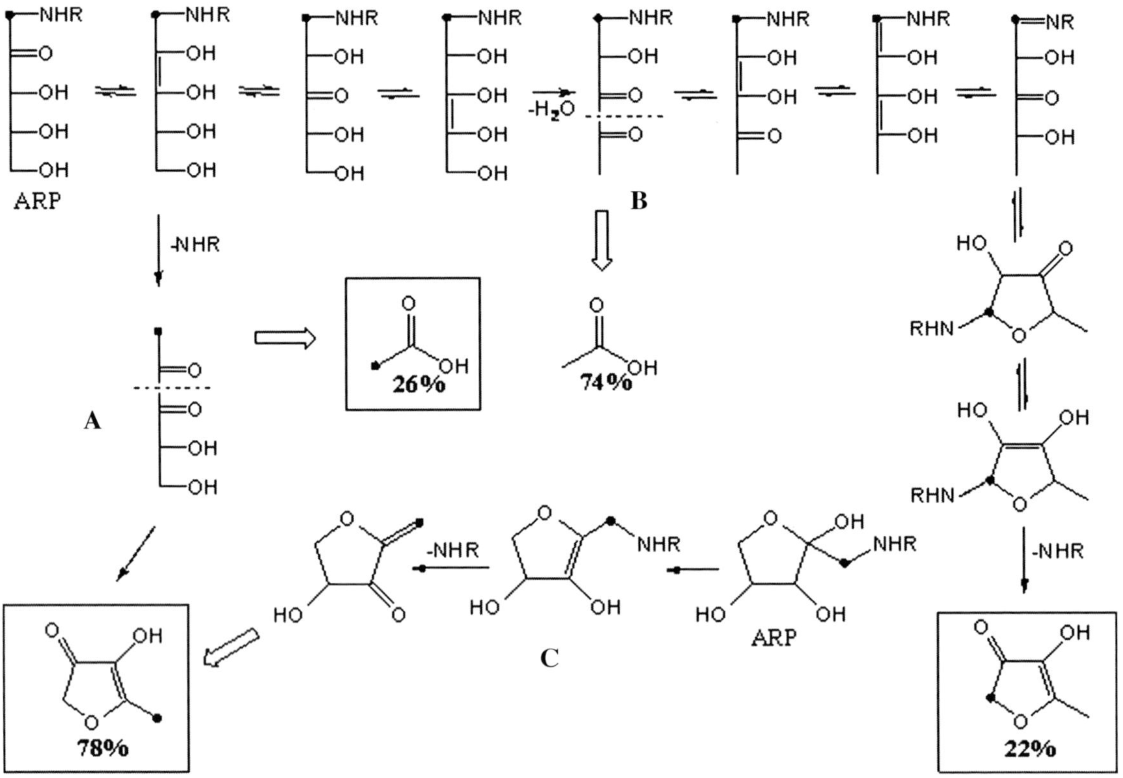

FIGURE 1. Formation pathway of 1-^{13}C-norfuraneol and 13C acetic acid from 1-^{13}C arabinose and model heptapeptide Nα-Acetyl-Lys-Lys-β-Ala-Lys-β-Ala-Lys-Gly. (**A**) 1-Deoxyosone pathway; (**B**) Reverse 1-deoxyosone pathway; (**C**) o-Elimination pathway.

group along the carbon backbone, resulting in an ARP-bound inverse 1-deoxyosone as shown in FIGURE 1; subsequently, keto enol tautomerism and deamination of the intermediate results in the formation of ^{13}CH$_2$ norfuraneol. C) o-Elimination of the cyclic sugar–ARP adduct can also form ^{13}CH$_3$ norfuraneol. Pathways A and C cannot be distinguished by tracing the m/z 43/m/z 44 mass label. However, acetic acid was characterized ^{12}C:^{13}C with 74%:26%. Therefore, the inverse 1-deoxyosone product obviously ends up in acetic acid, whereas only 22% of norfuraneol is formed by pathway B.

The carbonyl mobility at the sugar moiety of the ARP has previously been shown by Lederer *et al.*,[7]

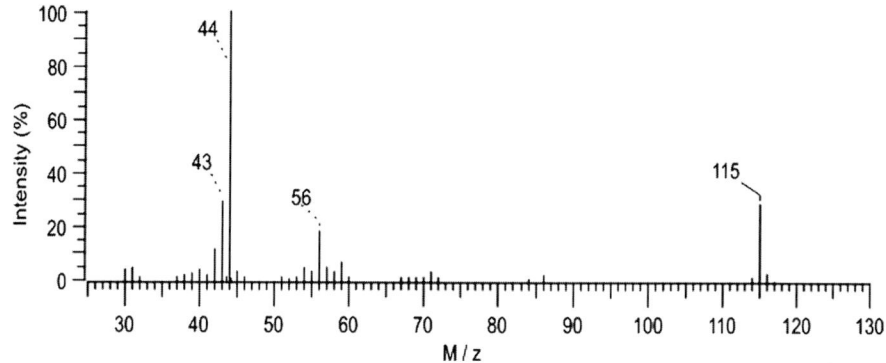

FIGURE 2. Electron impact mass spectrometry (low resolution) of 13C-labeled norfuraneol isolated from the reaction of 1-^{13}C arabinose and heptapeptide.

fully supporting our results within peptide- or protein-catalyzed Maillard reaction.

As a consequence, we suggest a slightly modified Maillard reaction pathway catalyzed by peptides or proteins instead of small amines (e.g., γ-aminobutyrate, GABA). A possible explanation for the different results of peptide, protein, and amino acid reactions are different shelf lives of peptide or protein ARP and amino acid ARP, resulting in the migration of the carbonyl group within the sugar carbon backbone.

One important and stringent way to obtain information about the Maillard reaction pathways is to introduce isotopically labeled educts and to characterize the position of ^{13}C in the labeled products. For this reason, EI–MS is an essential tool for pathway exploration. Maillard reaction studies *in vivo* require the characterization of formation pathways of volatiles, as shown in the present paper, as well as the identification of the peptide or protein modification. This will be the subject of a further publication.

Conflict of Interest

The authors declare no conflicts of interest.

References

1. MAILLARD, L.-C. 1912. Action des acides amines sur les sucres: formation des melanoidines par voie méthodique. C.R. Hedb. Seances Acad. Sci. **154:** 66–68.
2. HODGE, J.E. 1953. Dehydrated foods, chemistry of Browning reactions in model systems. J. Agric. Food Chem. **1:** 928–943.
3. LEDL, F. & E. SCHLEICHER. 1990. New aspects of the Maillard reaction in foods and in the human body *Angew.* Chem. Int. Ed. **29:** 565–594.
4. KENNEDY, L. & J.W. BAYNES. 1984. Nonenzymatic glycosylation and the chronic complications of diabetes: an overview. Diabetologia **26:** 93–98.
5. VLASSARA, H., R. BUCALA & L. STRIKER. 1994. Pathogenic effects of advanced glycosylation: biochemical, biologic and clinical implications for diabetes and aging. Lab. Invest. **70:** 138–151.
6. TRESSL, R., G.T. WONDRAK, L.-A. GARBE, *et al.* 1998. Pentoses and Hexoses as sources of new melanoidin-like Maillard polymers. J. Agric. Food Chem. **46:** 1765–1776.
7. BIEMEL, K.M., J. CONRAD & M.O. LEDERER. 2002. Unexpected carbonyl mobility in aminoketoses: the key to major Maillard crosslinks. Angew. Chem. Int. Ed. **41:** 801–804.

Model Studies on Protein Glycation

Influence of Cysteine on the Reactivity of Arginine and Lysine Residues toward Glyoxal

UWE SCHWARZENBOLZ, SUSANN MENDE, AND THOMAS HENLE

Institute of Food Chemistry, Technische Universität Dresden, Dresden, Germany

Mixtures of N^α-hippurylarginin, N^α-hippuryllysine, and glyoxal were incubated in the absence and presence of N^α-acetylcysteine in order to assess the individual reactivity of these nucleophilic amino acid residues. The incubations were performed under atmospheric and high hydrostatic pressure (400 MPa), and, at the same time, β-casein was reacted with glyoxal. The results showed that arginine is the main partner for glyoxal in the absence of cysteine, whereas a lysine derivatization was not apparent. In the presence of cysteine, however, arginine was almost completely protected from the reaction, whereas a noticeable formation of lysine derivatives, mainly carboxymethyllysine, was observed. Based on these findings, a reaction mechanism is proposed to explain the influence of cysteine on the reaction.

Key words: carboxymethyllysine; glyoxal; nucleophilic amino acids; high hydrostatic pressure

Introduction

Although glyoxal is well able to target nucleophilic amino acid residues (Lys, Arg, Cys) when reacting with proteins, these reactions have mainly been studied individually with the emphasis on elucidating reaction products.[1–3] Apart from lysine derivatization accompanied by the formation of N^ε-(carboxymethyl)lysine (CML),[4] reactions of arginine side chains with glyoxal have been found to be of major importance,[5,6] and recently the nucleophilic amino acid cysteine is reported as being a scavenger of glyoxal.[7] In food and other biological systems, the functional groups of these three amino acids would not occur separately from each other and therefore would compete for glyoxal based mainly on their nucleophilicity. For first insights into the underlying reaction mechanisms and the resulting spectrum of reaction products, we assessed the reactivity of lysine and arginine residues when competing for glyoxal in the absence and presence of cysteine and analyzed the amount of the resulting lysine and arginine products.

Address for correspondence: Uwe Schwarzenbolz, Institute of Food Chemistry, Technische Universität Dresden, D-01062 Dresden, Germany. Voice: +49 351 463 34465; fax: +49 351 463 34138.
Uwe.Schwarzenbolz@chemie.tu-dresden.de

Materials and Methods

Chemicals

N^α-hippurylarginine (Hiparg), N^α-hippuryllysine (Hiplys), N^α-acetylcysteine (Accys), and glyoxal (GO) were obtained from Sigma (Deisenhofen, Germany). Hydrochloric acid (HCl) was obtained from J.T. Baker (Deventer, the Netherlands), and cation-exchange resin Dowex 50W-X8 was obtained from Bio-Rad (Munich, Germany). All other chemicals were from Merck (Darmstadt, Germany).

Sample Preparation

Reaction mixtures, each containing 50 μmol/mL of Hiparg, Hiplys, and GO, were prepared in phosphate-buffered saline (pH 7.4) and reacted either in the absence or presence of 50 μmol/mL Accys at 40°C for 8 h. During this incubation, samples were taken and analyzed for hippuryl derivatives as well as for residual glyoxal. Similar experiments were performed in imidazole buffer (0.2 mol/L, pH 7.0). The incubation was carried out at atmospheric pressure as well as at 400 MPa in a high-pressure plant (Dieckers GmbH und Co., Willich, Germany).

Analysis of Hippuryl Derivatives

A reversed-phase (RP)-HPLC system with an Aqua 3 μC 18 125 A (150 × 4.6 mm) column (Phenomenex, Aschaffenburg, Germany) on an Agilent 1100 Series chromatography system with UV detection at 227 nm

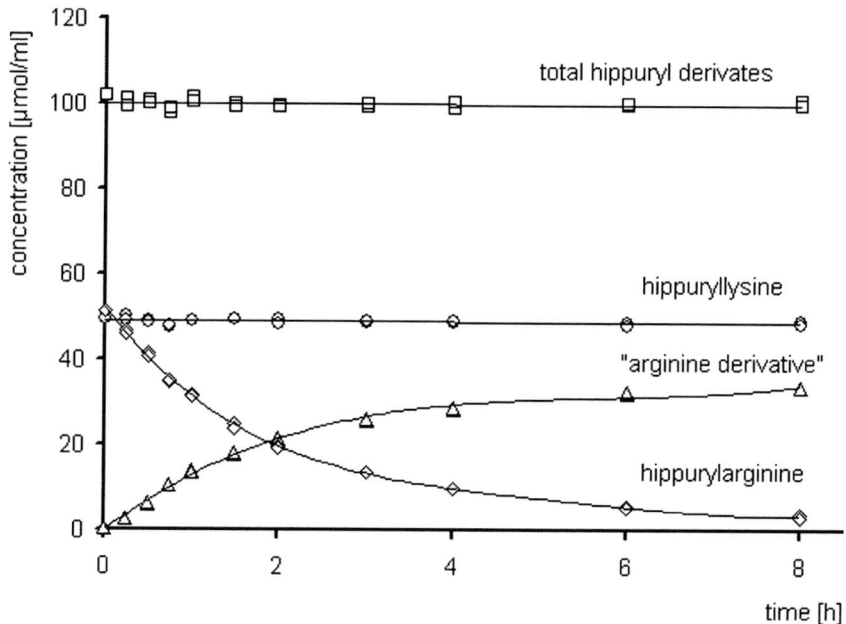

FIGURE 1. Concentration of hippuryl derivatives in incubation mixtures containing hippuryllysine, hippurylarginine, and glyoxal (molar ratios 1:1:1) in phosphate-buffered saline, pH 7.4.

was used. A gradient consisting of 10 mmol/L sodium phosphate (pH 7.0) (eluent A) and methanol (eluent B) at a flow rate of 0.6 mL/min was used. Initial conditions were 7% B which increased to 16% B within 22 min and subsequently to 25% B after an additional 3-min elution time. For regeneration, the column was flushed with 100% B for 10 min and equilibrated for 15 min at the initial conditions.

RP-HPLC Analysis of Glyoxal

Prior to analysis, sample mixtures were derivatized with o-phenylenediamine in 0.5 mol/L phosphate buffer (pH 6.5) under deaerated conditions for 12 h in the dark. An RP-HPLC system with an Aqua 3 μC 18 125 A (150 × 4.6 mm) column (Phenomenex) on an Agilent 1100 Series chromatography system with UV detection at 227 nm was used. The eluents consisted of 50 mmol/L acetic acid (eluent A) and methanol mixed with eluent A in a 1:1 ratio (eluent B). The HPLC was performed with a flow of 0.7 mL/min and a linear gradient from 100% eluent A to 25% eluent B after 40 min. The column was then equilibrated with 100% eluent A for 10 min.

Liquid Chromatography–Mass Spectrometry

Liquid chromatography–mass spectrometry (LC-MS) was performed on the HPLC system as described above, coupled to a Mariner time-of-flight (TOF) mass spectrometer (Perseptive Biosystems, Framingham, MA). Electrospray ionization was used. Eluents consisted of 10 mmol/L ammonia acetate (pH 5.0) (eluent A), and 10 mmol/L ammonia acetate in 84% acetonitrile (eluent B); the flow rate was 0.4 mL/min. The gradient started with 100% A for 2 min, linearly decreased to 75% A within 9 min, to 50% A in an additional 14 min, and further to 25% A after another 15 min. Prior to consecutive analysis, the column was equilibrated with the initial conditions.

Amino Acid Analysis

Acid hydrolysis prior to amino acid analysis was performed according to Henle et al.[8] Amino acids were analyzed with an Alpha Plus amino acid analyzer (LKB Biochrom, Cambridge, UK) using a PEEK column (150 × 4 mm, Alltech, Unterhaching, Germany) filled with ion-exchange resin DC4a-spec (sodium form, Benson, Reno, NV). The composition of the buffers and reagent as well as the running conditions are described in Henle et al.[8]

Results and Discussion

As the quantitative significance of individual derivatives resulting from the reaction of glyoxal and nucleophilic amino acid side chains (Arg, Lys, Cys) is still unclear, the main targets of a glyoxal derivatization on proteins are unknown. To address this, we incubated mixtures consisting of N^α-hippurylarginine, N^α-hippuryllysine, and glyoxal at 40°C up to 8 h in the

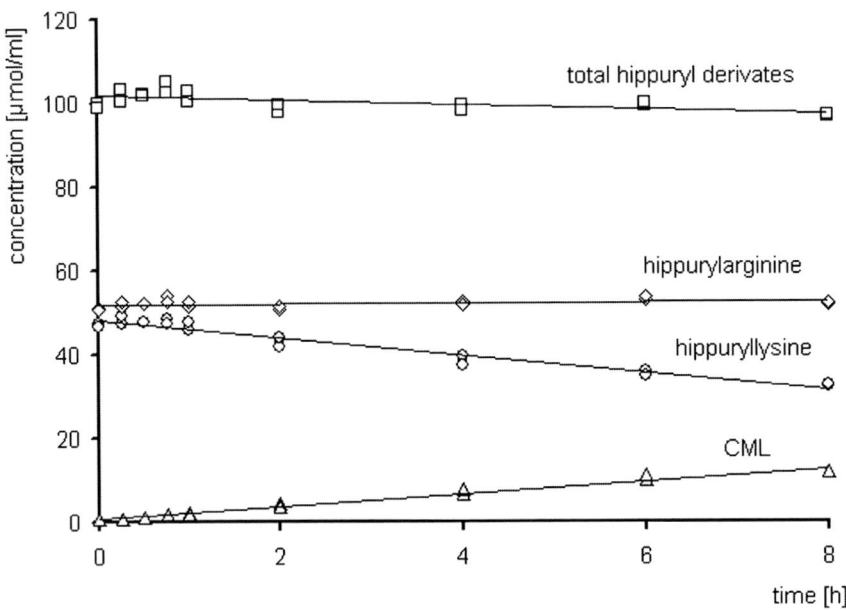

FIGURE 2. Concentration of hippuryl derivatives in incubation mixtures containing hippuryllysine, hippurylarginine, N^α-acetylcysteine, and glyoxal (molar ratios 1:1:1:1) in phosphate-buffered saline, pH 7.4.

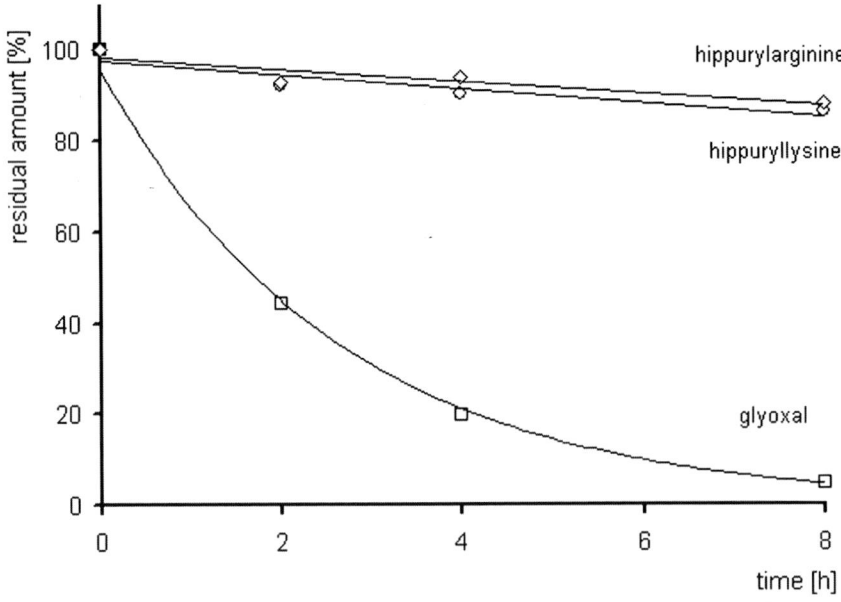

FIGURE 3. Residual amounts of hippuryl derivatives in incubation mixtures containing hippuryllysine, hippurylarginine, N^α-acetylcysteine, and glyoxal (molar ratios 1:1:1:1) in imidazole buffer, pH 7.0.

absence and presence of N^α-acetylcysteine. The resulting decrease in the hippuryl amino acid content was measured by RP-HPLC.

Phosphate-buffered mixtures of Hiparg, Hiplys, and Go that did not contain N^α-acetylcysteine showed an arginine derivatization up to 85%, while the lysine content remained almost unchanged (FIG. 1). As the arginine decreased, the formation of an "arginine derivative" became apparent. This compound, identified by MS, turned out to be dihydroimidazolinone.

The transfer of these experiments to a protein model system was performed with β-casein-glyoxal incubation mixtures. Amino acid analysis after acid

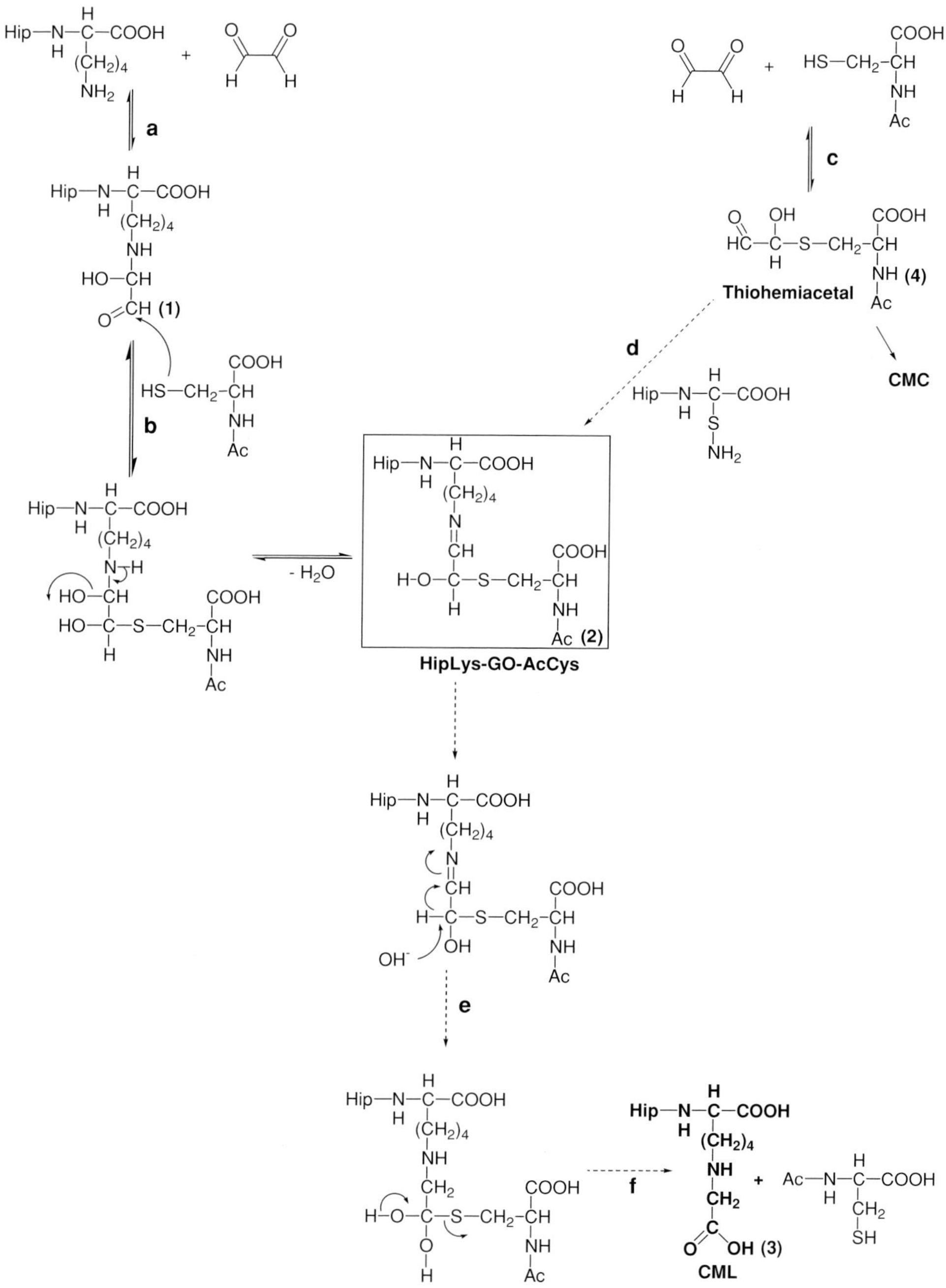

FIGURE 4. Proposed mechanism for a cysteine-mediated N^ϵ-(carboxymethyl)lysine formation.

hydrolysis showed that predominantly arginine residues reacted with glyoxal, whereas no lysine derivatization was apparent.

After the addition of N^α-acetylcysteine, the results changed drastically. An arginine derivatization was no longer visible, but the amount of unmodified lysine dropped to less than 70% after 8 h of incubation (FIG. 2), accompanied by CML formation. Although a high reactivity of cysteine toward glyoxal and products such as carboxymethylcysteine are described in the literature,[9] we were not able to unequivocally identify components by mass spectroscopy that are generated solely from cysteine and glyoxal under our study conditions.

High hydrostatic pressure represents a new technique for food preservation. To take this physical parameter into account, incubations were carried out in imidazole buffer at pH 7.0. In the absence of cysteine, arginine once again represented the main target for a reaction with glyoxal and was derivatized to about 50% of its initial amount. In contrast, lysine, except for a pressure-induced decrease even without glyoxal, did not react. A simultaneous chromatographical estimation of the residual glyoxal showed a decrease that was in line with the arginine derivatization and pointed to a 1:1 reaction between glyoxal and arginine.

The addition of N^α-acetylcysteine to the imidazole buffer incubation mixtures resulted in the same effects as in phosphate buffer but not as pronounced (FIG. 3). The reactivity of lysine was promoted, while that of arginine was markedly reduced, leading to almost identical derivatization rates. Owing to the cysteine content, the residual glyoxal content in these mixtures was reduced to almost zero, but despite the application of MS, we currently cannot link this effect to certain molecular structures.

These results show that the three different amino acid residues do not react separately with glyoxal in quantities that are linked to their nucleophilicity. The thiol is obviously able to affect the product spectrum, especially with respect to CML formation.

Based on these findings, we are proposing a reaction mechanism that leads to CML formation with the participation of cysteine (FIG. 4). Initially, glyoxal can either react with lysine to form a hemiaminal (route a, FIG. 4), which is stabilized by cysteine (route b, FIG. 4); or the primary step is the formation of a thiohemiacetal (route c, FIG. 4), which then subsequently reacts with lysine. In both initial steps, the key intermediate would be a product, "HipLys-GO-AcCys," with a theoretical molecular mass of 510.18 g/mol. This compound further degrades to result in CML under release of the thiol, which therefore acts as a catalyst. As all experiments were accompanied by LC–MS measurements at all different reaction stages, we were able to find a peak with the predicted protonated species (511.2 g/mol), which gave us evidence for the proposed pathway.

In order to find out which of the two possible initial steps predominantly leads to CML formation, we separately preincubated lysine and cysteine with glyoxal for 4 h at $40°C$ and then added the respective third component for an additional 4 h incubation. During that second period, samples were taken at certain time points and analyzed for CML. This experiment showed a CML content of up to 6% of the initial lysine in samples in which lysine was preincubated, representing a threefold higher CML content compared to samples with cysteine preincubation. Therefore, we suggest an initial lysine adduct of glyoxal (route a, FIG. 4) subsequently stabilized by cysteine (route b, FIG. 4) to be of particular importance for CML formation.

Our results show, that arginine is much more reactive toward glyoxal in thiol-free systems. Because cysteine-containing peptides (e.g., gluthathion) and proteins occur in physiological systems as well as in foods, their driving influence on the reaction of glyoxal with proteins cannot be neglected.

Conflict of Interest

The authors declare no conflicts of interest.

References

1. GLOMB, M.A. & V.M. MONNIER. 1995. Mechanism of protein modification by glyoxal and glycolaldehyde, reactive intermediates of the Maillard reaction. J. Biol. Chem. **270:** 10017–10026.
2. GLOMB, M.A. & G. LANG. 2001. Isolation and characterization of glyoxal-arginine modifications. J. Agric. Food Chem. **49:** 1493–1501.
3. BAYNES, J.W., N. FRIZZELL, R. NAGAI, et al. 2006. Chemical modification of protein thiols in diabetes. Abstracts of Papers, 232nd ACS National Meeting, San Francisco, CA, United States, Sept. 10–14.
4. ZYZAK, D.V., K.J. WELLS-KNECHT, J.A. BLACKLEDGE, et al. 1994. Pathways of the Maillard Reaction in vitro and in vivo. In Maillard Reactions in Chemistry, Food and Health. T.P. Labuza, G.A. Reineccius, et al., Eds.: 274–280. Royal Chemical Society. Bodmin, UK.
5. SCHWARZENBOLZ, U., T. HENLE, R. HAEßNER & H. KLOSTERMEYER. 1997. On the reaction of glyoxal with proteins. Z. Lebens. Unters. Forsch. A **205:** 121–124.
6. THORNALLEY, P.J. 2005. Dicarbonyl intermediates in the Maillard reaction. Ann. N.Y. Acad. Sci. **1043:** 111–117.
7. ZENG, J. & M.J. DAVIES. 2006. Protein and low molecular mass thiols as targets and inhibitors of glycation reactions. Chem. Res. Toxicol. **19:** 1668–1676.
8. HENLE, T., H. WALTER, I. KRAUSE & H. KLOSTERMEYER. 1991. Efficient determination of individual Maillard compounds. Int. Dairy J. **1:** 125–135.
9. ZENG, J. & M.J. DAVIES. 2005. Evidence for the formation of adducts and S-(Carboxymethyl)cysteine on reaction of a-dicarbonyl compounds with thiol groups on amino acids, peptides, and proteins. Chem. Res. Toxicol. **18:** 1232–1241.

Analysis of Amadori Peptides Enriched by Boronic Acid Affinity Chromatography

ANDREJ FROLOV AND RALF HOFFMANN

Institute of Bioanalytical Chemistry, Center for Biotechnology and Biomedicine, Faculty of Chemistry and Mineralogy, Leipzig University, Leipzig, Germany

Glycation of peptides and proteins by D-glucose is a universal, nonenzymatic reaction with important implications for the pathogenesis and diagnosis of many diseases, including diabetes mellitus. Whereas some modification sites have been identified in serum albumin and hemoglobin, a general approach to map glycation sites for nonabundant proteins present in complex mixtures, such as serum, is still missing. Here, we describe a universal enrichment procedure for glycated peptides using boronic acid affinity chromatography in the first dimension followed by reversed-phase chromatography, coupled either online to electrospray ionization mass spectrometry (ESI–MS) or offline to matrix-assisted laser desorption/ionization (MALDI) MS. This two-dimensional approach was optimized for high recoveries and low cross reactivities. For bovine serum albumin, a total of 31 Amadori peptides were identified in a tryptic digest corresponding to 26 different glycation sites.

Key words: m-aminophenylboronic acid; ESI; glycation; MALDI; mass spectrometry

Introduction

Glycation or nonenzymatic glycosylation refers to the reaction of reducing sugars (aldoses and ketoses), oxoaldehydes, or dicarbonyls with amino groups present in proteins at the N terminus or the side chains of lysine and arginine residues. This reaction is part of the Maillard reaction that finally leads to a series of colored and fluorescent compounds, termed advanced glycation end products (AGEs).[1,2] D-Glucose is a well-known, *in vivo*, glycation agent that modifies especially serum proteins because of its high concentration in blood, which is even more pronounced in diabetic patients.[3] Although it was suggested that these compounds contribute to diabetic complications,[4] their complexity, heterogeneity, and often unknown molecular structures limit their diagnostic value. However, the intermediately formed and relatively stable Amadori-modified proteins might be valuable diagnostic markers for both diagnosis and prognosis. Glycated hemoglobin (HbA1c), for example, has been used as a clinical marker to monitor long-term (3 months) glucose levels in diabetic patients.[5]

Glycated peptides can be identified by both matrix-assisted laser desorption/ionization (MALDI) and electrospray ionization (ESI) mass spectrometry (MS) by the mass shift of the modified peptides in an enzymatic digest of a purified protein,[6] although it can be ambiguous to identify modified peptides simply by their mass. The sensitivity and selectivity of this peptide mass fingerprint (PMF) can be further improved by coupling it offline or online to reversed-phase chromatography (RPC).[7] The inherent limitations of the PMF can be overcome by liquid chromatography–tandem mass spectrometry (LC–MS/MS), which provides, in addition to the peptide mass, characteristic fragmentation patterns to identify the modification site as well as the sugar that originally reacted with the peptide.[8] The glycation site can be identified by both collision-induced dissociation[8] and electron transfer dissociation.[9]

As a result of the typically low glycation levels at a certain position and the low concentrations of most proteins present in serum, it is not possible to directly analyze the glycation sites in a complex protein mixture. Thus, proteins of interest are typically isolated before analysis. This was shown for HbA_{1C}-based glycemic control in diabetic patients using m-aminophenylboronic acid affinity chromatography[10] and recently extended to proteins glycated *in vitro* and digested by trypsin prior to LC–MS/MS.[11] Here, we optimize the m-aminophenylboronic acid affinity chromatography to specifically enrich glycated

Address for correspondence: Professor Dr. Ralf Hoffmann, Institute of Bioanalytical Chemistry, Center for Biotechnology and Biomedicine, Faculty of Chemistry and Mineralogy, Leipzig University, Deutscher Platz 5, 04103 Leipzig, Germany. Voice: +49-0-341 9731331; fax: +49-0-341 9731330.

Hoffmann@chemie.uni-leipzig.de

peptides from digests of complex protein mixtures. The eluates were further separated by RP-HPLC and analyzed either online by nano-ESI–MS or offline by MALDI–MS. This two-dimensional separation scheme was successfully applied to identify 31 glycation sites in bovine serum albumin (BSA).

Experimental

Glycation of BSA

We incubated 1 mg protease-free, lyophilized albumin bovine fraction V (Serva GmbH, Heidelberg, Germany) in 1 mL phosphate buffer (100 mmol/L, pH 7.4) in the absence (control) or presence of D-glucose (0.5 mol/L; Carl Roth GmbH, Karlsruhe, Germany) at 50°C. After 4 days, the samples were dialyzed against aqueous NH_4HCO_3 buffer (3 mmol/L, pH 8.0) using an 8/32 dialysis tube (Serva, GmbH) two times for 24 h and concentrated under vacuum to a final volume of 100 μL. The proteins were reduced by addition of 2 μL aqueous dithiothreitol (Carl Roth GmbH) solution (100 mmol/L) at 50°C for 30 min, cooled to room temperature, and incubated with 2.1 μL iodoacetamide (Fluka, Sigma-Aldrich Chemie GmbH, Taufkirchen, Germany) solution (150 mmol/L) for 60 min to alkylate all thiol groups. The sample was digested with trypsin (Promega Corporation, Mannheim, Germany) at a substrate to enzyme ratio of 25 at 37°C for 8 h. The reaction was stopped by addition of 1 μL formic acid, dried, and stored at −20°C.

Chromatography

m-Aminophenylboronic acid-agarose (Sigma-Aldrich Chemie GmbH, Seelze, Germany) was suspended in washing buffer (250 mmol/L aqueous ammonium acetate buffer, 50 mmol/L $MgCl_2$, pH 8.0) and packed into a 1 mL polypropylene column (85 × 7.5 mm; Qiagen GmbH, Hilden, Germany). The column was equilibrated with washing buffer before 200 μL of the tryptic digest dissolved in washing buffer (1 μg/μL BSA) were loaded at 4°C. After 1 h, the column was washed with 10 mL washing buffer at room temperature and the glycated proteins or peptides were eluted with 5 mL elution buffer (200 mmol/L D-sorbitol or 250 mmol/L acetic acid, pH 2.8). The eluate was collected in 5 mL fractions.

The fractions were further separated by RP-HPLC using an Äkta™ purifier HPLC system (Amersham Biosciences Europe GmbH, Freiburg, Germany) containing an Aqua C-18 column (150 × 2 mm, particle size 3 μm; Phenomenex Inc., Aschaffenburg, Germany). Eluent A was 0.1% aqueous trifluoroacetic acid (TFA), and eluent B was acetonitrile containing 0.1% TFA. The column was equilibrated with 3% eluent B before the fractions were loaded and was washed for 10 min; the peptides were then eluted by a linear gradient from 3% to 60% eluent B for 57 min using a flow rate of 200 μL/min. The peptides were detected by their absorption at 220 nm. Two hundred microliter fractions were collected in 96-well microtiter plates for MS.

Mass Spectrometry

The RPC fractions were dried in vacuum and reconstituted in 10 μL aqueous acetonitrile (1:1 [v/v]). An aliquot of 0.5 μL of this solution was mixed with 0.5 μL matrix solution, which was prepared by dissolving 4 mg/mL α-cyano-4-hydroxy-cinnamic acid (CHCA; Bruker Daltonics GmbH, Bremen, Germany) in a mixture of 0.1% aqueous TFA and acetonitrile (1:1 [v/v]), on the target and air-dried. Samples were analyzed in positive-ion MALDI–time-of-flight (TOF) and MALDI–TOF/TOF–MS modes (4700 proteomic analyzer; Applied Biosystems GmbH, Darmstadt, Germany).

Alternatively, an aliquot of the fractions containing approximately 100 ng protein was separated by nano-RP-HPLC using a HPLC 1100 system (Agilent Technologies GmbH, Waldbronn, Germany) coupled online to the nano-ESI-source (Protana, Odense, Denmark) of a QqTOF hybrid mass spectrometer consisting of a quadrupole and an orthogonal TOF analyzer (QSTAR Pulsar I; Applied Biosystems GmbH). The tandem mass spectra were automatically acquired in positive-ion mode using an information-dependent acquisition experiment, as described recently.[8]

Results and Discussion

It is a well-known fact that boronic acids strongly bind to cis-diols forming a relatively stable five-membered ring under mild alkaline conditions, which can be used in affinity chromatography to enrich glycoproteins.[12] This interaction can also improve the separation of glycoproteins in SDS-PAGE, as addition of borate to the Laemmli buffer system results in negatively charged borate–sugar complexes that accelerate glycoproteins relative to their unglycosylated forms and thereby sharpen the bands.[13] Similar effects can be obtained for glycated peptides and proteins because of the high structural similarities compared to glycoproteins, i.e., the presence of cis-diols. Indeed, Yue et al. used m-aminophenylboronic acid-based affinity chromatography in the early 1980s to analyze the glycation pattern of hemoglobin.[10] The glycated or glycosylated proteins can be eluted either with acids (pH below 3.0) or D-sorbitol buffers.[14]

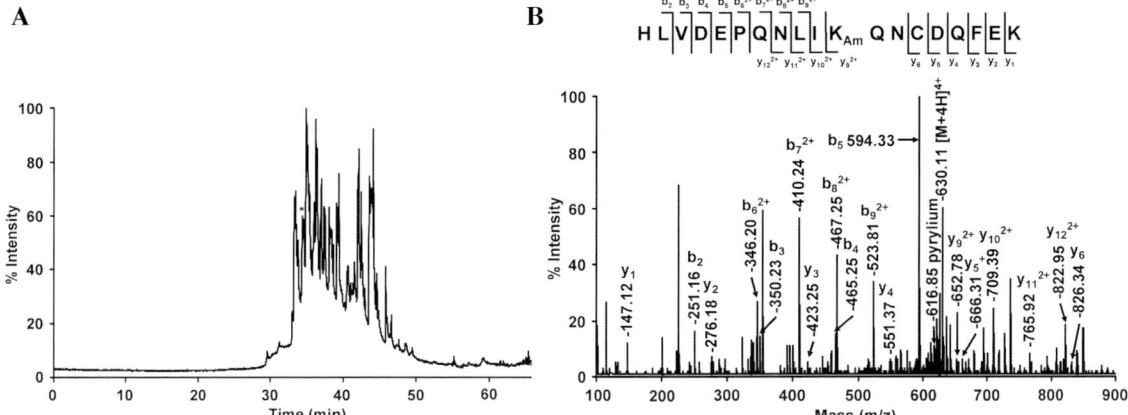

FIGURE 1. Reversed-phase HPLC separation of a tryptic digest of glycated bovine serum albumin enriched on m-aminophenylboronic acid agarose (**A**, total ion current chromatogram) coupled online to electrospray ionization–mass spectrometry and the tandem mass spectrum (**B**) for an early eluting peptide (marked by an asterisk) at m/z 839.8, including the sequence retrieved from the b- and y-ions.

In general, glycation sites could be identified in serum samples by two different strategies: enrichment of (i) glycated proteins by boronic acid affinity chromatography followed by mass spectrometrical analysis of the enzymatic digest or (ii) glycated peptides from a protein digest analyzed by MS. In our hands, the first strategy was not very effective, as almost 60% of all serum proteins bound to the m-aminophenylboronic acid phase, which is probably a result of unspecific binding. Consecutive MALDI–TOF/TOF analysis of the tryptic digest yielded only unglycated peptides, mostly from serum albumin. Because of these inherent disadvantages, we investigated the second strategy in more detail, optimizing the conditions for the affinity chromatography combined to RP-HPLC with glycated peptides synthesized on solid phase.[15] All studied glycated peptides bound quantitatively to the stationary phase at Mg^{2+} concentrations above 30 mmol, whereas unmodified peptides did not bind even at higher concentrations. Thus, 50 mmol/L $MgCl_2$ was used in all following experiments. Elution with 200 mmol/L D-sorbitol was very efficient, but the high viscosity of the solution caused a high back pressure at the RP-HPLC, requiring low flow rates with long loading times. Furthermore, polar peptides binding only weakly to the C_{18}-phase were not retarded and were lost for the consecutive analysis. Elution with acetic acid (pH 2.8) was superior with respect to the peptide recoveries in the second chromatographic dimension. Moreover, the elution of the glycated peptides was as efficient as with D-sorbitol based on the peak areas in RP-HPLC. A second elution step with D-sorbitol following the acetic acid procedure released only minor peptide amounts, i.e., less than 0.2% relative to the first acid elution step.

These optimized conditions enriched glycated peptides from spiked protein digests. Thus, BSA was glycated *in vitro*, digested with trypsin, and analyzed by the optimized two-dimensional chromatography strategy followed by online nano-ESI–MS (FIG. 1) or offline MALDI–MS. The online nano-RP-HPLC–ESI–MS/MS approach revealed a complex elution pattern with many peaks in the total ion current chromatogram (FIG. 1A). As the glycated position is not cleaved by trypsin but still represents a basic residue, the obtained Amadori peptides were relatively long and the spectra typically displayed the triply charged species as dominant ions. A total of 17 Amadori peptides, representing 17 different glycation sites, were identified by their fragmentation patterns, i.e., singly or doubly charged b- and y-ion series. The glycation sites were identified by their characteristic pyrylium (-54 u, $-3x\ H_2O$) and furylium (-84 u, $-3x\ H_2O$-HCHO) ions (FIG. 1B).[8] Only two unmodified peptides were detected, whereas no glycated peptides were detected in the control-BSA sample (incubated without D-glucose).

By MALDI–MS, a total of 27 Amadori peptides corresponding to 25 glycation sites were identified by their neural loss of 162 u besides only six unmodified peptides. Additionally, a mass loss of 120 u was obtained for most parent ions, which can be explained by the loss of $C_4O_4H_8$, as reported recently by Lapolla *et al.* for glycated human serum albumin-derived peptides analyzed by MALDI–MS.[7] Interestingly, we have not observed this neutral loss for synthetic Amadori peptides. The glycation sites of all peptides were determined by the characteristic reporter ions reported recently for glycated b- and y-ions.[8] Again, no peptides were detected for the tryptic digest of control BSA. The higher number of glycated peptides identified by MALDI–MS is only partially related to the 10 times higher sensitivity. Most likely, the major advantage was the shallower gradient and the longer analysis times allowing the operator to identify weaker signals in MS/MS mode.

In conclusion, we could show that glycated peptides can be quantitatively enriched from protein digests by m-aminophenylboronic acid-based affinity chromatography. This strategy allowed identification of 26 lysine-derived glycation sites in BSA by MALDI–MS and ESI–MS coupled offline or online to RP-HPLC. Among the 60 lysine residues, the following positions were glycated in BSA: 28, 36, 88, 117, 156, 160, 204, 228, 235, 256, 266, 285, 346, 374, 399, 401, 412, 420, 437, 455, 495, 498, 523, 548, 561, and 568.

Acknowledgments

We thank the Deutsche Forschungsgemeinschaft (DFG, Graduiertenkolleg 378) and the Free State Saxony for financial support.

Conflict of Interest

The authors declare no conflicts of interest.

References

1. FRYE, E.B., T.P. DEGENHARDT, S.R. THORPE & J.W. BAYNES. 1998. Role of the Maillard reaction in ageing of tissue proteins. J. Biol. Chem. **273:** 18714–18719.
2. SCHALKWIJK, C.G. et al. 1999. Amadori albumin in type 1 diabetic patients: correlation with markers of endothelial function, association with diabetic nephropathy, and localization in retinal capillaries. Diabetes **48:** 2446–2453.
3. PEPPA, M., J. URIBARRI & H. VLASSARA. 2003. Glucose, advanced glycation end products, and diabetes complications. Clin. Diabetes **21:** 186–187.
4. AHMED, N. & P.J. THORNALLEY. 2007. Advanced glycation endproducts: what is their relevance to diabetic complications. Diabetes Obes. Metab. **9:** 233–245.
5. KRISHNAMURTI, U. & M.W. STEFFES. 2001. Glycohemoglobin: primary predictor of the development of reversal of complications of diabetes mellitus. Clin. Chem. **47:** 1157–1165.
6. LAPOLLA, A. et al. 2004. Enzymatic digestion and mass spectrometry in the study of advanced glycation end products/peptides. J. Am. Soc. Mass Spectrom. **15:** 496–509.
7. LAPOLLA, A. et al. 2007. Off-line liquid chromatography-MALDI by with various matrices and tandem mass spectrometry for anaysis of glycated human serum albumin tryptic peptides. Mol. Nutr. Food Res. **51:** 456–461.
8. FROLOV, A., P. HOFFMANN & R. HOFFMANN. 2006. Fragmentation behavior of glycated peptides derived from D-glucose, D-fructose and D-ribose in tandem mass spectrometry. J. Mass Spectrom. **41:** 1459–1469.
9. ZHANG, Q. et al. 2007. Application of electron transfer dissociation mass spectrometry in analyses of non-enzymatically glycated peptides. Rapid Commun. Mass Spectrom. **21:** 661–666.
10. YUE, D.K. et al. 1982. The measurement of glycosylated hemoglobin in man and animals by aminophenylboronic acid affinity chromatography. Diabetes **31:** 701–705.
11. ZHANG, Q. et al. 2007. Enrichment and anaysis of nonenzymatically glycated peptides: boronate affinity chromatography coupled with electron-transfer dissociation mass spectrometry. J. Proteome Res. **6:** 2323–2330.
12. XIAO-CHUAN, L. 2004. Boronic acids for affinity chromatography. Chin. J. Chromatogr. **24:** 73–80.
13. KOENIG, R. et al. 1970. Protein subunits in the potato virus X group. Determination of the molecular weights by polyacrylamide electrophoresis. Biochim. Biophys. Acta. **207:** 184–189.
14. SHIMA, K. et al. 1988. High-performance liquid chromatographic assay of serum glycated albumin. Diabetologia **31:** 627–631.
15. FROLOV, A., D. SINGER & R. HOFFMANN. 2006. Site-specific synthesis of Amadori-modified peptides on solid-phase. J. Pept. Sci. **12:** 389–95.

Induction of Heat Shock Proteins and the Proteasome System by Casein-N^ε-(Carboxymethyl)lysine and N^ε-(Carboxymethyl)lysine in Caco-2 Cells

KAROLINE SCHMID,[a] MARTIN HASLBECK,[b] JOHANNES BUCHNER,[b] AND VERONIKA SOMOZA[a]

[a]*German Research Center for Food Chemistry, Garching, Germany*
[b]*Department of Chemistry, Technische Universität München, Munich, Germany*

Repeated mild heat shock treatment has been shown to have anti-aging effects on cellular mechanisms *in vitro*. Among these, the age-associated accumulation of advanced glycation end products (AGEs), such as N^ε-(carboxymethyl)lysine (CML), has been demonstrated to be effectively prevented in glyoxal-exposed human skin fibroblasts following mild heat shock treatment. The biochemical mechanism responsible for this inhibition is not yet known. However, the involvement of heat shock proteins (HSPs) and the misfolded proteins degrading the ubiquitin–proteasome system have been hypothesized. As AGE-modified proteins are likely to be conformationally modified, we investigated whether treatment of human intestinal cells with casein-linked CML or nonprotein-linked CML affects the expression of HSPs and the ubiquitin–proteasome system by using matrix-assisted laser desorption/ionization–time-of-flight tandem mass spectroscopy (after protein separation by two-dimensional gel electrophoresis) and by Western blotting. Compared to nontreated control cells, expression of HSP90, HSP60, HSP70 chaperones, and the proteasome S26 ATPase subunit 2 were significantly upregulated in casein-CML and in CML-treated cells. Exposure of Caco-2 cells to β-amyloid, a nonglycation product, revealed similar results. In conclusion, the results indicate that CML and casein-linked CML activate the expression of HSPs as well as the proteasome system, which are involved in the degradation of misfolded and possibly glycated proteins. Whether this mechanism is based on binding to cell surface receptors, such as the receptor for AGE, has to be clarified in future studies.

Key words: N^ε-(carboxymethyl)lysine; casein-CML; heat shock proteins

Introduction

N^ε-(Carboxymethyl)lysine (CML) is classified as a Maillard Reaction Product (MRP) as it is formed upon the reaction between glucose and lysine during heat treatment of foods.[1] On the other hand, CML is also termed an advanced glycation end product (AGE) as its *in vivo* formation has also been demonstrated during nonenzymatic glycation reactions, which are associated with the progression of aging and hyperglycemia.[2] The pathogenic role of AGEs is thought to be caused by glycation of proteins, which alters their native conformation and cellular function. Also, AGEs, such as CML, and also non-AGE compounds (e.g., amyloid beta peptides) have been shown to interact with cellular receptors that are involved in signaling pathways of inflammation and in the formation of reactive oxygen species.[3–6]

One of the enzymatic defense mechanisms to counteract the release of superoxide anions, a highly reactive oxygen species, is represented and achieved by the enzyme superoxide dismutase. During cellular processing, this enzyme cooperates with several chaperones.[7] Chaperones are a family of proteins[8] that arrange the correct folding or refolding of proteins, stabilize unfolded or misfolded proteins, prevent aggregation of proteins, or lead them to proteolytic degradation.

Chaperones are classified after their molecular weights. The heat shock protein 90 (HSP90) proteins stabilize non-native proteins and possess ATPase activity, which is essential for their function.[9] HSP70 proteins translocate proteins through membranes[10] and

Address for correspondence: P.D. Dr. Veronika Somoza, German Research Center for Food Chemistry, Lichtenbergstrasse 4, 85748 Garching, Germany. Voice: +49-89-289-14170; fax: +49-89-289-13248.
somoza@wisc.edu

refold aggregated proteins[10–12] in the cytosol. Inside the mitochondria, HSP60 proteins also help to fold proteins, such as actin and tubulin.[13]

An increased protein expression of HSPs in response to oxidative stress or protein glycation has already been demonstrated in *Xenopus laevis* oocytes.[14] Under physiological conditions, this stress response is beneficial and helps the cell to maintain normal function. Under pathological conditions, such as during the aging process, HSP expression is decreased.[15] In contrast, overexpression of HSPs in human fibroplast through repeated mild heat shock (RMHS) increases survival and has, therefore, anti-aging effects on cellular mechanims.[16]

In this work, we hypothesized that glycation of proteins or an amino acid results either in conformational changes or the release of reactive oxygen species and, thus, may increase the expression of HSPs. Therefore, the impact of CML and casein-linked CML on the expression of HSPs was studied in human intestinal Caco-2 cells by Western blot analysis and two-dimensional gel electrophoresis–matrix-assisted laser desorption/ionization (MALDI)– time-of-flight (TOF) mass spectroscopy (MS) techniques.

Methods

Chemicals

Unless stated otherwise, reagents and chemicals were purchased from Sigma (Deisenhofen, Germany). CML and β-amyloid were obtained from NeoMPS (Strasbourg, France) and Bachem (Weil am Rhein, Germany, H 6466), respectively. Human polyclonal antibody anti-HSP90 was purchased from Pineda (AK-service, Berlin, Germany); human polyclonal antibody anti-GAPDH was obtained from Dianova (Hamburg, Germany); and the secondary antibody anti-rabbit was purchased from New England Biolabs (Ipswitch, England).

Cell Culture

Human colon adenocarcinoma cells (Caco-2) were provided by the German Collection of Microorganisms and Cell Cultures (DSMZ, Braunschweig, Germany) and maintained in 90% Dulbecco's Eagle minimal essential medium with nonessential amino acids and Earle's balanced salt solution supplemented with 10% fetal calf serum, 2% L-glutamine, and 2% penicillin and streptomycin (PAA Laboratories, Wein, Austria, no. A15-151, M11-004, P11-010). Prior to each experiment, cells were treated with cell culture medium without fetal calf serum at 37°C and 5% CO_2 for 24 h. Afterward, time course experiments were carried out at the cell's confluence in a minimum of three biological replications. The compounds to which the cells were exposed to comprised casein-CML (3.7 mg/mL), CML (23 μg/mL), and β-amyloid (4.32 μg/mL). Except for casein-CML, which was prepared as described by Faist et al.,[17] all compounds were commercially obtained. Concentrations for casein-CML were chosen in accordance with previous experiments.[5] CML concentrations were adjusted to the CML content of casein-CML (70.5 g/kg [CML/casein-CML]). β-amyloid was used as another AGE compound-positive control because β-amyloid is also a ligand for the receptor for AGE (RAGE). The concentrations of β-amyloid were chosen in accordance with other studies[18] where an activation of nuclear factor kappa B (NF-κB) was shown.

After the cells were harvested, the samples were centrifuged (16,000g, 10 min), and the cell pellets were stored at −80°C until further analysis. After storage, pellets were resuspended in 500 μL lysis buffer and centrifuged again (16,000g, 10 min). Protein content was then determined in the supernatants by using the 2D-Quant Kit (GE Healthcare, Munich, Germany).

Sample preparation for MALDI–MS was performed using ZipTips (Qiagen, Hamburg, Germany) following the manufacturer's protocol.

Western Blotting

Equal amounts of protein were mixed with 3 × sample buffer (New England Biolabs), separated on 10–15% SDS-polyacrylamide gel, and transferred onto a polyvinylidene difluoride membrane (BioRad, Munich, Germany) via electrophoresis. The membrane was blocked with 10% blocking agent in Tris-buffered saline solution (TBS). Primary antibodies used for 1 h incubation were human polyclonal anti-GAPDH (diluted at 1:1000) and human polyclonal anti-HSP90 (1:2000). After washing with TBS and 0.05% Tween-20, membranes were incubated for 1 h with the horseradish peroxidase-conjugated secondary antibodies (1:1000), and washed. The signal was detected using a chemoluminescence detection kit (GE Healthcare) following the manufacture's protocol and analyzed by a Kodak Image station 2000R applying Kodak 1D software (Version 3.6; Munich, Germany).

Two-dimensional Electrophoresis

Two-dimensional analysis was performed as described by Görg et al.[19] Briefly, isoelectric focusing was carried out using 24-cm-fixed ampholyte gradients from pH 4.0 to 7.0 (GE Healthcare). In the second dimension, the proteins were separated by SDS-PAGE on 24 × 18 cm gels containing 13% acrylamide in an

TABLE 1. Percent changes of HSP90 protein expression in Caco-2 cells exposed to either casein-N^ϵ-(carboxymethyl)lysine (Casein-CML), CML, or β-amyloid for 0.5, 1.0, 2.0, 4.0, or 8.0 h (nontreated control cells = 100%)

	Time (h)					
	0.5	1.0	1.5	2.0	4	8
Casein-CML	−8.2 (±7.66)	−25.3* (±8.55)	+20.0 (±18.8)	+19.7*** (±2.57)	−15.4** (±4.48)	−28.9*** (±3.76)
CML	+71.0*** (±0.43)	+50.4** (±46.3)	+61.8** (±0.43)	+51.5* (±12.7)	−1.67 (±10.1)	−48.6 (±6.83)
β-amyloid	−35.8 (±4.12)	−0.40 (±10.2)	+18.9 (±14.6)	+19.0*** (±3.56)	−37.6** (±10.8)	−32.0*** (±6.91)

Data from at least triplicate Western blotting experiments ($n = 3$–5) are expressed as mean (± SEM) and were statistically analyzed by Student's t-test (*$P < 0.05$; **$P < 0.01$; ***$P < 0.001$).

Ettan Dalt chamber (GE Healthcare) and stained with Coomassie brilliant blue. For quantitative analysis, the software Melanie (ImageMaster 2D Platinum Software, Version 5.0 und 6.0; GE Healthcare) was used. Analyzed protein spots with a color intensity that differed from control cell protein spots by a factor of ≥1.3 were regarded as different from control cells.

MALDI–MS and MALDI–MS/MS Analysis

For MALDI–MS/MS identification, spots were excised and digested following the protocol of Schaefer et al.[20] and analyzed using an Ultraflex I ToF/ToF mass spectrometer (Bruker Daltonik, Rheinstetlen, Germany). Data analysis was performed using the BioTools (Bruker Daltonik) and Mascot (Matrix Science, London, UK) software packages.

Statistical Analysis

The unpaired Student's t-test was used to identify statistically significant differences ($P < 0.05$). Data are presented as mean values ± standard error mean.

Results and Discussion

Glycation of proteins and of amino acids has been demonstrated to be associated with the progression of various diseases, such as Diabetes mellitus[2] and Alzheimer's disease,[21] or aging.[2] Since glycation products are formed *in vivo* and are also absorbed from the habitual diet in considerable amounts,[22] a lifelong exposure to these compounds is inevitable. Therefore, we hypothesized that accumulation of glycated proteins or amino acids might induce HSPs as a defense system to counteract either the glycated and presumably conformationally changed proteins themselves or the increased formation of reactive oxygen species as cellular consequences of AGE's interaction with cell surface receptors.

First of all, the expression of HSP90 was quantified in human intestinal Caco-2 cells by Western blotting. HSP90 is the quantitatively predominant HSP in human cells that protects non-native proteins from misfolding.[9] The time course experiments revealed that exposure of Caco-2 cells to all of the tested compounds—casein-CML, CML, and β-amyloid—resulted in an increased HSP90 expression (TABLE 1). The maximum HSP90 expression for each of these compounds was reached after 2 h of treatment, at which point CML showed the strongest effect. The same result was obtained for shorter exposure times of 0.5 and 1.0 h, where the HSP90 expression increased by 71% and 50%, respectively.

Interestingly, treatment of the cells with β-amyloid, a nonglycated compound, showed a comparable effect as glycated casein-CML, whereas nonglycated casein did not change the HSP90 protein expression at a statistically significant level (data not shown). These results indicate that the induction of HSP90 protein expression might not only be a result of the presence of glycated compounds but might also be a consequence of binding to cell surface receptors and downstream signaling pathways.

In order to get a more comprehensive picture of the total of regulated HSPs, protein expression was then analyzed by two-dimensional gel electrophoresis and MALDI–TOF-MS. Again, cells were treated with casein, casein-CML, CML, or β-amyloid. As in the previous experiment, the maximum HSP90 expression was reached at 2 h, and the experiment was carried out applying the same exposure time and concentrations. FIGURE 1 shows a gel with protein spots for which a ≥1.3-fold change in intensity in treated cells compared to nontreated control cells was analyzed. Among

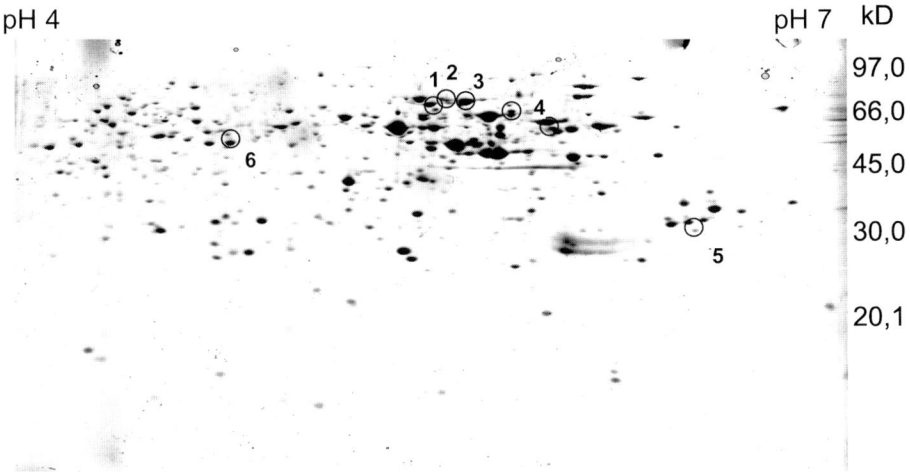

FIGURE 1. Picture of a two-dimensional gel of Caco-2 cells with selected identified HSPs and proteins of the proteasome system. The spot numbers have been identified to refer to the following proteins: 1, HSP70 binding protein; 2, HSP70 isoform 1; 3, HSP70 isoform 2; 4, HSP60; 5, HSP27; 6, S26 proteasome subunit 9.

TABLE 2. Relative changes of HSP and proteasome protein expression in Caco-2 cells exposed to either Casein-CML, CML, or β-amyloid for 2.0 h (nontreated control cells = 1.0)

	HSP27	HSP60	HSP-0 isoform 2	Proteasome S26 subunit 6	Proteasome S26 ATPase subunit 2
Casein-CML	3.40*	2.38***	1.41*	1.67*	1.80*
	(±0.74)	(±0.12)	(±0.16)	(±0.10)	(±0.27)
CML	1.38	2.12*	1.67*	1.34	2.11*
	(±0.40)	(±0.35)	(±0.27)	(±0.07)	(±0.36)
β-amyloid	1.96	2.58**	+1.91*	1.35	0.23***
	(±1.09)	(±0.08)	(±0.35)	(±0.56)	(±0.06)

Data from at least triplicate two-dimensional gel electrophoresis–matrix-assisted laser desorption/ionization–mass spectroscopy experiments ($n = 3$–5) are expressed as mean (± SEM) and were statistically analyzed by Student's t-test (*$P < 0.05$; **$P < 0.01$; ***$P < 0.001$).

all regulated proteins identified, the strongest effects were demonstrated for HSP27, HSP60, HSP70, and two proteasome subunits: S26, the ATP-independent subunit 6, and the ATP-dependent subunit 2. These proteasome subunits are involved in the degradation of misfolded proteins. Protein expression of the ATP-dependent proteasome S26 ATPase subunit 2 was induced by CML and casein-CML and inhibited by β-amyloid (TABLE 2). The stronger effect observed for CML compared to casein-CML is in agreement with the results obtained for the expression of HSP90, for which the ATP-activity is essential. In contrast, the ATP-independent proteasome S26 subunit 6 was only affected by casein-CML.

An induction of proteasome subunits in general indicate an increased proteolytic activity of the proteasome system. However, the present data do not provide any proof that CML or casein-CML is able to induce the degradation of glycated proteins or amino acids. This hypothesis has to be studied in future experiments.

Among the HSPs identified, expression of HSP27, HSP60, and HSP70 was upregulated by casein-CML, whereas treatment with the cells with CML and β-amyloid only induced the protein expression of HSP60 and HSP70 at a level of statistical significance (TABLE 2).

Interestingly, HSP27 was mainly upregulated by casein-CML, whereas HSP60 and HSP70 were predominantly affected by β-amyloid. Nonglycated casein, however, did not show any effect (data not shown).

The finding that β-amyloid, a nonglycated compound, also shows similar effects on the protein expression of selected HSPs as casein-CML and CML might lead to the speculation that these results are based on binding to the cell surface receptors RAGE. All of the tested compounds have been described as

RAGE ligands in earlier studies,[4–6] and it has also been demonstrated that binding to RAGE elicits an acute-phase response, including the activation of mitogen-activated kinases[6] and the NADPH-oxidase,[21] which is one of the main intracellular sources of superoxide anion radicals. As superoxide anion radicals are the substrate of the enzyme superoxide dismutase for which cooperation with HSP90 and HSP70 is discussed, the increase in their expression would support this theory.

Also, the remaining question of whether activation of HSPs and the proteasome system by nonenzymatic glycation products leads to their degradation cannot be answered by the results presented herein. Verbeke et al.[23] demonstrated the formation of CML to be effectively prevented in glyoxal-exposed human skin fibroblasts following RMHS treatment. Assuming an induction of HSPs during the RMHS treatment, one might speculate that this defense system is involved, although changes in the expression of HSPs or the proteasome system were not analyzed in this study.

In summary, it was shown for the first time that CML as a free molecule and as a protein-bound amino acid induces the protein expression of HSPs and proteasomes in Caco-2 cells. Whether these effects are based on binding to cell surface receptors, such as RAGE, and/or result in a degradation of glycated proteins has to be clarified in future studies.

Conflict of Interest

The authors declare no conflicts of interest.

References

1. BÜSER, W. & H.F. ERBERSDOBLER. 1986. Carboxymethyllysine, a new compound of heat damage in milk products. Milchwiss **41:** 780–785.
2. SCHLEICHER, E.D. et al. 1999. Increased accumulation of the glyoxidation product Nε-Carboxymethyllysine in human tissues in diabetes and aging. J. Clin. Invest. **99:** 457–468.
3. KISLINGER, T. et al. 1999. Nε-(carboxymethyl)lysine adducts of proteins are ligands for receptor for advanced glycation end products that activate cell signaling pathways and modulate gene expression. J. Biol. Chem. **274:** 31740–31749.
4. YAN, S.-D. et al. 2000. Receptor-depenednet cell stress and amyloid accumulation in systemic amyloidosis. Nat. Med. **6:** 643–651.
5. ZILL, H. et al. 2001. RAGE expression and AGE-induced MAP kinase activation in Caco-2-cells. Biochem. Biophys. Res. Comm. **288:** 1108–1111.
6. ZILL, H. et al. 2003. RAGE-mediated MAPK activation by food-derived AGE and non-AGE products. Biochem. Biophys. Res. Comm. **300:** 311–315.
7. MATSUMOTO, G. et al. 2006. Huntingtin and mutant SOD1 form aggregate structures with distinct molecular properties in human cells. J. Biol. Chem. **281:** 4477–4485.
8. HENDRICK, J.P. & F.U. HARTL. 1993. Molecular chaperone functions of heat-shock proteins. Annu. Rev. Biochem. **62:** 349–384.
9. OBERMANN, W.M. et al. 1998. In vivo function of HSP90 is dependent on ATP binding and ATP hydrolysis. J. Cell. Biol. **143:** 901–910.
10. LANGER, T. & W. NEUPERT. 1991. Heat shock proteins hsp60 and hsp70: their roles in folding, assembly and membrane translocation of proteins. Curr. Top. Microbiol. Immunol. **167:** 3–30.
11. HASLBECK, M. et al. 2005. Disassembling protein aggregates in the yeast cytosol. The cooperation of Hsp26 with Ssa1 and Hsp104. J. Biol. Chem. **280:** 23861–23868.
12. GOLOUBINOFF, P. et al. 1999. Sequential mechanism of solubilization and refolding of stable protein aggregates by a bichaperone network. Proc. Natl. Acad. Sci. USA **96:** 13732–13737.
13. RÜDIGER, S. et al. 2001. Its substrate specificity characterizes the DnaJ cochaperone as a scanning factor for the DnaK chaperone. Embo. J. **20:** 1042–1050.
14. MIFFLIN, L.C. & R.E. COHEN. 1994. Characterization of denatured protein inducers of the heat shock (stress) response in Xenopus laevis oocytes. J. Biol. Chem. **269:** 15710–15717.
15. FEDER, M.E. & G.E. HOFMANN. 1999. Heat-shock proteins, molecular chaperones, and the stress response: evolutionary and ecological physiology. Annu. Rev. Physiol. **61:** 243–282.
16. RATTAN, S.I.S. 1998. Repeated mild heat shock delays ageing in cultured human skin fibroplasts. Biochem. Mol. Biol. Int. **45:** 753–759.
17. FAIST, V. et al. 2001. Selective fortification of lysinoalanine, fructoselysine, and Nε-carboxymethyllysine in casein model systems. Nahrung **45:** 218–221.
18. KALTSCHMIDT, B. et al. 1997. Transcription factor NF-κB is activated in primary neurons by amyloid β peptides and in neurons surrounding early plaques from patients with Alzheimer disease. Proc. Natl. Acad. Sci. USA **94:** 2642–2647.
19. GÖRG, A. et al. 2000. The current state of two-dimensional electrophoresis with immobilized pH gradients. Electrophoresis **21:** 1037–1053.
20. SCHAEFER, H. et al. 2001. Identification of peroxisomal membrane proteins of Saccharomyces cerevisiae by mass spectrometry. Electrophoresis **22:** 2955–2968.
21. MUENCH, G. et al. 1998. Alzheimer's disease-synergistic effects of glucose deficit, oxidative stress and advanced glycation endproducts. J. Neural. Transm. **105:** 439–461.
22. SOMOZA, V. et al. 2006. Dose-dependent utilisation of casein-linked lysinoalanine, N(epsilon)-fructoselysine and N(epsilon)-carboxymethyllysine in rats. Mol. Nutr. Food Res. **50:** 833–841.
23. VERBEKE, P. et al. 2002. Hormetic action of mild heat stress decreases the inducibility of protein oxidation and glycoxidation in human fibroblasts. Biogerontology **3:** 117–120.

Reversal of Hyperglycemia-Induced Angiogenesis Deficit of Human Endothelial Cells by Overexpression of Glyoxalase 1 In Vitro

USMAN AHMED,[a,b] DARIN DOBLER,[a] SARAH J. LARKIN,[a,b] NAILA RABBANI,[a,b] AND PAUL J. THORNALLEY[a,b]

[a]*Department of Biological Sciences, University of Essex, Colchester, United Kingdom*

[b]*Clinical Sciences Research Institute, Warwick Medical School, University of Warwick, Coventry, United Kingdom*

Dicarbonyl glycation of RGD and GFOGER sites in type IV collagen has been associated with decreased angiogenesis. In this study, we investigated whether overexpression of glyoxalase 1 to decrease dicarbonyl glycation would prevent the angiogenesis deficit induced by hyperglycemia *in vitro*. Transfection of human microvascular endothelial cells resulted in a four-fold increase in glyoxalase 1 activity compared with controls. Incubation of human microvascular endothelial cells in model hyperglycemia produced a 32% decrease in formation of tube structures that was prevented by glyoxalase 1 overexpression. We conclude that increased protection against dicarbonyl glycation of endothelial cell protein protects hyperglycemia-induced angiogenesis deficit.

Key words: glyoxalase; angiogenesis; endothelial cells; diabetes; glycation; methylglyoxal

Introduction: Angiogenesis Deficit in Early Stages of Microvascular Complications of Diabetes

Early stages of retinopathy, neuropathy, and nephropathy and impaired wound healing in clinical diabetes are associated with decreased angiogenesis. According to the "ischemia-angiogenesis" hypothesis, the resulting ischemia, arising from decreased blood flow to surrounding tissue, may lead to later excessive angiogenesis in response to the expression of angiogenic growth factors including vascular endothelial-derived growth factor (VEGF), transforming growth factor-β, basic fibroblast growth factor, and connective tissue growth factor (CTGF). In the retina, the process starts in isolated capillaries with the loss of pericytes and endothelial cells (ECs) forming acellular capillaries. The capillary structure is thereby weakened severely and becomes occluded. This spreads to groups of capillaries and arterioles.[1]

Glycation of proteins, forming advanced glycation endproduct (AGE) residues, has been associated with both decreased and increased angiogenesis. Impaired collagenolysis following AGE residue formation has been linked to decreased angiogenesis and induction of increased angiogenic growth factors by AGE-modified proteins has been linked to increased angiogenesis. AGE formation may lead to decreased degradation of vascular collagen despite activation of matrix metalloproteinases,[2,3] and hence decreased angiogenesis. In one study, collagenolysis was impaired in diabetic mice in ischemia-induced angiogenesis and was restored by treatment with aminoguanidine,[2] a scavenger of dicarbonyl glycating agents.[4] Glycation of angiogenic growth factors has also been implicated in the angiogenesis deficit in diabetes.[5,6] Increased angiogenesis in response to proteins highly modified by AGEs *in vitro* is mediated by induction of VEGF and CTGF expression.[7,8]

Recent research has shown that protein glycation by α-oxoaldehydes, such as methylglyoxal (MG)—formed mainly by the degradation of cellular glycolytic intermediates—forms quantitatively important AGEs of cellular and extracellular proteins *in vivo*. Glycation by MG is mainly arginine residue-directed.[9] The plasma concentration of MG is increased in

Address for correspondence: Paul J. Thornalley, Clinical Sciences Research Institute, Warwick Medical School, University of Warwick, Coventry CV2 2DX, UK. Voice: +4424 7696 8594; fax +4424 7696 8595.
P.J.Thornalley@warwick.ac.uk

hyperglycemia associated with clinical diabetes,[10] giving rise to increased modification of vascular basement membrane type IV collagen. MG glycation forms arginine-derived hydroimidazolone residues at hotspot modification sites in RGD and GFOGER integrin-binding sites of collagen, causing EC detachment, anoikis, and inhibition of angiogenesis. ECs incubated in model hyperglycemia *in vitro* and experimental diabetes *in vivo* have produced the same modifications of vascular collagen, inducing similar responses.[11] Pedchenko et al. found similar impaired adhesion of ECs, mesangial cells, and podocytes to extracellular matrix (ECM) proteins glycated by MG.[12] These EC–ECM interactions are essential for angiogenesis.[13,14]

Glycation of the cellular protein of ECs by dicarbonyls—particularly integrins—may also lead to functional impairment. Glyoxalase 1 (Glo1) is the major cellular enzyme that catalyzes the metabolism of MG and thereby protects against dicarbonyl glycation.[15] To assess the role of dicarbonyl glycation of EC proteins in the hyperglycemia-induced angiogenesis deficit, we overexpressed Glo1 in ECs and investigated the effect on hyperglycemia-induced impairment of angiogenesis *in vitro*.

Methods

Glo1 cDNA in the pUC19 plasmid was provided by Dr. S. Ranganathan (FoxChase Cancer Center, Philadelphia, PA). PCR with primers containing appropriate restriction enzyme sites for the destination plasmid was performed to enable insertion of Glo1 cDNA into pIRES2-EGFP plasmid vector to produce pIRES2-EGFP-Glo1. An analytical digest was performed and indicated correct insertion of the Glo1 cDNA (approximately 622 bp). An empty plasmid served as a control (sham). Human dermal microvascular ECs (HMEC-1 cells) were seeded at 1.25×10^5 cells per well (per 9.5 cm^2) in 6-well microplates. After incubation for 24 h, a mixture of DNA/lipofectamine 2000 (1:1) was added and the cells were incubated for a further 24 h. The transfection efficiency was assessed by green fluorescent protein positive fluorescence. Glo1 activity of transfected cells was assayed by the spectrophotometric method[16] and protein content by the Bradford assay. Cells transfected with pIRES2-EGFP-Glo1 or pIRES2-EGFP were grown in model normoglycemia and hyperglycemia (5 mM and 20 mM glucose, respectively) for 3 days and then transferred to 5 mM glucose and plated onto Matrigel (ECM from the Engelbreth-Holm-Swarm tumor; Sigma Poole, UK) and incubated at 37 °C for 24 h. EC tube formation was quantified by image analysis of 100 x phase contrast images of eight random fields using NIH Image software (developed at the U.S. National Institutes of Health, USA, and available on the Internet at http://rsb.info.nih.gov/nih-image/).[11]

Results

The transfection efficiency of ECs with pIRES2-EGFP-Glo1 and pIRES2-EGFP vectors was 60–70%. ECs transfected with pIRES2-EGFP-Glo1 had a fourfold increase in Glo1 activity compared with controls (447.9 ± 44.8 mU/mg vs. 114.5 ± 20.4 mU/mg; data are expressed as mean ± SD, n = 3, P < 0.01). ECs incubated in hyperglycemia prior to assessment of the formation of tube-like structures showed an angiogenesis deficit that was corrected by overexpression of Glo1. Tube formation after 24 h (total tube length per field, μm; mean ± SD, n = 8) was: sham transfectant cells in normoglycemia 216 ± 61, Glo1 transfectant cells in normoglycemia 220 ± 59, sham transfectant cells in hyperglycemia 146 ± 39 ($p \leq 0.05$, with respect to sham transfectant in normoglycemia), and Glo1 transfectant in hyperglycemia 207 ± 44 ($p \leq 0.05$, with respect to sham transfectant in hyperglycemia).

Discussion

Impairment of angiogenesis by glycation in diabetes may contribute to tissue ischemia, impairment of wound healing, and coronary collateral vessel development.[1] Decreased formation of tube-like structures by ECs incubated under hyperglycemic conditions and prevention of this by overexpression of Glo1 suggests that: (i) dysfunction of ECs in hyperglycemia may contribute to decreased angiogenesis, and (ii) this effect may be mediated by glycation of cellular proteins by Glo1 substrates—glyoxal and MG. Increased dicarbonyl glycation of ECM proteins is also involved.[11,17] Arg-261 of integrin $\alpha_V\beta_3$ is a possible site of dicarbonyl glycation leading to impaired engagement of integrins in angiogenesis.[18]

We conclude that increased protection against dicarbonyl glycation of EC protein by overexpression of Glo1 protects against hyperglycemia-induced angiogenesis deficit. This implicates dicarbonyl glycation of integrins or related proteins in ECs in mediating decreased angiogenesis in hyperglycemia. Similar processes may contribute to the anti-angiogenic activity of cell permeable Glo1 inhibitors and their antitumor activity *in vivo*.[19]

Acknowledgments

We thank the Wellcome Trust and Cancer Research UK for support for our research.

Conflict of Interest

The authors declare no conflicts of interest.

References

1. MARTIN, A., M.R. KOMADA & D.C. SANE. 2003. Abnormal angiogenesis in diabetes mellitus. Medicinal Res. Revs. **23:** 117–145.
2. TAMARAT, R., J.S. SILVESTRE, M. HUIJBERTS, et al. 2003. Blockade of advanced glycation end-product formation restores ischemia-induced angiogenesis in diabetic mice. Proc. Natl. Acad. Sci. USA **100:** 8555–8560.
3. KUZUYA, M., S. SATAKE, S. AI, et al. 1998. Inhibition of angiogenesis on glycated collagen lattices. Diabetologia **41:** 491–499.
4. THORNALLEY, P.J., A. YUREK-GEORGE & O.K. ARGIROV. 2000. Kinetics and mechanism of the reaction of aminoguanidine with the a-oxoaldehydes, glyoxal, methylglyoxal and 3-deoxyglucosone under physiological conditions. Biochem. Pharmacol. **60:** 55–65.
5. FACCHIANO, F., A. LENTINI, V. FOGLIANO, et al. 2002. Sugar-induced modification of fibroblast growth factor 2 reduces its angiogenic activity in vivo. Am. J. Pathol. **161:** 531–541.
6. GIARDINO, I., D. EDELSTEIN & M. BROWNLEE. 1994. Nonenzymatic glycosylation in-vitro and in bovine endothelial cells alters basic fibroblast growth-factor activity—a model for intracellular glycosylation in diabetes. J. Clin. Invest. **94:** 110–117.
7. YAMAGISHI, S., Y. YONEKURA, Y. YAMAMOTO, et al. 1997. Advanced glycation end products-driven angiogenesis *in vitro*. J. Biol. Chem. **272:** 8723–8730.
8. TWIGG, S.M., M.M. CHEN, A.H. JOLY, et al. 2001. Advanced glycosylation end products up-regulate connective tissue growth factor (insulin-like growth factor-binding protein-related protein 2) in human fibroblasts: a potential mechanism for expansion of extracellular matrix in diabetes mellitus. Endocrinology **142:** 1760–1769.
9. AHMED, N., D. DOBLER, M. DEAN & P.J. THORNALLEY. 2005. Peptide mapping identifies hotspot site of modification in human serum albumin by methylglyoxal involved in ligand binding and esterase activity. J. Biol. Chem. **280:** 5724–5732.
10. MCLELLAN, A.C., P.J. THORNALLEY, J. BENN & P.H. SONKSEN. 1994. The glyoxalase system in clinical diabetes mellitus and correlation with diabetic complications. Clin. Sci. **87:** 21–29.
11. DOBLER, D., N. AHMED, L.J. SONG, et al. 2006. Increased dicarbonyl metabolism in endothelial cells in hyperglycemia induces anoikis and impairs angiogenesis by RGD and GFOGER motif modification. Diabetes **55:** 1961–1969.
12. PEDCHENKO, V., R. ZENT & B.G. HUDSON. 2004. $\alpha_v\beta_3$ and $\alpha_v\beta_5$ integrins bind both the proximal RGD site and non-RGD motifs within noncollagenous (NC1) domain of the $\alpha3$ chain of type IV collagen: implication for the mechanism of endothelial cell adhesion. J. Biol. Chem. **279:** 2772–2780.
13. SWEENEY, S.M., G. DILULLO, S.J. SLATER, et al. 2003. Angiogenesis in collagen I requires alpha(2)beta(1) ligation of a GFP* GER sequence and possibly p38 MAPK activation and focal adhesion disassembly. J. Biol. Chem. **278:** 30516–30524.
14. HYNES, R.O. 2002. A reevaluation of integrins as regulators of angiogenesis. Nat. Med. **8:** 918–921.
15. THORNALLEY, P.J. 2003. Glyoxalase I—structure, function and a critical role in the enzymatic defence against glycation. Biochem. Soc. Trans. **31:** 1343–1348.
16. MCLELLAN, A.C. & P.J. THORNALLEY. 1989. Glyoxalase activity in human red blood cells fractionated by age. Mech. Ageing Dev. **48:** 63–71.
17. PEDCHENKO, V.K., S.V. CHETYRKIN, P. CHUANG, et al. 2005. Mechanism of perturbation of integrin-mediated cell-matrix interactions by reactive carbonyl compounds and its implication for pathogenesis of diabetic nephropathy. Diabetes **54:** 2952–2960.
18. XIONG, J.P., T. STEHLE, B. DIEFENBACH, et al. 2001. Crystal structure of the extracellular segment of integrin alpha V beta 3. Science **294:** 339–345.
19. THORNALLEY, P.J., L.G. EDWARDS, Y. KANG, et al. 1996. Antitumour activity of *S-p*-bromobenzylglutathione cyclopentyl diester *in vitro* and *in vivo*. Inhibition of glyoxalase I and induction of apoptosis. Biochem. Pharmacol. **51:** 1365–1372.

Pathophysiological Role of the Glyoxalase System in Renal Hypoxic Injury

TAKANORI KUMAGAI, MASAOMI NANGAKU, AND REIKO INAGI

Division of Nephrology and Endocrinology, University of Tokyo School of Medicine, Tokyo, Japan

Methylglyoxal (MG), a reactive dicarbonyl compound mainly produced by metabolic pathways, such as glycolysis, binds to proteins or nucleic acids and forms advanced glycation end products. MG is efficiently metabolized by the glyoxalase system where MG is converted by glyoxalase I (GLO I) to S-D-lactoylglutathione. Although the glyoxalase system has been shown to play a pathological role in various diseases, including diabetic complications, its detailed pathophysiological function remains to be elucidated. We are interested in renal hypoxic diseases, but very little information is available regarding the association between the glyoxalase system and renal hypoxic diseases. Therefore, we investigated the biological role of GLO I in renal hypoxic diseases by using the rat ischemia/reperfusion (I/R) injury model. I/R induced the reduction of renal GLO I activity associated with morphological changes and renal dysfunction. Interestingly, the rats that overexpress human GLO I (GLO I Tg rats) showed amelioration of these manifestations in renal I/R (e.g., improvement of the tubulointerstitial injury and renal function). Accumulation of renal MG adducts, carboxyethyllysine, induced by I/R also decreased in GLO I Tg rats compared to wild-type rats. These results demonstrate that GLO I has renoprotective effects in I/R injury via reduction of protein modification by MG.

Key words: methylglyoxal; glyoxalase I; renal ischemia/reperfusion; tubulointerstitial injury; proximal tubular cells; acute renal failure

Methylglyoxal, a Precursor of Advanced Glycation End Products

Chemical modification of proteins by reactive aldehydes and ketones is a well-recognized phenomenon in diabetes. Such modifications occur through a reaction known as the Maillard reaction. Through a series of intermediates, this reaction leads to the formation of irreversible adducts on proteins that are known as advanced glycation end products (AGEs).[1] AGEs are thought to be responsible, in part, for diabetic complications, such as retinopathy, nephropathy, and vascular injury. α-Dicarbonyl compounds are thought to be the major precursors of AGEs, and, among these, methylglyoxal (MG) has recently received considerable attention as a mediator to form AGEs. A recent study on the formation of AGEs in endothelial cells cultured under hyperglycemic conditions indicated that MG was the major precursor of AGEs.[2] It has been reported that MG primarily reacts with arginine residues to form N^δ-(5-hydroxy-4,6-dimethylpyrimidine-2-yl)-L-ornithine (argpyrimidine)[3] and also reacts with lysine residues to generate N^ε-(carboxyethyl)lysine (CEL).[4] In addition, MG is thought to be a toxic substance.[5a]

Glyoxalase System for Detoxification of Methylglyoxal

The glyoxalase system detoxifies MG and is composed of two enzymes: glyoxalase I (GLO I), which metabolizes MG to S-D-lactoylglutathione, and GLO II, which converts S-D-lactoylglutathione to D-lactate.[5] The glyoxalase system was originally described in 1913, but the complete function remains to be elucidated. Shinohara *et al.* demonstrated that overexpression of GLO I in bovine endothelial cells reduced intracellular AGEs when the cells were cultured in the presence of high glucose.[2] Moreover, a series of studies indicated that certain tumor primary cultures and cell lines overexpress GLO I, suggesting that increased amounts of this enzyme prevent tumor cell apoptosis, possibly by limiting MG production.

Address for correspondence: Reiko Inagi, Ph.D., Division of Nephrology and Endocrinology, University of Tokyo School of Medicine, 7-3-1 Hongo, Bunkyo-ku, Tokyo 113–8655, Japan. Voice: +81-3-3815-5411 (ext. 33128); fax: +81-3-5800-8806.

inagi-npr@umin.ac.jp

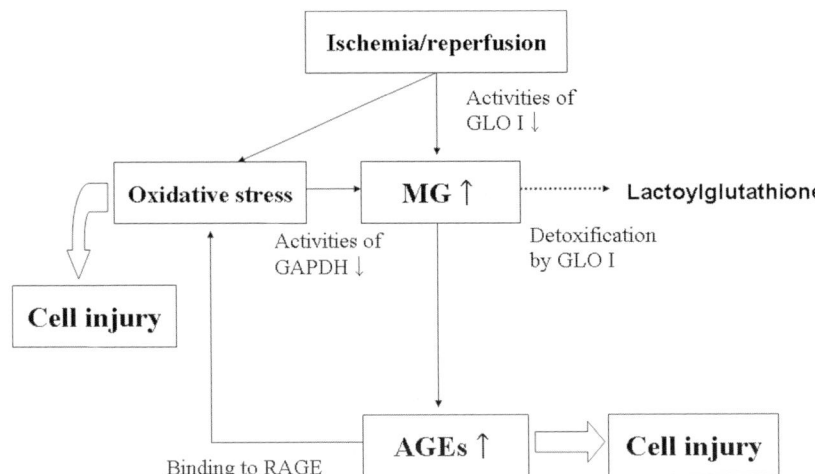

FIGURE 1. Proposed mechanism for improvement of renal ischemia/reperfusion injury by overexpressing glyoxalase I (GLO I). Ischemia/reperfusion leads to elevation of methylglyoxal (MG) from the decrease in GLO I activity. MG reacts with proteins to form advanced glycation end products (AGEs). The interaction of AGEs with the receptor for AGE (RAGE) leads to oxidative stress. Oxidative stress may further increase MG. Overexpression of GLO I detoxifies MG and can cut off the vicious cycle. GAPDH, glyceraldehyde-3-phosphate dehydrogenase.

Pathophysiological Functions of the Glyoxalase System in Diseases

MG level is increased in various oxidative stress-related diseases. The glyoxalase system-related pathological conditions include diabetic nephropathy,[6] diabetic retinopathy,[7] Alzheimer's disease,[8] anxiety,[9] and resistance to anticancer drugs.[10] In addition, GLO I deficiency was associated with unusually high plasma levels of AGEs in a hemodialysis patient.[11] The list of the glyoxalase system-related diseases continues to expand.

Glyoxalase System and Kidney Diseases

Regarding the relationship between the glyoxalase system and kidney diseases, diabetic nephropathy has been the focus of intensive research. For example, Beisswenger *et al.* demonstrated that MG levels are elevated in diabetic subjects with more rapid progression of nephropathy.[6] On the other hand, the relationship between acute kidney injury and the glyoxalase system remains to be elucidated despite acute kidney injury being a significant health care concern with high mortality rates.

Role of the Glyoxalase System in Renal Hypoxic Disease

Ischemia from hypotension or sepsis is the most common cause of human acute kidney injury, and ischemia/reperfusion (I/R) injury is a significant problem in kidney transplantation, indicating that renal I/R injury is clinically important. The mechanisms underlying renal I/R injury are most likely multifactorial and interdependent and involve hypoxia, inflammatory responses, and free radical damage. Recently, several reports demonstrated that MG is increased in cardiac or cerebral I/R. For example, in cardiac I/R, MG is increased in heart homogenates,[12] and in cerebral I/R, MG–arginine adducts (argpyrimidine) have been detected in arterial walls within the infracted zone after 24 h reoxygenation.[13] These findings spurred us to investigate the relationship between the glyoxalase system and renal I/R injury.

We therefore investigated the biological role of GLO I in the rat renal I/R injury model. I/R was induced by occlusion of the left renal vessels for 45 min, followed by 24 h reperfusion. I/R induced the reduction of GLO I activity in the kidney, which was associated with morphological damage and renal dysfunction. Interestingly, in the rats overexpressing human GLO I (GLO I Tg rats),[14] these manifestations were significantly ameliorated (e.g., reduction of tubulointerstitial injury and apoptosis of tubular cells and improvement of renal function). Accumulation of renal MG adduct, CEL, induced by I/R was also decreased in GLO I Tg rats compared with the wild type as shown by immunohistochemical techniques. In an *in vitro* study, we exposed immortalized rat proximal tubular cells (IRPTC) or GLO I knocked-down IRPTC with small interfering (si)RNA to hypoxia (0.2% O_2) followed by reoxygenation (95% air). Knock-down of GLO I exacerbates the cellular injury estimated by the lactate dehydrogenase assay.

Taken together, GLO I has renoprotective effects in I/R injury via reduction of protein modification by MG in the tubular cells. The proposed mechanism of improvement of renal I/R injury by overexpressing GLO I is shown in FIGURE 1.

Conclusions

GLO I has renoprotective effects in renal I/R injury by reducing AGEs in the tubular cells. Further studies are required to elucidate a mechanism for the reduction of GLO I activity by I/R. Our findings provide an important insight in understanding the pathophysiology of renal I/R injury and suggest a possible therapeutic approach targeting the glyoxylase system.

Acknowledgments

This research was supported by Grants-in-Aid for Scientific Research from the Japan Society for the Promotion of Science (19590939 to R.I. and 19390228 to M.N.).

Conflict of Interest

The authors declare no conflicts of interest.

References

1. BROWNLEE, M., A. CERAMI & H. VLASSARA. 1988. Advanced glycosylation end products in tissue and the biochemical basis of diabetic complications. N. Engl. J. Med. **318:** 1315–1321.
2. SHINOHARA, M. *et al.* 1998. Overexpression of glyoxalase-I in bovine endothelial cells inhibits intracellular advanced glycation endproduct formation and prevents hyperglycemia-induced increases in macromolecular endocytosis. J. Clin. Invest. **101:** 1142–1147.
3. SHIPANOVA, I.N., M.A. GLOMB & R.H. NAGARAJ. 1997. Protein modification by methylglyoxal: chemical nature and synthetic mechanism of a major fluorescent adduct. Arch. Biochem. Biophys. **344:** 29–36.
4. AHMED, M.U. *et al.* 1997. N-epsilon-(carboxyethyl)lysine, a product of the chemical modification of proteins by methylglyoxal, increases with age in human lens proteins. Biochem. J. **324:** 565–570.
5a. KALAPOS, M.P. 1999. Methylglyoxal in living organisms: chemistry, biochemistry, toxicology and biological implications. Toxicol. Lett. **22:** 145–175.
5. THORNALLEY, P.J. 1990. The glyoxalase system: new developments towards functional characterization of a metabolic pathway fundamental to biological life. Biochem. J. **269:** 1–11.
6. BEISSWENGER, P.J. *et al.* 2005. Susceptibility to diabetic nephropathy is related to dicarbonyl and oxidative stress. Diabetes **54:** 3274–3281.
7. MILLER, A.G. *et al.* 2006. Glyoxalase I is critical for human retinal capillary pericyte survival under hyperglycemic conditions. J. Biol. Chem. **281:** 11864–11871.
8. KUHLA, B. *et al.* 2005. Methylglyoxal, glyoxal, and their detoxification in Alzheimer's disease. Ann. N.Y. Acad. Sci. **1043:** 211–216.
9. HOVATTA, I. *et al.* 2005. Glyoxalase 1 and glutathione reductase 1 regulate anxiety in mice. Nature **438:** 662–666.
10. SAKAMOTO, H. *et al.* 2000. Glyoxalase I is involved in resistance of human leukemia cells to antitumor agent-induced apoptosis. Blood **95:** 3214–3218.
11. MIYATA, T. *et al.* 2001. Glyoxalase I deficiency is associated with unusual level of advanced glycation end products in hemodialysis patient. Kidney Int. **60:** 2351–2359.
12. LOREDANA, G.B. *et al.* 2006. Receptor for advanced-glycation end products: key modulator for myocardial ischemic injury. Circulation **113:** 1226–1234.
13. OYA, T. *et al.* 1999. Methylglyoxal modification of protein. Chemical and immunochemical characterization of methylglyoxal-arginine adducts. J. Biol. Chem. **274:** 18492–18502.
14. INAGI, R. *et al.* 2002. Efficient in vitro lowering of carbonyl stress by the glyoxalase system in conventional glucose peritoneal dialysis fluid. Kidney Int. **62:** 679–687.

A419C (E111A) Polymorphism of the Glyoxalase I Gene and Vascular Complications in Chronic Hemodialysis Patients

MARTA KALOUSOVÁ,[a] ALEXANDRA GERMANOVÁ,[a] MARIE JÁCHYMOVÁ,[a] OTO MESTEK,[b] VLADIMÍR TESAŘ,[c] AND TOMÁŠ ZIMA[a]

[a]*Institute of Clinical Chemistry and Laboratory Diagnostics, First Faculty of Medicine and General University Hospital, Charles University, Prague, Czech Republic*

[b]*Institute of Chemical Technology, Prague, Czech Republic*

[c]*Department of Nephrology, First Faculty of Medicine and General University Hospital, Charles University, Prague, Czech Republic*

Advanced glycation end products (AGEs) take part in the pathogenesis of vascular, diabetic, and uremic complications. Their precursors are detoxified by the glyoxalase system. Our aim was to study A419C (E111A) single nucleotide polymorphism (SNP) of the glyoxalase I gene in hemodialysis (HD) patients. A419C SNP, several laboratory parameters including soluble receptor for AGEs (sRAGE), and clinical data were studied in 214 HD patients and 89 controls. Allelic and genotypic frequencies did not differ between HD patients and controls. A419C SNP was significantly linked with serum sRAGE, which sensitively reflects the AGE burden of the organism (3986 ± 1638 pg/mL in the CC variant versus 3277 ± 1398 pg/mL in the AC variant and 3297 ± 1445 pg/mL in the AA variant, $P < 0.01$). In the CC variant, significantly higher prevalence of cardiovascular disease and peripheral vascular disease was found, while the prevalence of hypertension, diabetes mellitus, and dyslipidemia did not differ between genotypes. In summary, in this study we demonstrate for the first time the association of A419C polymorphism of the glyoxalase I gene with sRAGE levels and show the genetic predisposition to vascular complications in HD patients.

Key words: advanced glycation end products; cardiovascular; glyoxalase; hemodialysis; polymorphisms; sRAGE

Introduction

Advanced glycation end products (AGEs) take part in the pathogenesis of diabetic, vascular, as well as uremic complications. Glyoxalase I is a zinc-binding enzyme that has an outstanding role in the metabolism of the major AGE precursors: methylglyoxal and glyoxal. The gene for glyoxalase I is located on chromosome 6 (locus 6p21,3–6p21,2), close to the major histocompatibility complex HLA-DR.[1] Seventy single nucleotide polymorphisms (SNPs) of the glyoxalase I gene were identified.[2] The A419C polymorphism (also E111A or Glu111Ala polymorphism)[3] is the only polymorphism within the coding sequence.

Since AGEs accumulate in patients with decreased renal function and cardiovascular risk is increased in these patients compared to the general population, our aim was to study A419C SNP of the glyoxalase I gene in hemodialysis (HD) patients in relation to their clinical status and laboratory characteristics.

Subjects and Methods

Study Population

The study group was comprised of 214 unrelated, Caucasian, chronic HD patients from six centers, who were in stable clinical status; 89 unrelated, Caucasian, healthy subjects served as controls. The study was approved by the Ethical Committee of the First Faculty of Medicine and General University Hospital, Prague, Czech Republic. All patients gave their informed consent prior to entering the study. The

Address for correspondence: Marta Kalousová, M.D., Ph.D., Institute of Clinical Chemistry and Laboratory Diagnostics, 1st Faculty of Medicine and General University Hospital, Charles University, Karlovo nám. 32, 121 11 Prague 2, Czech Republic. Voice: +420-224966620; fax: +420-2249642848.

marta.kalousova@seznam.cz, mkalousova@hotmail.com

TABLE 1. Laboratory characteristics of hemodialysis (HD) patients and healthy controls

Parameter	HD patients	Controls
Number of patients (men/women)	214 (119/95)	89 (31/58)*
Age (years)	63.1 ± 13.4	56.6 ± 8.3**
Hemoglobin (g/L)	106 ± 12.9	142 ± 10.4**
Leukocytes ($\times 10^9$/L)	6.91 ± 1.95	6.25 ± 1.56*
Creatinine (μmol/L)	757 ± 204	83 ± 14**
Albumin (g/L)	38.1 ± 3.7	45.1 ± 2.5**
Cholesterol (mmol/L)	4.8 ± 1.1	5.5 ± 0.9**
CRP (mg/L)	9.3 ± 15.3	3.2 ± 2.5**
Glucose (mmol/L)	6.2 ± 2.3	5.2 ± 0.9**
HbA1c (%) ($n = 64$)	5.0 ± 1.8	not assessed
AGE-related fluorescence (AU)	11.2 ± 2.7 × 10^5	3.8 ± 1.9 × 10^5**
sRAGE (pg/mL)	3427 ± 1508	1758 ± 637**
Zinc (μmol/L)	10.4 ± 2.2	13.4 ± 2.2**

Data are expressed as mean ± SD. *$P < 0.005$, **$P < 0.001$, HD versus controls.

study is registered as a clinical trial in The Cochrane Renal Group Registry <http://www.cochrane-renal.org/dbsearch.php> CRG110500022. The group of HD patients consisted of 56% men and 44% women (mean age 63.1 ± 13.4 years). Their dialysis treatment lasted, on average, for 2 years, and arteriovenous fistule was used for dialysis. Thirty-one percent of patients suffered from diabetes mellitus, and 41% had dyslipidemia. The case history of the patients included hypertension (87%), cardiovascular disease (61%), cerebrovascular diseases (21%), and peripheral vascular disease (26%). TABLE 1 lists the basic laboratory characteristics of the studied patients.

Samples

Before starting the dialysis session, blood was collected via puncture of the arteriovenous fistule in the HD patients and via puncture of the cubital vein in the controls. Tubes containing EDTA were used for DNA analysis. For biochemical analysis, blood was centrifuged for 10 min at 1450 g and serum was frozen at −80°C.

Glyoxalase I Polymorphism Genotyping

SNP E111A in exon 4 is caused by the change of adenine–cytosine in mRNA position 419. The region containing E111A polymorphism (203 bp long) was amplified by PCR with primers (forward primer 5′GCA GGG GTT AGG CCA ATT AT3′ and reverse primer 5′CAG GCA AAC TTA CCG AAT CC3′) with the initial denature at 92°C for 5 min followed by 30 cycles of 94°C for 30 s, 60°C for 30 s, and 72°C for 1 min and additionally 68°C for 5 min. The primers were predicted by Primer3 Input (http://frodo.wi.mit.edu/). For restriction analysis, restriction endonuclease Bsm AI at 37°C overnight was used. Fragment sizes were assessed using NebCutter V2.0 (http://tools.neb.com/NEBcutter2) as follows: 143 bp and 60 bp for the wild-type allele 419A and 203 bp for the mutant allele 419C. Products were separated by electrophoresis in 3% agarose gel and visualized in UV light after ethidium bromide staining. The restriction fragment length polymorphism (RFLP) method for detection of E111A polymorphism was confirmed by direct sequencing (CEQ^{TM} 8000, Genetic Analysis System, Beckman Coulter, Fullerton, CA) of the PCR product.

Biochemical Analyses

The soluble receptor for AGEs (sRAGE) was measured with enzyme-linked immunosorbent assay (ELISA) using standard kits (Quantikine, RD Systems, Minneapolis, MN). AGEs-related fluorescence was measured spectrofluorimetrically. Other parameters were measured by standard methods. CRP was measured turbidimetrically; HbA1c with high performance liquid chromatography (VARIANT II; Biorad, Hercules, CA) and zinc with atomic absorption spectroscopy (SPECTR AA ZZO FS; Varian, Palo Alto, CA). Other parameters were measured by standard clinical-chemistry methods recommended by the IFCC (International Federation of Clinical Chemistry). Blood count was measured with an automated hematological analyzer.

Statistical Analysis

The results of the biochemical parameters are expressed as mean ± standard deviation. Comparison of groups was performed with analysis of variance

TABLE 2. Genotype frequencies of A419C SNP of the glyoxalase I gene in HD patients HD and controls

Patients/gynotype	AA	AC	CC
HD patients (HWE)	32.7% (26.4%)	37.4% (50.0%)	29.9% (23.6%)**
Controls (HWE)	32.6% (27.3%)	39.3% (49.9%)	28.1% (22.8%)*

Determined frequencies and (expected frequencies) according to the Hardy–Weinberg equilibrium (HWE). *$P < 0.05$, **$P < 0.001$ versus HWE.

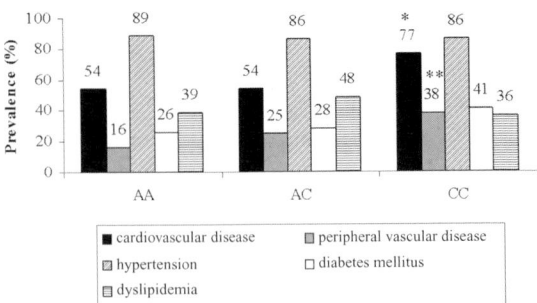

FIGURE 1. Prevalence of comorbidities in each genotype of A419C polymorphism of the glyoxalase I gene in hemodialysis patients. *$P < 0.01$ CC versus AA and AC, **$P < 0.005$ CC versus AA.

(ANOVA), the Kruskal-Wallis test, unpaired t-test, and Wilcoxon signed rank test. Results of the genetic analysis were evaluated using the χ^2 test. All results were considered as statistically significant at $P < 0.05$.

Results

Allelic as well as genotypic frequencies did not differ between HD patients and controls but were not in Hardy–Weinberg equilibrium (TABLE 2). The A allele was found in 51.4% of HD patients and in 52.2% of controls while the C allele was present in 48.6% of HD patients and 47.8% of controls. The studied SNP was not associated with serum levels of glucose, AGE-related fluorescence, and zinc and with HbA1c. However, it was significantly linked with serum sRAGE levels, which sensitively reflect the AGE burden of the organism (highest in the CC variant, 3986 ± 1638 pg/mL versus 3277 ± 1445 pg/mL in AC and 3297 ± 1398 pg/mL in AA, $P < 0.01$) and with the AGEs to sRAGE ratio (lowest in the CC variant, $P < 0.05$). In the CC variant, a significantly higher prevalence of cardiovascular disease ($P < 0.01$ CC versus AA and AC) and peripheral vascular disease ($P < 0.005$ CC versus AA) was found, while the prevalence of hypertension, diabetes mellitus, and dyslipidemia did not differ between genotypes (FIG. 1).

Discussion

This is the first study addressing A419C SNP of the glyoxalase I gene in HD patients. A possible role of glyoxalase in vascular damage was suggested by Miyata et al.[4] They reported a case of a Japanese woman on chronic HD with multiple cardiovascular complications in absence of predisposing risk factors but with a deficiency of glyoxalase I and an unusually high level of AGEs (pentosidine and carboxymethyllysine). Studying A419C SNP of the glyoxalase I gene, we demonstrate a higher prevalence of both cardiovascular and peripheral vascular disease in the mutated homozygotes (CC), although the prevalence of risk factors (hypertension, diabetes mellitus, and dyslipidemia) does not differ among genotypes. Our analysis revealed a significant deviation of allele distribution from the Hardy–Weinberg equilibrium. In order to be sure that it is not a methodological problem, RFLP–PCR was confirmed by DNA sequencing. It is in line with findings for other both old and young populations,[5] suggesting the existence of selection forces.

The studied SNP was significantly linked to the levels of sRAGE, which sensitively reflect the AGE burden of the organism. Indeed, glyoxalase I expression and activity is mediated by RAGE.[6] On the other hand, neither pentosidine nor carboxymethyllysine levels were correlated with glyoxalase I activity.[4]

In conclusion, in this study we demonstrate for the first time the association of A419C polymorphism of glyoxalase I gene with sRAGE levels and show the genetic predisposition to vascular complications in HD patients.

Acknowledgments

This study was supported by the research project MSM 0021620807. The authors are thankful to physicians and nurses from cooperating dialysis centers for technical assistance, especially to Dr. Kazderová, Professor Dusilová-Sulková and Dr. Hodková (Prague), Dr. Gorun (Ústí nad Orlicí), Dr. Hobzek and Mrs. Žďánská (Písek), Dr. Suchanová and Dr. Křížová (Tábor), and Dr. Nýdlová (Strakonice). The authors are equally thankful to laboratory staff for their excellent laboratory skills, especially to Dr. Soukupová, Dr. Vinglerová, Mrs. Miškovská, Mrs. Dományová, Mrs. Medová, Mrs. Řeháková, and Mrs. Pourová.

Conflict of Interest

The authors declare no conflicts of interest.

References

1. THORNALLEY, P.J. 1990. The glyoxalase system: new developments towards functional characterization of a metabolic pathway fundamental to biological life. Biochem. J. **269:** 1–11.
2. GALE, C.P. & P.J. GRANT. 2004. The characterisation and functional analysis of the human glyoxalase-1 gene using methods of bioinformatics. Gene **340:** 251–260.
3. KIM, N.S. et al. 1995. cDNA cloning and characterization of human glyoxalase I isoforms from HT-1080 cells. Biochem. **117:** 359–361.
4. MIYATA, T. et al. 2001. Glyoxalase I deficiency is associated with an unusual level of advanced glycation end products in a hemodialysis patient. Kidney Int. **60:** 2351–2359.
5. CHEN, F. et al. 2004. Role for glyoxalase I in Alzheimer's disease. Proc. Natl. Acad. Sci. **101:** 7687–7692.
6. BIERHAUS, A. & P. NAWROTH. 2007. Role of AGEs in pain and loss of pain perception. Abstract Book 9th International Symposium on the Maillard Reaction. 2007; 41.

Succination of Proteins by Fumarate

Mechanism of Inactivation of Glyceraldehyde-3-Phosphate Dehydrogenase in Diabetes

MATTHEW BLATNIK,[a] SUZANNE R. THORPE,[b] AND JOHN W. BAYNES[a,b,c]

Department of [a]Chemistry and Biochemistry, and [b]Exercise Science, University of South Carolina, Columbia, South Carolina 29208, USA

S-(2-succinyl)cysteine (2SC) is a chemical modification of proteins formed by a Michael addition reaction between the Krebs cycle intermediate, fumarate, and thiol groups in protein—a process known as *succination* of protein. Succination causes irreversible inactivation of glyceraldehyde-3-phosphate dehydrogenase (GAPDH) *in vitro*. GAPDH was immunoprecipitated from muscle of diabetic rats, then analyzed by ultra-performance liquid chromatography–electrospray ionization–mass spectroscopy. Succination of GAPDH was increased in muscle of diabetic rats, and the extent of succination correlated strongly with the decrease in specific activity of the enzyme. We propose that 2SC is a biomarker of mitochondrial and oxidative stress in diabetes and that succination of GAPDH and other thiol proteins may provide the chemical link between glucotoxicity and the pathogenesis of diabetic complications.

Key words: protein; chemical modification; cysteine; diabetes; fumarate; glyceraldehyde-3-phosphate dehydrogenase; oxidative stress; mitochondrial stress; succination

Introduction

In a *Unifying Hypothesis* on the origin of diabetic complications, Brownlee and colleagues proposed that inactivation of glyceraldehyde-3-phosphate dehydrogenase (GAPDH) is a critical, early step in the metabolic derangements that lead to the development of diabetic complications.[1,2] Inactivation of this enzyme causes an increase in intracellular concentrations of glucose and glycolytic intermediates, leading to enhanced activity of the polyol pathway, the hexosamine pathway, and the protein kinase C pathways and accelerated formation of advanced glycation end products (AGE). The original proposal, that GAPDH was inactivated by reactive oxygen species, such as superoxide,[1] was followed by a later study indicating that GAPDH was inactivated by poly-ADP-ribosylation.[3] However, the Krebs cycle intermediate, fumarate, reacts with cysteine residues in proteins, producing S-(2-succinyl)cysteine (2SC)—a process known as *succination* of protein (FIG. 1)[4]; succination of GAPDH *in vitro* causes inactivation of this enzyme. When GAPDH was immunoprecipitated from skeletal muscle of control and streptozotocin-induced (type 1) diabetic rats and tryptic peptides analyzed by matrix-assisted laser desorption/ionization (MALDI)–time-of-flight (TOF) mass spectroscopy (MS), we observed that succination of GAPDH was significantly increased in muscle of diabetic rats.[5] While MALDI–TOF is not considered a quantitative technique, we concluded that the extent of succination was consistent with the decrease in specific activity of GAPDH in muscle of diabetic rats. In the present study, we provide additional evidence that succination contributes to inactivation of GAPDH in muscle of diabetic rats.

Results

To obtain additional information regarding the extent of modification of GAPDH, we isolated the enzyme by immunoprecipitation from muscle of control and diabetic rats, then applied ultra-performance liquid chromatography (UPLC)–electrospray ionization (ESI$^+$)–MS for analysis of the native (vinylpyridine [VP]-modified) and 2SC peptides. GAPDH has four cysteine residues, of which only Cys-149 and Cys-244, in peptides #17 and #26, respectively, are

Address for correspondence: Dr. John W. Baynes, Department of Exercise Science, PHRC, 921 Assembly St., University of South Carolina, Columbia, SC 29208. Voice/Fax: +1-803-777-7272.

john.baynes@sc.edu

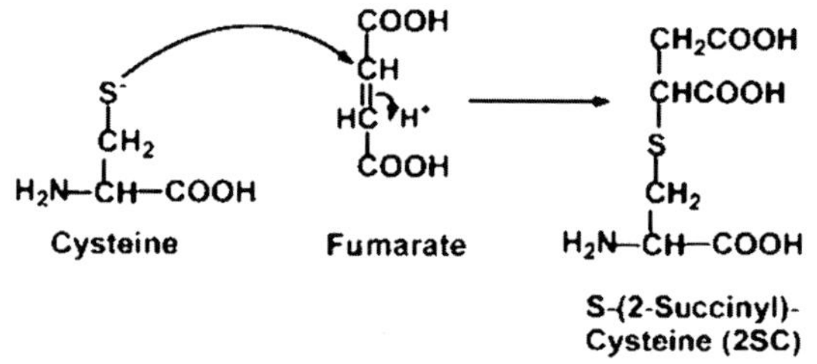

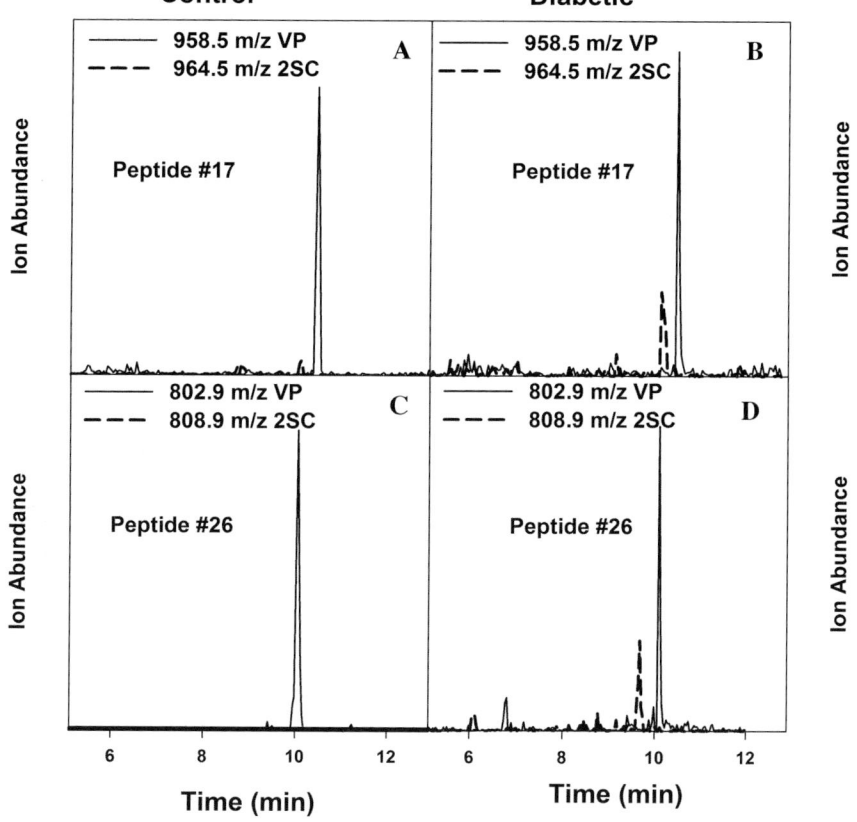

FIGURE 1. Mechanism of formation of S-(2-succinyl)cysteine (2SC). Nucleophilic addition of cysteine to fumarate yields 2SC by a Michael addition reaction.

subject to succination; Cys-149 is the active site cysteine, and Cys-244 is a peripheral, nucleophilic, cysteine residue. Chemical modification of either of these thiol groups is known to cause inactivation of the enzyme.[5]

As shown in FIGURE 2, only traces of 2SC peptides are detected in muscle from control rats (FIG. 2A & C), while the 2SC peptides are readily detected in GAPDH from diabetic rats (FIG. 2B & D). As reported previously,[5] the ratio of 2SC:VP peptide was significantly increased in muscle of diabetic compared to control rats ($P < 0.01$ for both peptides), and there was a strong correlation between the ratio of 2SC:VP peptides and the decrease in specific activity of GAPDH in muscle of diabetic rats ($P < 0.01$ for peptide #17, and $P < 0.05$ for peptide #26). As a more quantitative approach to estimating the extent of succination of peptides, we used the relative area (RA) technique of Brock et al.,[6] in which the sum of the area units for the several charge states of the native (VP) peptide is compared to the sum of the area units of a reference peptide in native enzyme. A decrease in the RA of the VP

FIGURE 2. Increased 2SC modification of glyceraldehyde-3-phosphate dehydrogenase (GAPDH) peptides isolated from gastrocnemius muscle of diabetic versus control rats. GAPDH immunoprecipitates were reduced with vinylpyridine (VP), fractionated by 1D-SDS-PAGE, stained with Coomassie blue, digested in-gel with trypsin, then analyzed by ultra-performance liquid chromatography–electrospray ionization–mass spectroscopy; details of methods are presented elsewhere.[5] **(A)** Peptide #17 from control rat; **(B)** peptide #17 from diabetic rat; **(C)** peptide #26 from control rat; **(D)** peptide #26 from diabetic rat. Solid lines, VP peptide; dotted lines, 2SC peptide.

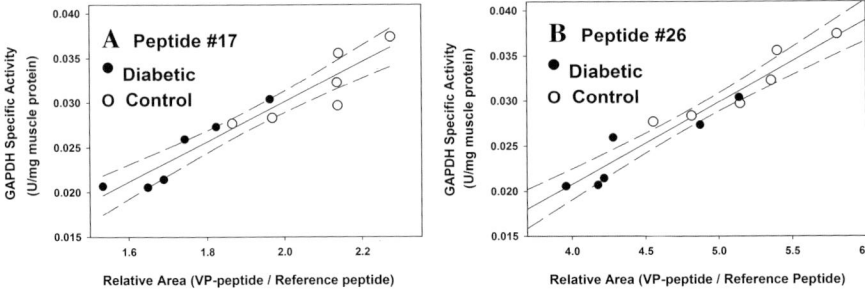

FIGURE 3. Correlation between extent of modification of peptides and specific activity of GAPDH in control and diabetic rat gastrocnemius muscle. **(A)** Correlation between relative area (RA) of 2SC peptide #17 and specific activity of GAPDH ($P < 0.0001$; $r^2 = 0.90$). **(B)** Correlation between RA of 2SC peptide #26 and specific activity of GAPDH ($P < 0.0001$; $r^2 = 0.92$).

peptide versus reference peptide is indicative of modification of the peptide. Peptide #2 (VGVNGFGR) was chosen as the reference peptide because it lacks a cysteine residue and yields a strong, well-resolved signal on ESI$^+$–liquid chromatography (LC)/MS. While the RA method does not establish that all of the loss of native peptides is caused by succination, the mean RA value decreased significantly in diabetic compared to control rats ($P < 0.01$ for peptide #17; $P < 0.05$ for peptide #26). In addition, there was a strong correlation between the RA of the native peptides and the specific activity of GAPDH (FIG. 3)—as the RA decreased, there was a corresponding decrease in the specific activity of GAPDH.

Discussion

The studies described here with UPLC–ESI$^+$–MS complement the results of MALDI–TOF analysis of tryptic peptides from GAPDH isolated from muscle of control and diabetic rats.[5] Based on correlations between the extent of modification and the specific activity of the enzyme (FIG. 3), we conclude that succination is a major cause of the decrease in specific activity of GAPDH in diabetic muscle. In other work, we have shown that succination of proteins is increased in adipocytes grown in cell culture media at high (30 mmol/L) glucose concentration[7] and in adipose tissue of *db/db* diabetic mice, a model of type 2 diabetes (unpublished observations). Overall, it appears that succination of proteins is increased in the presence of higher-than-normal glucose concentration, both *in vitro* and *in vivo*, and that succination may be an indication of mitochondrial stress resulting from glucotoxicity in both type 1 and type 2 diabetes.

Mechanistically, we propose that succination is an early step in the response to excess glucose concentration in cell culture or *in vivo* in diabetes. The abundant fuel supply and accumulation of ATP would lead to an increase in mitochondrial nicotinamide adenine dinucleotide (NADH) and then to hyperpolarization of the inner mitochondrial membrane, as seen in adipocytes.[8] The increase in NADH would then lead to accumulation of Krebs cycle intermediates, including fumarate, causing an increase in succination of proteins. The accumulation of NADH may also trigger the production of reactive oxygen species.[9] Eventually, damage to the integrity of the mitochondrial membranes would open the mitochondrial permeability transition pore, precipitating a series of events culminating in apoptosis.[10] Studies on succination of protein may permit early detection of mitochondrial stress, perhaps even in obesity and metabolic syndrome, and trigger effective clinical intervention at the earliest stages in development of diabetes and its complications.

Acknowledgments

This work was supported by the United States Public Health Service Research Grant DK-19971 from the National Institutes of Diabetes, and Kidney and Digestive Diseases.

Conflict of Interest

The authors declare no conflicts of interest.

References

1. BROWNLEE, M. 2005. The pathobiology of diabetic complications: a unifying mechanism. Diabetes **54:** 1615–1625.
2. NISHIKAWA, T., D. EDELSTEIN, X.L. DU, *et al.* 2000. Normalizing mitochondrial superoxide production blocks three pathways of hyperglycaemic damage. Nature **404:** 787–790.

3. Du, X., T. Matsumura, D. Edelstein, et al. 2003. Inhibition of GAPDH activity by poly(ADP-ribose) polymerase activates three major pathways of hyperglycemic damage in endothelial cells. J. Clin. Invest. **112:** 1049–1057.
4. Alderson, N.L., Y. Wang, M. Blatnik, et al. 2006. S-(2-succinyl)cysteine: a novel chemical modification of tissue proteins by a Krebs cycle intermediate. Arch. Biochem. Biophys. **450:** 1–8.
5. Blatnik, M., N. Frizzell, S.R. Thorpe & J.W. Baynes. 2008. Inactivation of glyceraldehyde-3-phosphate dehydrogenase by fumarate in diabetes: formation of S-(2-succinyl)cysteine: a novel chemical modification of protein and possible biomarker of mitochondrial stress. Diabetes **57:** 41–49.
6. Brock, J.W.C., D.J.S. Hinton, W.E. Cotham, et al. 2003. Proteomic analysis of the site specificity of glycation and carboxymethylation of ribonuclease. J. Proteome Res. **2:** 506–513.
7. Nagai, R., J.W.C. Brock, M. Blatnik, et al. 2007. Succination of proteins in adipocytes: S-(2-succinyl)cysteine is a biomarker of oxidative stress during maturation of adipocytes. J. Biol. Chem. **282:** 34219–34228.
8. Lin, Y., A.H. Berg, P. Iyengar, et al. 2005. The hyperglycemia-induced inflammatory response in adipocytes: the role of reactive oxygen species. J. Biol. Chem. **280:** 4617–4626.
9. Kohanski, M.A., D.J. Dwyer, B. Hayete, et al. 2007. A common mechanism of cellular death induced by bactericidal antibiotics. Cell **130:** 797–810.
10. Belizario, J.E., J. Alves, J.M. Occhiucci, et al. 2007. A mechanistic view of mitochondrial death decision pores. Braz. J. Med. Biol. Res. **40:** 1011–1024.

Dietary Advanced Glycation Endproducts and Oxidative Stress

In Vivo Effects on Endothelial Function and Adipokines

ALIN STIRBAN,[a] MONICA NEGREAN,[a] CHRISTIAN GÖTTING,[b] JAIME URIBARRI,[c] THOMAS GAWLOWSKI,[a] BERND STRATMANN,[a] KNUT KLEESIEK,[b] THEODOR KOSCHINSKY,[d] HELEN VLASSARA,[e] AND DIETHELM TSCHOEPE[a]

[a]*Diabetes Center and* [b]*Institute for Laboratory and Transfusion Medicine, Heart and Diabetes Center North Rhine–Westphalia, Bad Oeynhausen, Germany*

[c]*Division of Nephrology, Department of Medicine and* [e]*Division of Diabetes and Aging Research, The Brookdale Department of Geriatrics, Mount Sinai School of Medicine, New York, USA*

[d]*Professor emeritus, Heinrich-Heine University Duesseldorf, Duesseldorf, Germany*

Advanced glycation endproducts (AGEs) and oxidative stress (OS) contribute to the development and progression of diabetic complications. We have reported that dietary AGEs and OS induce acute endothelial dysfunction *in vivo*, but little is known about their effects on adipokines. Twenty inpatients with type 2 diabetes mellitus (mean age: 55.9; range: 32–71 years), received a standard diabetes diet for 6 days. On days 4 and 6, the acute effects of a high-AGE (HAGE) or a low-AGE (LAGE) meal (15.100 vs. 2.750 kU AGE) were studied in a randomized, cross-over, investigator-blinded design. Measurements were performed after an overnight fast, at baseline (B) and at 2, 4, and 6 h after the HAGE or LAGE meals. Both meals had the same ingredients and differed only by the cooking method. Two h following HAGE, a significant decrease from baseline occurred in adiponectin (−10%*‡ vs. +0%) and leptin (−22%*‡ vs. −13%*), and a significant increase occurred in vascular cell adhesion molecule 1 (+19%*‡ vs. −5%) and thiobarbituric acid reactive substances (+23%*‡ vs. +6%). These changes did not occur, or occurred to a lesser extent, following LAGE. At 4 h following HAGE, an increase in methylglyoxal (+20%‡ vs. −5%) and E-selectin (+54%*‡ vs. −3%) occurred. Urinary AGEs increased only after HAGE (+51%*‡ vs. −2%; values presented as HAGE vs. LAGE; *$P < 0.05$ vs. baseline, ‡$P < 0.05$ vs. LAGE). The postprandial excursions in glucose, insulin, and triglycerides were similar between both meals. A meal rich in AGEs induces acute endothelial and adipocyte dysfunction. These effects were prevented by changing the cooking method.

Key words: endothelium; postprandial; adipokines; advanced glycation endproducts; oxidative stress

Introduction

Postprandial metabolic changes play an important role in the development of cardiovascular disease (CVD) and diabetic complications.[1] Endothelial dysfunction precedes by decades the onset of atherosclerosis and CVD, and is characterized by imbalances between relaxing and constricting factors, procoagulant and anticoagulant substances, and proinflammatory and antiinflammatory mediators.[2] Nitric oxide (NO), mainly secreted by endothelial cells, is a potent vasodilatator and has marked antiatherogenic properties. A decrease in NO-mediated effects can have three main causes: reduction in NO synthesis (endothelial dysfunction), increased NO scavenging (e.g., by exacerbated oxidative stress [OS] or advanced glycation endproducts [AGEs]), or decreased sensitivity of target organs (e.g., smooth muscle cells) to NO.[3] NO is partly inactivated by peroxynitrites, forming metabolites that react with the amino acid tyrosine. The result is the production of nitrotyrosine (NT), a parameter that reflects NO inactivation and OS.[4] Endothelial dysfunction can

Address for correspondence: Alin Stirban, Heart and Diabetes Center NRW, Georgstrasse 11, 32545 Bad Oeynhausen, Germany. Voice: +49-5731-973724; fax: +49-5731-972145.

astirban@hdz-nrw.de

be indirectly assessed by measuring NO metabolites, but several specific laboratory markers of endothelial dysfunction have also been proposed, among them vascular cell adhesion molecule 1 (VCAM-1) and E-selectin (endothelial-leukocyte adhesion molecule-1).[5]

Adipocytes, by releasing adipokines (e.g., adiponectin, leptin, resistin, and visfatin), play an important role in the modulation of insulin sensitivity, inflammation, endothelial function, and satiety.[6] In obesity, adiponectin decreases while leptin increases and leptin resistance develops.[6] Even though adipokine levels have been extensively investigated in diabetes, and a diurnal variation has been described, little is known about postprandial adipokine regulation, especially in people with type 2 diabetes mellitus (T2DM).[7]

AGEs constitute a heterogeneous group of compounds formed by the nonenzymatic glycation of proteins, lipids, or nucleic acids.[8] They can be generated endogenously, but an important source consists of food AGEs,[8] which are absorbed (about 10%) and partly retained in the body or eliminated in the urine.[9] *In vitro* data have shown that AGEs impair endothelial function[10] and leptin secretion from adipocytes.[11] We have recently suggested that dietary AGEs and OS induce endothelial dysfunction[3,12,13] and decrease postprandial adiponectin levels[7] *in vivo*. In the present article, we offer an overview on the effects of food AGEs and OS on circulating markers of endothelial and adipocyte function.

Research Design and Methods

Twenty inpatients with T2DM (age: 55.9 ± 2.0 years; diabetes duration: 8.7 ± 1.7 years; body mass index: 29.5 ± 0.8 kg/m^2; hemoglobin A1c: $8.9 \pm 0.4\%$ [mean $\pm$ SEM]; male/female: 13/7; smokers/nonsmokers: 4/16; treatment: oral/oral + insulin/insulin alone: 15/3/2; angiotensin receptor blockers: 14; statins inhibitors: 8; betablockers: 7; diuretics: 7; calcium channel blockers: 3; aspirin: 16 [number]) and without a history of cardiovascular events were investigated after approval of the institutional review board and individual written consent. Patients on a standard diabetes diet for the 6-day study period were studied on two occasions, following an overnight fast and after a 12-h withdrawal of any other medication. On days 4 and 6, the effects of a high-AGE (HAGE) and low-AGE (LAGE) meal on endothelial and adipocyte function were studied in an investigator-blinded, randomized, cross-over design (n = 10 began with HAGE, n = 10 with LAGE) after an overnight fast (7:00 AM) as well as at 2, 4, and 6 h postprandially. We also obtained timed samples of urine at baseline and 6 h following the meal intake.

The two meals were isocaloric (580 kcal), had identical ingredients, and differed only by the temperature and time of cooking (HAGE: frying or broiling at 230 °C for 20 min; LAGE: steaming or boiling at 100 °C for 10 min).[12] The two meals differed in calculated AGE content (HAGE: 15100 kU AGE; LAGE: 2750 kU AGE).[8]

Blood samples were analyzed for serum glucose, cholesterol, triglycerides, and low-density lipoprotein and high-density lipoprotein cholesterol (Architect ci8200 analyzer, Abbott Diagnostics, Wiesbaden, Germany), thiobarbituric acid reactive substances (TBARS; Alexis Biochemicals, Gruenberg, Switzerland), serum adiponectin (R&D Systems, Wiesbaden, Germany), leptin (R&D Systems, Wiesbaden, Germany), NT (Oxis Research, Portland, OR), insulin (DPC Biermann, Bad Nauheim, Germany), and serum methylglyoxal (MG)-derivatives (enzyme-linked immunosorbent assay [ELISA], monoclonal anti-MG-BSA antibody -MG3D11; Dr. Y. Al-Abed, The Picower Institute, Manhasset, NY). Urinary AGEs were measured by ELISA using an antibody against carboxymethyllysine (4G9 mab, Alteon, Inc, Northvale, NJ).

Statistical significance of postprandial changes was analyzed by a two-way ANOVA for repeated measurements, followed by a two-tailed paired *t*-test with Bonferroni's correction for multiple testing. To compare the effects between the two meals, differences at certain time points from baseline were analyzed by a paired *t*-test. Data are expressed as mean $\pm$ SEM unless stated otherwise. The level of significance was defined as $P < 0.05$.

Several aspects of this study have been previously reported[3,7,12] and the leptin data have been submitted for publication.

Results

A more pronounced decrease in leptin occured after HAGE compared with LAGE (-22%*‡ vs. -13%*; *$P < 0.05$ vs. baseline, ‡$P < 0.05$ vs. LAGE). Further data are presented in TABLE 1.

We found a significant correlation between blood glucose and TBARS change at 2 h (r = 0.868) and at 4 h (r = 0.681, $P < 0.001$ for both). Postprandial excursions in glucose, insulin, triglycerides, and NT were comparable between meals.

TABLE 1. Study parameters (mean ± SEM)

Parameter	Meal	Baseline	2 h	4 h	6 h
Adiponectin (ng/mL)	HAGE	4102 ± 543	3707 ± 476*‡	3769 ± 466	3830 ± 494
	LAGE	3856 ± 399	3859 ± 441	3668 ± 464	3875 ± 445
VCAM-1 (ng/mL)	HAGE	538 ± 41	642 ± 61*‡	577 ± 59	535 ± 40
	LAGE	567 ± 46	540 ± 47	485 ± 33	499 ± 35
E-selectin (ng/mL)	HAGE	35 ± 2		54 ± 9*‡	
	LAGE	37 ± 3		36 ± 3	
TBARS (nmol/mL)	HAGE	7.4 ± 0.3	9.1 ± 0.7*‡	6.7 ± 0.4	6.5 ± 0.3
	LAGE	8.1 ± 0.4	8.6 ± 0.6	6.5 ± 0.3	6.3 ± 0.3
MG (nmol/mL)	HAGE	3.0 ± 0.3	2.8 ± 0.3	3.6 ± 0.3‡	2.6 ± 0.3
	LAGE	3.1 ± 0.3	2.5 ± 0.3	2.9 ± 0.3	2.5 ± 0.3
Urinary AGE (U/mL)	HAGE	35 ± 4			53 ± 7*‡
	LAGE	43 ± 7			42 ± 5
Nitrotyrosine (nmol/L)	HAGE	45 ± 3	46 ± 3	46 ± 3	46 ± 3
	LAGE	45 ± 3	44 ± 3	44 ± 2	44 ± 3
Glucose (mg/dL)	HAGE	153 ± 9	182 ± 13	130 ± 9	114 ± 6
	LAGE	154 ± 10	175 ± 11	118 ± 6	106 ± 6
Insulin (pmol/L)	HAGE	13 ± 4	44 ± 15	27 ± 13	16 ± 6
	LAGE	11 ± 4	40 ± 14	17 ± 7	11 ± 3
Triglycerides (mg/dL)	HAGE	143 ± 16	152 ± 17	164 ± 19	163 ± 18
	LAGE	171 ± 28	174 ± 27	189 ± 30	184 ± 26

*$P < 0.05$ vs. baseline, ‡$P < 0.05$ vs. LAGE

Discussion

Following a meal with high AGE content (HAGE), we found a marked increase in circulating markers of endothelial dysfunction (VCAM-1 and E-selectin) and decreases in adiponectin and leptin levels, suggesting acute postprandial changes in endothelial and adipocyte function. These effects were prevented by changing the cooking method (LAGE).

Previous reports suggested a postprandial increase in NT.[4] But in our study, NT did not change postprandially during either intervention. The balanced nutrition that our subjects received before and during the study period may have protected them against a postprandial NT increase because NT levels have been shown to remain stable in the face of increased OS in subjects with good antioxidant intake.[14]

We found a more pronounced increase in AGEs and OS following HAGE, as shown by the postprandial increase in MG, urinary AGE, respectively TBARS. Hyperglycemia and hypertriglyceridemia increase OS[4] and AGEs[15] and induce endothelial dysfunction.[4] But postprandial glucose and triglyceride excursions were similar in HAGE and LAGE; therefore, the differences between meals have to be explained by other mechanisms. We suggest the increase in OS and AGE as responsible pathomechanisms. Because OS and AGEs are known to potentiate each other,[10] and because both can induce endothelial[4,10] and adipocyte[11] dysfunction, it is difficult to conclude from our model the extent to which OS and AGEs separately contributed to the effects seen.

Interestingly, the amount of AGEs administered during the HAGE intervention was similar to the average estimated daily intake by the general population.[9] Because similar amounts of AGEs are consumed regularly by many people, we assume that similar endothelial and adipocyte insults may occur quite frequently and may, over time, promote CVD and insulin resistance, especially in populations at risk.

In summary, we have shown that a meal with a high AGE content can acutely induce significant endothelial and adipocyte dysfunction. We believe that the postprandial increases of circulating AGEs and OS represent the two main pathomechanisms. Further studies are warranted to investigate the clinical importance of postprandial adiponectin and leptin decrease for the modulation of insulin sensitivity and food intake.

Conflict of Interest

The authors declare no conflicts of interest.

References

1. CERIELLO, A. 2005. Postprandial hyperglycemia and diabetes complications: is it time to treat? Diabetes **54:** 1–7.
2. POREDOS, P. 2001. Endothelial dysfunction in the pathogenesis of atherosclerosis. Clin. Appl. Thromb. Hemost. **7:** 276–280.
3. STIRBAN, A., M. NEGREAN, B. STRATMANN, et al. 2006. Benfotiamine prevents macro- and microvascular endothelial dysfunction and oxidative stress following a meal rich in advanced glycation end products in individuals with type 2 diabetes. Diabetes Care **29:** 2064–2071.
4. CERIELLO, A., C. TABOGA, L. TONUTTI, et al. 2002. Evidence for an independent and cumulative effect of postprandial hypertriglyceridemia and hyperglycemia on endothelial dysfunction and oxidative stress generation: effects of short- and long-term simvastatin treatment. Circulation **106:** 1211–1218.
5. MEIGS, J.B., F.B. HU, N. RIFAI & J.E. MANSON 2004. Biomarkers of endothelial dysfunction and risk of type 2 diabetes mellitus. JAMA **291:** 1978–1986.
6. FANTUZZI, G. 2005. Adipose tissue, adipokines, and inflammation. J. Allergy Clin. Immunol. **115:** 911–919.
7. STIRBAN, A., M. NEGREAN, B. STRATMANN, et al. 2007. Adiponectin decreases postprandially following a heat processed meal in people with type 2—an effect prevented by benfotiamine and cooking method. Diabetes Care **30:** 2514–2516.
8. GOLDBERG, T., W. CAI, M. PEPPA, et al. 2004. Advanced glycoxidation end products in commonly consumed foods. J. Am. Diet. Assoc. **104:** 1287–1291.
9. KOSCHINSKY, T., C.J. HE, T. MITSUHASHI, et al. 1997. Orally absorbed reactive glycation products (glycotoxins): an environmental risk factor in diabetic nephropathy. Proc. Natl. Acad. Sci. USA **94:** 6474–6479.
10. PEPPA, M., J. URIBARRI & H. VLASSARA 2004. The role of advanced glycation end products in the development of atherosclerosis. Curr. Diab. Rep. **4:** 31–36.
11. UNNO, Y., M. SAKAI, Y. SAKAMOTO, et al. 2004. Advanced glycation end products-modified proteins and oxidized LDL mediate down-regulation of leptin in mouse adipocytes via CD36. Biochem. Biophys. Res. Commun. **325:** 151–156.
12. NEGREAN, M., A. STIRBAN, B. STRATMANN, et al. 2007. Effects of low- and high-advanced glycation endproduct meals on macro- and microvascular endothelial function and oxidative stress in patients with type 2 diabetes mellitus. Am. J. Clin. Nutr. **85:** 1236–1243.
13. URIBARRI, J., A. STIRBAN, D. SANDER, et al. 2007. Single oral challenge by advanced glycation end products acutely impairs endothelial function in diabetic and nondiabetic subjects. Diabetes Care Accepted, in press.
14. BO, S., R. SANDER, S. GUIDI, et al. 2005. Plasma nitrotyrosine levels, antioxidant vitamins and hyperglycaemia. Diabet. Med. **22:** 1185–1189.
15. BEISSWENGER, P.J., S.K. HOWELL, R.M. O'DELL, et al. 2001. Alpha-dicarbonyls increase in the postprandial period and reflect the degree of hyperglycemia. Diabetes Care **24:** 726–732.

Preparation of Nucleotide Advanced Glycation Endproducts—Imidazopurinone Adducts Formed by Glycation of Deoxyguanosine with Glyoxal and Methylglyoxal

THOMAS FLEMING,[a] NAILA RABBANI,[a,b] AND PAUL J. THORNALLEY[a,b]

[a]*Department of Biological Sciences, University of Essex, Colchester, United Kingdom*
[b]*Clinical Sciences Research Institute, Warwick Medical School, University of Warwick, Coventry, United Kingdom*

An analytical procedure was developed for nucleotide advanced glycation endproducts formed by the reaction of glyoxal and methylglyoxal with deoxyguanosine under physiological conditions. For this, the imidazopurinone derivatives, 3-(2′-deoxyribosyl)-6,7-dihydro-6,7-dihydroxyimidazo[2,3-b]purin-9(8)one (dG-G) and 3-(2′-deoxyribosyl)-6,7-dihydro-6,7-dihydroxy-6-methylimidazo-[2,3-b]purine-9(8)one (dG-MG), were prepared. Authentic standard and stable isotope-substituted standard adducts were prepared and an isotopic dilution analysis assay methodology was developed using liquid chromatography with tandem mass spectrometry and optimized DNA extraction and nuclease digestion procedures. Analysis of dG-G, dG-MG, and the oxidative marker 8-hydroxydeoxyguanosine in the DNA of cultured human cells and mononuclear leukocytes showed that nucleotide advanced glycation endproducts are major markers of DNA damage in human cells.

Key words: glycation; DNA; methylglyoxal; glyoxal; deoxyguanosine; 8-hydroxydeoxyguanosine; imidazopurinone

Introduction

Cellular DNA suffers continuous damage from oxidation, deamination, and other processes.[1] One such further process leading to DNA damage is glycation, particularly by the physiological reactive dicarbonyl glycating agents, glyoxal and methylglyoxal, forming nucleotide advanced glycation endproducts (AGEs). Although glycation damage to DNA is associated with mutagenesis and carcinogenesis, the mutagenic potential is low while the cellular protection against glycation is functioning.[2]

Dicarbonyl glycation of DNA may give rise to significant steady-state levels of nucleotide AGEs relative to the DNA oxidation marker 8-hydroxydeoxyguanosine (8-HOdG). The methylglyoxal-derived imidazopurinone nucleotide AGE, 3-(2′-deoxyribosyl)-6,7-dihydro-6,7-dihydroxy-6-methylimidazo-[2,3-b]purine-9(8)one (dG-MG), has been detected by a ^{32}P-post-labelling technique[3] in cellular DNA, and N_2-(1-carboxyethyl)-2-deoxyguanosine has been detected by liquid chromatography with tandem mass spectrometry (LC–MS/MS) in an alkaline digest of DNA glycated by methylglyoxal.[4] In this study, we prepared authentic standard imidazopurinone adducts, 3-(2′-deoxyribosyl)-6,7-dihydro-6,7-dihydroxyimidazo[2,3-b]purin-9(8)one (dG-G) and dG-MG; FIG. 1), for use in an isotopic dilution analysis LC–MS/MS assay methodology for quantitation of the major nucleotide AGEs derived from glyoxal and methylglyoxal with deoxyguanosine.

Methods

Materials: 2′-Deoxyguanosine monohydrate and glyoxal and methylglyoxal solutions (40%) were purchased from Sigma (Poole, Dorset, UK). U-[^{13}C,^{15}N]-2′-Deoxyguanosine (all >98% isotopic purity) was purchased from Cambridge Isotope Laboratories (Andover, MA).

Preparation of dG-G: A mixture of 2′-deoxyguanosine monohydrate (300 mg, 1.05 mmol)

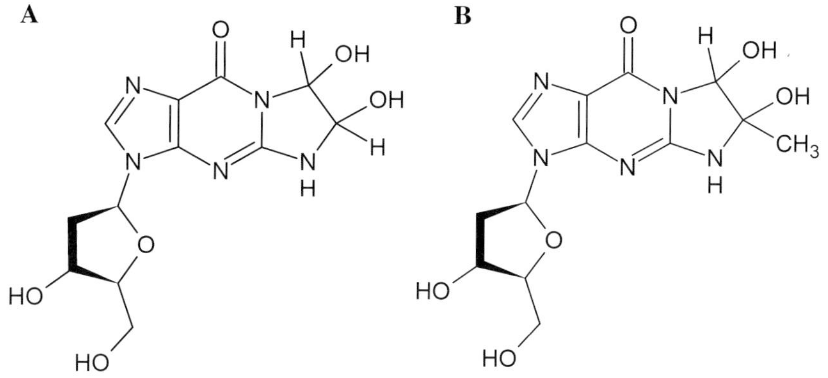

FIGURE 1. Imidazopurinone nucleotide AGEs, (**A**) 3-(2′-deoxyribosyl)-6,7-dihydro-6,7-dihydroxyimidazo[2,3-b]purin-9(8)one (dG-G) and (**B**) 3-(2′-deoxyribosyl)-6,7-dihydro-6,7-dihydroxy-6-methylimidazo-[2,3-b]purine-9(8)one (dG-MG).

and glyoxal (40% solution in water, 185 µL, 92 mg, 1.59 mmol) in water (30 mL) was stirred for 4 days at room temperature. The solution was lyophilized to dryness. The nucleotide AGE product was then purified by reversed-phase high-performance liquid chromatography (HPLC) using a Waters 25 × 100 mm NOVAPAK C18 (Waters, Elstree, UK), 6-µm pore size cartridge and monitoring the eluate by absorbance at 254 nm. The mobile phase was 50 mM ammonium formate (pH 4.6) with a linear gradient of 0–5% acetonitrile over 30 min. Eluate fractions containing the product were collected, pooled, and lyophilized to dryness.

Preparation of dG-MG: A mixture of 2′-deoxyguanosine monohydrate (300 mg, 1.05 mmol) and methylglyoxal (40% solution in water, 253 µL, 120 mg, 1.67 mmol) in water (30 mL) was stirred for 4 days at room temperature. The solution was lyophilized to dryness and purified by reversed-phase HPLC as described above.

Results

dG-G: The major product isolated and purified from the reaction of glyoxal with deoxyguanosine was characterized by ^{1}H and ^{13}C NMR, high-resolution mass spectrometry, and UV and infrared absorbance spectroscopy. ^{1}H NMR (270 MHz, dimethylsulfoxide [DMSO]-d$_6$) yielded the following chemical shift δ_H (ppm) and coupling constant J (Hz) values: 8.93 (bs, 1H, N^2-H); 8.07 (s, 1H, 8-H); 7.36 (d, 1H, c-H, J = 6.10); 6.59 (d, 1H, b-H, J = 7.34); 6.23 (d, 1H, H1′, 1H, J = 7.03); 5.58 (d, 1H, d-H, J = 6.72); 5.42 (d, 3′-OH, 1H, J = 3.67); 5.07 (t, 1H, 5′-OH, J = 5.51); 4.97 (d, 1H, a-H, J = 7.32); 4.44 (m, 1H, H3′); 3.92 (m, 1H, H4′); 3.63 (m, 2H, H5′/H5″); 2.35 (m, 1H, H2″). ^{13}C NMR (68 MHz, DMSO-d$_6$) gave the following chemical shift δ_C (ppm) values: 154.8 (C6); 154.6 (C2); 150.5 (C4); 135.6 (C8); 117.5 (C5); 87.6 (C4′); 84.0 (C-a); 83.6 (C-b); 82.8 (C1′); 70.6 (C3′); 61.5 (C5′). High-resolution mass spectrometry gave a molecular ion with m/z 348.0909 (calculated for $C_{12}H_{15}N_5O_6Na$, 348.0915). UV spectrophotometry yielded the following absorbance maxima at wavelength λ_{max} (nm) with extinction coefficient ε ($M^{-1}cm^{-1}$) values: λ_{max} = 249 and ε = 14,319 ± 214; and λ_{max} = 275 and ε = 6,787 ± 84. Infrared spectroscopy (Nujol) gave the following λ_{max} (cm^{-1}): 1713 $V_{C=O}$ (s); 1600–1553 $V_{C=C}$ (m); 1600 δ_{N-H} (m); 1331–1298 δ_{O-H} (s); 1229–1050 V_{C-O} (s); 1331–1050 V_{C-N} (m); 1102–1078 V_{C-O} (s; C-O-C). The percentage yield was 87% (299 mg), based on deoxyguanosine.

dG-MG: The major product isolated and purified from the reaction of methylglyoxal with deoxyguanosine was characterized by ^{1}H and ^{13}C NMR, high-resolution mass spectrometry, and UV and infrared absorbance spectroscopy. ^{1}H NMR (270 MHz, DMSO-d$_6$) yielded the following chemical shift δ_H (ppm) and coupling constant J (Hz) values: 8.70 (bs, 1H, N^2-H); 7.95 (s, 1H, 8-H); 7.18 (d, 1H, c-OH, J = 6.72); 6.21 (d, 1H, b-OH, J = 3.67); 6.12 (t, 1H, H1′, J = 7.02); 5.36 (d, 1H, d-H, J = 7.32); 5.31 (d, 1H, 3′-OH, J = 4.27); 4.97 (t, 1H, 5′-OH, J = 3.67); 4.34 (m, 1H, H3′); 3.82 (m, 1H, H4′); 3.17 (m, 2H, H5′/H5″); 2.22 (m, 1H, H2″); 1.39 (s, 3H, a-CH3). ^{13}C NMR (68 MHz, DMSO-d$_6$) gave the following chemical shift δ_C (ppm) values: 155.0 (C6); 154.1 (C2); 150.4 (C4); 135.4 (C8); 117.6 (C5); 87.6 (C4′); 87.2 (C-c); 84.6 (C-b); 82.8 (C1′); 70.6 (C3′); 61.6 (C5′); 21.0 (C-a). High-resolution mass spectrometry gave a molecular ion with m/z 362.1065 (calculated for $C_{13}H_{17}N_5O_6Na$, 362.1071). UV spectrophotometry yielded the following absorbance maxima at wavelength λ_{max} (nm) with extinction coefficient ε ($M^{-1}cm^{-1}$) values: $\lambda_{max} = 250$ and $\varepsilon = 11,202 \pm 327$; and $\lambda_{max} = 275$ nm and $\varepsilon = 7,099 \pm 78$. Infrared spectroscopy (Nujol) gave the following λ_{max} (cm^{-1}): 1694 $V_{C=O}$ (s); 1602–1553 $V_{C=C}$ (m); 1602 δ_{N-H} (m); 1297 δ_{O-H} (s); 1230–1053 V_{C-O} (s); 1297–1053 V_{C-N} (m); 1096 V_{C-O} (s; C-O-C). The percentage yield was 62 (222 mg), based on deoxyguanosine.

Stable isotopic standards of dG-G and dG-MG: U-[$^{13}C,^{15}N$]-2′-deoxyguanosine (98% isotopic purity in each isotope) were used to prepare the isotopically labelled standards of dG-G and dG-MG.

Discussion

In the preparation of glycation adducts of deoxyguanosine with glyoxal and methylglyoxal, we found that imidazopurinone adducts were the major nucleotide AGEs formed. Similar reaction of deoxyguanosine with glyoxal and methylglyoxal under physiological conditions of 37 °C and pH 7.4 gave similar results.

The authentic standard adducts of dG-G and dG-MG, together with stable isotopomers, were used in an isotopic dilution LC–MS/MS assay for quantitation of these adducts in calf thymus DNA, glycated by glyoxal and methylglyoxal, and in DNA extracts of human cells—peripheral mononuclear leukocytes, human leukemia 60 cells, and hepatoma G2 cells. Analysis of dG-G, dG-MG, and the oxidative marker 8-hydroxydeoxyguanosine in DNA of cultured human cells and mononuclear leukocytes showed that nucleotide AGEs are major markers of DNA damage in human cells.

Acknowledgments

We thank the Wellcome Trust and Cancer Research UK for support for our research.

Conflict of Interest

The authors declare no conflicts of interest.

References

1. FRIEDBERG, E.C. 2003. DNA damage and repair. Nature **421:** 436–440.
2. THORNALLEY, P.J. 2003. Protecting the genome: defence against nucleotide glycation and emerging role of glyoxalase I over expression in multidrug resistance in cancer chemotherapy. Biochem. Soc. Trans. **31:** 1372–1377.
3. VACA, C.E., J.-L. FANG, M. CONRADI & S.-M. HOU. 1994. Development of a ^{32}P-postlabelling technique for the analysis of 2′-deoxyguanosine-3′-monophosphate and DNA of methylglyoxal. Carcinogenesis **15:** 1887–1894.
4. FRISCHMANN, M., C. BIDMON, J. ANGERER & M. PISCHETSRIEDER. 2005. Identification of DNA adducts of methylglyoxal. Chemical Research in Toxicology **18:** 1586–1592.

Maillard Products as Biomarkers in Cancer

BEATRICE E. BACHMEIER,[a] ANDREAS G. NERLICH,[b] HELMUT ROHRBACH,[b] ERWIN D. SCHLEICHER,[c] AND ULRICH FRIESS[c]

[a]*Department of Clinical Chemistry and Clinical Biochemistry, Ludwig-Maximilians-University Munich, D-80336 Munich, Germany*

[b]*Institute of Pathology, Academic Hospital Munich-Bogenhausen, D-81925 Munich, Germany*

[c]*Division of Clinical Chemistry, University of Tübingen, D-72076 Tübingen, Germany*

Because tumors exert increased glycolysis rates, a high intracellular carbonyl stress with the formation of Maillard products may evolve. Therefore, we studied the presence of N^ε-(carboxymethyl)lysine (CML) modification in breast cancer tissues from 20 patients and found significant cytoplasmatic staining in tumor cells that was independent of the tumor stage, tumor type, and microanatomic localization. Studying breast cancer cell lines, we also found strong cytoplasmatic CML staining that was again independent of their invasive or metastatic behavior. Our results reveal that tumor cells show a strong cytoplasmatic immunoreactivity to CML without evident association with breast carcinoma type, differentiation, tumor stage, or intratumoral localization. We conclude that CML formation is a general tumor cell-associated process.

Key words: N^ε-(carboxymethyl)lysine; breast cancer; oxidative stress; metastasis; invasion

Introduction

Previous studies revealed an increased formation of Maillard–advanced glycation end products (AGEs) in aging, in diseases with disturbed carbohydrate metabolism, and in inflammation.[1,2] The enhanced formation of AGEs may not only be a biomarker but may also indicate a causal involvement of AGEs in the progression of these diseases because receptors for AGEs (RAGE)[3] were characterized activating the nuclear factor kappa B (NFκB) signaling pathway, thus leading to cytokine expression.[4] Since RAGE expression is also under the control of NFκB, chronic stimulation of RAGE leads to a vicious cycle, thus perpetuating an inflammatory response to an initiating event.[5] Increased AGE formation and upregulation of RAGE has been particularly demonstrated in tissues affected by diabetes (e.g., kidney,[6] nerve,[7] and eye[8]). The possible causal involvement of the AGE–RAGE pathway in the development of diabetic nephropathy was demonstrated by application of soluble RAGE, which reduced the structural and functional renal changes in a diabetic animal model.[9] Several hundred Maillard products have been characterized in food and more than 10 Maillard products have been identified *in vivo* with N^ε-(carboxymethyl)lysine (CML) as a major Maillard product[10,11] and as a ligand of RAGE.[12]

Previous studies showed that AGE products are formed via intracellularly generated dicarbonyl compounds, possibly derived from glycolysis intermediates yielding arginine- and lysine-modified proteins.[13] Because glucose transporter 1 expression and subsequently the glycolysis rate is substantially upregulated in cancer tissue and cancer cells as a result of an insufficient oxygen supply, we hypothesized that an increased formation of CML may occur in tumor tissues. To prove this hypothesis, clinically and histologically well-characterized breast cancer specimens from 20 individuals were investigated. We studied the occurrence and localization of CML in larynx and colon carcinomas for comparison. To evaluate the formation of CML in tumor cells, CML was studied in three breast cancer cell lines and three keratinocyte cell lines *in vitro*.

Materials and Methods

Study Population

This study was performed on a series of human breast cancer specimens obtained during routine

Address for correspondence: Erwin Schleicher, Department of Internal Medicine (Division of Clinical Chemistry), University of Tübingen, Otfried-Mueller-Str. 10, 72076 Tübingen, Germany. Voice: +49 7071 298 0602; fax: +49 7071 29 4696.

Erwin.Schleicher@med.uni-tuebingen.de

TABLE 1. Characteristics of the study population as represented by 10 out of 20 patients from breast cancer cases analyzed

Age	Type	Diff.[a]	Size (mm)	LN[b] status	CML-pos. tu-cells	CML-pos. stroma	CML-pos. endothel	CML-pos. pre-existent
67	IDC[c]	2	12	0	1	1	2	2
78	ILC[d]	2	9	0	3	1	1	2
66	IDC	3	19	0	1	1	1	1
75	ILC	2	12	0	3	1	1	2
81	IDC	3	30	0	2	1	2	3
63	ILC	2	12	0	3	1	1	1
82	IDC	3	18	0	1	1	1	3
77	IDC	2	19	1	1	2	1	2
87	IDC	3	36	3	1	1	1	3
61	IDC	3	27	0	2	2	2	3

[a]Diff, differentiation.
[b]LN, lymph node.
[c]IDC, invasive ductal carcinoma.
[d]ILC, invasive lobular carcinoma.

surgery. The characteristics of the study population and tumor staging are shown in TABLE 1. Similar specimens from a smaller series of larynx carcinoma and colon cancers were obtained for routine histology.

Tissue Processing and Immunohistochemical Staining Procedures

The presence and the extent of CML staining were determined by immunohistochemistry as described previously.[2,14] Briefly, appropriate paraffin sections were deparaffinized, pretreated as previously indicated, and exposed to a polyclonal CML antibody with proven specificity.[2] Antibody–antigen complexes were visualized with the avidin–biotin complex method using the peroxidase staining reaction as described.[2,14]

In Vitro Studies

For the *in vitro* analysis, we used the breast cancer cell lines MCF-7, MDA-MB 231, and MDA-MB 435[15] and the keratinocyte cell lines HaCaT, A5, and II-4 RT[16] as previously described. Immunocytochemical stainings were performed as described for the tissue sections with the modification that the cell preparations were only fixed for a short time (30 s, acetone:methanol 2:1 v/v) without further pretreatment. For Western blotting, protein lysates were prepared as described previously.[17] In short, tissue extracts were prepared using a liquid nitrogen-cooled vibration mill (Micro Dismembrator U; Braun, Melsungen, Germany) and a motor-driven Potter-Elvejem homogenizator (Braun, Melsungen, Germany). For cell lysates, cultured cells were grown at basal conditions to 80% density, harvested, washed three times in buffered isotonic saline, and resuspended in lysis buffer (10 mmol/L TRIS/HCl [pH 7.4], 320 mmol/L saccharose, 1 mmol/L EDTA). Lysis and subcellular fractionation were performed by the addition of digitonin (100 μg/mL) and differential centrifugation, and cytosolic and mitochondrial fractions were characterized by lactate dehydrogenase and citrate synthetase according to Minich *et al.*[18] Aliquots equivalent to 30 μg protein were separated by sodium dodecyl polyacrylamide gel electrophoresis (SDS-PAGE) on 7.5% gels, transferred to nitrocellulose membranes, and specifically stained with polyclonal anti-CML antiserum (rabbit, 1:8000), which has been characterized extensively.[2] CML-modified proteins were visualized by the enhanced chemiluminescense method.

Results

To evaluate the possible value of the major Maillard product CML as a biomarker for tumor cells and possibly their invasive and metastatic potential *in vivo*, we studied the detailed localization of CML by immunohistochemistry in breast carcinomas with different tumor type, staging, differentiation, and microanatomic localization. In parallel, the surrounding stroma and supplying vessels were investigated. As shown in FIGURE 1, we observed a significant cytoplasmatic staining in nearly all tumor cells and cell groups regardless of tumor stage, tumor type, and localization at the tumor front or the tumor center. Endothelial cells of the vessels were CML positive, although to very different extents. In addition, several peritumoral stromal cells were also positively labeled but to a significantly

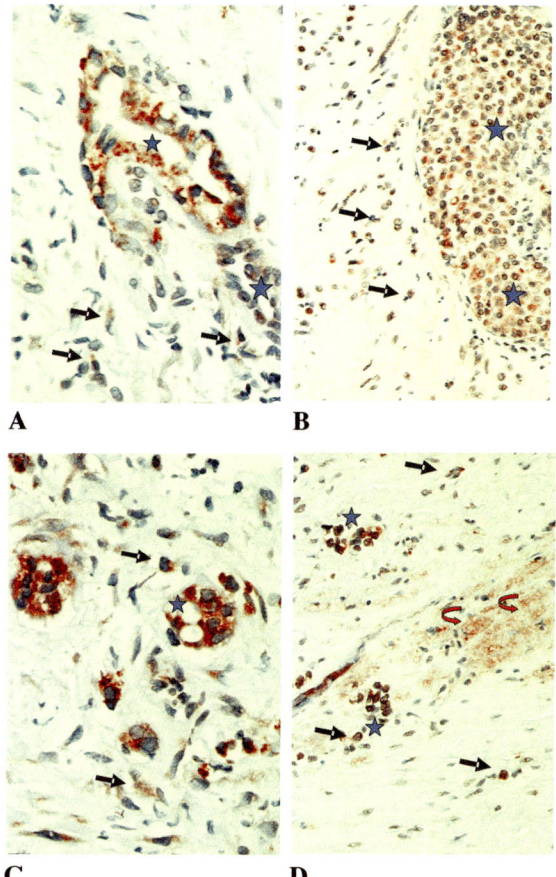

FIGURE 1. Immunolocalization of N^ε-(carboxymethyl)lysine (CML) staining in breast cancer tissue: **(A)** invasive ductal breast carcinoma with strong cytoplasmatic staining; **(B)** staining of invasive and intraductal carcinoma cells and adjacent stroma cells; **(C)** higher magnification of breast cancer tissue section with CML staining of tumor and stroma cells; **(D)** the pre-existing peritumoral matrix shows focal CML staining (curved arrows). The staining of stroma cells is indicated by (straight) arrows, and cytoplasmatic staining of tumor cells is indicated by stars (original magnification: A–D ×200).

lesser extent than the tumor cells. Furthermore, we observed a diffuse staining of the extracellular matrix in the surrounding matrix distant to the tumor zone in those cases of individuals of advanced age (more than 50-years old). This matrix staining was not seen in younger individuals (data not shown). The results of the individual subjects are summarized in TABLE 1. We also studied the tissue from 15 patients with larynx cancer and from 10 patients with colon cancer (both of different tumor stage and differentiation) and obtained similar results as described for breast cancer in detail above (data not shown).

To evaluate if the biological behavior of tumor cells influences CML formation, we studied three different breast cancer cell lines with different degrees of invasiveness and/or tendency for metastasis: MCF-7, MDA-MB 231, and MDA-MB 435 representative for noninvasive, nonmetastatic; invasive, low metastatic; or invasive metastatic behavior, respectively. We found strong cytoplasmatic CML staining of comparable extent in all cell lines (FIG. 2). Furthermore, we found no dependency of CML staining from the cell density. Similar results were obtained for three human keratinocytic cell lines with different biological behavior (HaCaT, nontumorigen; A5, tumorigen but benign growth; II-4 RT, tumorigen and invasively growing) (data not shown). Noteworthy, CML staining was not evenly distributed in the cytoplasma but concentrated in the perinuclear space (endoplasmatic reticulum).

To study the possible source of oxidative stress (CML), we separated mitochondrial and cytoplasmatic fraction of the breast cancer cell lines and found that distinct proteins were CML modified in mitochondria and others in cytoplasma, indicating the specific occurrence of CML modification in cellular compartments (FIG. 3A,B). Western blotting of tissue extracts from tumor and the surrounding nontumor tissue from four breast cancer samples showed similar CML-stained proteins in tumor and nontumor tissue (FIG. 3C).

Discussion

This is the first extensive study that investigates the occurrence and distribution of the AGE-modification CML in human tumor cells *in vitro* and *in vivo*. Using several series of tumor tissue from breast, larynx, and colon cancer and different cell lines of breast carcinomas and benign and malignant keratinocytes, we provide circumstantial evidence that tumor cells produce and/or accumulate CML-modified proteins. This is considerably higher than in peritumoral stroma cells or associated endothelial cells. Accordingly, we could identify the tumor cells as the main sources for CML-modified proteins during tumor growth.

In our experiments, we could not clarify the origin of CML formation in tumor cells. Because in previous studies overexpression of glucose transporter 1[17] (which leads to fivefold increased glycolysis rate without increasing oxidative stress) does not lead to increased CML modification of cellular plasma proteins, we speculate that an enhanced oxidative stress promotes the formation of CML in tumor cells and tissues.

There is only minimal information on the presence of well-characterized AGE products in human

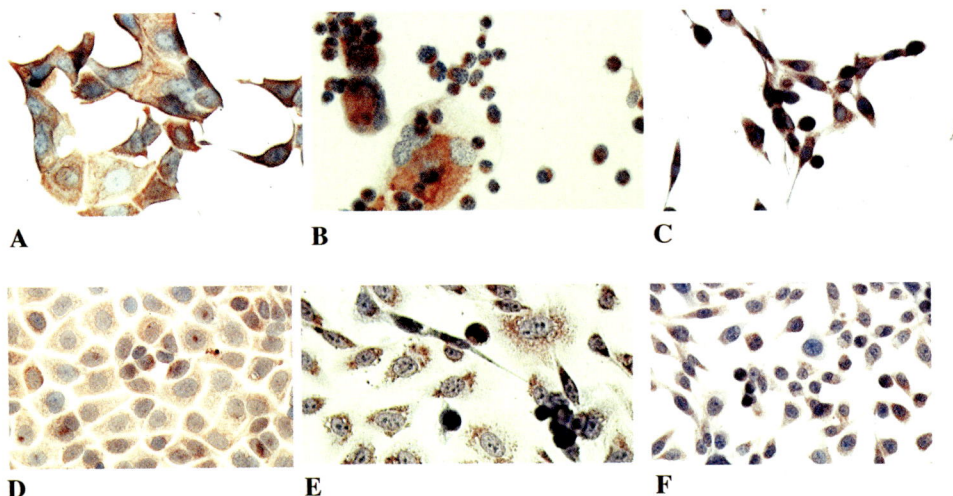

FIGURE 2. Immunolocalization of CML staining in breast cancer cell lines with different degrees of invasiveness and/or tendency for metastasis. (**A, D**) MCF-7 a noninvasive nonmetastatic cell line; (**B, E**) MDA-MB 231, an invasive low-metastatic cell line; and (**C, F**) MDA-MB 435, a cell line with invasive metastatic behavior. A–C shows low cell density; D–F reveals high cell density. There is significant cytoplasmic CML staining in all cells of different growth pattern and cell density (A–F×400).

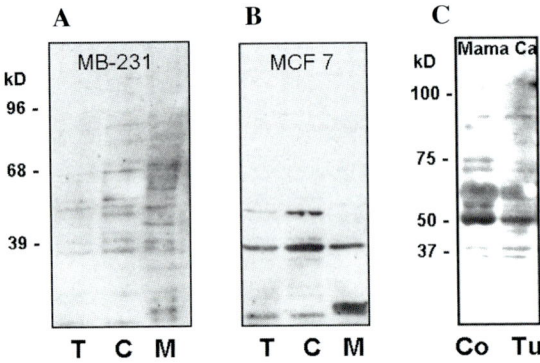

FIGURE 3. Western blotting of proteins extracted from (**A**) the invasive, low-metastatic, Mama carcinoma cell line MDA-MB 231; (**B**) from the noninvasive, nonmetastatic, cell line MCF-7; (**C**) and from a solid Mama carcinoma. For the cell lines, CML-immunoreactivity in the total cell lysate (T) is compared to the cytosolic (C) and mitochondrial (M) fraction. For the solid tumor, CML staining of surrounding tissue (control, Co) is compared to the tumor (Tu).

cancer. Schalkwijk et al. studied the localization of CML and argpyrimidine using specific antibodies.[19] They detected the presence of both antigens in five samples of breast and colon carcinomas but with different intensities. In two out of five cases of larynx carcinomas and leiomyosarcomas, no staining of either antigen was found. Furthermore, this group identified heat shock protein 27 as a major argpyrimidine-modified protein in human nonsmall cell lung cancer.[20] Since the antiapoptotic effect of this protein is increased upon modification, our results together with previous reports may indicate that tumor-induced modification of distinct proteins may lead to increased resistance to apoptosis in human carcinomas.

Many reports have confirmed the expression of RAGE in various human tumors since the first report.[21] The possible relevance of the activated AGE–RAGE pathway for tumor progression has been studied by the group of Schmidt et al.[22] They demonstrated that blockade of RAGE–amphoterin decreased growth and metastases of both implanted tumors and tumors developing spontaneously in susceptible mice. These data together with various other reports indicate that activation of the RAGE induces signal transduction pathways important for tumor progression and metastasis.

In conclusion, our in vivo studies reveal that tumor cells in all carcinomas studied show a strong cytoplasmatic immunoreactivity to CML with no further information on the characterization of carcinoma in respect to origin, differentiation, invasiveness, or tumor stage.

Conflict of Interest

The authors declare no conflicts of interest.

References

1. VLASSARA, H., R. BUCALA & L. STRIKER. 1994. Pathogenic effects of advanced glycosylation: biochemical, biologic,

and clinical implications for the diabetes and aging. Lab. Invest. **70:** 138–151.
2. SCHLEICHER, E.D., E. WAGNER & A. NERLICH. 1997. Increased accumulation of the glycoxidation product Nε-(carboxymethyl)lysine in human tissues in diabetes and aging. J. Clin. Invest. **99:** 457–468.
3. SCHMIDT, A.M., O. HORI & R. CAO, et al. 1996. RAGE: a novel cellular receptor for advanced glycation end products. Diabetes **45**(Suppl 3): S77–80.
4. LILIENSIEK, B., M.A. WEIGAND & A. BIERHAUS, et al. 2004. Receptor for advanced glycation end products (RAGE) regulates sepsis but not the adaptive immune response. J. Clin. Invest. **113:** 1641–1650.
5. ANDRASSY, M., J. IGWE & F. AUTSCHBACH, et al. 2006. Post-translationally modified proteins as mediators of sustained intestinal inflammation. Am. J. Pathol. **169:** 1223–1237.
6. TANJI, N., G.S. MARKOWITZ, C. FU, et al. 2000. Expression of advanced glycation end products and their cellular receptor RAGE in diabetic nephropathy and nondiabetic renal disease. J. Am Soc. Nephrol. **11:** 1656–1666.
7. HASLBECK, K.M., B. NEUNDORFER, U. SCHLOTZER-SCHREHARDTT, et al. 2007. Activation of the RAGE pathway: a general mechanism in the pathogenesis of polyneuropathies? Neurol. Res. **29:** 103–110.
8. HAMMES, H.P., H. HOERAUF, A. ALT, et al. 1999. Nε(Carboxymethyl)Lysin and the AGE Receptor RAGE colocalize in age-related macular degeneration. Invest. Ophth. Vis. Sci. **40:** 1855–1859.
9. WENDT, T.M., N. TANJI, J. GUO, et al. 2003. RAGE drives the development of glomerulosclerosis and implicates podocyte activation in the pathogenesis of diabetic nephropathy. Am. J. Pathol. **162:** 1123–1137.
10. LEDL, F. & E. SCHLEICHER. 1990. New aspects of the maillard reaction in foods and in the human body. Angew. Chem. Int. Ed. Engl. **29:** 565–594.
11. REDDYS, S., J. BICHLER, K.J. WELLS-KNECHT, et al. 1995. Nε(Carboxymethyl)Lysin is a dominant advanced glycation end product (AGE) antigen in tissue proteins. Biochemistry **34:** 10872–10878.
12. KISLINGER, T., C. FU, B. HUBER, et al. 1999. Nε(Carboxymethyl)Lysin adducts of proteins are ligands for receptor for advanced gylcation end products that activate cell signalling pathways and modulate gene expression. J. Biol. Chem. **274:** 31740–31749.
13. THORPE, S.R. & J.W. BAYNES. 2003. Maillard reaction products in tissue proteins: new products and new perspectives. Amino Acids **25:** 275–281.
14. NERLICH, A.G., B.E. BACHMEIER, E. SCHLEICHER, et al. 2007. Immunomorphological analysis of RAGE receptor expression and NF-kappaB activation in tissue samples from normal and degenerated intervertebrals discs of various ages. Ann. N. Y. Acad. Sci. **1096:** 239–248.
15. BACHMEIER, B.E., A.G. NERLICH, et al. 2007. The chemopreventive polyphenol curcumin prevents hematogenous breast cancer metastases in immunodeficient mice. Cell. Physiol. Biochem. **19:** 137–152.
16. BACHMEIER, B.E. & A.G. NERLICH. 2002. Immunohistochemical pattern of cytokeratins and MMPs in human keratinocyte cell lines of different biological behaviour. Int. J. Oncol. **20:** 495–499.
17. SCHLEICHER, E., C. WEIGERT, H. ROHRBACH, et al. 2005. Role of glucoxidation and lipid oxidation in the development of atherosclerosis. Ann. N. Y. Acad. Sci. **1043:** 343–354.
18. MINICH, T., S. YOKOTA, R. DRINGEN, et al. 2003. Cytosolic and mitochondrial isoforms of NADP+dependent isocitrate dehydrogenases are expressed in cultured rat neurons, astrocytes, oligodendrocytes and microglial cells. J. Neurochem. **86:** 605–614.
19. VAN HEIJST, J.W., H.W. NIESSEN, K. HOEKMAN, et al. 2005. Advanced glycation end products in human cancer tissue. Ann. N.Y. Acad. Sci. **1043:** 725–733.
20. VAN HEIJST, J.W., H.W. NIESSEN, R.J. MUSTERS, et al. 2006. Argpyrimidine-modified heat shock protein 27 in human non-small cell lung cancer: a possible mechanism for evasion of apoptosis. Cancer Lett. **241:** 309–319.
21. HSIEH, H.L., B.W. SCHAFER, N. SASAKI, et al. 2003. RAGE is expressed in human tumour tissue. Biochem. Biophys. Res. Com. **307:** 375–381.
22. TAGUCHI, A., D.C. BLOOD, G. DEL TORO, et al. 2000. Blockade of RAGE-amphoterin signalling suppresses tumour growth and metastases. Nature **405:** 354–360.

Glycation of Plasma Lipoprotein Lipid Membrane and Screening for Lipid Glycation Inhibitor

KIYOTAKA NAKAGAWA, DAIGO IBUSUKI, SHINJI YAMASHITA, AND TERUO MIYAZAWA

Food and Biodynamic Chemistry Laboratory, Graduate School of Agricultural Science, Tohoku University, Sendai, Japan

We recently reported that phosphatidylethanolamine (PE)-linked Amadori product (Amadori-PE) increased abnormally in diabetic plasma. However, the glycation mechanism of human plasma low-density lipoprotein (LDL) is still unclear. Moreover, lipid glycation inhibitors have yet to be discovered. In this study, we compared the glycation kinetics of LDL lipid and LDL protein *in vitro* and screened lipid glycation inhibitors. LDL-PE was converted to Amadori-PE followed by LDL protein (apoB) glycation. Pyridoxal 5′-phosphate could easily react with PE before the glucose–PE reaction, and the PE–pyridoxal 5′-phosphate adduct was detected in human red blood cells. Pyridoxal 5′-phosphate can be used in diabetes prevention.

Key words: Amadori product; diabetes; glycation; lipid peroxidation; low-density lipoprotein; Maillard reaction; phosphatidylethanolamine; pyridoxal 5′-phosphate

Introduction

Lipid peroxidation plays a role in the pathophysiology of diabetes.[1] To detect lipid hydroperoxides as the primary oxidation product, we used the previously established chemiluminescence–high performance liquid chromatography (CL–HPLC) method to selectively measure lipid hydroperoxides.[2] We found that the level of plasma phosphatidylcholine hydroperoxide (PCOOH) is increased in type 2 diabetic patients,[3] and we therefore postulate that PCOOH-mediated cytotoxicity is involved in the pathophysiology of diabetes. Recently, while investigating the factors responsible for PCOOH increases in diabetic plasma, we found that glycated phospholipids significantly accumulated in diabetic plasma.[4,5] Glucose reacts with the amino group of phosphatidylethanolamine (PE) to form an unstable Schiff base that undergoes an Amadori rearrangement to yield the stable PE-linked Amadori product (Amadori-PE) deoxy-D-fructosyl PE. Since Amadori-PE triggers the oxidative modification of lipids via the generation of superoxides, Amadori-PE and PCOOH could play a role in the development of diabetes.

We developed an analytical method for the estimation of Amadori-PE using liquid chromatography combined with quadrupole/linear ion-trap tandem mass spectrometry (LC–MS/MS) and determined human plasma Amadori-PE concentrations. The amount of Amadori-PE in the plasma of diabetic patients was higher than that in the control group. Plasma Amadori-PE was positively correlated with plasma PCOOH.[6,7]

However, despite the potential significance of Amadori-PE in pathological signaling, the glycation of human plasma low-density lipoprotein (LDL) membrane, a PE-rich component in blood, is still unclear. Moreover, lipid glycation inhibitors have not yet been discovered because of the absence of a lipid glycation model for inhibitor screening. In this study, we analyzed and estimated LDL Amadori-PE using LC–MS/MS and compared the glycation kinetics of LDL lipid and LDL protein *in vitro*. Subsequently, we optimized reaction conditions between PE and glucose and developed a lipid glycation model for screening lipid glycation inhibitors.

Comparing Glycation Kinetics of LDL Lipid and LDL Protein

We compared the glycation kinetics of LDL lipid and LDL protein. Plasma from healthy human subjects was collected and subjected to centrifugation to prepare LDLs. After incubating LDL with glucose,

Address for correspondence: Teruo Miyazawa, Food and Biodynamic Chemistry Laboratory, Graduate School of Agricultural Science, Tohoku University, 1-1, Tsutsumidori-Amamiyamachi, Aoba-ku, Sendai, 981-8555, Japan. Voice: +81 22 717 8904; fax: +81 22 717 8905.
miyazawa@biochem.tohoku.ac.jp

LDL Amadori-PE was estimated using LC combined with LC–MS/MS. LDL lipids were extracted using the Folch method and analyzed using LC–MS/MS as in the previous study.[6,7] The protein glycation marker Glycated apoB was measured using a Glycacor kit (Exocell, Philadelphia, PA)-included antibody recognizing fructosyl-adducts.

LC–MS/MS with multiple reaction monitoring enabled the separation and the detection of predominant LDL Amadori-PE molecular species (e.g., 16:0–18:2, 16:0–20:4, and 16:0–22:6 Amadori-PE) with high sensitivity and selectivity. The Amadori-PE concentration of nonglycated LDL was approximately 0.8 mol% of LDL-PE. After incubation of LDL with glucose, the nonglycated PE peak decreased and the Amadori-PE peak significantly increased. We found that lipid glycation increased along with an increase in glucose concentration and incubation time. In addition, the glycated LDL concentration did not change as rapidly as the Amadori-PE concentration. Therefore, it was hypothesized that when LDL was incubated with glucose, LDL-PE was glycated and converted to Amadori-PE, followed by LDL protein (apoB) glycation. The number of glycatable ε-amino residues in apoB (approximately 20–136)[8] is similar to that in the total content of PE (approximately 137) per LDL particle. These results demonstrate that LDL-PE is susceptible to glycation, suggesting the involvement of LDL lipid glycation in the pathophysiology of diabetes and demonstrating the usefulness of Amadori-PE as a glycation marker *in vivo*.

Development of Lipid Glycation Model and Lipid Glycation Inhibitor Screening

PE glycation involves the formation of a Schiff base and its rearrangement to Amadori-PE. Therefore, to develop a lipid glycation model, we optimized the reaction conditions (glucose concentration, temperature, buffer type, and pH) for Schiff-PE and Amadori-PE in the reaction between PE and glucose. Schiff-PE and Amadori-PE were detected by reversed-phase LC combined with evaporative light scattering detection-MS (LC–ELSD-MS) as in the previous study.[9]

High glucose concentration (300–500 mmol/L) stimulated Schiff-PE formation but temperature did not. Methanolic buffer (70–80% methanol) enhanced Schiff-PE formation, but excess methanol (80%) in the buffer caused pH instability. Based on composition and pH, phosphate buffer (pH 7.0–7.4) was found to be the most effective for Schiff-PE formation. The accelerating effect of phosphate anions on the Maillard reaction is well known.[10–12] It is likely that phosphate anions enhance lipid glycation by its binding and function as a neighboring group catalyst of glycation. Based on these results, we selected the following lipid glycation system: dioleoyl-PE (0.3 mmol/L) incubated with glucose (500 mmol/L) in 1 mL of 70% methanolic phosphate buffer (0.1 mmol/L pH, 7.4) at 37°C. In this system, Schiff-PE formation reached its maximum level (0.18 mmol/L) at 6 h after incubation and then decreased gradually. During 12–48 h incubation, Amadori-PE formation occurred consistently. These findings can be explained by the Maillard theory that PE reacts with glucose to form Schiff-PE, followed by rearrangement to Amadori-PE. Therefore, compounds that can reduce Schiff-PE formation can be used as lipid glycation inhibitors.

Using this model, various protein glycation inhibitors (aminoguanidine, pyridoxamine, aspirin, and carnosine), antioxidants (ascorbic acid, α-tocopherol, quercetin, and rutin), and other food compounds (L-lysine, L-cysteine, pyridoxine, pyridoxal and pyridoxal 5′-phosphate) were evaluated for their antiglycation properties. Unexpectedly, some components considered as protein glycation inhibitors and antioxidants had little inhibitory effect on lipid glycation. The most active inhibitors were pyridoxal 5′-phosphate and pyridoxal (vitamin B_6 derivatives). The inhibitory mechanism involved condensation of the aldehyde group of pyridoxals with the amino group of PE. However, low-molecular-weight aldehydes (glyoxal, methylglyoxal, and DL-glyceraldehyde) had no inhibitory effect on lipid glycation. These pyridoxals could easily be condensed with PE before the glucose–PE reaction occurred. In this manner, pyridoxals strongly inhibited lipid glycation. Further, the PE–pyridoxal 5′-phosphate adduct was detected in human red blood cells.[9] These findings suggest a new role of pyridoxal 5′-phosphate in preventing lipid glycation *in vivo*.

Conflict of Interest

The authors declare no conflicts of interest.

References

1. WITZTUM, J.L. & D. STEINBERG. 1991. Role of oxidized low density lipoprotein in atherogenesis. J. Clin. Invest. **88:** 1785–1792.
2. MIYAZAWA, T. 1989. Determination of phospholipid hydroperoxides in human blood plasma by a chemiluminescence-HPLC assay. Free Radic. Biol. Med. **7:** 209–217.
3. NAGASHIMA, T., S. OIKAWA & Y. HIRAYAMA, et al. 2002. Increase of serum phosphatidylcholine hydroperoxide

dependent on glycemic control in type 2 diabetic patients. Diabet. Res. Clin. Pract. **56:** 19–25.
4. OAK, J.H., K. NAKAGAWA & T. MIYAZAWA. 2000. Synthetically prepared Amadori-glycated phosphatidylethanolamine can trigger lipid peroxidation via free radical reactions. FEBS Lett. **481:** 26–30.
5. MIYAZAWA, T., J.H. OAK & K. NAKAGAWA. 2005. A convenient method for preparation of high-purity, Amadori-glycated phosphatidylethanolamine and its prooxidant effect. Ann. N.Y. Acad. Sci. **1043:** 276–279.
6. NAKAGAWA, K., J.H. OAK & O. HIGUCHI, et al. 2005. Ion-trap tandem mass spectrometric analysis of Amadori-glycated phosphatidylethanolamine in human plasma with or without diabetes. J. Lipid Res. **46:** 2514–2524.
7. MIYAZAWA, T., J.H. OAK & K. NAKAGAWA. 2005. Tandem mass spectrometry analysis of amadori-glycated phosphatidylethanolamine in human plasma. Ann. N.Y. Acad. Sci. **1043:** 280–283.
8. SEGREST, J.P., M.K. JONES & H.D. LOOF, et al. 2001. Structure of apolipoprotein B-100 in low density lipoproteins. J. Lipid Res. **42:** 1346–1367.
9. HIGUCHI, O., K. NAKAGAWA & T. TSUZUKI, et al. 2006. Aminophospholipid glycation and its inhibitor screening system: a new role of pyridoxal 5′-phosphate as the inhibitor. J. Lipid Res. **47:** 964–974.
10. TESSIER, F. & I.B. ARAGON. 1998. Effect of pH, phosphate and copper on the interaction of glucose with albumin. Glycoconj. J. **15:** 571–574.
11. WATKINS, N.G., C.I. NEGLIA-FISHER & D.G. DYER, et al. 1987. Effect of phosphate on the kinetics and specificity of glycation of protein. J. Biol. Chem. **262:** 7207–7212.
12. BURTON, H.S. & D.J. MCWEENY. 1963. Non-enzymic browning reactions: consideration of sugar stability. Nature **197:** 266–268.

Analysis of Amadori-glycated Phosphatidylethanolamine in the Plasma of Healthy Subjects and Diabetic Patients by Liquid Chromatography–Tandem Mass Spectrometry

TERUO MIYAZAWA, DAIGO IBUSUKI, SHINJI YAMASHITA, AND KIYOTAKA NAKAGAWA

Food and Biodynamic Chemistry Laboratory, Graduate School of Agricultural Science, Tohoku University, Sendai, Japan

Peroxidized phospholipid-mediated cytotoxity, the abnormal increase in the levels of phosphatidylcholine hydroperoxide (PCOOH) found in the plasma of type 2 diabetic patients, is involved in the pathophysiology of many diseases. PCOOH accumulation may be related to Amadori-glycated phosphatidylethanolamine (deoxy-D-fructosyl PE, or Amadori-PE) because Amadori-PE causes oxidative stress. However, the occurrence of lipid glycation products, including Amadori-PE, *in vivo* remains unclear. We developed a method to analyze Amadori-PE by using quadrupole/linear ion-trap mass spectrometry, the Applied Biosystems 4000 Q TRAP. We found that pyridoxals could easily be condensed with PE before the glucose–PE reaction occurred. The PE-pyridoxal 5′-phosphate adduct was detectable in human red blood cells, and the increased plasma Amadori-PE concentration in streptozotocin-induced diabetic rats was decreased by dietary supplementation with pyridoxal 5′-phosphate. Therefore, it is likely that pyridoxal 5′-phosphate acts as a lipid glycation inhibitor *in vivo*, and this may contribute to diabetes prevention.

Key words: glycation; phosphatidylethanolamine; Amadori product; tandem mass spectrometry; diabetes; glycation inhibitor; pyridoxal 5′-phosphate

Introduction

Lipid oxidative stress plays a role in the pathophysiology of atherogenesis, diabetes, aging, and other conditions.[1] To detect the primary oxidation product, lipid hydroperoxide, we established the chemiluminescence detection–liquid chromatography method.[2] Using this method, we confirmed that the level of plasma phosphatidylcholine hydroperoxide (PCOOH) increases in patients with hyperlipidemia[3] and type 2 diabetes.[4] Hence, we hypothesized that plasma PCOOH formation is closely involved in the pathophysiology of such diseases.

Maillard reactions, which occur between an amino group and a reducing sugar, are the most important among the chemical and oxidative reactions occurring in foods and biological samples; they contribute to food deterioration and the pathophysiology of human diseases. Although protein glycation has been investigated thoroughly, little attention has been paid to lipid glycation. Some studies showed that phosphatidylethanolamine (PE) reacts with glucose, forming a PE-linked Amadori product (Amadori-PE) through the formation of an unstable Schiff base.[5] Our subsequent studies demonstrated that Amadori-PE is capable of generating reactive oxygen species, and thereby triggers lipid peroxidation in the presence of ferrous ions[6]; moreover, it is an important compound that promotes vascular disease as a result of its angiogenic activity on endothelial cells.[7]

Despite the potential significance of Amadori-PE in pathological signaling, its characteristics and a quantitative method for its detection *in vivo* have not been established. Accordingly, we developed a liquid chromatography method by using an ultraviolet-labeling reagent for the detection of Amadori-PE in foods.[8] However, we were unable to detect Amadori-PE in biological samples because of insufficient sensitivity of the method.

Address for correspondence: Teruo Miyazawa, 1-1, Tsutsumidori-Amamiyamachi, Aoba-ku, Sendai, 981-8555, Japan, Voice: +81 22 717 8904; fax: +81 22 717 8905.

miyazawa@biochem.tohoku.ac.jp

A recently developed hybrid quadrupole/linear ion trap (QqLIT) spectrometer, the Applied Biosystems 4000 Q TRAP, is particularly advantageous as a liquid chromatography–tandem mass spectrometry (LC–MS/MS) detector for the analysis of biomolecules.[9–11] The development of the 4000 Q TRAP has allowed both triple quadrupole and ion-trap scans to be performed simultaneously in a single stage. The product ion scan, neutral loss scan, and multiple reaction monitoring (MRM) can provide useful structural information regarding the analyte even in the presence of major background contaminants from complex biological matrices. Recently, the analysis of lipids (i.e., sphingolipids and phospholipids) using 4000 Q TRAP has been reported.[9]

Based on this knowledge, we developed a 4000 Q TRAP LC–MS/MS method for detecting Amadori-PE in the plasma of human subjects with or without diabetes. Using this method, we confirmed that Amadori-PE is present at higher levels in the plasma of diabetic patients compared with that of healthy subjects; we discuss here the possible peroxidative role of Amadori-PE *in vivo*. Because no lipid glycation inhibitor has been discovered thus far, we used LC–MS/MS to conduct *in vitro* screening of the possible therapeutic roles of the vitamin B_6 derivative, pyridoxal 5′-phosphate, *in vivo*.

Quantitative Analysis of Amadori-PE by LC–MS/MS

MS/MS analysis was carried out in the positive ion mode; the PE standard *(m/z* 744) yielded a product ion *(m/z* 603) corresponding to the neutral loss of the ethanolaminephosphate head group (141 amu) from the precursor ion. The same product ion *(m/z* 603) was also observed when the Amadori-PE standard *(m/z* 906) was analyzed by MS/MS. This suggested that the presence of Amadori-PE molecular species can be verified by identifying the specific neutral loss (303 amu) using MS/MS in the neutral loss scan mode. Using this mode, we successfully determined that Amadori-PE species exist in the plasma of diabetic patients undergoing chronic hemodialysis. To identify whether these species have the specific structural characteristics of Amadori-PE, three principal species *(m/z* 880, 906, and 926) in the plasma of diabetic patients undergoing chronic hemodialysis were examined by MS/MS analysis. Consequently, the specific product ion *(m/z* 603), corresponding to the neutral loss of the glucose-adducted ethanolaminephosphate head group from Amadori-PE, was observed in each of the three species.

Therefore, we are certain that lipid glycation occurs in the human body.[12]

The three predominant Amadori-PE species (16:0–18:1, 16:0–20:4, and 16:0–22:6) in the plasma of healthy subjects and diabetic patients were individually quantified using LC–MS/MS with MRM. All calibration curves showed good linearity ($R^2 = 0.995$–0.999), and the detection limits were ∼0.1 pmol/injection at a signal-to-noise ratio of 3. Similarly, the parameters for LC–MS/MS with MRM could be optimized for detecting nonglycated PE standards. When these conditions were fully optimized, the native PE species (16:0–18:1, 16:0–20:4, and 16:0–22:6) and their Amadori products appeared as clear peaks in MRM chromatograms of the plasma of diabetic patients. The plasma 16:0–18:1-Amadori-PE concentration was ∼3 pmol/mL in healthy subjects and 6–11 pmol/mL in diabetic patients. In the calculation of the 16:0–18:1-Amadori-PE/16:0–18:1 PE mol%, the amounts of Amadori-PE in the plasma of diabetic patients (0.15 mol%), diabetic patients on chronic hemodialysis (0.29 mol%), and nondiabetic patients on chronic hemodialysis (0.13 mol%) were higher than those of the control group (0.08 mol%). Similarly, the average glycation rate of 16:0–20:4-Amadori-PE/16:0–20:4 PE and of 16:0–22:6-Amadori-PE/16:0–22:6 PE in the plasma of diabetic patients was ∼0.13–0.30 mol%, which was higher than the rates in the plasma of the healthy control subjects (0.05–0.10 mol%). It is notable that the plasma Amadori-PE levels were positively correlated with the plasma levels of PCOOH (an oxidative stress marker).

Occurrence of the PE-pyridoxal 5′-phosphate Adduct in Human Red Blood Cells

Using this lipid glycation system, we investigated the inhibitory effects of several compounds on Schiff-PE formation. Screening revealed that pyridoxal 5′-phosphate and pyridoxal (vitamin B_6 derivatives) were the most appropriate antiglycative compounds.[13,14] Despite the wide distribution of pyridoxal 5′-phosphate in human tissues and body fluids, the occurrence of the PE-pyridoxal 5′-phosphate adduct *in vivo* has never been investigated. In the positive ion mode, collision-induced dissociation of a reduced form of the PE-pyridoxal 5′-phosphate adduct produced a diglyceride ion ($[M+H-372]^+$), permitting neutral loss scanning and MRM. When $NaBH_4$-treated red blood cell (RBC) extract was infused into the 4000 Q TRAP, molecular species of PE-pyridoxal 5′-phosphate adduct (reduced forms) could be screened out by neutral loss scanning. The predominant molecular species of PE-pyridoxal 5′-phosphate adduct in RBCs were 16:0–18:1 (m/z 949.9), 16:0–18:2 (m/z 947.9), 16:0–20:4 (m/z 971.9), 16:0–22:6 (m/z 995.9), 18:0–18:2 (m/z 975.9), 18:0–20:4 (m/z 999.9), and 18:0–22:6 (m/z 1023.8) as diacyl species and 16:0–20:4 (m/z 955.8), 16:0–22:6 (m/z 1007.8), 18:0–20:4 (m/z 983.8), 18:0–22:6 (m/z 1007.8), and 18:1–20:4 (m/z 981.9) as the alkenyl-acyl species (plasmalogen). The neutral loss spectra of the PE-pyridoxal 5′-phosphate adduct and PE itself indicated that the molecular species of PE were randomly condensed with pyridoxal 5′-phosphate. These results provided the first direct structural evidence for the existence of the PE-pyridoxal 5′-phosphate adduct *in vivo*.

Interfacing LC with the 4000 Q TRAP enabled the separation and detection of the predominant molecular species of the PE-pyridoxal 5′-phosphate adduct (reduced forms) in RBCs. The concentration of the PE-pyridoxal 5′-phosphate adduct (16:0–18:2 [m/z: 947.6], one of the predominant molecular species) was ∼78 ± 12 pmol/mL of packed cells. In contrast, the concentration of the PE-pyridoxal adduct was below the detection limit (<1 pmol/mL of packed cells). These findings suggest a new role of pyridoxal 5′-phosphate in preventing lipid glycation *in vivo*.

Dietary Pyridoxal 5′-phosphate Prevents Lipid Glycation *In Vivo*

The effect of dietary pyridoxal 5′-phosphate on lipid glycation was investigated using streptozotocin (STZ)-induced diabetic rats. The amount of Amadori-PE (18:0–22:6-Amadori-PE, a predominant species in plasma) in the plasma of control rats (not treated with STZ; $n = 6$) was 44.5 ± 25.7 pmol/mL. In STZ-induced diabetic rats (80 mg/kg of body weight; $n = 6$), a higher concentration of plasma Amadori-PE (18:0–22:6-Amadori-PE) was observed (945.7 ± 429.0 pmol/mL). The concentration of plasma Amadori-PE was proportional to that of plasma PCOOH (control rats, 8.6 ± 2.8 pmol/mL; STZ-induced diabetic rats, 68.9 ± 33.7 pmol/mL). The increments in the levels of both plasma Amadori-PE and PCOOH were significantly suppressed by 38% and 39%, respectively, in STZ-induced diabetic rats that received dietary supplementation with 2 mM pyridoxal 5′-phosphate (300 mg/kg of body weight per day) for 10 weeks. Although no significant difference was observed between the plasma lipids of the STZ-induced diabetic rats fed pyridoxal 5′-phosphate and the plasma lipids of rats that did not receive dietary supplementation, dietary pyridoxal 5′-phosphate tended to improve the plasma cholesterol and triglyceride levels of the STZ-induced diabetic rats.

Conflict of Interest

The authors declare no conflicts of interest.

References

1. WITZTUM, J.L. & D. STEINBERG. 1991. Role of oxidized low density lipoprotein in atherogenesis. J. Clin. Invest. **88:** 1785–1792.
2. MIYAZAWA, T. 1989. Determination of phospholipid hydroperoxides in human blood plasma by a chemiluminescence-HPLC assay. Free Radic. Biol. Med. **7:** 209–217.
3. KINOSHITA, M., S. OIKAWA, K. AYASAKA, *et al.* 2000. Age-related increases in plasma phosphatidylcholine hydroperoxide concentrations in control subjects and patients with hyperlipidemia. Clin. Chem. **46:** 822–828.
4. NAGASHIMA, T., S. OIKAWA, Y. HIRAYAMA, *et al.* 2002. Increase of serum phosphatidylcholine hydroperoxide dependent on glycemic control in type 2 diabetic patients. Diabet. Res. Clin. Pract. **56:** 19–25.
5. BUCALA, R., Z. MAKITA, T. KOSCHINSKY, *et al.* 1993. Lipid advanced glycosylation: pathway for lipid oxidation in vivo. Proc. Natl. Acad. Sci. USA **90:** 6434–6438.
6. OAK, J., K. NAKAGAWA & T. MIYAZAWA. 2000. Synthetically prepared Aamadori-glycated phosphatidylethanolamine can trigger lipid peroxidation via free radical reactions. FEBS Lett. **481:** 26–30.
7. OAK, J., K. NAKAGAWA, S. OIKAWA, *et al.* 2003. Amadori-glycated phosphatidylethanolamine induces angiogenic differentiations in cultured human umbilical vein endothelial cells. FEBS Lett. **555:** 419–423.
8. OAK, J.H., K. NAKAGAWA & T. MIYAZAWA. 2002. UV analysis of Amadori-glycated phosphatidylethanolamine in foods and biological samples. J. Lipid Res. **43:** 523–529.
9. HOUJOU, T., K. YAMATANI, H. NAKANISHI, *et al.* 2004. Rapid and selective identification of molecular species in phosphatidylcholine and sphingomyelin by conditional neutral loss scanning and MS3. Rapid Commun. Mass Spectrom. **18:** 3123–3130.
10. MIGLIORANÇA, L.H., R.E. BARRIENTOS-ASTIGARRAGA, B.S. SCHUG, *et al.* 2005. Felodipine quantification in human plasma by high-performance liquid chromatography coupled to tandem mass spectrometry. J. Chromatogr. B Analyt. Technol. Biomed. Life Sci. **814:** 217–223.
11. YAO, M., H. ZHANG, S. CHONG, *et al.* 2003. A rapid and sensitive LC/MS/MS assay for quantitative determination of digoxin in rat plasma. J. Pharm. Biomed. Anal. **32:** 1189–1197.
12. NAKAGAWA, K., J.H. OAK, O. HIGUCHI, *et al.* 2005. Ion-trap tandem mass spectrometric analysis of Amadori-glycated phosphatidylethanolamine in human plasma with or without diabetes. J. Lipid Res. **46:** 2514–2524.

13. HIGUCHI, O., K. NAKAGAWA, T. TSUZUKI, et al. 2006. Aminophospholipid glycation and its inhibitor screening system: a new role of pyridoxal 5′-phosphate as the inhibitor. J. Lipid Res. **47:** 964–974.
14. RYBAK, M.E. & C.M. PFEIFFER. 2004. Clinical analysis of vitamin B_6: determination of pyridoxal 5′-phosphate and 4-pyridoxic acid in human serum by reversed-phase high-performance liquid chromatography with chlorite postcolumn derivatization. Anal. Biochem. **333:** 336–344.

Nonenzymatically Glycated Lipoprotein ApoA-I in Plasma of Diabetic and Nephropathic Patients

ANNUNZIATA LAPOLLA,[a] MAURA BRIOSCHI,[b,c] CRISTINA BANFI,[b,c] ELENA TREMOLI,[b,c] CHIARA COSMA,[a] LUCIANA BONFANTE,[a] SIMONE CRISTONI,[b,d] ROBERTA SERAGLIA,[e] AND PIETRO TRALDI[e]

[a]*Dipartimento di Scienze Mediche e Chirurgiche, Cattedra di Malattie del Metabolismo, Università di Padova, Padova, Italy*

[b]*Centro Cardiologico Monzino IRCCS, Milano, Italy*

[c]*Dipartimento di Scienze Farmacologiche, Università degli studi di Milano, Milano, Italy*

[d]*ISB, Ion Source & Biotechnologies, Milan, Italy*

[e]*CNR-ISTM, Corso Stati Uniti 4, Padova, Italy*

ApoA-I, which constitutes 70% of the apolipoprotein content of high-density lipoproteins, acts as an acceptor for the transfer of phospholipids and free cholesterol from peripheral tissues and transports cholesterol in the liver and other tissue for excretion and steroidogenesis. In order to verify its possible structural alteration in pathological states, plasma samples from healthy, diabetic, and nephropathic subjects have been analyzed by two-dimensional gel electrophoresis. By this approach, clear differences among the three classes of subjects become evident and, in the case of diabetic and nephropathic patients, intense spots are present. The matrix-assisted laser desorption/ionization mass spectrometry analysis of their digestion products shows that an overexpression of unglycated and glycated ApoA-I is present in both pathological states, reasonably affecting the efficiency in cholesterol transport.

Key words: nonenzymatic glycation; diabetes; nephropathy; MALDI/MS; 2D gel electrophoresis

Introduction

Data from epidemiologic studies have shown that high plasma levels of high-density lipoproteins (HDL) protect against the development of atherosclerosis.[1] This is a result of the actions of HDL, the most important of which is the ability to promote the efflux of cholesterol from cells present in the arterial wall to the liver, a process called reverse cholesterol transport.[2] In this context, ApoA-I, which constitutes 70% of the apolipoprotein content of HDL, acts as an acceptor for the transfer of phospholipids and free cholesterol from peripheral tissues and transports cholesterol in the liver and other tissues for excretion and steroidogenesis.[3]

Atherosclerotic vascular disease is a major complication of diabetes, and among the known risk factors of atherosclerosis, such as hyperlipoproteinemia, obesity, hypertension, hyperinsulinemia, and inflammation,[4] low levels of HDL play an important role.[5]

Recently it has been shown that patients with end-stage renal disease have decreased levels of ApoA-I from increased catabolism compared to control subjects.[6,7] In addition to the reduction in ApoA-I levels, a series of studies have shown that post-translational modification of this apolipoprotein, including nonenzymatic glycation, can contribute to its impaired action.[8–11]

In the present investigation, a proteomic approach based on two-dimensional gel electrophoresis (2DE) and matrix-assisted laser desorption/ionization (MALDI) mass spectrometry (MS) has been employed to study plasma samples from healthy, diabetic, and nephropathic subjects with the main aim being to identify possible post-translational modifications of ApoA-I resulting from nonenzymatic glycation processes.

Address for correspondence: Professor Annunziata Lapolla, DPT Medical and Surgical Sciences, Padova University, Via Giustiniani n2, I-35127 Padova, Italy. Voice: +0039-049-8216857; fax: +0039-049-8216838.
annunziata.lapolla@unipd.it

TABLE 1. Metabolic parameters evaluated in 10 diabetic patients, 10 nephropathic patients, and 10 normal controls

Subjects	Fasting plasma glucose (mg/dL)	HbA1c (%)	Cholesterol (mg/dL)	HDL cholesterol (mg/dL)	Triglycerides (mg/dL)	Creatinine (mg/dL)
Diabetics	212 ± 12	9.0 ± 0.2	165 ± 31	53 ± 10	118 ± 34	1.1 ± 0.2
Nephropathic	92 ± 5	5.8 ± 0.2	195 ± 52	47 ± 11	193 ± 90	8.2 ± 1.2
Controls	89 ± 4	5.5 ± 0.3	212 ± 30	53 ± 15	110 ± 45	0.87 ± 0.02

Values are mean ± SD. Hb1Ac = glycated β-globulin; HDL = high-density lipoprotein.

Experimental

Subjects

Ten type 2-diabetic patients (mean age ± SD, 63 ± 7 years; mean duration of disease, 18 ± 9 years), 10 patients affected by end-stage renal disease and subjected to peritoneal dialysis (mean age ± SD, 60 ± 6 years; mean duration of disease, 4 ± 2 years), and 10 normal controls (mean age ± SD, 62 ± 4 years) were evaluated. Their metabolic parameters are reported in TABLE 1. All subjects gave their informed consent to the study that was carried out, following the Helsinki Declaration rules and the study was approved by the local ethics committee.

Plasma Sample Preparation and Two-dimensional Gel Electrophoresis

The plasma samples (10 mL), obtained by mixing the samples of each group of subjects, were diluted four times with 20% (v/v) acetonitrile and applied onto a Centriplus centrifugal concentrator membrane with MWCO 30000 (Millipore, Billerica, MA). The centrifugal filter membranes were rinsed and used according to manufacturer's instruction. The samples were centrifuged at 3000 g until >90% of the input plasma had passed through the membranes.[12] The filtrates were lyophilized to dryness and resuspended in water before desalting and delipidating by the trichloroacetic/acetone procedure. With this method we obtained a recovery of 150 μg of low-molecular-weight proteins from 600 mg of total plasma proteins. Desalted and delipidated proteins were resuspended in a buffer containing 8 mol/L urea, 2 mol/L thiourea, 4% 3-[(3-cholamiopropyl)dimethylammonio]-1-propanesulfonate, 2% v/v carrier ampholytes (pH 3–10), 20 mmol/L Tris, 55 mmol/L dithiothreitol (DTT), and bromophenol blue. 2DE was carried out on the Protean IEF cell (Bio-Rad Laboratories, Segrate, Milano, Italy) as previously described.[13] IPG ready strips (7 cm, pH 3–10 nonlinear gradient) (Bio-Rad) were actively rehydrated at 50 V for 24 h. Proteins were loaded at the cathode using the cup loading tray for the Protean IEF cell and focused for a total of 10 kVh. After focusing, the strips were first equilibrated 15 min with a solution containing 50 mmol/L Tris-HCl, 6 mol/L urea, 30% v/v glycerol, 2% SDS, and 2% DTT and then equilibrated again with the same buffer containing 4.5% iodoacetamide instead of DTT. The focused proteins were then separated according to size by SDS-PAGE on 17% Tricine gel and stained with colloidal blue stain. The protein patterns were digitalized with a scanner and compared with PDQUEST 6.0 (Bio-Rad).

Protein Digestion

The 2DE protein spot was digested, slightly modifying the method employed by Shevchenko et al.[14] Briefly, after washing the spot with 200 μL of NH_4HCO_3 (50 mmol)/CH_3CN 1:1 in order to completely remove the colloidal blue stain, the gel was dried and rehydrated with 20 μL of 6 ng/mL trypsin solution for 30 min at 4°C. Then, 5 μL of NH_4HCO_3 was added on the gel protein spot, and the sample was incubated at 37°C overnight. After enzymatic digestion, the sample was centrifuged and 2 μL of H_2O + 1% trifluoroacetic acid solution (TFA) solution was added to the obtained peptide mixture. The sample was sonicated using an ultrasound bath for 30 min and was centrifuged at 13,000 g/min for 1 min, and 1 μL of the surnatant was characterized by MALDI.

Mass Spectrometry

Spectra were acquired by means of a Voyager-DE STR MALDI-TOF instrument (Applied Biosystem, Foster City, CA), operating in reflectron positive ion mode. A nitrogen laser ($\lambda = 337$ nm) was used, operating at 20 Hz with adjustable laser energy. α-4-hydroxycinnamic acid was used as the matrix. We deposited 1 μL of the digested protein peptide mixture on the MALDI steel plate and dried it at room temperature. We then deposited 0.5 μL of matrix solution (10 mg/mL in a H_2O/CH_3CN 1/1 +0.1% TFA solution) directly on the dried spot and allowed it to dry at room temperature. The operative conditions used were the following: accelerating voltage, 25 kV; grid

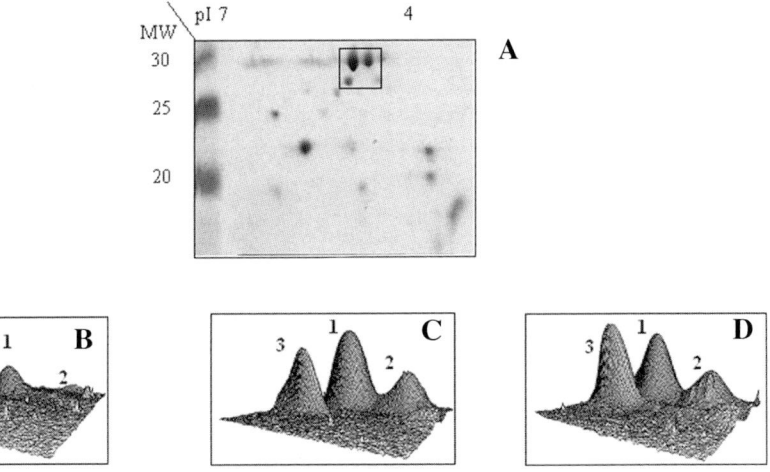

FIGURE 1. (A) Two- and three-dimensional plots of 2DE protein spots of **(B)** control, **(C)** diabetic, and **(D)** nephropathic subjects.

voltage, 75%; delay time, 100 ns. The trypsin autolysis fragments at m/z 842.4059 and 2211.152 were used in order to calibrate the mass spectra of the digested proteins. Post-source decay (PSD)[15] experiments were used to obtain the protein sequence, identifying the modified amino acids.

Data Analysis
2DE Image Analysis

PDQuest software (Bio-Rad) was used for spot detection, spot quantification, gel matching, and statistical analysis of differences between the experimental groups. The cut-off level for a differentially expressed protein was defined as at least a twofold increase or decrease in spot intensity. Statistically significant differences between groups for each protein were computed by analysis of variance. The level of significance of difference was set at $P < 0.05$.

MALDI Data

Aldente[16] and Profound[17] peptide mass fingerprint tools were used to identify the analyzed digested proteins, while the algorithm GlycoMod[18] was used to identify the modified glycated peptide. Modified peptide sequences were confirmed by the PSD approach. In particular the on-line available InSpecT software,[19] based on the tag sequencing approach,[20] was used to obtain the peptide sequence from PSD fragmentation spectra. Each modified sequence was assigned by means of a statistical score, expressed as a p value, provided by the software.

Results

The results of 2DE analysis of the three groups of plasma samples are reported in FIGURE 1, visualized in the three-dimensional views of the area of interest. These data show that, in the case of healthy subjects, essentially only one peak is present, while in the case of diabetic and nephropathic patients, three different peaks are detected in the same region.

The MALDI data of the samples obtained by enzymatic digestion of the differentially expressed spots indicate that spots 1 and 2 correspond to ApoA-I, while spot 3 is a result of retinol-binding proteins (P value from 3.6×10^{-24} to 1×10^{-7}). MALDI analysis indicates that spot 2 corresponds to glycated ApoA-I, which is also present in normal subjects to a lesser extent.

The glycation sites have been individuated by the analysis of the digestion fragments. Mono and diglycated peptides are evidenced in the case of spot 2. These data have been further confirmed by the PSD approach followed by peptide sequence tags to filter the database. The modified peptide sequences are reported in TABLE 2 together with their probability score, expressed as a p value (between 6×10^{-7} and 5×10^{-5}).

Discussion

Plasma proteomics is a complex task because of the copresence of many proteins of strongly different relative abundance. In particular, albumin and immunoglobulin hide all the other low-abundant

TABLE 2. Modified peptides identified by accurate mass measurement in spots 1, 2, and 3 of control, diabetic, and nephropathic subjects

Sample (Spot - Name)	Spot density (OD·mm²)	m/z values (mass accuracy in ppm)	Modification	Modified sequence	Sequence position
Control (1 - Apo A-I)	0.35 ± 0.07	-	-	-	-
Control (2 - Not found)	0.10 ± 0.02	-	-	-	-
Control (3 - Not Found)	0.12 ± 0.02	-	-	-	-
Diabetic (1 - Apo A-I)	2.01 ± 0.08*	-	-	-	-
Diabetic (2 Apo A-I)	0.51 ± 0.05*	1838.908 (4)	GLUC-GLUC	VSFLSALEEYTKK	251–263
		2922.342 (58)	GLUC-GLUC	QGLLPVLESFKVSFLSALEEYTK	240–262
		2239.102 (21)	GLUC-PHOS	DYVSQFEGSALGKQLNLK	52–69
Diabetic (3- RBP)	0.62 ± 0.07*	-	-	-	-
Nephropathic (1 - Apo A-I)	1.32 ± 0.10*	-	-	-	-
Nephropathic (2 - Apo A-I)	0.43 ± 0.08*	1838.908 (4)	GLUC-GLUC	VSFLSALEEYTKK	251–263
		2922.342 (58)	GLUC-GLUC	QGLLPVLESFKVSFLSALEEYTK	240–262
		2239.102 (21)	GLUC-PHOS	DYVSQFEGSALGKQLNLK	52–69
Nephropathic (3 RBP)	0.77 ± 0.12*	-	-	-	-

The modification type is also reported. Densitometric analysis: *$P < 0.05$ versus control.

proteins. To overcome this problem, in recent years different plasma sample treatments have been proposed; these are mainly based on purification and prefractionation techniques.

In the present investigation, the 2DE approach was used. Ten plasma samples for each of the groups under investigation (healthy subjects, type 2-diabetic patients, and nephropathic patients) were mixed and each of the obtained plasma pools was analyzed. Our attention was focused on the low-molecular-weight protein fraction (MW < 30,000 Da), which was purified by centrifugal ultrafiltration before 2DE. Albumin is known to bind and transport small molecules and peptides; for this reason the centrifugal ultrafiltration was conducted with solvents in order to disrupt the protein–protein interactions.

The results of our study provide evidence of a clear difference between the samples of healthy subjects and those of diabetic and nephropathic patients. In particular, overexpression of ApoA-I protein is observed in the pathological cases. Furthermore, spot 2, analyzed by MALDI and PSD experiments, corresponds to glycated ApoA-I. The glycated peptide sequences indicate that the preferential glycation sites are the ε-amino group of lysines belonging to the protein chain. These data imply that in plasma samples of both diabetic and nephropathic subjects glycated ApoA-I is as abundant as unglycated protein. Furthermore, considering that equal amounts of the plasma samples have been analyzed, both unglycated and glycated proteins are overexpressed in these groups compared to control subjects.

These findings are partially in agreement with the published literature on ApoA-I levels. In fact, the overexpression of ApoA-I in the diabetic subjects is in agreement with the data of Calvo et al.,[8–10] while the same behavior in the nephropathic subjects conflicts with the data published by Batista et al.[6] and Okubo et al.[7] on the basis of stable isotope studies. This discrepancy can be attributed to an increased catabolism of triglycerides occurring in patients with end-stage renal disease.

The glycation of ApoA-I in the diabetic patients is in line with what we have observed for other circulating proteins, such as human serum albumin, hemoglobin, and immunoglobulin G.[21] The present investigation presented an undescribed behavior: in the case of nephropathic subjects, a glycation level analogous to that of diabetics was observed. This could be a result of carbonyl stress and/or glucose from peritoneal dialysis.

The data obtained indicate that the evaluation of ApoA-I and glycated ApoA-I can be a valid diagnostic tool to assess the metabolic state of diabetic and/or nephropathic patients. Furthermore, glycated ApoA-I levels could be related with the glyco-oxidation stress experienced by the patient during the protein half-life. Its possible change in functionality could reflect a different cholesterol transport efficiency, which would lead atherosclerosis development.

The same trend is also observed for end-stage renal disease patients, but it necessarily originates by a different mechanism, reasonably related to the efficiency of glycated ApoA-I clearance. These results can explain the occurrence of macrovascular disease in both types of patients.

Conflict of Interest

The authors declare no conflicts of interest.

References

1. CASTELLI, W.P., R.J. GARRISON, P.W. WILSON, et al. 1986. Incidence of coronary heart disease and lipoprotein cholesterol levels. The Framingham study. JAMA **256:** 2835–2838.
2. LEWIS, G.F. & D.J. RADER. 2005. New insights into the regulation of HDL metabolism and reverse cholesterol transport. Circ. Res. **96:** 1221–1232.
3. RADER, D. 2003. Regulation of reverse cholesterol transport and clinical implications. Am. J. Cardiol. **92:** 42J–49J.
4. ROHER, L., M. HERSBERGER & A. VON ECKARDSTEIN. 2004. High density lipoproteins in the intersection of diabetes mellitus, inflammation and cardiovascular disease. Curr. Opin. Lipidol. **15:** 269–278.
5. QUINTAO, E.C., W.L. MEDINA & M. PASSARELLI. 2000. Reverse cholesterol transport in diabetes mellitus. Diabetes Metab. Res. Rev. **16:** 237–250.
6. BATISTA, M.C., F.K. WELTY, M.R. DIFFENDERFER, et al. 2004. Apolipoprotein A-I, B-100 and B-48 metabolism in subjects with chronic kidney disease, obesity and the metabolic syndrome. Metabolism **53:** 1255–1261.
7. OKUBO, K., K. IKEWAKI, S. SAKAI, et al. 2004. Abnormal HDL apolipoprotein A-I and A-II kinetics in hemodialysis patients: a stable isotope study. J. Am. Soc. Nephrol. **15:** 1008–1015.
8. CALVO, C., G. PONSIN & F. BERTHEZENE. 1988. Characterization of the non enzymatic glycation of high density lipoprotein in diabetic patients. Diabete Metab. **14:** 264–269.
9. CALVO, C., C. TALUSSOT, G. PONSIN, et al. 1988. Nonenzymatic glycation of apolipoprotein A-I. Effects on its self-association and lipid binding properties. Biochem. Biophys. Res. Commun. **153:** 1060–1067.
10. CALVO, C. & C. VERDUGO. 1992. Association in vivo of glycated apolipoprotein A-I with high density lipoproteins. Eur. J. Clin. Chem. Clin. Biochem. **30:** 3–5.
11. SHISHINO, K., M. MURASE, S. MAKINO, et al. 2000. Glycated apolipoprotein A-I assay by combination of affinity chromatography and latex immunoagglutination. Ann. Clin. Biochem. **37:** 498–506.
12. HARPER, R.G., S.R. WORKMAN, S. SCHUETZNER, et al. 2004. Low-molecular-weight human serum proteome using ultrafiltration, isoelectric focusing and mass spectrometry. Electrophoresis **25:** 1299–1306.
13. BANFI, C., M. BRIOSCHI, R. WAIT, et al. 2006. Proteomic analysis of membrane microdomains derived from both failing and non-failing human hearts. Proteomics **6:** 1976–1988.
14. SHEVCHENKO, A., M. WILM, O. VORM, et al. 1996. Mass spectrometric sequencing of proteins from silver stained polyacrylamide gels. Anal. Chem. **68:** 850–858.
15. MONETI, G., F. FRANCESE, G. MASTROBUONI, et al. 2007. Do collisions inside the collision cell play a relevant role in CID-LIFT experiments? J. Mass Spectrom. **42:** 117–126.
16. http://www.expasy.org/tools/aldente/
17. http://prowl.rockefeller.edu/
18. http://www.expasy.ch/tools/glycomod/
19. http://peptide.ucsd.edu/
20. TANNER, S., H. SHU, A. FRANK, et al. 2005. InsPecT: Fast and accurate identification of post-translationally modified peptides from tandem mass spectra. Anal. Chem. **77:** 4626–4639.
21. LAPOLLA, A., D. FEDELE, R. SERAGLIA, et al. 2006. The role of mass spectrometry in the study of non-enzymatic protein glycation in diabetes: an update. Mass Spectrom. Rev. **25:** 775–797.

Evaluating the Extent of Protein Damage in Dairy Products

Simultaneous Determination of Early and Advanced Glycation-induced Lysine Modifications

JÖRG HEGELE,[a] VÉRONIQUE PARISOD,[a] JANIQUE RICHOZ,[a] ANKE FÖRSTER,[b] SARAH MAURER,[b] RENÉ KRAUSE,[b] THOMAS HENLE,[b] TIMO BÜTLER,[a] AND THIERRY DELATOUR[a]

[a]*Nestlé Research Centre, Nestec Ltd., CH-1000 Lausanne 26, Switzerland*
[b]*Institute of Food Chemistry, Technische Universität Dresden, D-01062 Dresden, Germany*

An isotope dilution liquid chromatography-tandem mass spectrometry (LC-MS/MS) method was developed to determine lysine (Lys), N^ε-fructosyllysine (FL), N^ε-carboxymethyllysine (CML), and pyrraline (Pyr) in dairy products. The presented approach entails protein cleavage via enzymatic digestion to liberate the aforementioned compounds, which were then quantified using a stable isotope dilution assay. LC-MS/MS analysis was performed by positive electrospray ionization recording two transition reactions per analyte in selected reaction monitoring mode. The CML and Lys values obtained with enzymatic digestion were compared to those acquired with acid hydrolysis HCl (6 mol/L), and the two proteolysis methods yielded comparable quantifications. Allowing for the fact that the investigated compounds are formed during different stages of the glycation process, the method is able to reveal the progress of protein glycation in dairy products.

Key words: glycation; mass spectrometry; N^ε-carboxymethyllysine; N^ε-fructosyllysine; pyrraline; dairy products

Introduction

Thermal treatment of food induces a complex network of chemical reactions. A part of this network is termed the Maillard reaction, or nonenzymatic browning. This reaction is a cascade of various reactions initiated by the condensation of an amino group, most notably stemming from lysine (Lys) and arginine residues of proteins, and a carbonyl group of reducing sugars to yield a labile Schiff base. The latter is subsequently rearranged to give rise to the so-called Amadori product during the "early" stage of the Maillard reaction, which may undergo further degradation steps and eventually lead to a smorgasbord of different "advanced" products.[1] At present, a plethora of different methods are available for quantifying glycation-induced protein modifications. N^ε-carboxymethyllysine (CML), for example, has been measured in various matrices by HPLC with fluorescence detection,[2] enzyme-linked immunosorbent assay (ELISA),[3] or in model proteins by means of gas chromatography–mass spectrometry (GC/MS).[4] Regarding N^ε-fructosyllysine (FL), the Amadori product of glucose and lysine, very little information is available as the Amadori product is not stable during acid hydrolysis. Direct quantification after enzymatic digestion was achieved using ion-exchange chromatography.[5] The common approach, however, consists of an indirect quantification of FL as the furosine derivative, which is formed in constant amounts during acid hydrolysis.[6] Another major advanced modification, termed pyrraline (Pyr), a pyrrole derivative containing the N^ε-amino group of Lys, was quantified in various foods using ion-exchange chromatography[7] or reversed-phase HPLC,[8] both in combination with photodiode array detection. During recent years, liquid chromatography-tandem mass spectrometry (LC-MS/MS) has become a well-established and powerful analytical technique that represents an attractive alternative for sensitive and selective detection of low levels of compounds with a high degree of certainty,

Address for correspondence: T. Delatour, Nestlé Research Centre, Vers-chez-les-Blanc, 1000 Lausanne 26, Switzerland. Voice: +41 21 785 9220; fax: +41 21 785 8553.
thierry.delatour@rdls.nestle.com

especially when combined with stable isotope dilution. This approach has recently been used by Ahmed et al.[9] to assess CML levels of raw, pasteurized, and sterilized bovine milk as well as cola drinks.

In the present paper, an LC-MS/MS method is reported for the simultaneous quantification of Lys, FL, CML, and Pyr in dairy products by isotope dilution. The methodology consists of protein cleavage employing either acid hydrolysis (HCl, 6 mol/L) or enzymatic digestion to release free amino acids prior to sample cleanup by solid phase extraction (SPE) and analysis with an LC-MS/MS system.

Materials and Methods

Chemicals, Reagents, and Enzymes

The following products were used: nonafluoropentanoic acid (NFPA) 97% (Acros Organics, Geel, Belgium); ammonium carbonate and triethylphosphine (Fluka, Buchs, Switzerland); Tris (Merck, Darmstadt, Germany); iodethanol, pepsin, leucine aminopeptidase, and prolidase (Sigma, Steinheim, Germany); pronase E (Serva, Heidelberg, Germany); ($^{13}C_6$,$^{15}N_2$)-Lys-dihydrochloride (Euriso-Top GmBH, Saarbruecken, Germany).

Synthesis of Isotopomers

($^{13}C_2$)-CML: The synthesis was performed as described previously by Delatour et al.[10] Briefly, N^α-acetyllysine was incubated with 1,2-$^{13}C_2$-glyoxylic acid in the presence of sodium cyanoborohydride prior to purification by HPLC-UV after conversion of the N^α-acetyl derivative of CML into the di-*n*-butyl ester. The release of CML was achieved with an incubation for 24 h in HCl (6 mol/L) at 110°C. *($^{13}C_6$,$^{15}N_2$)-FL:* The synthesis of ($^{13}C_6$,$^{15}N_2$)-FL was adapted from Heyns and Noack,[11] and the purification was carried out as reported by Finot and Mauron[12] with some modifications.[13] Briefly, ($^{13}C_6$,$^{15}N_2$)-Lys dihydrochloride was refluxed in dry methanol in the presence of glucose. The solution was purified using a cation-exchange column (120 × 25 mm) AG 50W-X8 (H+ form, 20–50 mesh, Bio-Rad). The isolation of ($^{13}C_6$,$^{15}N_2$)-FL was achieved by cation-exchange chromatography using a 480 × 15 mm column filled with AG 50W-X8 (pyridinium form, 100–200 mesh, Bio-Rad). *($^{13}C_6$,$^{15}N_2$)-Pyr:* The synthesis was performed according to the method of Henle and Bachmann.[14] Briefly, 3-deoxyglucosulose and N^α-Boc-($^{13}C_6$,$^{15}N_2$)-Lys were heated in the presence of cellulose in a solid state. Following the extraction of the mixture, the extracts were purified by liquid chromatography using C18-reverse-phase column material and UV detection at 280 nm. The resulting N^α-Boc-Pyr was deprotected using acetic acid and the solution subsequently evaporated to dryness.

Sample Preparation

The protein content in dairy products was evaluated with the nitrogen content using the Kjeldahl method and applying the conversion: protein content (mass %) = 6.3 × nitrogen content (mass %).[15] A portion of sample equivalent to 100 mg of protein was dissolved or diluted in water to yield a final volume of 10 mL. An aliquot of 50 µL corresponding to 500 µg of protein was either enzymatically digested or hydrolyzed by hydrochloric acid.

Enzymatic Digestion

Sample portions with 500 µg of protein were diluted with 40 µL of water, 10 µL of ammonium carbonate (1 mol/L, ph 10.5), and 100 µL of reduction/alkylation reagent (final volume: 200 µL). Samples were incubated at 37°C for 1 h and subsequently evaporated to dryness. Dry residues were reconstituted in 1 mL of HCl (0.02 mol/L) and subjected to multienzymatic digestion as described elsewhere.[16]

Acid Hydrolysis

Sample portions containing 500 µg of protein were mixed with 1.5 mL of sodium borate (0.2 mol/L, ph 9.5). One milliliter of sodium borohydride (1.0 mol/L in NaOH 0.1 mol/L) was added to the mixture followed by incubation for 4 h at room temperature. After addition of 2.6 mL of HCl (37%), the samples were hydrolyzed for 24 h at 110°C. Hydrolysates were further evaporated to dryness and reconstituted in 1380 µL of water prior to the cleanup by solid phase extraction.

Solid Phase Extraction

An aliquot of 200 µL of enzymatic hydrolysate or 345 µL of acid hydrolysate was diluted in 1 mL of water. After the addition of 100 µL of NFPA and isotopically labeled internal standards, i.e., ($^{13}C_2$)-CML, ($^{13}C_6$,$^{15}N_2$)-FL, ($^{13}C_6$,$^{15}N_2$)-Lys, and ($^{13}C_6$,$^{15}N_2$)-Pyr, the samples were loaded onto preconditioned Oasis HLB 6 cc (200 mg) cartridges (Waters, Milford, MA) and eluted. Fractions were evaporated to dryness and resuspended in 40 µL (enzymatic digestion) or 100 µL (acid hydrolysis) of water. Ten microliters (10 µL) was injected into the LC-MS/MS system for the analysis.

Liquid Chromatography–Electrospray Tandem Mass Spectrometry

The HPLC system consisted of an Agilent 1100 series (Agilent Technologies, Waldbronn, Germany) coupled to the TSQ Quantum mass spectrometer from ThermoFinnigan (San Jose, CA). The column was an

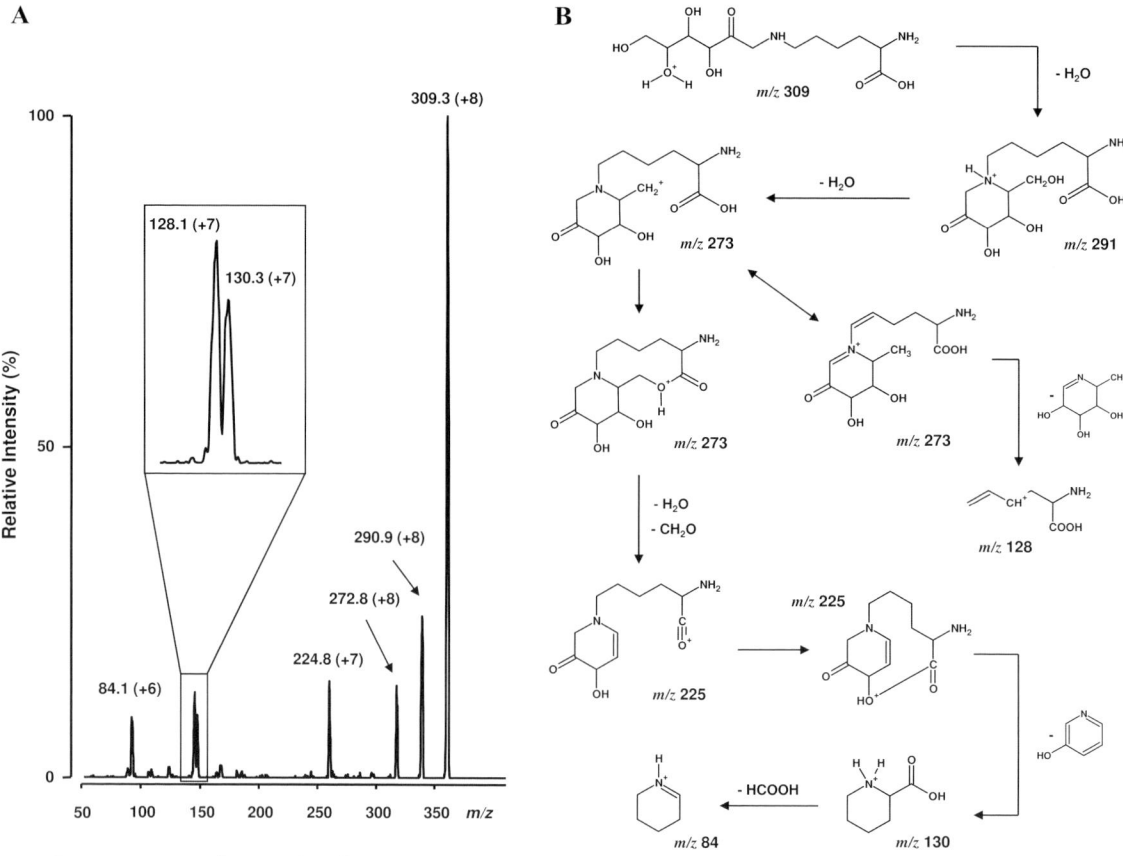

FIGURE 1. (A) Collision-induced dissociation spectrum of N^ε-fructosyllysine (FL). The upmass shifts of fragment ions derived from the ($^{13}C_6, ^{15}N_2$)-isotopomer are indicated in brackets. **(B)** Postulated fragmentation pathway of FL.

Atlantis dC18, 3 μm, 2.1 × 150 mm from Waters, and the flow rate was set at 200 μL/min. The separation of CML, FL, Lys, and Pyr was achieved with the following gradient using NFPA (5 mmol/L) in water as solvent A and NFPA (5 mmol/L) in acetonitrile in solvent B (t in [min]/[% B]): (0/15), (15/85), (20/85), (21/15), (50/15). The detection was carried out by positive electrospray ionization, and the source parameters were selected as follows: spray voltage, 4.5 kV; capillary temperature, 300°C; collision pressure, 1.2 mTorr. The data acquisition was performed with a scan time of 50 ms. The chromatographic profile was recorded in the selected reaction monitoring mode, and two characteristic transitions were monitored per compound in order to improve the selectivity.

Results and Discussion

Fragmentation of FL and Pyr

The assignment of all fragments was confirmed by MS/MS analysis of FL and Pyr and the ($^{13}C_6, ^{15}N_2$)-labeled compounds. *FL:* Upon fragmentation of the protonated molecule ([M+H]$^+$ at m/z 309) with a low collision energy, six product ions formed via two fragmentation pathways in the collision-induced dissociation (CID)-spectrum (FIG. 1). The protonation of the alcohol in the γ-position of the keto group of the fructose moiety led to a loss of water and a possible formation of a six-membered ring (m/z 291) by a nucleophilic attack of the N^ε-nitrogen of the lysine moiety on the corresponding carbocation. A further loss of water involving the protonated nitrogen of the ring and the alcohol function of the –CH$_2$OH group generated the product ion at m/z 273. The subsequent elimination of water and formaldehyde yielded the fragment ion at m/z 225, which, following elimination of 3-hydroxypyridine, led to the formation of piperidinum-2-carboxylic acid (m/z 130). The product ion at m/z 84 was generated after elimination of formic acid. The remaining product ion at m/z 128 was obtained from the fragment ion at m/z 273 as a result of a second possible fragmentation pathway that comprises rearrangements of hydrogen atoms and the elimination of a substituted six-membered ring,

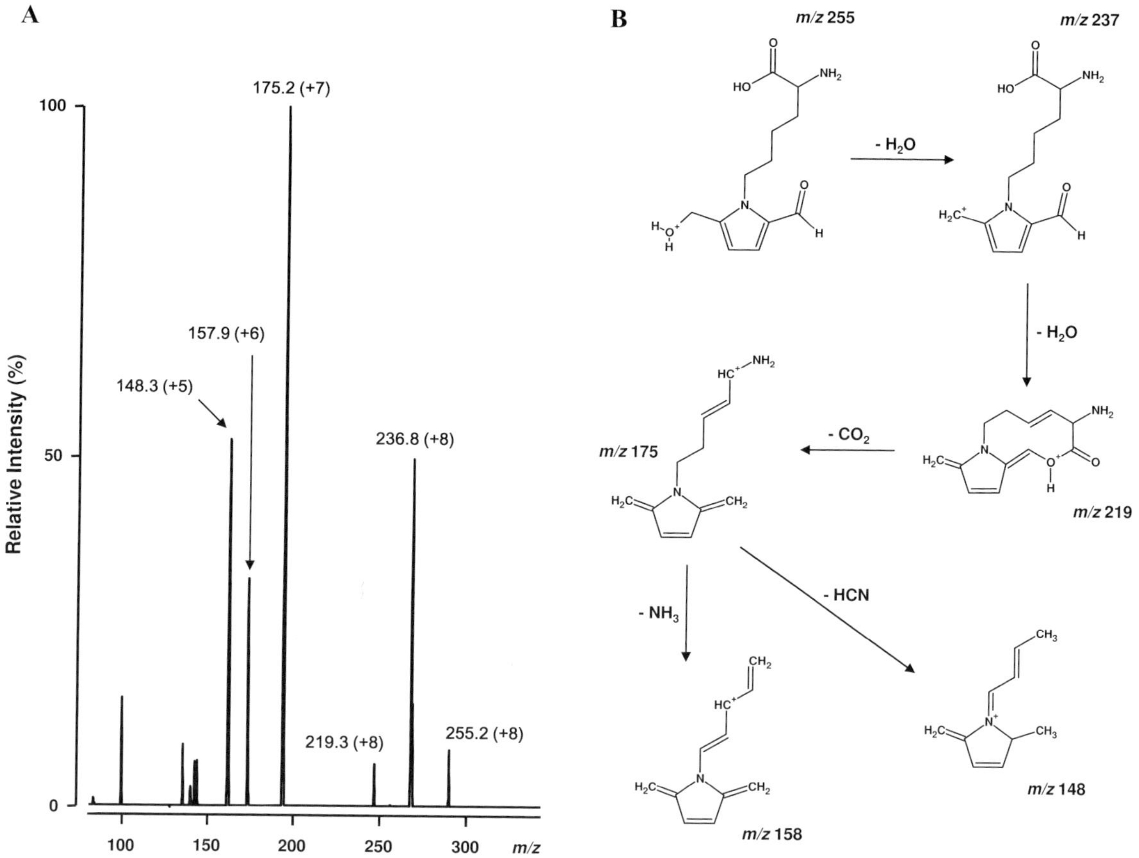

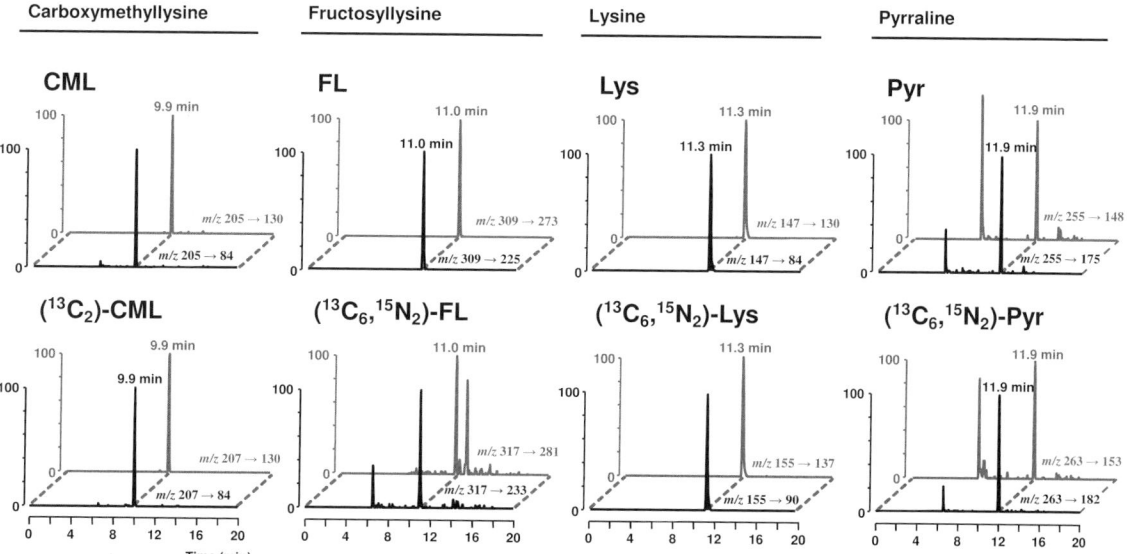

FIGURE 2. (A) CID-spectrum of pyrraline (Pyr). The upmass shifts of fragment ions derived from the ($^{13}C_6$,$^{15}N_2$)-isotopomer are indicated in brackets. (B) Postulated fragmentation pathway of Pyr.

FIGURE 3. Chromatographic profile of a dairy product containing N^ϵ-carboxymethyllysine (CML), FL, lysine (Lys), and Pyr. For each analyte and its corresponding isotopomer, two transition reactions resembling quantifier and qualifier are depicted.

including the N^ε-nitrogen of the Lys moiety. *Pyr:* The fragmentation of the protonated molecule ion $[M+H]^+$ at m/z 255 was initiated by a loss of water from the protonated $-CH_2OH$ side chain of the pyrrole substructure to form a fragment ion at m/z 237 (FIG. 2). A subsequent delocalization of the positive charge to the aldehyde group, followed by rearrangements of hydrogen atoms and a final loss of water, gave rise to the product ion at m/z 219. The signal observed at m/z 175 could be assigned to a further loss of CO_2. Additional elimination of either NH_3 or HCN from the latter fragment generated ions at m/z 158 and 148, respectively.

MS Detection

The outstanding reliability of the method is based upon several distinct criteria. Besides the corresponding retention time, two characteristic transition reactions were monitored per analyte (FIG. 3). In this context, the peak area of the most intense fragment ion (1st transition reaction) was used for quantification, while the second transition reaction, yielding a product ion with a lower signal intensity, served as a qualifier. Furthermore, the area ratio between the two recorded fragments constitutes a supplementary characteristic which was used for confirming the presence of each of the respective analytes. Single- and double-transition LC-MS/MS-based methods for the analysis of glycation-derived modifications in biological and food matrices have been used by Thornalley et al.[17] and Teerlink et al.,[18] respectively, in order to try to overcome certain inconveniences of other methods (e.g., the lack of reliability of immunochemical methods often indicating the results in arbitrary units rather than providing absolute values or the necessity to derivatize the analytes prior to GC/MS analysis). Regarding the quantification of CML, a recent comparison of GC/MS and ELISA[19] clearly disclosed an alarming inconsistency of the observed data depending on the sample type. Whereas CML values obtained for powdered infant formulas match very closely (i.e., 24 ± 12 ng/mg protein [ELISA], 21 ± 10 ng/mg protein [GC/MS]) a huge discrepancy in the measurements was observed for liquid infant formulas (e.g., 801 ± 420 ng CML/mg protein [ELISA] and 47 ± 17 ng CML/mg protein [GC/MS] in the case of hydrolyzed formulas). This example strongly demonstrates the need for accurate and reliable methods for the quantification of glycation-induced protein modifications.

Artifactual Conversion of FL into CML

Ahmed et al.[20] established that FL decomposes into CML under certain conditions. Therefore, we con-

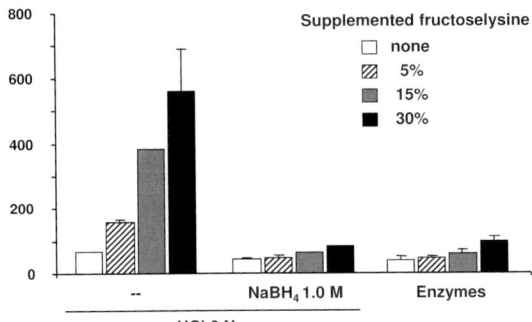

FIGURE 4. Artifactual conversion of FL into CML. Bovine serum albumin (BSA) was fortified with FL at 0, 5, 15, and 30% (mol/mol) relative to the Lys content of BSA. The protein was cleaved using acid hydrolysis (with and without previous reduction mediated by $NaBH_4$) or enzymatic hydrolysis.

ducted some preliminary tests to investigate the artifactual formation of CML during the sample preparation and hence prevent an overestimation of this analyte. Bovine serum albumin (BSA) was fortified with FL at 0, 5, 15, and 30% (mol/mol) relative to the Lys content of BSA. A significant dose-dependent conversion of FL into CML was observed after acid hydrolysis with HCl (6 mol/L) (FIG. 4). However, the generation of this artifact was drastically diminished by introducing a reduction step with sodium borohydride (1 mol/L) prior to hydrolysis. The values obtained here were in the same range as those observed after enzymatic digestion of the samples, where again only a residual induction of CML was detected.

Measurements in Dairy Products

As can easily be inferred from FIGURE 5A, the results obtained for the quantification of CML in dairy products by means of enzymatic digestion and acid hydrolysis match very closely (slope $= 0.97$). When CML content was expressed in relation to unmodified Lys (cf. FIG. 5B), the slope was 1.06, close to the calculated regression curve with an ideal slope at $\alpha = 1$ (broken line), showing that the Lys contents measured with both proteolysis methods were consistent. Moreover, a good correlation ($r^2 = 0.683$ and 0.613, respectively) between the values obtained by enzymatic digestion and acid hydrolysis was observed. However, it seemed that the CML concentration was slightly overestimated by enzymatic digestion when compared to acid hydrolysis (positive y-intercept). Drawing a further comparison between enzymatic digestion and the furosine method with respect to the determination of Lys, the correlation uncovered a trend towards an overestimation

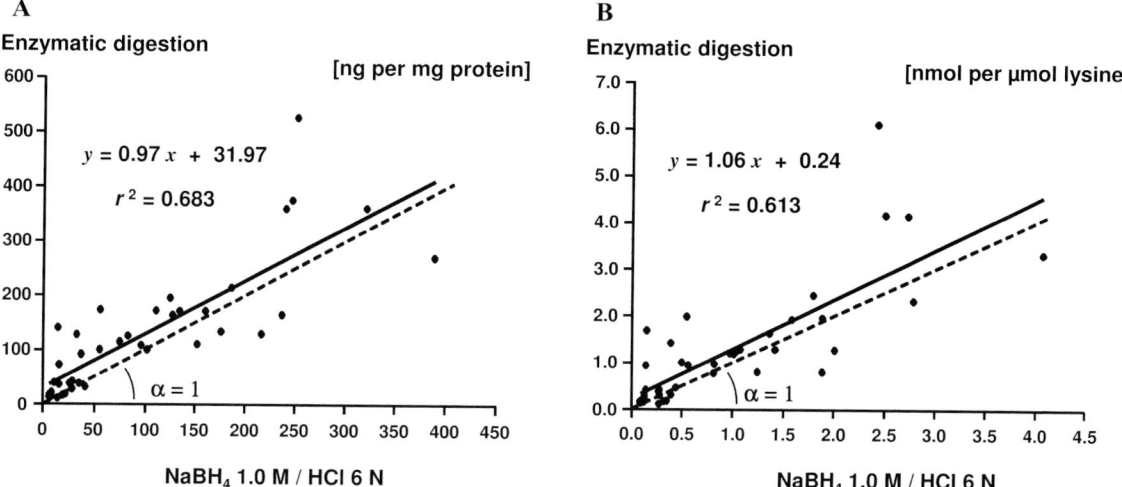

FIGURE 5. Comparison of CML measurements in dairy products applying enzymatic digestion or acid hydrolysis (HCl, 6 mol/L) for protein cleavage. Each measurement represents a mean of two independent analyses. **(A)** Results are indicated as ng CML per mg protein. **(B)** The CML content is referred to as nmol CML per µmol Lys. The *broken line* indicates an ideal regression line with a slope at α = 1.

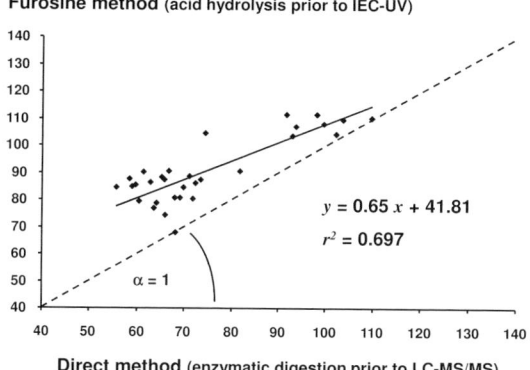

FIGURE 6. Comparison of an indirect and a direct methodology for the determination of Lys in proteins: furosine method versus enzymatic digestion. Results are given as µg Lys per mg protein. The *broken line* indicates an ideal regression line with a slope at α = 1. IEC, ion exchange chromatography.

of Lys using the furosine method (FIG. 6). This is in good accordance with the findings published by Henle et al.[5] who proved that the furosine method underestimated modified Lys. Finally, it should also be mentioned that a good correlation was also observed between CML and Pyr (data not shown). Owing to the existence of several pathways which give rise to CML, whereas Pyr, based on the current state of knowledge, is exclusively formed during the glycation process, one could presume a lack of correlation between the two analytes. Nevertheless, the coefficient of determination turned out to be $r^2 = 0.778$.

Conflict of Interest

The authors declare no conflicts of interest.

References

1. FRIEDMAN, M. 1996. Food browning and its prevention: an overview. J. Agric. Food Chem. **44:** 631–653.
2. HARTKOPF, J. *et al.* 1994. Determination of N$^\varepsilon$-carboxymethyllysine by a reversed-phase high-performance liquid chromatography method. J. Chromatogr. A. **672:** 242–246.
3. GOLDBERG, T. *et al.* 2004. Advanced glycoxidation end products in commonly consumed foods. J. Am. Diet. Assoc. **104:** 1287–1291.
4. LECLÈRE, J., I. BIRLOUEZ-ARAGON & M. MELI. 2002. Fortification of milk with iron-ascorbate promotes lysine glycation and tryptophan oxidation. Food Chem. **76:** 491–499.
5. HENLE, T., H. WALTER & H. KLOSTERMEYER. 1991. Evaluation of the extent of the early Maillard-reaction in milk products by direct measurement of the Amadori-product lactuloselysine. Z. Lebensm. Unters. Forsch. **193:** 119–122.
6. HENLE, T., G. ZEHETNER & H. KLOSTERMEYER. 1995. Fast and sensitive determination of furosine. Z. Lebensm. Unters. Forsch. **200:** 235–237.
7. HENLE, T. & H. KLOSTERMEYER. 1993. Determination of protein-bound 2-amino-6-(2-formyl-2-hydroxymethyl-1-pyrrolyl)-hexanoic acid ("pyrraline") by ion-exchange chromatography and photodiode array detection. Z. Lebensm. Unters. Forsch. **196:** 1–4.
8. RUFIAN-HENARES, J.A., E. GUERRA-HERNANDES & B. GARCIA-VILLANOVA. 2004. Pyrraline content in enteral formula processing and storage and model systems. Eur. Food Res. Technol. **219:** 42–47.

9. AHMED, N. et al. 2005. Assay of advanced glycation endproducts in selected beverages and food by liquid chromatography with tandem mass spectrometric detection. Mol. Nutr. Food Res. **49:** 691–699.
10. DELATOUR, T. et al. 2006. Synthesis, tandem MS- and NMR-based characterization, and quantification of the carbon 13-labeled advanced glycation endproduct, 6-N-carboxymethyllysine. Amino Acids **30:** 25–34.
11. HEYNS, K. & H. NOACK. 1962. Die Umsetzung von D-Fructose mit L-Lysin und L-Arginin und deren Beziehung zu nichtenzymatischen Bräunungsreaktionen. Chem. Ber. **95:** 720–727.
12. FINOT, P.A. & J. MAURON. 1969. Le blocage de la lysine par la reaction de Maillard. I. Synthese de N-(desoxy-I-D-fructosyl-I) et N-desoxy-I-u-lactulosyl-1)-L-lysines. Helv. Chim. Acta **52:** 1488–1495.
13. KRAUSE, R., K. KNOLL & T. HENLE. 2003. Studies on the formation of furosine and pyridosine during acid hydrolysis of different Amadori products of lysine. Eur. Food Res. Technol. 216: 277–283.
14. HENLE, T. & A. BACHMANN. 1996. Synthesis of pyrraline reference material. Z. Lebensm. Unters. Forsch. **202:** 72–74.
15. MCKENZIE, H.A. & W.H. MURPHY. 1970. General methods in elemental analysis. In Milk Proteins Chemistry and Molecular Biology. Vol. 1: 127–262. Academic Press. New York.
16. DELATOUR, T. et al. 2007. A comparative study of proteolysis methods for the measurement of 3-nitrotyrosine residues: enzymatic digestion versus hydrochloric acid-mediated hydrolysis. J. Chromatogr. B. 851: 268–276.
17. THORNALLEY, P.J. et al. 2003. Quantitative screening of advanced glycation endproducts in cellular and extracellular proteins by tandem mass spectrometry. Biochem. J. 375: 581–592.
18. TEERLINK, T. et al. 2004. Measurement of N$^\varepsilon$-(carboxymethyl)lysine and N$^\varepsilon$-(carboxyethyl)lysine in human plasma protein by stable-isotope-dilution tandem mass spectrometry. Clin. Chem. **50:** 1222–1228.
19. CHARISSOU, A., L. AÏT-AMEUR & I. BIRLOUEZ-ARAGON. 2007. Evaluation of a gas chromatography/mass spectrometry method for the quantification of carboxymethyllysine in food samples. J. Chromatogr. A **1140:** 189–194.
20. AHMED, M.U., S.R. THORPE & J.W. BAYNES. 1986. Identification of N$^\varepsilon$-carboxymethyllsyine as a degradation product of fructoselysine in glycated protein. J. Biol. Chem. 261: 4889–4894.

A Novel Yellow Pigment, Furpipate, Derived from Lysine and Furfural

MASATSUNE MURATA,[a] HANA TOTSUKA,[a] AND HIROSHI ONO[b]

[a]*Department of Nutrition and Food Science, Ochanomizu University, Tokyo, Japan*
[b]*National Food Research Institute Tsukuba, Japan*

Furfural is an important intermediate compound of the Maillard reaction of pentose or ascorbic acid. We examined the browning of furfural and lysine by heating these compounds in solution and found two yellow compounds. The first, furpipate, is a novel pipecolic acid derivative and was (Z)-3-(2-furylmethylidene)-3H, 4H, 5H, 6H-pyridine-2-carboxylic acid. The second was decarboxylated-furpipate. Furpipate shows absorption maxima at 370 nm and 310 nm under acidic and alkaline conditions, respectively. This compound was the major colored compound of the heated solution containing lysine and furfural.

Key words: Maillard reaction; furfural; lysine; pipecolic acid; browning; pigment

Foods derived from plants contain pentose in addition to hexose. It is well known that pentose contributes more to browning by the Maillard reaction than hexose. Hydroxymethylfurfural (HMF) is one of the major decomposed products of hexoses, such as glucose and fructose, under acidic conditions, while furfural is the corresponding one of pentoses and is also formed by the decomposition of ascorbic acid. The decomposition or reaction products of hexoses, especially glucose, by the Maillard reaction have been intensively examined, while there are insufficient data on the reaction products of pentoses or furfural.

Our group has shown that furfural is a good indicator of the browning of orange juice, which is rich in ascorbic acid, during storage.[1,2] Several colored reaction products of furfural[3,4] (FIG. 1[1, 2]) or pentoses[5–7] (FIG. 1[3-5]) have been reported. For example, a yellow compound (FIG. 1[1])[3] was formed by the reaction between furfural and proline, while red compounds (FIG. 1[2]) were formed by the reaction between furfural and alanine.[3,4] However, the data on the browning of furfural and amino acids and their reaction products are insufficient. In this present study, we examined the browning reaction between lysine and furfural and found two yellow compounds (FIG. 1[6, 7]). One of these (FIG. 1[6]) is a novel yellow compound, named furpipate, that was formed during the reaction as the major low-molecular-weight colored compound in the solution.[8] The second yellow compound was decarboxylated-furpipate (FIG. 1 [7]).

Methods

A solution (100 mL) containing 30 mmol/L Boc-Lys-OH and 90 mmol/L furfural was autoclaved at 121°C for 30 min before being cooled and washed with ethyl acetate. The water layer was then concentrated *in vacuo* and applied to a column of octadecylsilane (ODS). The column was developed with a mixture of 0.1% trifluoroacetic acid and MeOH (7:3). The yellow solution was collected and concentrated *in vacuo* before being applied to preparative HPLC (column, Mightysil RP-18 (Kanto Chemical, Tokyo, Japan); eluent, 0.1% trifluoroacetic acid and MeOH (5:1); wavelength for detection, 360 nm). A peak at a retention time of about 26 min was collected and concentrated *in vacuo*. Furpipate (6.6 mg) was obtained as a greenish amorphous powder.

Results and Discussion

We first compared the browning of pentoses, such as xylose and arabinose, glucose, HMF, and furfural, under weakly acidic conditions. Each compound was dissolved in a 0.5 mol/L acetate buffer (pH 5.0) because, in general, foods derived from plants are weakly acidic. Each solution was mixed with lysine and heated. The solution containing xylose or arabinose turned intense brown. The solution containing furfural turned less brown than that containing xylose or arabinose but more brown than that containing glucose or HMF.

Address for correspondence: Masatsune Murata, 2-1-1 Otsuka, Bunkyo-ku, Tokyo 112-8610, Japan. Voice: +81-3-5978-5753; fax: +81-3-5978-5755.
murata.masatsune@ocha.ac.jp

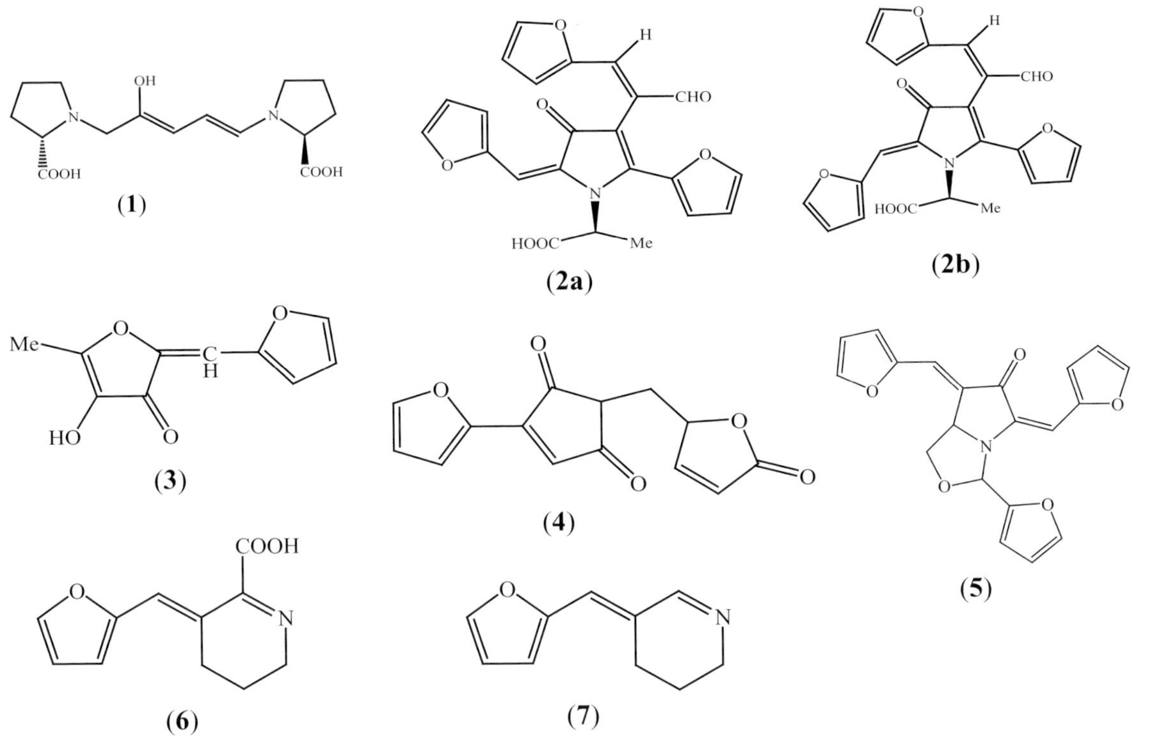

FIGURE 1. Chemical structures of the colored compounds derived from furfural (**1**,[3] **2a** and **2b**,[3,4] **6**,[8] and **7**[10]) and pentose (**3**,[5] **4**,[6] and **5**[7]) with amino acids.

Next, the effect of amino acids on the browning of furfural was examined. When the browning was evaluated by the absorbance at 450 nm, the absorbance was very strong in the solution containing tryptophan or histidine. Next in strength of absorbance were the solutions containing lysine, cysteine, asparagine, phenylalanine, proline, arginine, and glycine. The browning index[9] or the L*, a*, and b* values of a solution calculated from its transmittances (445, 495, 550, and 625 nm) was nearer to our sensory evaluation of browning than the absorbance at 450 nm. The tendency for browning estimated by the browning index was slightly different from that estimated by the absorbance at 450 nm. Histidine showed the most intensive browning, followed by cysteine, tryptophan, proline, lysine, and asparagine.

The reaction product of furfural and lysine was then examined. N^α-(*tert*-butoxycarbonyl)-(L)-lysine (Boc-Lys-OH) was first used instead of lysine because we expected a simple reaction product.

The solution of furpipate was yellow, while its powder was slightly greenish. The absorption maxima were 375 nm and 310 nm under acidic and alkaline conditions, respectively. The mass spectrometry (MS) data showed its molecular weight and molecular formula to be 205 and $C_{11}H_{11}NO_3$, respectively. The nuclear magnetic resonance (NMR) data enabled the chemical structure of this compound to be determined as (*E*)-3-(2-furylmethylidene)-3*H*, 4*H*, 5*H*, 6*H*-pyridine-2-carboxylic acid (FIG. 1 [6]).[8] This compound is a novel pipecolic acid derivative and was named furpipate.

As furpipate does not contain a *tert*-butoxycarbonyl group, we examined if furpipate was formed from lysine and furfural. The retention time and the absorption spectrum of the major peak of heated solution of lysine and furfural coincided with those of authentic furpipate, which strongly suggest that furpipate was the major pigment of the heated solution. To confirm this, the peak corresponding to furpipate was isolated and analyzed by high resolution ^{1}H-NMR and was found to coincide with the data of authentic furpipate. We concluded that furpipate had been formed from lysine and furfural as well as from Boc-Lys-OH and furfural.

We observed another yellow peak, other than furpipate, on the HPLC. We then isolated this compound and identified it by instrumental analyses, such as MS and NMR, as decarboxylated-furpipate (FIG. 1 [7]). This compound had been synthesized from furfural and 2,3,4,5-tetrahydropyridine.[10]

Finally, the detection limits and color contributions[11] of furpipate and decarboxylated-furpipate were evaluated. The detection limits for furpipate and decarboxylated-furpipate were 5.5 μg/mL and 93.2 μg/mL, respectively. The color contributions of these two compounds were 25.8% and 2.6%, respectively.

In conclusion, a novel pipecolic acid derivative, furpipate, was isolated and identified as the major low-molecular-weight yellow pigment in a heated solution of furfural and lysine.

Conflict of Interest

The authors declare no conflicts of interest.

References

1. SHINODA, Y. *et al*. 2005. Browning of model orange juice solution: factors affecting the formation of decomposition products. Biosci. Biotechnol. Biochem. **69:** 2129–2137.
2. SHINODA, Y. *et al*. 2004. Browning and decomposed products of model orange juice. Biosci. Biotechnol. Biochem. **68:** 529–536.
3. HOFMANN, T. 1998. Characterization of the chemical structure of novel colored Maillard reaction products from furan-2-carboxaldehyde and amino acids. J. Agric. Food Chem. **46:** 932–940.
4. HOFMANN, T. 1997. Determination of the chemical structure of novel colored 1H-pyrrol-3(2H)-one derivatives formed by Maillard-type reactions. Helv. Chim. Acta **80:** 1843–1856.
5. AMES, J.M. *et al*. 1993. Low molecular weight colored compounds formed in xylose-lysine model systems. Food Chem. **46:** 121–127.
6. ARNOLDI, A. *et al*. 1997. New colored compounds from the Maillard reaction between xylose and lysine. J. Agric. Food Chem. **45:** 650–655.
7. HOFMANN, T. 1998. Identification of novel colored compounds containing pyrrole and pyrrolinone structures formed by Maillard reactions of pentoses and primary amino acids. J. Agric. Food Chem. **46:** 3902–3911.
8. MURATA, M., H. TOTSUKA & H. ONO. 2007. Browning of furfural and amino acids, and a novel yellow compound, furpipate, formed from lysine and furfural. Biosci. Biotechnol. Biochem. **71:** 1717–1723.
9. MATIACEVICH, S.B. & M. P. BUERA. 2006. A critical evaluation of fluorescence as a potential marker for the Maillard reaction. Food Chem. **95:** 423–430.
10. MILLER, R. 1987. Synthesis and stereochemistry of (*E*)-5-(3, 4,5,6-tetrahydropyrid-3-ylidenemethyl)-2-furanmethaol, a product of the reaction between D-glucose and L-lysine. Acta Chem. Scand. **B41:** 208–209.
11. HOFMANN, T. 1998. Characterization of the most intense colored compounds from Maillard reactions of pentoses by application of color dilution analysis. Carbohydr. Res. **313:** 203–213.

Time-dependent Component-specific Regulation of Gastric Acid Secretion-related Proteins by Roasted Coffee Constituents

M. RUBACH,[a] R. LANG,[b] T. HOFMANN,[b] AND V. SOMOZA[a]

[a]*Deutsche Forschungsanstalt für Lebensmittelchemie, Garching, Germany*

[b]*Lehrstuhl für Lebensmittelchemie und Molekulare Sensorik, Technische Universität München, Freising-Weihenstephan, Germany*

Consumption of coffee beverages has been reported to cause gastric irritation in some consumers as a result of increased gastric acid secretion. In the complex mechanisms of gastric acid secretion, the activity and expression of the H^+,K^+-ATPase is regulated by transmitters, such as histamine, acetylcholine, gastrin, somatostatin, and their corresponding receptors. Here, we report the effect of three coffee constituents, chlorogenic acid, caffeine, and N-methyl pyridinium ions, on the expression of the histamine receptor H2, the acetylcholine receptor M3, the gastrin receptor, the somatostatin receptor, and the H^+,K^+-ATPase. Human gastric cancer cells were exposed to chlorogenic acid, caffeine, or N-methyl pyridinium in their coffee brew-representative concentrations as well as to physiological stimulators of gastric acid secretion. Gene expression levels of receptor proteins and those of the H^+,K^+-ATPase were measured at different time points by real-time PCR. Expression of prosecretory receptors significantly increased between one and one-half to twofold after treatment with chlorogenic acid or caffeine compared to control cells at the same time point. Chlorogenic acid and caffeine also increased the H^+,K^+-ATPase gene expression twofold higher compared to control cells. In contrast, N-methyl pyridinium downregulated the expression of the prosecretory gastrin receptor significantly, by −27%. In conclusion, chlorogenic acid, caffeine, and N-methyl pyridinium impair the expression of gastric acid secretion-related proteins in a time-dependent manner. Future work will be aimed at the elucidation of the cooperative interplay of individual components using recombinates of single coffee constituents.

Key words: coffee; gastric acid secretion; regulation

Introduction

Coffee is one of the most popular thermally processed beverages. Nevertheless, some consumers report gastric irritation from coffee consumption, likely as a result of increased gastric acid secretion.[1] To reduce coffee-induced stomach discomfort, stomach-friendly coffee is steam treated to remove compounds that are hypothesized to stimulate gastric acid secretion (e.g., 5-hydroxytryptamides and chlorogenic acid).[2]

In the complex process of gastric acid secretion, which takes place in gastric parietal cells, a number of different cell surface receptors, functional proteins, and signaling proteins are involved.[3] The key player at the functional side of acid secretion is the H^+,K^+-ATPase. This heterodimeric protein transports hydrogen ions into the gastric lumen in antiport to potassium ions. The activity of the H^+,K^+-ATPase can be regulated by the physiological transmitters histamine, acetylcholine, gastrin, or somatostatin that bind to their corresponding receptor.[4] On the other hand, pharmacological compounds, such as omeprazole, have been developed to irreversibly or reversibly inhibit H^+,K^+-ATPase activity.[5] Since pharmacological compounds mostly interact directly with the functional subunit of the H^+,K^+-ATPase, physiological hormones and transmitters bind to their corresponding receptors. The resulting intracellular signal either affects the target directly or can be transduced into the core following activation of transcription of related genes. In each case, different pathways are involved in signal transduction. For instance, phosphorylation of the kinases Akt, extracellular signal-regulated kinase, or glycogen synthase kinase represent signaling pathways that have been shown to be involved in gastric acid secretion.[6–8]

Address for correspondence: P.D. Dr. Veronika Somoza, German Research Center for Food Chemistry, Lichtenbergstrasse 4, 85748 Garching, Germany. Voice: +49-89-289-14170; fax: +49-89-289-13248.
somoza@wisc.edu

Taken together, gastric acid secretion by parietal cells is regulated by the activation of receptors, signaling pathways, gene transcription, and enzyme activities. In case of cell surface receptors, it was shown that the expression levels of certain receptors may change after activation in order to increase or decrease the susceptibility of the cell to extracellular signals.[9–12] Here, we hypothesize that coffee constituents, as small organic compounds, may also affect the regulation of cell surface receptors that are related to gastric acid secretion and of the H^+,K^+-ATPase. The experiments were carried out using the HGT-1 human gastric cancer cell line, which has been previously shown to express the receptors related to gastric acid secretion and the H^+,K^+-ATPase.[13,14] The coffee components tested were selected according to previous findings hypothesizing that caffeine and chlorogenic acid are potent activators of stomach acid secretion having an irritating potential.[2,15] In addition, we decided to study the effects of the N-methyl pyridinium ion, which is known as a thermal degradation product released from the coffee alkaloid trigonellin and has been recently identified as a key chemopreventively active component in coffee brew.[16]

Material and Methods

Cell Culture

HGT-1 cells were cultured at 37°C and with 5% CO_2 in Dulbecco's modified Eagle medium with 20% fetal calf serum, 5% glutamine, and 5% penicillin and streptomycin. Before each experiment, the cells were starved with medium containing only 5% glutamine and 5% penicillin and streptomycin for 24 h.

After synchronization, cells were exposed to caffeine (Sigma-Aldrich, Munich, Germany), chlorogenic acid (Sigma-Aldrich), and N-methyl pyridinium (synthesized following the protocol published by Somoza et al.[16]) in concentrations according to their contents quantified in a regular coffee beverage by HPLC techniques: chlorogenic acid, 1038 mg/L; caffeine, 618 mg/L; N-methyl pyridinium, 32 mg/L. After 5, 10, 15, or 20 min of exposure, cells were harvested for the analysis of gene expression.

Real-time PCR Assay Validation

Twelve reference genes were tested for their most constant expression levels under different conditions with GENORM (University of Gent, Gent Belgium). The human endogenous reference gene panel (Bioline, Luckenwalde, Germany) was used in combination with the Full Velocity SYBR Green Kit (Stratagene, Amsterdam, the Netherlands) on a Mx3000p cycler (Stratagene). Cycling conditions were chosen according to the manufacturer's protocol.

Expression Assays

Primers ATP4A, HRH2, CCKBR, SSTR2, and CHRM3 (TABLE 1) were designed with Beacon Designer 7.0 (PremierBiosoft, Palo Alto, CA) and validated by standard and melting curve analysis. The correct sequence of PCR products was verified by sequencing (Medigenomics, Martinsried, Germany). Real-time PCR assays were performed using the Brilliant SYBR Green Kit (Stratagene) on a Mx3000p cycler (Stratagene). Cycling conditions followed the manufacturer's protocol.

RNA Isolation and cDNA Synthesis

Total RNA was isolated using the Rneasy Midi Kit (Qiagen, Hilden, Germany). DNase I digest was performed with RNase free DNase Kit (Qiagen) on column. Prior to qPCR, total RNA contents were quantitated photometrically at 260 nm. The cDNA was synthesized using the cDNA High Capacity Synthesis Kit (Applied Biosystems, Munich, Germany) as described in the manufacturer's manual.

Statistical Analysis

Statistical analysis was performed with Excel 2003. Data sets generated by qPCR were transformed by

TABLE 1. Nucleotide sequences of primers designed for studying the gene expression of the cell surface receptors of histamine (HRH2), acetylcholine (CHRM3), gastrin (CCKBR), and somatostatin (SSTR2) as well as that of the H^+,K^+-ATPase (ATP4A)

HRH2:	Forward 3'-TGG GAG CAG AGA AGA AGC AAC C-5'
	Reverse 3'-GAT GAG GAT GAG GAC CGC AAG G-5'
CHRM3:	Forward 3'-AGC AGC AGT GAC AGT TGG AAC-5'
	Reverse 3'-CTT GAG CAC GAT GGA GTA GAT GG-5'
CCKBR:	Forward 3'-GAC TAC TCA TGG TGC CCT AC-5'
	Reverse 3' AGC AGA AGC AGC AGT ACG 5'
SSTR2:	Forward 3' TCC TCC GCT ATG CCA AGA TGA AG 5'
	Reverse 3' AGA TGC TGG TGA ACT GAT TGA TGC 5'
ATP4A:	Forward: 3'-CGGCCAGGAGTGGACATTCG-5'
	Reverse: 3'-ACACGATGGCGATCACCAGG-5'

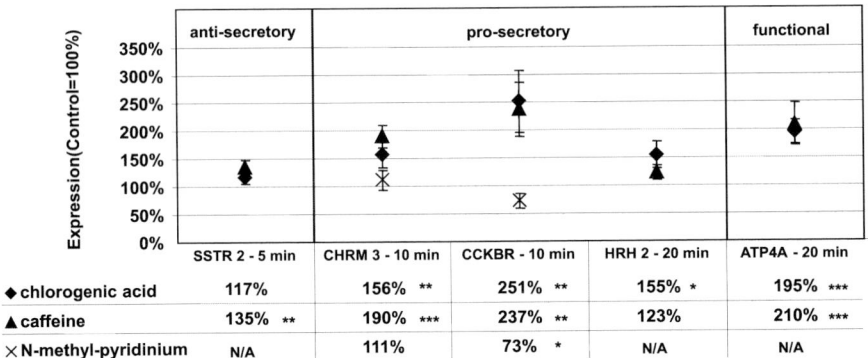

FIGURE 1. Maximum expression of antisecretory, prosecretory, and functional proteins after time-dependent exposure of HGT-1 cells to chlorogenic acid, caffeine, or N-methyl pyridinium compared to a nontreated control (=100%). Data are given as mean values with standard error mean ($n = 3–9$; ***$P \leq 0.001$, **$P \leq 0.01$, *$P \leq 0.01$).

logarithmic conversion to reach normality. Single comparisons between treated and control cells were done with the two-tailed Student's t-test for equal variances. Each experiment was performed with at least six independent biological samples ($n = 6$), except those for N-methyl pyridinium ($n = 3$). The Nalimov test was used to identify nonconfident values.

Results

Expression Analysis of Gastric Acid Secretion-related Genes

HGT-1 cells were treated with three different single compounds according to their concentrations quantified in a regular coffee brew by HPLC techniques. All genes showed different expression maxima during a time-dependent measurement (FIG. 1). After 5 min, the expression level of the somatostatin receptor 2 (SSTR2) was only impaired by caffeine. However, the increase of about +35% was significant ($P \leq 0.01$, $n = 6$). In contrast to these results obtained for an antisecretory receptor, the expression of all prosecretory receptors examined was upregulated either by chlorogenic acid, caffeine, or both compounds after 10 to 15 min of exposure to the cells. Chlorogenic acid significantly increased the expression of the acetylcholine receptor M3 (CHRM3) by about +56% after 10 min of exposure ($P \leq 0.01$, $n = 9$). Also, caffeine was shown to almost double the expression of CHRM3. This result was highly significant compared to control cells ($P \leq 0.001$, $n = 9$). Acetylcholine, as a physiological ligand of the acetylcholine receptor, slightly increased the expression by about +21% after 10 min of exposure. A similar result was obtained for the gastrin receptor (CCKBR). Treatment with chlorogenic acid or caffeine increased the CCKBR expression by about +151% or +135% ($P \leq 0.01$, $n = 9$), which corresponded to a threefold to fourfold higher stimulation than caused by human gastrin at a concentration of 0.07 mmol/L (+39%, $P \leq 0.05$, $n = 3$; result not shown). Interestingly, N-methyl pyridinium was the only compound that decreased the expression of CCKBR in HGT-1 cells after an exposure time of 10 min. Specifically, N-methyl pyridinium ions significantly decreased the expression of this prosecretory receptor by 73% compared to nontreated control cells (−27%, $P \leq 0.05$, $n = 3$). The histamine receptor H2 (HRH2) was shown to be upregulated significantly by chlorogenic acid after 20 min (+55%, $P \leq 0.05$, $n = 6$) and also by caffeine after 5 min of exposure (+32, $P \leq 0.05$, $n = 6$; result not shown). In comparison to treatment with the potent physiological stimulant histamine, which resulted in a maximum increase in HRH2 expression of +41% after 5 min ($P \leq 0.05$, $n = 4$), these increases were within the physiological range. At the same point of time, the functional subunit of the H^+,K^+-ATPase (ATP4A), a key player of gastric acid secretion, was impaired both by chlorogenic acid and caffeine. In contrast, exposure of HGT-1 cells to chlorgenic acid led to an increase in ATP4A expression of about +95% ($P \leq 0.001$, $n = 6$), whereas treatment with caffeine increased this expression by +110% ($P \leq 0.001$, $n = 6$).

Discussion

The present study was aimed at the elucidation of coffee components as potent stimulants or inhibitors of mechanisms associated with gastric acid secretion in parietal cells. Here, we were able to identify the impact of three coffee constituents on the regulation of

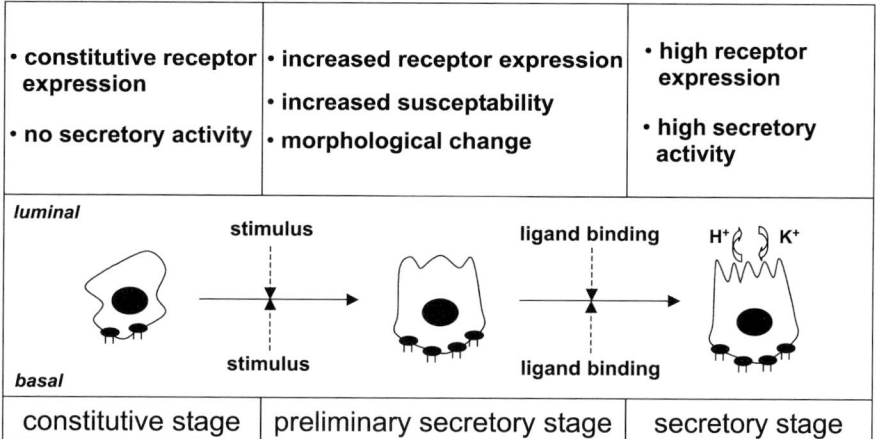

FIGURE 2. Different stages of parietal cells in the process of gastric acid secretion. The constitutive stage is characterized by the expression of antisecretory and prosecretory receptors at the basal level. After stimulation by extracellular signals, the parietal cell starts a morphological change. The expression of prosecretory receptors is increased. In the secretory stage, the morphological change has been completed and the parietal cell reaches its maximum secretory activity.

gene transcription of five proteins that are related to signaling and functional mechanisms of gastric acid secretion. In general, proteins involved in the regulation of gastric acid secretion can be subdivided into three classes: antisecretory and prosecretory cell surface receptor proteins as well as functional proteins. The gene expression of the antisecretory somatostatin receptor was shown to be regulated only by caffeine after 5 min of cell exposure. At this time, also the prosecretory histamine H2 receptor showed an upregulation by caffeine to an almost equivalent level. All of these changes were demonstrated to be comparable to those observed for the cell's treatment with histamine.

Taking into account that an increased gene expression of a certain cell surface receptor protein leads to an increased amount of receptor protein, a higher susceptibility for the specific ligand seems feasible. Applying this principle to the results presented herein, an equivalent increase of antisecretory and prosecretory receptors at the cell surface will end in two signals compensating each other. The following time points were only characterized by expression changes of prosecretory receptors and the H^+,K^+-ATPase. After 10 min of treatment, the cells' expression of the acetylcholine receptor and the gastrin receptor was strongly increased by caffeine and chlorogenic acid. Acetylcholine increased the expression of the acetylcholine receptor as well but twofold to fourfold less than caffeine or chlorogenic acid did. In the context of gastric acid secretion, this means the susceptibility of parietal cells after treatment with coffee constituents is much higher than under physiological conditions. The same regulation occurred for the gastrin receptor. Human gastrin caused a twofold to fourfold lower increase in expression than coffee constituents. At the same point of time, another coffee constituent, N-methyl pyridinium, decreased the expression of the gastrin receptor significantly, which would result in a decreased susceptibility. Finally, at the last measured time point (20 min of exposure), the expression of the HRH2 increased to a higher level. Summarizing the results, the expression levels of all prosecretory receptor proteins was increased in the presence of chlorogenic acid and caffeine after 10 to 20 min of the cell's exposure, whereas at these time points, the gene expression of the antisecretory somatostatin receptor protein was not upregulated anymore.

As receptor signaling and signal transduction quickly stimulate intracellular synthesis of functional proteins, the gene expression of the key player of gastric acid secretion, the H^+,K^+-ATPase, was also analyzed. The results demonstrated that after 20 min of exposure, gene expression of the H^+, K^+-ATPase was doubled by chlorogenic acid and by caffeine compared to nontreated control cells. Presumably, at this stage, the parietal cell is capable of a high secretory performance as a result of the upregulation of prosecretory receptor proteins and of the functional H^+,K^+-ATPase proteins.

Taking all together, two important aspects how coffee constituents influence the regulation of gastric acid secretion were shown in this work. First, the intragastric response of acid secretion follows a concerted reaction of the parietal cell in response to caffeine, chlorogenic acid, and N-methyl pyridinium ions. The cell's response mode is switched from a constitutive stage to a prosecretory stage (FIG. 2). This is reached by a temporary downregulation of antisecretory receptors with a subsequent upregulation of prosecretory receptors. In the preliminary secretory stage, those proteins needed for acid secretion are synthesized and expressed at the cell surface. This presumably results in the secretory stage, which is characterized by a high secretory activity. The second aspect concerns the interaction of single coffee brew components. As shown, every compound regulates the expression of each relevant gene to a certain extent. This can be an increase, a decrease, or no effect at all. Regarding the vast number of structurally diverse compounds in coffee brews, it would be very feasible that interactions might result in compensating or synergistic effects. In order to identify these diverse effects, bioinformatic modeling provides the possibility of evaluating the influence of single compounds and recombinates on regulative and functional aspects of gastric acid secretion based on experimental data.

In conclusion, the data presented here shows, for the first time, the impact of coffee constituents on the expression of gastric acid secretion-related proteins. From this model system, we hypothesize that the effect of coffee constituents and combinations thereof may lead to yet unknown regulative and functional activities in the parietal cell. The examination of the regulative and functional influence of coffee constituents on gastric acid secretion as recombinates and as whole beverage will be part of future work.

Acknowledgments

We thank Dr. C.L. Laboisse (Inserm 94-04, Faculte de Medicine, Nantes, France) for kindly providing the HGT-1 cells, clone 6. The financial support for this project was obtained from the FEI (Forschungskreis der Ernaehrungsindustrie e.V., Bonn, Germany), the Arbeitsgemeinschaft industrieller Forschungsvereinigungen (AiF), and the Ministry of Economics and Technology (Project no. AIF-FV 14042N).

Conflict of Interest

The authors declare no conflicts of interest.

References

1. BOEKEMA, P.J. et al. 1999. Coffee and gastrointestinal function: facts and fiction. A review. Scand. J. Gastroenterol. Suppl. **230:** 35–39.
2. DARBOVEN, A. 1997. Verfahren zur Qualitätsverbesserung von Rohkaffee durch Behandlung mit Wasserdampf und Wasser/95109295.6. Europäisches Patentblatt. 1997/05.
3. YAO, X. & J.G. FORTE. 2003. Cell biology of acid secretion by the parietal cell. Annu. Rev. Physiol. **65:** 103–131.
4. MITCHELL, S.L. 2004. Gastric secretion. Curr. Opin. Gastroenterol. **20:** 519–525.
5. MUNSON, K., R. GARCIA & G. SACHS. 2005. Inhibitor and ion binding sites on the gastric H,K-ATPase. Biochemistry **44:** 5267–5284.
6. TAKEUCHI, Y. et al. 1997. Functional role of extracellular signal-regulated protein kinases in gastric acid secretion. Am. J. Physiol. **273:** G1263–1272.
7. TODISCO, A. et al. 2001. Functional role of protein kinase B/Akt in gastric acid secretion. J. Biol. Chem. **276:** 46436–46444.
8. THIEL, A. et al. 2006. Expression of cyclooxygenase-2 is regulated by glycogen synthase kinase-3beta in gastric cancer cells. J. Biol. Chem. **281:** 4564–4569.
9. MIYOSHI, K. et al. 2006. Recent advances in molecular pharmacology of the histamine systems: regulation of histamine H1 receptor signaling by changing its expression level. J. Pharmacol. Sci. **101:** 3–6.
10. KOLIVAS, S. & A. SHULKES. 2004. Regulation of expression of the receptors controlling gastric acidity. Regul. Pept. **121:** 1–9.
11. KOLIVAS, S. et al. 2000. Quantitative measurement of mRNA coding for the receptors controlling acid secretion in the ovine fundus and antrum by using RT-PCR. J. Gastroenterol. Hepatol. **15:** 1257–1266.
12. YIP, L., H.C. LEUNG & Y.N. KWOK. 2004. Effect of omeprazole on gastric adenosine A1 and A2A receptor gene expression and function. J. Pharmacol. Exp. Ther. **311:** 180–189.
13. CARMOSINO, M. et al. 2000. The cultured human gastric cells HGT-1 express the principal transporters involved in acid secretion. Pflugers Arch. **440:** 871–880.
14. LABOISSE, C.L. et al. 1982. Characterization of a newly established human gastric cancer cell line HGT-1 bearing histamine H2-receptors. Cancer Res. **42:** 1541–1548.
15. COHEN, S. & G.H. BOOTH, JR. 1975. Gastric acid secretion and lower-esophageal-sphincter pressure in response to coffee and caffeine. N. Engl. J. Med. **293:** 897–899.
16. SOMOZA, V. et al. 2003. Activity-guided identification of a chemopreventive compound in coffee beverage using in vitro and in vivo techniques. J. Agric. Food Chem. **51:** 6861–6869.

Maillard Reaction versus Other Nonenzymatic Modifications in Neurodegenerative Processes

REINALD PAMPLONA, EKATERINA ILIEVA, VICTORIA AYALA, MARIA JOSEP BELLMUNT, DANIEL CACABELOS, ESTHER DALFO, ISIDRE FERRER, AND MANUEL PORTERO-OTIN

Department of Experimental Medicine, School of Medicine, University of Lleida-IRBLLEIDA, Lleida 25008, Spain

Nonenzymatic protein modifications are generated from direct oxidation of amino acid side chains and from reaction of the nucleophilic side chains of specific amino acids with reactive carbonyl species. These reactions give rise to specific markers that have been analyzed in different neurodegenerative diseases sharing protein aggregation, such as Alzheimer's disease, Pick's disease, Parkinson's disease, dementia with Lewy bodies, Creutzfeldt-Jakob disease, and amyotrophic lateral sclerosis. Collectively, available data demonstrate that oxidative stress homeostasis, mitochondrial function, and energy metabolism are key factors in determining the disease-specific pattern of protein molecular damage. In addition, these findings suggest the lack of a "gold marker of oxidative stress," and, consequently, they strengthen the need for a molecular dissection of the nonenzymatic reactions underlying neurodegenerative processes.

Key words: advanced glycation end products; advanced lipoxidation end products; Alzheimer's disease; amyotrophic lateral sclerosis; Creutzfeldt-Jakob disease; energy metabolism; free radicals; mitochondria; oxidative stress; Parkinson's disease; Pick's disease; reactive carbonyl species

Nonenzymatic Oxidative Protein Modification

As a rule, chemical reactions in living cells are under strict enzymatic control and conform to a tightly regulated metabolic program. One important factor implicit in evolution, from a biomolecular view, is the minimizing of unwanted side reactions. Nevertheless, uncontrolled and potentially deleterious reactions occur, even under physiological conditions. Free radicals (reactive oxygen species [ROS] and reactive nitrogen species [RNS]) are generated by both enzymatic and nonenzymatic sources and have been implicated in a multitude of physiological processes including aging and disease initiation and/or progression.[1] Oxidative stress occurs when the net flux of free radical production during normal aerobic metabolism exceeds the antioxidant defenses of the cell. Emerging evidence indicates that this stress causes specific protein modifications that may lead to a change in the structure and/or function of the oxidized protein.[2,3] Carbonylation is one of those changes, altering the conformation of the polypeptide chain and determining the partial or total inactivation of proteins. This can have a wide range of downstream functional consequences and may be the cause of subsequent cellular dysfunctions and tissue damage.

Structurally, carbonylation may arise from direct oxidation of amino acid side chains, mainly Pro, Arg, Lys, and Thr, resulting in the formation, among others, of glutamic semialdehyde (GSA) and aminoadipic semialdehyde (AASA), the main carbonyl products of metal-catalyzed oxidation of proteins. In addition, carbonyl groups may be introduced into proteins by secondary reaction of the nucleophilic side chains of Cys, His, and Lys residues with reactive carbonyl species (RCS) produced during lipid peroxidation (lipoxidation reactions) or generated as a consequence of the reaction with highly reducing sugars, such as glyoxal and methylglyoxal (an usual byproduct of glycolysis) or their oxidation products (glycation and glycoxidation reactions). Most of the biological effects of intermediate RCS are attributed to their capacity to react with the nucleophilic sites of proteins, forming advanced lipoxidation end products (ALEs) and advanced glycation end products (AGEs).[2] Compared to free radicals, RCS are stable and can diffuse within or even escape from the cell and attack targets far from the site of formation. Therefore, these soluble reactive intermediates are not only cytotoxic per se but

Address for correspondence: Manuel Portero-Otin, Departament de Medicina Experimental, Facultat de Medicina, Universitat de Lleida-IRBLLEIDA; Carrer Montserrat Roig, 2; 25008 Lleida, Spain. Voice: +34-973702408; fax: +34-973702426.

reinald.pamplona@cmb.udl.cat, manuel.portero@cmb.udl.cat

also behave as mediators and propagators of oxidative stress and tissue damage, acting as second cytotoxic messengers.[4]

Oxidative decomposition of polyunsaturated fatty acids (PUFAs) initiates chain reactions that lead to the formation of a variety of RCS which, by reacting with nucleophilic sites in proteins, generate specific ALEs, such as MDA-Lys, HNE-Lys, and Nε-(hexanoyl)lysine, among others. These adducts have been detected by chemical and immunohistochemical methods in a broad range of tissues and species during physiological aging[4] and specific pathological states. The involvement of toxic RCS as products and propagators of oxidative damage in neurodegenerative diseases is currently under study.

The residual aldehyde group in some ALEs can further react to give protein cross-links and fluorescent products that are very similar to AGEs. Lipofuscin, the nondegradable intralysosomal fluorescent pigment that accumulates with age in postmitotic cells, is a recognized hallmark of aging.[4] Other important toxic products formed during nonenzymatic modification of proteins in aging and disease are referred to as "either advanced glycation or lipoxidation end products," so named because they may be formed from either carbohydrates or lipids. Nε-(carboxymethyl)lysine (CML) and Nε-(carboxyethyl)lysine (CEL) are, on a molar basis, the major modifications that have been measured in tissue proteins among these mixed-origin products, emphasizing the importance of the intersection between carbohydrate and lipid chemistry.[2]

Protein Damage and Neurodegenerative Diseases

The nervous system is potentially sensitive to oxidative modifications because i) the particular fatty acid composition of neuronal tissues that is rich in PUFA (hence easily peroxidizable); ii) the high O_2 consumption; and iii) the relatively poor expression antioxidant systems.[5] Also, nervous tissue is considered a postmitotic tissue and therefore highly susceptible to aging.[6] Cells in all regions of the nervous system are affected by aging, as indicated by the decline of sensory, motor, and cognitive functions with time. As this process is involved as a risk factor in most neurodegenerative diseases—there is a dramatic increase in the probability of developing a neurodegenerative disorder (e.g., Alzheimer's disease [AD], Parkinson's disease [PD], or amyotrophic lateral sclerosis [ALS], among others) during the sixth, seventh, and eighth decades of life—and oxidative modifications play a key role in aging, it is often accepted that these diseases should have increased oxidative damage.

Cells in the nervous system are affected by, and respond to, aging much as cells in other organ systems do, and so cells in the brain experience increased amounts of oxidative stress, impaired mitochondrial function and perturbed energy homeostasis, accumulation of damaged proteins, and lesions in their nucleic acids.[5,6] These changes during normal aging are exacerbated in vulnerable populations of neurons in neurodegenerative disorders. Therefore, some diseases might be viewed as a syndrome of accelerated aging in selected neurons.

The interest in the molecular dissection of each of these three pathways (i.e., direct oxidative modification, glycoxidation, and lipoxidation) clearly exceeds an academic context. An appropriate knowledge in this sense could help to rationally design therapeutic approaches aimed either at diminishing oxidative damage in a nonselective way (provided each oxidative pathway is increased in a similar extent) or at pinpointing those processes selectively increased. With this goal in mind, tissues from human tauopathies (AD and Pick's disease [PiD]), synucleopathies (PD and dementia with Lewy bodies [DLB]), and other neurodegenerative processes linked to protein misfolding and/or deposits (Creutzfeldt-Jakob disease [CJD] and ALS) were studied (TABLE 1). The concentration of selected markers of each pathway of protein oxidative damage was analyzed by gas chromatography coupled to mass spectrometry by using authentic deuterated internal standards, according to previously described procedures.[7] Tissues located in "target" zones of the diseases (showing pathological abnormalities) and "control" zones (without morphological changes) were evaluated in order to offer a biochemical correlate of the disease (TABLE 1). These diseases were chosen by the fact that all share accumulation or involvement of structurally modified protein deposits, which have been detected as modified with oxidation products using immunohistochemical procedures.

Tauopathies: Alzheimer's Disease and Pick's Disease

Alzheimer's disease is the more studied and prevalent neurodegenerative disease and is associated with β-amyloid deposits either in neurofibrillary tangles or hyaline bodies. The measurement of oxidative protein modifications[7] reveals that GSA and AASA contents are higher in the frontal cortex (area 8) of AD patients than in age- and sex-matched healthy controls. These increases (around 50% over control values) are larger than those present for well-known

TABLE 1. Changes in specific markers from oxidation-, glycoxidation-, and lipoxidation-derived reactions in human neurodegenerative diseases: a comparative molecular pathology approach

Disease	Number of cases	Location	GSA	AASA	CEL	CML	PI	DHA	MDAL
AD[7]	13	FC	↑	↑↑	↑	↑	↑	↑	↑↑
PiD[8] (1)	7	FC	↑	↑↑	↓	↓	↓	↓	↑
	7	OC	↑	↑	↑	↑	↑	↑	↑↑
PD[10]	7	SN	=	=	=	=	↓	↓	↑
	7	FC	=	=	↓	=	↑	↑↑	↑↑
DLB[10]	4	FC	=	=	↓	=	↑	↑	=
CJD (2)	10	FC	↑	↑	↑	↑	↓	↓	↑
ALS[12]	11	SC	↑	↑	↑	↑	↓	↓↓	↑↑
	11	FC	↑	↑	↑	↑	↓	↑	↑

Abbreviations: AD, Alzheimer's disease; ALS, amyotrophic lateral sclerosis; CJD, Creutzfeldt-Jakob disease; DLB, dementia with Lewy bodies; PD, Parkinson's disease; PiD, Pick's disease. Location: FC, frontal cortex; OC, occipital cortex; SC, spinal cord; SN, substantia nigra. (1) unpublished results; (2) Pamplona, Naudi, Gavin, et al., University of Lleída, Lleída Spain AASA, aminoadipic semialdehyde; CEL, Nε-(carboxyethyl)lysine; CML, Nε-(carboxymethyl)lysine; DHA, docosahexaenoic acid; GSA, glutamic semialdehyde; MDAL, Nε-(malondialdehyde)lysine; PI, peroxidizability index.

AGE markers (CEL and CML), which roughly increase 15% over control values. As suggested by changes in peroxidizability, which is significantly increased in these AD patients (basically because of changed contents of docosahexaenoic acid [DHA]), the most affected marker of protein oxidative modification is Nε-(malondialdehyde)lysine (MDAL), which doubles its content in samples from AD patients. By using a combination of two-dimensional electrophoresis, Western blot, and peptide fingerprinting with matrix-assisted laser desorption/ionization time-of-flight (MALDI-TOF), several cytoesqueletal proteins, metabolic enzymes, and heat shock proteins were identified as modified by MDAL.[7]

To ascertain whether those phenomena are specific to AD, we analyzed PiD (unpublished results and Ref. 8). PiD is another tauopathy, characterized by the specific involvement of the frontotemporal cortex. In common with AD, significant increases were found in AASA, GSA, and MDAL, suggesting increased direct oxidative and lipoxidative damage, although at a lower extent than in AD. However, concentrations of lipoxidative and glycoxidative protein modifications were decreased (both CEL and CML), a fact that can be related to the loss of glycolytic potential, well described in this disease.

The occipital cortex is usually viewed as a location without morphological evidence of involvement of the disease, and hence morphological evidence serves as controls for measurements. In this case, we evidenced increased oxidative, glycoxidative, and lipoxidative damage in this location, supporting the fact that oxidative stress may be an early-stage change in the pathogenesis of this disease. Most interestingly, there was a direct and significant correlation between CEL concentration and DHA levels, suggesting that DHA is increased in response to neuronal stress. This would involve both oxidative stress and increased glycolysis, leading to increased CEL through potentially increased methylglyoxal efflux in the occipital cortex, whereas in the frontal cortex, because of neuronal loss and consequent decreased glycolysis, decreased values in both DHA and CEL content are present. Concerning the targets of oxidative damage, five different proteins exhibit increased anti-DNP staining: reduced form of nicotinamide adenine dinucleotide phosphate (NADPH) carbonyl-reductase, glial fibrillary acidic protein, heat shock protein 70, cathepsin D precursor, and vesicle-fusing ATPase (unpublished results).

Synucleopathies: Parkinson's Disease and Dementia with Lewy Bodies

Parkinson's disease is the most prevalent synucleopathy. Despite previous results describing increased immunoreactivity to anti-AGE and oxidative damage adducts,[9] no chemical evidence of protein oxidative damage was available. When evaluating the content of the above-mentioned markers in substantia nigra from incidental DLB,[10] MDAL was the only marker that was significantly increased in PD (approximately 100%). Similar increases were also found in the frontal cortex and amygdala, suggesting the importance of lipoxidation. In clear contrast, CEL levels were significantly decreased in both the amygdala and the frontal cortex, a fact that can be in accordance with a described loss of glycolysis in PD.[11] Targets of lipoxidative damage comprise several antioxidant enzymes, proteasome components, α-synuclein, and other proteins not shared with AD or PiD. To shed further light on the potential relationship of oxidative damage in synucleopathies, we analyzed cortex samples from patients with DLB, which showed a lack of increased MDAL content but, again, significant decreases in CEL content.[10]

Other Neurodegenerative Processes Linked to Protein Misfolding and/or Deposits

Creutzfeldt-Jakob disease is a neurodegenerative spongiform disease, linked to transmissible prionopathies. The protein oxidative profile in the frontal cortex shows similarities to AD: increased direct oxidative, glycoxidative, and lipoxidative damage, but in this case a decrease in n-3 fatty acids is present. (Pamplona, Naudi, Gavin, et al., University of Lleída, Lleída, Spain) All increases are in the same range, suggesting a general change in the modified protein turnover. This would be compatible with reported alterations in proteasome present in related prion-induced diseases. Concerning targets of glycoxidative damage, we evidenced two key enzymes in glycolysis to be highly modified in CJD samples: glyceraldehyde-3-phosphate dehydrogenase and fructose-1,6-bisphosphate aldolase. As the activity of this latter enzyme involves the formation of a Schiff's base in its active site, we hypothesize that some of those bases may be transformed, under increased oxidative conditions, to CEL.

Amyotrophic lateral sclerosis is characterized by the selective loss of motor neurons in spinal cord and in brain cortex, associated with highly ubiquitinated deposits of proteinaceous material. Spinal cord lysates from ALS patients showed significant increases in direct oxidative, glycoxidative, and lipoxidative damage.[12] Analogous to samples from the brain cortex in AD, the more sensible marker was MDAL, suggesting the importance of lipoxidative modification in this context. Similar to substantia nigra samples in PD and frontal cortex samples in PiD, lipoxidative damage was accompanied by a strong loss in the content of DHA. As observed from other locations without morphological evidence of pathology, samples from the brain cortex of ALS patients also showed increased oxidation, glycoxidation, and lipoxidation, associated with reactive increases in the content of n-3 fatty acids, particularly DHA. All these features are reproducible *in vitro* by the generation of chronic excitotoxicity—a mechanism linked to selective neuronal loss by disturbed intracellular Ca^{++} homeostasis—in a spinal cord organotypic culture, supporting the involvement of this neurodegenerative pathway *in vivo*.[12]

Final Remarks

In conclusion to these analyses, no single marker of protein oxidative modification (among those used here) can be viewed as a gold standard for assessment of oxidative damage in neurodegenerative processes. The same applies to targets of oxidative damage that show disorder-specific differences. Moreover, in some cases CEL levels decreased, supporting CEL relationship with glycolysis potential. There are significant associations between changes in fatty acid composition (especially DHA) and protein oxidative damage. Most interestingly, when tissue from pathologically preserved locations was available, the tissue analyses indicated that protein oxidative modifications take place before potential morphological and clinical changes appear, suggesting an early involvement of protein oxidative damage in neurodegenerative process.

Acknowledgments

This study was supported in part by I+D grants from the Spanish Ministry of Education and Science (BFU2006-14495/BFI), the Spanish Ministry of Health (ISCIII, Red de Envejecimiento y Fragilidad, RD06/0013/0012), and the Generalitat of Catalunya (2005SGR00101) to R.P.; the Spanish Ministry of Health (FIS 04-0355 and 05-2241), Spanish Ministry of Education and Science (AGL2006-12433), and "La Caixa" Foundation to M.P.O. and E.D., E.I. and D.C.; and Field-initiated Studies Program grants 020004 and 03-006, Spanish Ministry of Education and Science (CICYT SAF-2001-4681E), and European Union project Brain Net Europe II to I.F.

Conflict of Interest

The authors declare no conflicts of interest.

References

1. SANZ, A., R. PAMPLONA & G. BARJA. 2006. Is the mitochondrial free radical theory of aging intact? Antioxid. Redox Signal. **8:** 582–599.
2. THORPE, S.R. & J.W. BAYNES. 2003. Maillard reaction products in tissue proteins: new products and new perspectives. Amino Acids **25:** 275–281.
3. PORTERO-OTIN, M. & R. PAMPLONA. 2006. Is endogenous oxidative protein damage envolved in the aging process?. *In* Recent Research Developments in Pathological Biochemistry. Vol. 1 Protein Oxidation and Disease. J. Pietzsch, Ed.: 91–142. Research Signpost. Kerala, India.

4. HULBERT, A.J., R. PAMPLONA, R. BUFFENSTEIN & W.A. BUTTEMER. 2007. Life and death: metabolic rate, membrane composition and life span of animals. Physiol. Rev. doi: 10.1152/physrev.00047.2006.
5. MATTSON, M.P. & T. MAGNUS. 2006. Ageing and neuronal vulnerability. Nat. Rev. Neurosci. **7:** 278–294.
6. BARJA, G. 2004. Free radicals and aging. Trends Neurosci. **27:** 595–600.
7. PAMPLONA, R., E. DALFO, V. AYALA, et al. 2005. Proteins in human brain cortex are modified by oxidation, glycoxidation, and lipoxidation: Effects of Alzheimer's disease and identification of lipoxidation targets. J. Biol. Chem. **280:** 21522–21530.
8. MUNTANE, G., E. DALFO, A. MARTINEZ, et al. 2006. Glial fibrillary acidic protein is a major target of glycoxidative and lipoxidative damage in Pick's disease. J. Neurochem. **99:** 177–185.
9. CASTELLANI, R., M.A. SMITH, P.L. RICHEY, et al. 1996. Glycoxidation and oxidative stress in Parkinson disease and diffuse Lewy body disease. Brain Res. **737:** 195–200.
10. DALFO, E., M. PORTERO-OTIN, V. AYALA, et al. 2005. Evidence of oxidative stress in the neocortex in incidental Lewy body disease. J. Neuropathol. Exp. Neurol. **64:** 816–830.
11. HU, M.T., S.D. TAYOR-ROBINSON, K.R. CHAUDHURI, et al. 2000. Cortical dys function in non-demented Parkinson's disease patients: a combined (31) P-MRS and (18) FDG-PET study. Brain **123:** 340–352.
12. ILIEVA, E., V. AYALA, M. JOVE, et al. 2007. Oxidative and endoplasmic reticulum stress interplay in sporadic amyotrophic lateral sclerosis. Brain, doi: 10.1093/brain/awm190.

Suppression of Renal α-Dicarbonyl Compounds Generated following Ureteral Obstruction by Kidney-Specific α-Dicarbonyl/L-Xylulose Reductase

HIROKO ODANI,[a] JUN ASAMI,[b] AIKO ISHII,[b] KAYOKO OIDE,[b] TAKAKO SUDO,[b] ATSUSHI NAKAMURA,[b] NORIYUKI MIYATA,[b] NOBORU OTSUKA,[b] KENJI MAEDA,[c] AND JUNICHI NAKAGAWA[b,d]

[a]*Department of Nephrology, Fujita Health University School of Medicine, Toyoake 470-1192, Japan*

[b]*Medical Research Laboratories, Taisho Pharmaceutical Co., Ltd., Saitama 331-9530, Japan*

[c]*Daikou Medical Engineering Institute, Nagoya 461-0047, Japan*

[d]*Department of Food Science, Faculty of Bio-industry, Tokyo University of Agriculture, Abashiri, 099-2493, Hokkaido, Japan*

Renal unilateral ureteral obstruction (UUO) causes acute generation of α-dicarbonyl stress substances, such as glyoxal, 3-deoxyglucosone, and methylglyoxal, in the kidneys. These α-dicarbonyl compounds are prone to form advanced glycation end products (AGEs) via the nonenzymatic Maillard reaction. Using transgenic (Tg) mice overexpressing a kidney-specific short-chain oxidoreductase, α-dicarbonyl/L-xylulose reductase (DCXR), we measured generation of α-dicarbonyls following UUO by means of electrospray ionization/liquid chromatography/mass spectrometry in their kidney extracts. The accumulation of 3-deoxyglucosone was significantly reduced in the kidneys of the mice Tg for DCXR compared to their wild-type littermates, demonstrating 4.91 ± 2.04 vs. 6.45 ± 1.85 ng/mg protein ($P = 0.044$) for the obstructed kidneys, and 3.68 ± 1.95 vs. 5.20 ± 1.39 ng/mg protein ($P = 0.026$) for the contralateral kidneys. Despite the reduction in accumulated α-dicarbonyls, collagen III content in kidneys of the Tg mice and their wild-type littermates showed no difference as monitored by *in situ* hybridization. Collectively, DCXR may function in the removal of renal α-dicarbonyl compounds under oxidative circumstances, but it is not sufficient to suppress acute renal fibrosis during 7 days UUO.

Key words: α-dicarbonyls; α-dicarbonyl/L-xylulose reductase; ESI/LC/MS; unilateral ureteral obstruction; UUO animal model; renal failure; 3-deoxyglucosone; methylglyoxal

Introduction

It is well recognized that the Maillard reaction readily takes place in the tissue under hyperglycemic circumstances, and that the resulting advanced glycation end products (AGEs) can cause tissue dysfunction by cross-linking proteins. Pathological consequences follow not only through its complex tissue deformation but also through emission of inflammatory signals following activation of a scavenger receptor called receptor for AGE (RAGE).[1] Important intermediate compounds for the generation of AGEs are α-dicarbonyls, such as methylglyoxal (MGO), glyoxal (GO), and 3-deoxyglucosone (3-DG), all having α-dicarbonyl groups in their molecules. The Maillard reaction and the subsequent Amadori rearrangement trigger complicated nonenzymatic and enzymatic processes in tissues and generate AGEs, such as carboxymethyllysine (CML), often detected in the stressed portion of the tissues. Under highly oxidative conditions, as in renal dialysis, generation of α-dicarbonyls is enhanced, and, consequently, the serum concentration of α-dicarbonyls is elevated in nephritic patients as detected by electrospray ionization/liquid chromatography/mass spectrometry (ESI/LC/MS).[2,3] The fact

Address for correspondence: Hiroko Odani, Department of Nephrology, Fujita Health University School of Medicine, 1-98 Dengakugakubo, Kutsukake-cho, Toyoake, 470-1192, Japan. Voice/fax: +81-52-837-0210.
hirokood@kjps.net

that augmented renal generation of α-dicarbonyls occurs even without a diabetic background implies the renal detoxifying ability is at least in part responsible for the removal of α-dicarbonyls. One of the candidate enzymes for the detoxification process is α-dicarbonyl/L-xylulose reductase (DCXR).[4] DCXR is highly expressed in the kidney where it is specifically localized in the microvilli protruding from the inner walls of the proximal tubules, as demonstrated at the ultrastructural level. This suggests that DCXR may readily encounter reabsorbed α-dicarbonyls in the primary urines coming from glomeruli. We generated transgenic (Tg) mice overexpressing DCXR[5] and examined DCXR's relevance in α-dicarbonyl reduction under an oxidative condition, namely unilateral ureteral obstruction (UUO) treatment.[6] We measured MGO, GO, and 3-DG content in the kidneys by ESI/LC/MS.

Methods

Standardization of Dicarbonyls and ESI/LC/MS Analysis

The content of dicarbonyl compounds was measured after resolution of derivatized dicarbonyl compounds by reverse-phase high performance liquid chromatography followed by mass spectrometry analysis as described previously.[2] Briefly, formation of the derivatized products of 3-DG (Dojindo Labs, Kumamoto, Japan), MGO (Sigma-Aldrich, St. Louis, MO), and GO (Nakalai Tesque, Kyoto, Japan) were made through reaction with 2,3-diaminonaphthalene (Dojindo Labs) and 3,4-hexanedione was used as the internal standard. Quantification of the three α-dicarbonyl compounds was done according to each protonated molecular ion peak area ratio obtained by ESI/LC/MS in the selected ion monitoring mode. The correction coefficients between the added three dicarbonyl standard concentration and the peak area ratio were 0.972 (regression equation of the standard curve: $y = 0.69154 + 0.03024x$) for 3-DG, 0.990 ($y = -2.1462 + 0.08555x$) for MGO, and 0.996 ($y = -11.45 + 0.2455x$) for GO, with a standard concentration range from 1 nmol/mL to 600 nmol/mL. The data were normalized by protein concentration.

Animals

Wistar male rats (7-weeks old) and male mice (12–14-weeks old) Tg for DCXR protein (line #4001) previously generated by introducing mouse *dcxr* cDNA into FVB mice[5] were used. The animals were housed in an approved animal care facility at Taisho Pharmaceutical Co., Ltd. (Saitama, Japan) and had free access to food and water.

UUO Operation and Kidney Extraction

DCXR Tg mice and their control littermates received a UUO operation at the left urinary tract obstructed by ligation. After 7 days, the mice were sacrificed and the kidney extracts were prepared as follows: The kidney was minced with scissors and soaked in a buffer containing 10 mmol Tris-HCl (pH 8.0), 0.5 mmol EDTA/EGTA, 1 mmol $MgCl_2$, 1 mmol β-mercaptoethanol, and 0.2% TritonX100 supplemented with one tablet of the protease inhibitor cocktail (Roche/Boehringer Mannheim, Basel, Switzerland) per 50 mL of buffer and then homogenized with a homogenizer Hitachi HG30 (Hitachi, Tokyo, Japan). The homogenate was ultracentrifuged at 15,000 g for 30 min in a TLA55 rotor (Beckman Coulter, Fullerton, CA).

Biochemical Analysis

Western blotting was performed using 6D12 monoclonal antibody against CML (25 μg/mL) (Dojin Chemicals, Kumamoto, Japan) and α-DCXR antibody (2 μg/mL).[4]

Immunochemical Analysis of Renal Fibrosis and DCXR Expression

The kidney was fixed in a neutral buffer containing 10% formalin and then embedded in paraffin. General staining, fibrosis, and collagen accumulation were examined with hematoxylin and eosin, Masson trichrome, and Sirius red, respectively. For immunohistochemical analysis, 3-mm-thick sections were incubated with α-DCXR peptide antibody (0.5 μg/mL), α-smooth-muscle actin antibody (1A4, 1:5 dilution) (DAKO, Glostrup, Denmark), and α-CML monoclonal antibody 6D12 (2 μg/mL). Immunostaining was performed using the Ventana NX Automated Immunohistochemistry System (Ventana Japan, Tokyo) according to the manufacturer's protocol. For *in situ* hybridization (ISH) of collagen type III, a probe comprising nucleotide no. 2673–3263 (591 base-pairs long) was generated from the cDNA (accession no. XML 343563), and digoxigenin (DIG)-labeled RNA probes were synthesized. ISH was done using the Ventana HX Automated ISH System (Ventana Japan, Yokohama). Images were taken under a light microscope (Carl Zeiss, Oberkochen, Germany).

Results

DCXR Expression in UUO-treated Tg Mice

Previously we generated Tg mice with the *dcxr* gene using a universal promoter consisting of a

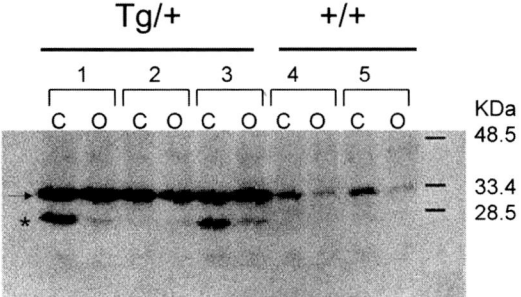

FIGURE 1. Behavior of α-dicarbonyl/L-xylulose reductase (DCXR) protein in Tg mice during unilateral ureteral obstruction (UUO). Kidney extracts of DCXR Tg mice and their wt littermates were prepared following UUO, and Western blotting was performed. A representative blotting is shown. Tg/+, dcxr Tg, heterozygous (1–3); +/+, wild-type littermate (4 and 5). C, contralateral (normal side kidney); O, obstructed. Numbers indicate individual animal naming. Note that the lower bands (28.5 kDa, marked with an asterisk) should be the degradation product of the 33.4 kDa mature DCXR bands (arrow) [5]. Reprinted from Ref. 6.

cytomegaloviral (CMV) enhancer and a rabbit β-globin promoter.[5] These mice overexpressed DCXR throughout the body, and homozygous and heterozygous mice expressed sixfold and threefold DCXR protein, respectively, compared to the wild type as measured by quantifying Western blots of the kidney extracts.[5] Using Tg mice, we aimed to compare the renal dicarbonyl metabolism under oxidative conditions in the DCXR Tg mice. We first examined the abundance of Tg DCXR protein following UUO treatment in the kidneys of these mice. As shown in FIGURE 1, expression of the Tg DCXR protein remained constant in the course of UUO (lanes 1–3; compare C and O) while the endogenous DCXR obviously decreased with increasing time of UUO (lanes 4 and 5; compare C and O), indicating that expression of the DCXR gene under the CMV enhancer and β-globin promoter in the mouse kidney was not repressed by the mechanical or oxidative stress generated during UUO treatment, unlike that of the endogenous gene. The smaller bands marked with an asterisk were previously estimated to be degradation products in the course of sample preparation,[5] and these bands tended to be more susceptible to UUO (lanes 1 and 3; compare C and O).

Reduction of Accumulated α-Dicarbonyls in the UUO-treated Kidneys of DCXR Tg Mice

The fact that transgene-encoded DCXR maintained elevated expression during UUO treatment indicated a potential to reduce UUO-derived dicarbonyls, we next examined the generation of α-dicarbonyls in these Tg mice following UUO. The accumulation of renal dicarbonyls was measured following 7 days of UUO treatment of DCXR Tg mice. As expected, the content of dicarbonyls in the obstructed kidneys was significantly higher than in the contralateral kidneys (FIG. 2). As for 3-DG, the content in both the obstructed and the contralateral kidneys was significantly lower in the Tg mice compared to their wild-type (wt) littermates (FIG. 2, left panel). The accumulation of 3-DG in the kidneys of Tg mice compared to their wt littermates was 4.91 ± 2.04 vs. 6.45 ± 1.85 ng/mg protein ($P = 0.044$) for the obstructed kidneys and 3.68 ± 1.95 vs. 5.20 ± 1.39 ng/mg protein ($P = 0.026$) for the contralateral kidneys. As for MGO, there was no significant difference between Tg mice and their wt littermates in the obstructed or contralateral kidneys. On the other hand, the MGO content in the obstructed kidneys was significantly higher than in the contralateral kidneys in Tg mice (FIG. 2, middle panel). GO displayed similar tendencies (right panel). We used heterologous Tg mice in order to ensure a sufficient number of animals to allow statistical evaluation.

Immunohistochemical Examination of DCXR and Type III Collagen Expression

Kidney sections were prepared from UUO-treated mice and general appearance and the expression of DCXR were examined. The UUO-operated kidneys showed significant dilation of the renal pelvis, thinning of the cortex, and dilation of the tubules. It was also noted that the appearance of regenerated renal tubules were frequently associated with thick basement membrane and interstitial infiltration of inflammatory cells and fibrosis in their surroundings (data not shown).[6] UUO treatment caused a reduction in endogenous DCXR protein (FIG. 3, A versus B), but the Tg DCXR protein expressed in the Tg mice remained at a constant level (FIG. 3, C versus D). DCXR expression in the Tg mice covered virtually all the area, with the most remarkable expression in the medullary collecting ducts (FIG. 3, C and D).

In order to examine fibrotic changes during UUO treatment, we did ISH using a probe for type III collagen mRNA as ISH is a sensitive method for visualization of the initial stages of fibrosis. The results clearly showed that UUO treatment upregulated collagen gene transcription (FIG. 3, E versus F and G versus H) and that the extent of collagen transcriptional elevation did not appear to be significantly reduced in the Tg mice (FIG. 3, F versus H).

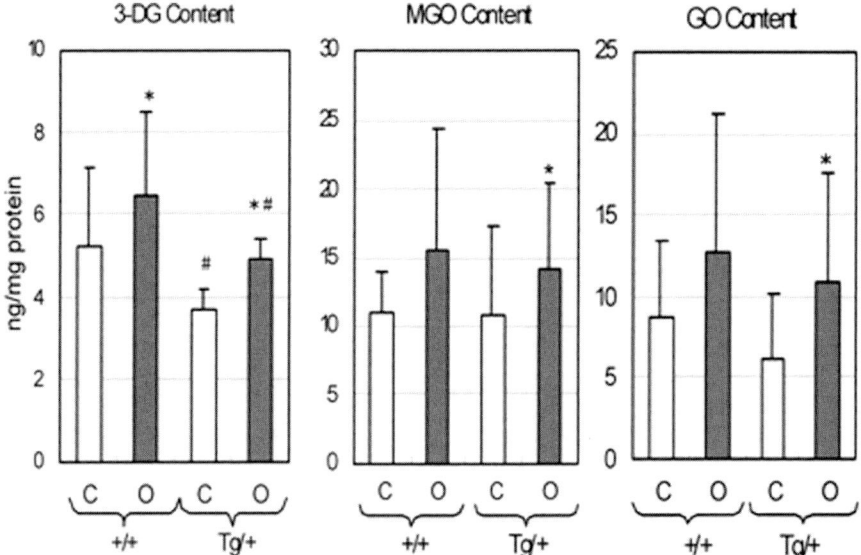

FIGURE 2. Effect of UUO and DCXR overexpression on the dicarbonyl contents in mouse kidneys. The content of 3-DG (*left*), MG (*middle*), and GO (*right*) in kidney extracts of DCXR Tg mice was calculated following normalization to the protein content. Values are the means with standard deviations from 13 wild-type and 16 Tg mice. C, contralateral; O, obstructed. (*$P < 0.05$ in comparison of normal and occlusion within wild-type or Tg mice and #$P < 0.05$ in comparison of wild-type and Tg mice within normal side or obstructed side.) Reprinted from Ref. 6.

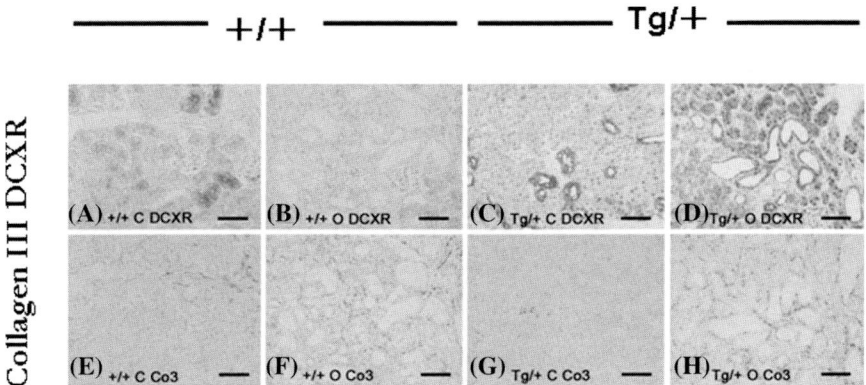

FIGURE 3. Fibrotic changes and DCXR expression of UUO-treated kidney. UUO-operated kidneys of DCXR Tg and wt mice were examined immunochemically with α-DCXR antibody (*upper panel*) and by *in situ* hybridization of collagen type III cDNA (*lower panel*). Wild-type (+/+) and DCXR Tg (Tg/+) kidneys are shown. C, contralateral right kidneys; O, obstructed left kidneys. Magnification: bars, 50 μm.

Discussion

In this study, the UUO animal model, known to generate carbonyl stress acutely, was used to study the function of DCXR in the regulation of accumulation of AGEs and α-dicarbonyls in kidneys. Within the renal tubules, DCXR expression was most prominent at the collecting tubules and ducts and had moderate expression in other parts of the tubules. Expression in the interstitial and mesangial regions was scarce.

Previously we observed that expression of DCXR protein in rat kidneys decreased with increasing duration of UUO treatment and the accumulation of CML increased in the obstructed kidney, suggesting that insufficient reducing capacity of the obstructed kidney might be attributable to a decrease in the amount of DCXR.[6] On the other hand, the transgenically introduced *dcxr* gene expressed by a promoter consisting of the CMV enhancer and the β-globin promoter[5] in mice did not decrease during

UUO treatment. If DCXR is indeed involved in the reduction of α-dicarbonyls in the kidney, constitutively elevated expression of the *dcxr* transgene should protect animals from carbonyl stress. UUO did indeed induce carbonyl compounds and initial renal fibrotic changes in mice used here. A significant difference in the metabolism of dicarbonyls in the Tg mice from that in wt mice was found upon comparison of the accumulation of dicarbonyls. The kidneys of Tg mice contained significantly lower amounts of 3-DG than their wt littermates, both in obstructed and in contralateral kidneys (FIG. 3, left panel). A tendency for a similar reduction in MGO and GO in the Tg mouse kidneys was observed, albeit without statistical significance. These observations indicate that the overall dicarbonyl content in the kidney tended to be reduced in the Tg mice compared to their littermates, probably as a result of an amplification of DCXR activity. Taking into account the specific distribution of endogenous DCXR in the renal tubules, DCXR might play a role in the detoxification of nephritic dicarbonyls generated during chronic renal disorder and under hyperglycemic pressure as well, both of which cause accumulation of renal carbonyl stress.

Although a significant reduction in renal dicarbonyl accumulation was observed in the DCXR Tg mice, we were not able to detect any difference in the reduction of renal fibrosis in the Tg mice from their wt littermates in this acute model. A longer continuation of fibrotic insult might be necessary to detect such effects as the chronic renal fibrosis occurring in diseases like diabetic nephropathy.

Conflict of Interest

The authors declare no conflicts of interest.

References

1. BOHLENDER, J. M., *et al.* 2005. Advanced glycation end products and the kidney. Am. J. physiol. Renal Physiol. **289:** 645–659.
2. ODANI, H. & K. MAEDA. 1999. Increase in three α,β-dicarbonyl compound levels in human uremic plasma: specific in vivo determination of intermediates in advanced Maillard reaction. Biochem. Biophys. Res. Commun. **256:** 89–93.
3. ODANI, H. & IIJIMA, K. 2001. Identification of N (omega)-carboxymethylarginine, a new advanced glycation end-product in serum proteins of diabetic patients: possibility of a new marker of aging and diabetes. Biochem. Biophys. Res. Commun. **285:** 1232–1236.
4. NAKAGAWA, J. & A. HARA. 2002. Molecular characterization of mammalian dicarbonyl/l-xylulose reductase and its localization in kidney. J. Biol. Chem. **277:** 17883–17891.
5. SUDO, T. & J. NAKAGAWA. 2005. Transgenic mice overexpressing dicarbonyl/L-xylulose reductase gene crossed with KK-Ay diabetic model mice: an animal model for the metabolism of renal carbonyl compounds. Exp. Anim. **54:** 385–394.
6. ASAMI, J. & J. NAKAGAWA. 2006. Suppression of AGE precursor formation following unilateral ureteral obstruction in mouse kidneys by transgenic expression of α-dicarbonyl/L-xylulose reductase. Biosci. Biotechnol. Biochem. **70:** 2899–2905.

Comparison of Pharmacokinetics between Highly and Mildly Modified AGE Proteins in Mice

RYOJI NAGAI,[a] KATSUMI MERA,[a,b] YUKIO FUJIWARA,[a] MIME NAGAI,[a] AND MASAKI OTAGIRI[b]

[a]*Department of Medical Biochemistry, Faculty of Medical and Pharmaceutical Sciences, Kumamoto University, Kumamoto 860-8556, Japan*

[b]*Department of Biopharmaceutics, Graduate School of Pharmaceutical Sciences, Kumamoto University, Kumamoto 862-0973, Japan*

We previously demonstrated that RAW 264.7 cells (murine macrophage cell line) recognize highly modified advanced glycation end products (AGE)-bovine serum albumin (BSA) (high-AGE-BSA), which was prepared by incubating BSA with 1600 mmol/L glucose for 40 weeks. In the present study, we prepared mildly modified AGE-BSA (mild-AGE-BSA) and conducted an endocytic uptake study using human monocyte-derived macrophages and Chinese hamster ovary cells which overexpressed such scavenger receptors as CD36, SR-BI (scavenger receptor class B type-I), and LOX-1 (lectin-like oxidized low-density lipoprotein receptor-1). Although high-AGE-BSA was significantly recognized by these cells, mild-AGE-BSA did not show any ligand activity to these cells. Furthermore, when ^{111}In-labeled mild- or high-AGE-BSA was injected into the tail vein of male ddY mice, ^{111}In-high-AGE-BSA was rapidly cleared from the circulation, with about 80% of the injected ^{111}In-high-AGE-BSA being eliminated within 5 min. In contrast, the clearance rate of ^{111}In-mild-AGE-BSA was very slow, similar to the ^{111}In-native BSA. Taken together, our results indicate that the ligand activity of AGE-BSA to scavenger receptors and those pharmacokinetic properties depend on their rate of modification by AGEs.

Key words: advanced glycation end products; macrophage; scavenger receptor

Introduction

Cellular interaction with advanced glycation end product (AGE)-modified proteins is believed to induce several biological responses that are involved in the development of diabetic vascular complications.[1] These cellular interactions are thought to be mediated by AGE receptors, such as scavenger receptor (SR)-A (class A scavenger receptor types I and II),[2] CD36,[3] SR-BI (scavenger receptor class B type-I),[4] LOX-1 (lectin-like oxidized-low-density lipoprotein receptor-1),[5] and RAGE (receptor for AGE).[6] However, AGE proteins are prepared independently by research groups using different protocols, and all researchers, including our group, use excessively high concentrations of glucose and aldehydes. For instance, we prepared high-AGE-bovine serum albumin (BSA) by incubating BSA with 1600 mmol/L glucose for 40 weeks or 33 mmol/L glycolaldehyde for 7 days and demonstrated that high-AGE-BSA is recognized by SR-A.[2] Schmidt *et al.* demonstrated that AGE-BSA, prepared by incubating BSA with 250 mmol/L glucose-6-phosphate for 4 weeks, is recognized by RAGE.[6] Taken together, those reports demonstrate that the AGE proteins were prepared with unphysiologically high concentrations of aldehydes and then were used for the cellular experiments. In the present study we prepared mildly modified AGE-BSA (mild-AGE-BSA) by incubating BSA with 50 mmol/L glucose for 24 weeks and then compared mild-AGE-BSA ligand activity to high-AGE-BSA ligand activity against the scavenger receptors.

Materials and Methods

Preparation of AGE-modified BSA

High-AGE-BSA was prepared as described previously.[7] Mild-AGE-BSA was prepared by incubating

Address for correspondence: Ryoji Nagai, Ph.D., Department of Medical Biochemistry, Faculty of Medical and Pharmaceutical Sciences, Kumamoto University, Honjo 1-1-1, Kumamoto 860-8556, Japan. Fax: +81-96-364-6940.

nagai-883@umin.ac.jp

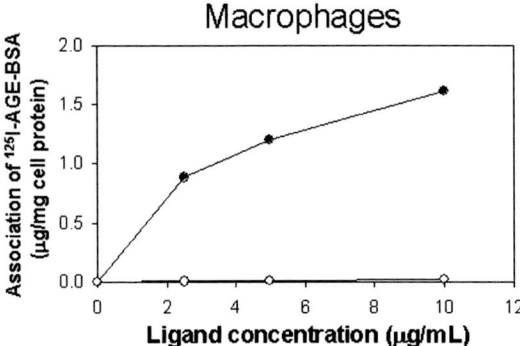

FIGURE 1. Ligand activity of advanced glycation end products (AGE)-bovine serum albumin (BSA) to human monocyte-derived macrophages. Human monocyte-derived macrophages were incubated with the indicated concentration of ^{125}I-high-AGE-BSA (closed circles) or ^{125}I-mild-AGE-BSA (open circles).

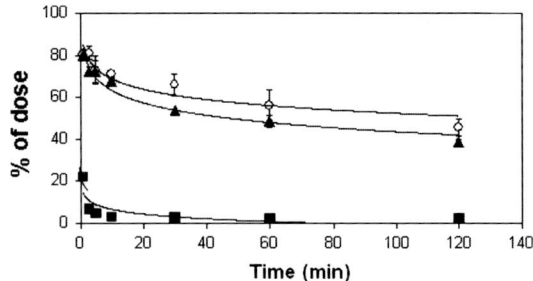

FIGURE 2. Plasma clearance of AGE-BSA after intravenous administration to mice. BSA (open circles), mild-AGE-BSA (closed triangles), and high-AGE-BSA (closed squares) were labeled with ^{111}In, injected as a bolus through the tail vein of mice, and the relative radioactivity plotted against the time after injection. Each data point represents the mean ± SD for three mice.

0.05 g/mL of fatty acid-free BSA with 50 mmol/L of glucose in 0.05 mol sodium phosphate buffer (pH 7.4) at 37°C for 24 weeks, followed by dialysis against phosphate-buffered saline solution.

Cellular and Clearance Assays

Human monocyte-derived macrophages were prepared as described previously.[8] Chinese hamster ovary (CHO) cells overexpressing CD36,[3] SR-BI,[4] and OX-1[5] were prepared as described previously. Modified BSA preparations were radiolabeled with ^{125}I using Iodo-Gen (Thermo Scientific, Rockford, IL), and the specific association of ^{125}I-AGE-BSA with the cells was determined as described previously.[2] The BSA, mild-AGE-BSA, and high-AGE-BSA were radiolabeled with ^{111}In. The mice received tail-vein injections of ^{111}In-labeled proteins in saline at a dose of 1 mg/kg. At the appropriate intervals after the injection, blood was collected from the vena cava under anesthesia and plasma was obtained by centrifugation.

Results and Discussion

As shown in FIGURE 1, the ^{125}I-high-AGE-BSA was specifically associated with human monocytes-derived macrophages in a dose-dependent manner, whereas these changes were not observed in the ^{125}I-mild-AGE-BSA. Similar tendencies were observed in the CHO overexpressing scavenger receptors. Therefore, the ^{125}I-high-AGE-BSA was specifically associated with the CHO cells overexpressing SR-BI, CD36, and LOX-1, whereas the association of the ^{125}I-mild-AGE-BSA to these cells was negligible (data not shown), demonstrating that only the high-AGE-BSA showed ligand activity to the scavenger receptors. Furthermore, the plasma clearance rates of ^{111}In-mild-AGE-BSA were very slow, similar to the ^{111}In-BSA, whereas the radioactivity of ^{111}In-high-AGE-BSA was rapidly cleared from the circulation, with about 90% of the injected ^{111}In-high-AGE-BSA being eliminated within 5 min after the intravenous administration (FIG. 2). Thornalley et al.[9] demonstrated that end-stage renal disease is associated with a significant increase in the molecular mass of human serum albumin (HSA) (+255 Da, relative to control subjects). However, our study using MALDI TOF mass analysis demonstrated that the molecular mass of mild-AGE-BSA was 658 Da larger than native BSA (data not shown), indicating that our experimentally prepared mild-AGE-BSA is already more profoundly modified than physiological HSA under (patho)physiological conditions. These results indicate that endocytic uptake of AGE proteins through scavenger receptors is negligible or unlikely to occur in vivo. This study demonstrates that the ligand activity of the AGE proteins to the scavenger receptors and those pharmacokinetic properties are dependent on their rate of modification by the AGEs, and we should carefully prepare the AGE proteins in vitro to clarify the physiological significance of the interaction between AGE receptors and AGE proteins.

Acknowledgments

This work was supported in part by Grants-in-Aid for Scientific Research (No. 18790619 to R.N) from the Ministry of Education, Science, Sports and Cultures of Japan.

Conflicts of Interest

The authors declare no conflicts of interest.

References

1. VLASSARA, H., R. BUCALA & L. STRIKER. 1994. Pathogenic effects of advanced glycosylation: biochemical, biologic, and clinical implications for diabetes and aging. Lab Invest. **70:** 138–151.
2. NAGAI, R. *et al.* 2000. Glycolaldehyde, a reactive intermediate for advanced glycation end products, plays an important role in the generation of an active ligand for the macrophage scavenger receptor. Diabetes **49:** 1714–1723.
3. OHGAMI, N. *et al.* 2001. Cd36, a member of the class b scavenger receptor family, as a receptor for advanced glycation end products. J. Biol. Chem. **276:** 3195–3202.
4. OHGAMI, N. *et al.* 2001. Scavenger receptor class B type I-mediated reverse cholesterol transport is inhibited by advanced glycation end products. J. Biol. Chem. **276:** 13348–13355.
5. JONO, T. *et al.* 2002. Lectin-like oxidized low density lipoprotein receptor-1 (LOX-1) serves as an endothelial receptor for advanced glycation end products (AGE). FEBS Lett. **511:** 170–174.
6. SCHMIDT, A.M. *et al.* 1992. Isolation and characterization of two binding proteins for advanced glycosylation end products from bovine lung which are present on the endothelial cell surface. J. Biol. Chem. **267:** 14987–14997.
7. HIGASHI, T. *et al.* 1997. The receptor for advanced glycation end products mediates the chemotaxis of rabbit smooth muscle cells. Diabetes **46:** 463–472.
8. NAGAI, R. *et al.* 2007. Investigation of pathways of advanced glycation end-products accumulation in macrophages. Mol. Nutr. Food Res. 51: 462–467.
9. THORNALLEY, P.J. *et al.* 2000. Mass spectrometric monitoring of albumin in uremia. Kidney Int. **58:** 2228–2234.

Modification of Vimentin

A General Mechanism of Nonenzymatic Glycation in Human Skin

THOMAS KUEPER,[a] TILMAN GRUNE,[b] GESA-MEIKE MUHR,[a] HOLGER LENZ,[a] KLAUS-PETER WITTERN,[a] HORST WENCK,[a] FRANZ STÄB,[a] AND THOMAS BLATT[a]

[a]R & D, Beiersdorf AG, Hamburg, Germany

[b]Department of Biological Chemistry and Nutrition, University of Hohenheim, Stuttgart, Germany

In a recent study, we were able to show that the intermediate filament protein vimentin aggregates in human dermal fibroblasts because of modification by the advanced glycation endproduct carboxymethyllysine (CML). In this work, we investigated the formation of intracellular CML in relation to the concentration of glucose in the culture medium. The natural degradation product of glucose, methylglyoxal, was able to induce the aggregation of vimentin. This dicarbonyl leads to the formation of the modifications MG-H1 and carboxyethyllysine (CEL) as a result of the reaction with arginine and lysine residues of proteins. Furthermore, we found that the protein vimentin was modified, not only by CML and CEL, but also by pentosidine and pyrraline. These findings underline the special position of vimentin as a preferential target of the Maillard reaction in human skin.

Key words: glycation; AGEs; vimentin; carboxymethyllysine; CML; carboxyethyllysine; CEL; skin; Maillard; fibroblasts

The formation of advanced glycation endproducts (AGEs) is based on the reaction between reducing sugars and protein residues,[1] resulting in a heterogeneous group of structures with varying chemical and physical properties.[2] These modifications accumulate in various tissues during the process of aging, leading to the loss of elasticity of connective tissues and vessel walls.[3-8] In patients with renal failure, adducts of the Maillard reaction increase dramatically in plasma as a result of inefficient renal clearance.[9] Furthermore, several AGE modifications have been postulated as biomarkers for different diseases. These include the serum level of pentosidine as a marker for heart failure,[10] succination of proteins as a biomarker for mitochondrial stress in adipocytes,[11] or the level of carboxymethyllysine (CML) as marker for diabetic complications.[12,13] Additionally, many reports in the literature link development of AGEs to age-related diseases, such as Alzheimer's disease, arthritis, atherosclerosis, or pulmonary fibrosis.[14] The formation of intracellular glycation products also plays an important physiological role

because of the rapid reaction of dicarbonyls with protein residues.[15,16] However, less information is available about the contribution of intracellular AGEs to aging. Glyoxal and methylglyoxal are regarded as key players among these intracellular dicarbonyls, finally leading to the formation of carboxyethyllysine (CEL), CML, and methylglyoxal-derived hydroimidazo one (MG-H1).[17,18] In the experiments presented here, we were able to prove that the comparable glycation of vimentin by various AGEs is found in dermal fibroblasts with relevance for human skin aging.

Results

Enhancement of Glucose in Medium Correlates with Intracellular CML

Because we were able to show that the cultivation of fibroblasts with glyoxal—which reacts with the lysine resulting in CML—leads to the redistribution of vimentin in fibroblasts, we decided to test the intracellular level of CML following incubation with different concentrations of glucose. Therefore, primary human fibroblasts were incubated for one week with varying amounts of glucose, and the level of CML was measured using flow cytometry. Fibroblasts cultured with 200 mM glucose showed a 43% higher intracellular

Address for correspondence: Thomas Kueper, Beiersdorf AG, Unnastrasse 48, 20253 Hamburg, Germany. Voice: 49 0 1799206938; fax: 49 0 4018014917.

Thomas.Kueper@beiersdorf.com

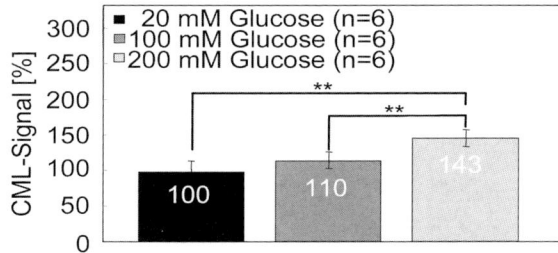

FIGURE 1. Glucose in culture medium induces intracellular CML formation. Fibroblasts were cultured for 1 week with different concentrations of glucose. Afterwards, the level of CML was determined by flow cytometry using an antibody specific for this modification. The CML level of fibroblasts cultured with 20 mM glucose was set to 100%. Data are based on six independent measurements; Mean ± SD, **$P < 0.01$.

CML signal compared with fibroblasts cultured in the presence of 20 mM glucose (FIG. 1). This clearly reflects the importance of the breakdown products of glucose—such as glyoxal and methylglyoxal—in the formation of intracellular AGE modifications.

Methylglyoxal Induces Redistribution of Vimentin

Because of the chemical similarity of glyoxal and methylglyoxal, it appeared reasonable to test whether cultivation of fibroblasts with methylglyoxal induces a redistribution of vimentin, as was previously shown for cultivation with glyoxal.[19] Human dermal fibroblasts were cultured with glyoxal (FIG. 2A) and methylglyoxal (FIG. 2B) to mimic the perennial modification by these glycation agents. Remarkably, both dicarbonyls induced the rearrangement of vimentin as a result of its glycation. In contrast, the protein actin did not indicate any visible alterations in its structure.

Glycation of Vimentin with Various AGEs

Proteins of primary human fibroblast lysates were separated by 2-dimensional gel electrophoresis and blotted onto nitrocellulose for subsequent Western blot analyses with antibodies specific for the AGEs pentosidine,[20] pyrraline,[21] CEL,[22] and the intermediate filament protein vimentin,[23] which was shown to be affected by exceptionally high glycation with CML in a previous study.[19] Therefore, we decided to investigate whether these additional AGEs target vimentin as well. FIGURE 3 exhibits a clear colocalization of vimentin with all modifications tested, supporting the finding that vimentin is an important target of AGEs.

Experimental Procedures

Cell Culture

Human facial skin biopsies were isolated from healthy donors. Primary dermal cells were enzymatically prepared using a standardized dispase (Roche Applied Science, Mannheim, Germany) digestion technique. The dermal fraction was cultured at 37°C and 7% CO_2 (in air) in six-well plates containing Dulbecco's modified Eagle's medium (Invitrogen, Eggenstein, Germany) and supplemented with 10% fetal calf serum (Invitrogen) and penicillin and streptomycin (50 μg/mL, Invitrogen). After 5–6 weeks of incubation, confluent fibroblasts were seeded into the appropriate flasks. As required by the experimental design, fibroblasts were incubated for 7 days with 200 μM glyoxal[14,24] or methylglyoxal.

Immunoblotting Analysis

Immunoblotting was performed following standard procedures using chemiluminescence methods. Briefly, proteins were blotted on nitrocellulose or polyvinylidene difluoride sheets and visualized with the Lumilight Plus Western blot detection kit (Roche Applied Sciences). Imaging was performed using LUMI-Imager (Boehringer, Mannheim, Germany). Signals were quantified using LumiAnalyst (Roche Applied Sciences). The CML antibody used was clone 6D12.[25] CEL,[22] pentosidine,[20] and pyrraline[21] were purchased from TransGenic (Kumamoto, Japan, 1:200 dilution). Antibodies detecting vimentin (sc-32322, dilution 1:100,000) were purchased from Santa Cruz Biotechnology (Santa Cruz, CA). Secondary antibodies labeled with peroxidase were purchased from Sigma (St. Louis, MO).

Cell Monolayer Immunohistochemistry

Cells were fixed in 3% paraformaldehyde for 30 min at room temperature, washed with phosphate-buffered saline solution (PBS), and permeabilized with 0.5% Triton X-100 for 5 min, after which the cells were blocked with 3% bovine serum albumin (BSA) for 30 min. Cells were then incubated with primary antibody in 1% BSA for 1 h. Antibodies detecting

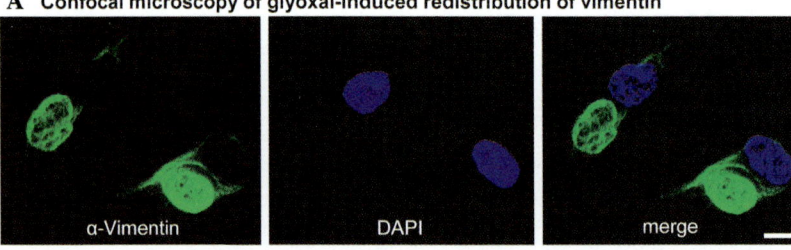

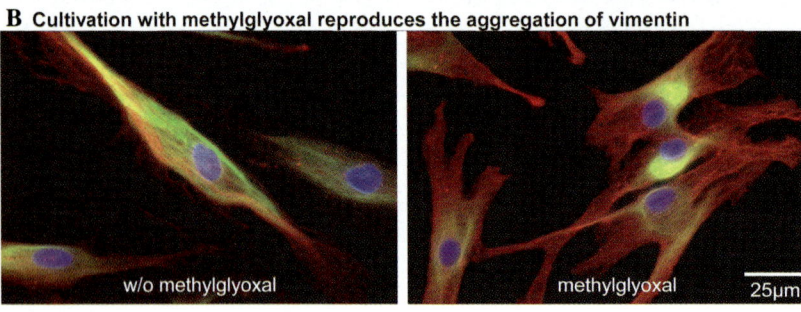

FIGURE 2. Glyoxal and methylglyoxal induce redistribution of vimentin. (**A**) The upper lane shows confocal images of glyoxal-cultured fibroblasts stained for vimentin (green) and DAPI (blue), bar = 31.5 μm. (**B**) Primary human skin fibroblasts were cultured with methylglyoxal (*right*) and stained for vimentin (green) and actin (red). The nucleus was visualized by DAPI-staining (blue). The control fibroblasts were cultured in medium without (w/o) methylglyoxal (*left*).

vimentin (sc-32322, dilution 1:200) and actin (sc-1615, dilution 1:100) were purchased from Santa Cruz Biotechnology. After washing the cells with 0.05% Nonidet P-40 and PBS they were incubated for 1 h with 1% BSA solution containing fluorescence-labeled secondary antibodies (AlexaFluor 488 chicken anti-mouse IgG, AlexaFluor 594 donkey anti-goat IgG, Molecular Probes, Eugene, OR) at 1:1000 to visualize the target protein and 4′,6-diamidin-2′-phenylindol (DAPI) (Molecular Probes, dilution 1:1000). After extensive washing with PBS, fluorescence images were recorded on a fluorescence microscope (Olympus, Hamburg, Germany) with an attached closed-circuit display camera. Confocal fluorescence images were recorded on an LSM 510 Meta (Carl Zeiss, Jena, Germany).

Flow Cytometry

Fibroblasts were trypsinized, washed twice with PBS, and subsequently fixed in 3% paraformaldehyde for 30 min at room temperature. After two washing steps, the cells were permeabilized with 0.5% Triton X-100 (Sigma) in PBS, washed again, and blocked with 3% BSA (Roth, Karlsruhe, Germany) in PBS for 30 min at room temperature. Fluorescein isothiocyanate-labeled antibody detecting CML[25] (Transgenic Inc., Kumamoto, Japan, 1:100 dilution) was added in 1% BSA and incubated for 1 h at room temperature. After three washing steps with PBS, the fluorescence was measured using FACSCanto (BD Biosciences, San Jose, CA) and analyzed by the software BD FACSDiva 4.0.

Two-dimensional Electrophoresis

One-hundred microgram protein was precipitated from cell lysates using ice cold trichloroacetic acid. The pellet was resuspended in rehydration buffer (8 M urea, 2% 3-[(3-cholamiopropyl)dimethylammonio]-1-propanesulfonate), 0.002% bromphenol blue, 18 mM dithiothreitol), supplemented with 0.2% Bio-Lyte (Bio-Rad, Hercules, CA). The pH 3–10 IPG-strips were covered with this solution and silicon oil. The rehydration was performed using PROTEAN_IEF-Cell (50 V, 12 h) followed by the isoelectric focusing (IEF). The immobilized pH gradient (IPG)-strip was equilibrated for 15 min in equilibration buffer (50 mM Tris-HCl, 6 M urea, 30% glycerol, 2% sodium dodecyl sulfate (SDS), 0.002% bromphenol blue) supplemented with 65 mM dithiothreitol followed by 15 min incubation in equilibration buffer with 135 mM iodoacetamide. The separation was performed in 10% polyacrylamide gels using a standard electrophoresis buffer (250 mM Tris-HCL, 1.9 M glycine, 1% SDS).

Statistical Analysis

Data were analyzed statistically using Statistica 7.1 (StatSoft, Tulsa, OK). Normality was tested

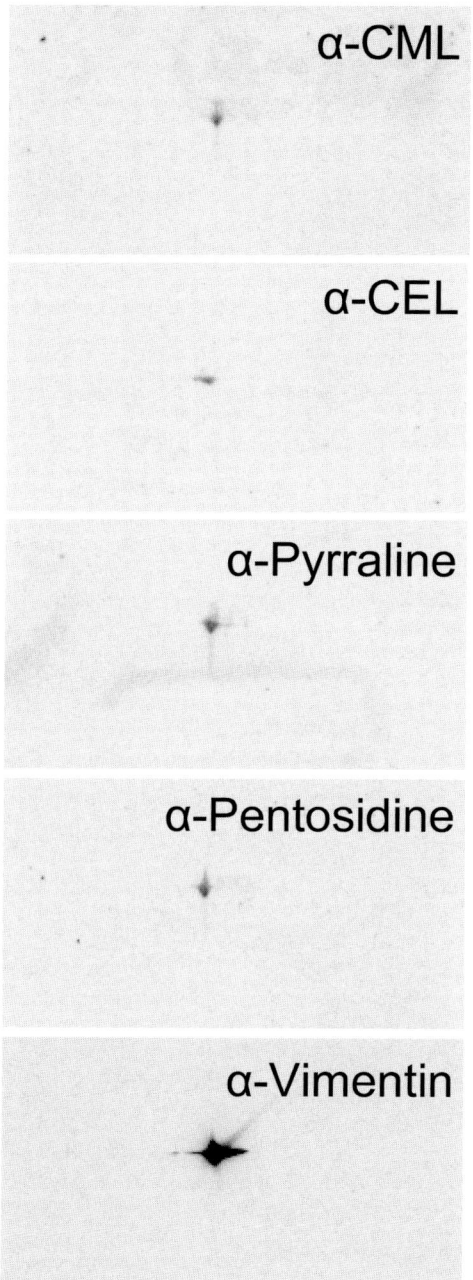

FIGURE 3. Vimentin is the target of various AGEs. Lysates of primary human fibroblasts were applied to 2-dimensional electrophoresis (pH 3–10) followed by Western blot analyses with antibodies specific for different AGEs and the filament protein vimentin.

performing the Lilliefors-test. Statistical significance was determined using the two-tailed Student's unpaired t-test.

Discussion

This study clearly shows the modification of the intermediate filament protein vimentin by several AGEs. These include not only CML and CEL, but also pyrraline and pentosidine, highlighting the physiological relevance of vimentin glycation. Furthermore, the incubation of human dermal fibroblasts with methylglyoxal leads to the same structural breakdown and reorganization of vimentin as observed for glycation with glyoxal in a previous study.[19] The concentration of methylglyoxal is approximately 1 μM in human blood, but it can increase up to 5 μM in diabetic patients.[26] Methylglyoxal is detoxified by the glyoxalase system[27] of the human body, and glyoxalase is upregulated in tumors,[28] which show a higher glycolytic rate in general.[29] This is interesting in that we were able to show that the intermediate filament protein vimentin— a protein expressed by various tumors[30,31]— is affected by CML-induced rearrangement in nonpathogenic skin.[19] Glyoxalase might therefore protect cancer cells from methylglyoxal-induced aggregation of vimentin, preventing the loss of carcinoma cell migration and adhesion.[31] In congruence with the glucose-mediated increase in methylglyoxal in diabetes, we were able to show that cultivation of primary human fibroblasts with high levels of glucose enhances the intracellular amount of glyoxal-derived CML modifications. Notably, the accumulation of CML in human skin during normal aging[3–8] might result partly from ineffective glucose use in old skin.[32] The finding that glyoxal and methylglyoxal induce a rearrangement of vimentin in cultured fibroblasts raises the question whether diabetic skin—a system with impaired glucose metabolism—shows a higher number of vimentin aggregates compared with nonpathogenic skin. This question becomes even more important considering our finding of aggregated vimentin in sections of nonpathogenic skin *in vivo*.[19] Further studies are needed to investigate this issue.

Acknowledgments

We thank Dr. Ulrich Hahn for helpful suggestions regarding the realization of this study. Katja Farhat provided technical help concerning flow cytometry.

Conflict of Interest

The authors declare no conflicts of interest.

References

1. JEANMAIRE, C. et al. 2001. Glycation during human dermal intrinsic and actinic ageing: an in vivo and in vitro model study. Br. J. Dermatol. **145:** 10–18.
2. BUCCIARELLI, L.G. et al. 2002. RAGE is a multiligand receptor of the immunoglobulin superfamily: implications for homeostasis and chronic disease. Cell. Mol. Life Sci. **59:** 1117–1128.
3. VLASSARA, H. 2005. Advanced glycation in health and disease: role of the modern environment. Ann N.Y. Acad Sci. **1043:** 452–460.
4. CLOOS, P.A. & S. CHRISTGAU. 2002. Non-enzymatic covalent modifications of proteins: mechanisms, physiological consequences and clinical applications. Matrix Biol. **21:** 39–52.
5. SCHMIDT, A.M. et al. 1995. The dark side of glucose. Nat. Med. **1:** 1002–1004.
6. DUNN, J.A. et al. 1991. Age-dependent accumulation of N epsilon-(carboxymethyl)lysine and N epsilon-(carboxymethyl)hydroxylysine in human skin collagen. Biochemistry **30:** 1205–1210.
7. DUNN, J.A. et al. 1989. Oxidation of glycated proteins: age-dependent accumulation of N epsilon-(carboxymethyl) lysine in lens proteins. Biochemistry **28:** 9464–9468.
8. BAYNES, J.W. 2001. The role of AGEs in aging: causation or correlation. Exp. Gerontol. **36:** 1527–1537.
9. THORNALLEY, P.J. 2006. Advanced glycation end products in renal failure. J. Ren. Nutr. **16:** 178–184.
10. KOYAMA, Y. et al. 2007. High serum level of pentosidine, an advanced glycation end product (AGE), is a risk factor of patients with heart failure. J. Card Fail. **13:** 199–206.
11. NAGAI, R. et al. 2007. Succination of protein thiols during adipocyte maturation - a biomarker of mitochondrial stres. J. Biol. Chem **282:** 34219–34228.
12. MONNIER, V.M. et al. 2005. Glycation products as markers and predictors of the progression of diabetic complications. Ann. N.Y. Acad. Sci. **1043:** 567–581.
13. GENUTH, S. et al. 2005. Glycation and carboxymethyllysine levels in skin collagen predict the risk of future 10-year progression of diabetic retinopathy and nephropathy in the diabetes control and complications trial and epidemiology of diabetes interventions and complications participants with type 1 diabetes. Diabetes **54:** 3103–3111.
14. KASPER, M. & R.H. FUNK. 2001. Age-related changes in cells and tissues due to advanced glycation end products (AGEs). Arch. Gerontol. Geriatr. **32:** 233–243.
15. DEGENHARDT, T.P. et al. 1998. Chemical modification of proteins by methylglyoxal. Cell. Mol. Biol. (Noisy-le-grand) **44:** 1139–1145.
16. SHINOHARA, M. et al. 1998. Overexpression of glyoxalase-I in bovine endothelial cells inhibits intracellular advanced glycation endproduct formation and prevents hyperglycemia-induced increases in macromolecular endocytosis. J. Clin. Invest. **101:** 1142–1147.
17. THORNALLEY, P.J. 2007. Endogenous alpha-oxoaldehydes and formation of protein and nucleotide advanced glycation endproducts in tissue damage. Novartis Found Symp. **285:** 229–243; discussion 243–246.
18. AGALOU, S. et al. 2005. Advanced glycation end product free adducts are cleared by dialysis. Ann. N.Y. Acad. Sci. **1043:** 734–739.
19. KUEPER, T. et al. 2007. Vimentin is the specific target in skin glycation. Structural prerequisites, functional consequences, and role in skin aging. J. Biol. Chem. **282:** 23427–23436.
20. MIYATA, T. et al. 1996. Identification of pentosidine as a native structure for advanced glycation end products in beta-2-microglobulin-containing amyloid fibrils in patients with dialysis-related amyloidosis. Proc. Natl. Acad. Sci. USA **93:** 2353–2358.
21. MIYATA, S. & V. MONNIER. 1992. Immunohistochemical detection of advanced glycosylation end products in diabetic tissues using monoclonal antibody to pyrraline. J. Clin. Invest. **89:** 1102–1112.
22. AHMED, M.U. et al. 1997. N-epsilon-(carboxyethyl)lysine, a product of the chemical modification of proteins by methylglyoxal, increases with age in human lens proteins. Biochem. J. **324:** 565–570.
23. DRABEROVA, E. et al. 1986. A common antigenic determinant of vimentin and desmin defined by monoclonal antibody. Folia Biol. (Praha) **32:** 295–303.
24. CERVANTES-LAUREAN, D. et al. 2005. Nuclear proteasome activation and degradation of carboxymethylated histones in human keratinocytes following glyoxal treatment. Free Radic Biol. Med. **38:** 786–795.
25. IKEDA, K. et al. 1996. N (epsilon)-(carboxymethyl)lysine protein adduct is a major immunological epitope in proteins modified with advanced glycation end products of the Maillard reaction. Biochemistry **35:** 8075–8083.
26. CHAN, W.H. & H.J. WU. 2007. Methylglyoxal and high glucose co-treatment induces apoptosis or necrosis in human umbilical vein endothelial cells. J. Cell Biochem. In press.
27. THORNALLEY, P.J. 1990. The glyoxalase system: new developments towards functional characterization of a metabolic pathway fundamental to biological life. Biochem. J. **269:** 1–11.
28. ANTOGNELLI, C. et al. 2006. Overexpression of glyoxalase system enzymes in human kidney tumor. Cancer J. **12:** 222–228.
29. VAN HEIJST, J.W. et al. 2006. Argpyrimidine-modified Heat shock protein 27 in human non-small cell lung cancer: a possible mechanism for evasion of apoptosis. Cancer Lett. **241:** 309–319.
30. WU, M. et al. 2007. Proteome analysis of human androgen-independent prostate cancer cell lines: variable metastatic potentials correlated with vimentin expression. Proteomics **7:** 1973–1983.
31. MCINROY, L. & A. MAATTA. 2007. Down-regulation of vimentin expression inhibits carcinoma cell migration and adhesion. Biochem. Biophys. Res. Commun. **360:** 109–114.
32. LENZ, H. et al. 2005. The creatine kinase system in human skin: protective effects of creatine against oxidative and UV damage in vitro and in vivo. J. Invest. Dermatol. **124:** 443–452.

Erratum for Ann. N.Y. Acad. Sci. 1072: 386–388

Inflammatory Bowel Disease: Genetics, Barrier Function, Immunologic Mechanisms, and Microbial Pathways. Wolfram W. Domschke, Martin F. Kagnoff, Torsten F. Kucharzik, Lloyd F. Mayer, and Stephan R. Targan, Eds.

In the chapter cited in Reference 1, A. Di Sabatino's name was misspelled. Please see the list of authors below for the correct spelling.

SYLVIA L.F. PENDER,[a] C.K.F. LI,[b] A. DI SABATINO,[c] T.T. MACDONALD,[c] AND M.G. BUCKLEY[a]

We apologize for this error.

Reference

1. PENDER, S.L.F., C.K.F. LI, A.D.I. SABATINO, T.T. MACDONALD, and M.G. BUCKLEY. 2006. Role of macrophage metalloelastase in gut inflammation. Ann. N.Y. Acad. Sci. **1072:** 386–388.

Erratum for Ann. N.Y. Acad. Sci. 1088: A1–A10

Neuroendocrine and Immune Crosstalk. George P. Chrousos, Gregory A. Kaltsas, and George Mastorakos, Eds.

The chapter cited below in Reference 1 was omitted from the print version. The online version can be found at http://www.annalsnyas.org/cgi/content/full/1088/1/A1, and the print version is included here in the following pages.

We apologize for this error.

Reference

1. MORIKAWA, Y., T. HISAOKA, T. KITAMURA & E. SENBA. 2006. TROY, a novel member of the tumor necrosis factor receptor superfamily in the central nervous system. Ann. N.Y. Acad. Sci. **1088**: A1–A10.

TROY, a Novel Member of the Tumor Necrosis Factor Receptor Superfamily in the Central Nervous System

YOSHIHIRO MORIKAWA,[a] TOMOKO HISAOKA,[a] TOSHIO KITAMURA,[b] AND EMIKO SENBA[a]

[a]*Department of Anatomy and Neurobiology, Wakayama Medical University, Wakayama 641-8509, Japan*

[b]*Division of Cellular Therapy, Advanced Clinical Research Center, The Institute of Medical Science, The University of Tokyo, Minato-Ku, Tokyo 108-8639, Japan*

ABSTRACT: Using a signal sequence trap method, we isolated TROY, a novel member of the tumor necrosis factor receptor superfamily (TN-FRSF), from a mouse brain cDNA library. TROY mRNA is strongly expressed in brain and embryo. *In situ* hybridization analysis of the embryo showed that TROY mRNA was exclusively expressed in the epithelium of many tissues, including neuroepithelium. In the developing central nervous system, TROY mRNA was strongly expressed in the ventricular and subventricular zones, which contain neuronal and glial precursors during mouse embryogenesis that are both region-specific and stage-dependent. In addition, TROY mRNA was expressed in the developing olfactory bulb from embryonic day (E) 13.5 to neonate. Next, we focused on the detailed cellular characterization of TROY-expressing cells in the developing olfactory system. TROY mRNA was first detected in the olfactory nerve layer (ONL) of the olfactory bulb at E13.5 and was expressed most intensely in the inner ONL (ONL-i) during late embryogenesis. In the postnatal olfactory bulb, TROY-expressing cells were also detected in the glomerular layer (GL) and ONL-i. TROY was intensely expressed in olfactory ensheathing cells (OECs) of the ONL-i, which are positive for neuropeptide Y (NPY), but negative for S-100 or p75 low-affinity nerve growth factor receptor. Furthermore, TROY was also detected in glial fibrillary acidic protein (GFAP)-positive glial cells of the ONL-i and GL. Thus, TROY was expressed in some specific subsets of glial cells in the olfactory bulb, including OECs, and may play some roles in the developing and adult olfactory system.

KEYWORDS: tumor necrosis factor receptor superfamily; development; neuroepithelium; olfactory ensheathing cells; subventricular zone; mouse TROY

Address for correspondence: Yoshihiro Morikawa, Department of Anatomy and Neurobiology, Wakayama Medical University, 811-1, Kimiidera, Wakayama 641-8509, Japan. Voice/fax: +81-73-441-0617.

yoshim@wakayama-med.ac.jp

ISOLATION OF THE MOUSE TROY, A NEWLY IDENTIFIED MEMBER OF THE TNFR SUPERFAMILY

Tumor necrosis factor (TNF)-related cytokines form a large family of pleiotropic mediators of host defense, inflammation, apoptosis, autoimmunity, and organogenesis that act either locally as membrane proteins or on distant target cells as secreted proteins. Members of the tumor necrosis factor receptor superfamily (TNFRSF) mediate such actions of TNF-related cytokines.[1]

To identify the genes encoding proteins with signal sequence from mouse brain, we performed the signal sequence trap by retrovirus-mediated expression screening (SST-REX). In the SST-REX, Ba/F3 clones transduced with cDNAs containing signal sequences showed factor-independent growth through surface expression of a constitutively active receptor for thrombopoietin as a fusion protein.[2] One of the cDNA clones isolated by SST-REX from a mouse brain cDNA library showed homology with members of the TNFRSF at the amino acid sequence level.[3] We designated the gene as TROY (TNFRSF expressed on the mouse embryo), because mRNA of the gene was strongly expressed in mouse embryo. Full-length cDNA of TROY contains an open reading frame of 1251 nucleotides, which encodes a protein of 416 amino acids (FIG. 1A). Two hydrophobic regions are present in the protein, representing the signal sequence and the transmembrane region. Like other members of the TNFRSF, TROY contains the characteristic cysteine-rich motifs (FIG. 1A). TROY exhibited a significant and extensive homology (33%) with all three motifs of Edar, the *dl* gene product.[4,5] The cytoplasmic domain of TROY spans amino acids 194–416 of the precursor protein and contains a major TNFR-associated factor (TRAF) 2-binding consensus sequence, TLQE (amino acids 276–279). However, TROY does not have the death domain, which is contained in the cytoplasmic domains of CD95, TNFR, TNF-related apoptosis-inducing ligand receptor, and Edar.[1] A shorter clone was also identified from a cDNA library of mouse embryonic day (E) 17.5 skin, which was designated as dTROY. Although the extracellular and transmembrane domains of dTROY were identical to those of TROY, its cytoplasmic domain has only 21 amino acids and does not contain a TRAF2-binding consensus sequence (FIG. 1B). Transient transfection-based overexpression of TROY revealed TROY induced activation of nuclear factor κB, which was inhibited by dominant negative forms of TRAF2, TRAF5, and TRAF6.[3] In addition, Eby *et al.* have reported that TROY activated the c-Jun N-terminal kinase pathway.[6] To confirm that dTROY was not an artifact, a dTROY-specific reverse transcription-polymerase chain reaction was performed using a sense primer of the common sequence and an antisense primer of the unique sequence of dTROY. A band of dTROY was amplified from the total RNA of E17.5 skin but not from mouse brain or liver.

A human counterpart of TROY was obtained from a cDNA library derived from the human gliosarcoma cell line GI-1. This cDNA contains an open

A

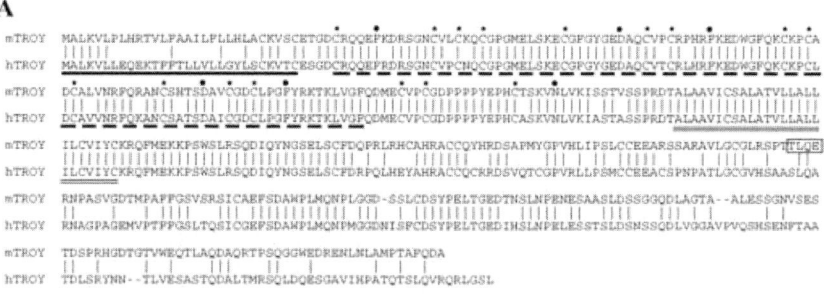

B

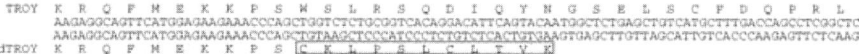

FIGURE 1. (**A**) Comparison between the protein sequence of mouse and human TROY. The signal sequence is *underlined* and transmembrane domain is *double-underlined*. *Vertical lines* indicate common amino acid residues. Two perfect TNFR cysteine-rich motifs are *broken-underlined*, conserved cysteine residues and conserved amino acid residues are indicated by the *asterisks* and *dots*, respectively. The *boxed region* indicates the TRAF2-binding domain. (**B**) Structure of dTROY, a putative decoy receptor. The region of dTROY that is structurally different from TROY is *boxed*.

reading frame of 1269 nucleotides and encodes a cysteine-rich protein of 423 amino acids with a calculated molecular mass of 46 kDa. The overall identity between human TROY and mouse TROY at the amino acid level is 75%, with a 92% identity in the extracellular and transmembrane domain. The cytoplasmic tail of human TROY was 234 amino acids long and had a 57% homology with mouse TROY (FIG. 1A).

Expression of TROY mRNA was examined by Northern blot analysis (FIG. 2A). TROY mRNA was strongly expressed in the brain, with an approximate molecular size of 4.5 kb. TROY mRNA was detected in most tissues, but not in the spleen. In the embryo, the expression level was periodically increased and was particularly strong in the skin (data not shown). TROY mRNA was also detected in human glioma U251 cells, GI-1 cells, A3–1 embryonic stem cells, and nullipotent embryonal carcinoma NF-1 cells (data not shown).

In situ hybridization in mouse E13.5 embryo revealed that TROY mRNA was detected exclusively in the epithelium (i.e., neuroepithelium in the frontal and lateral lobes, the epidermis of the skin, bronchiolar epithelium, epithelium of the tongue, gastric epithelium, conjunctiva, and cochlea). In neonatal mice, however, TROY mRNA was mainly detected in hair follicles like Edar and in neural cells in the cerebrum but not in the epidermis of the skin (FIG. 2B).

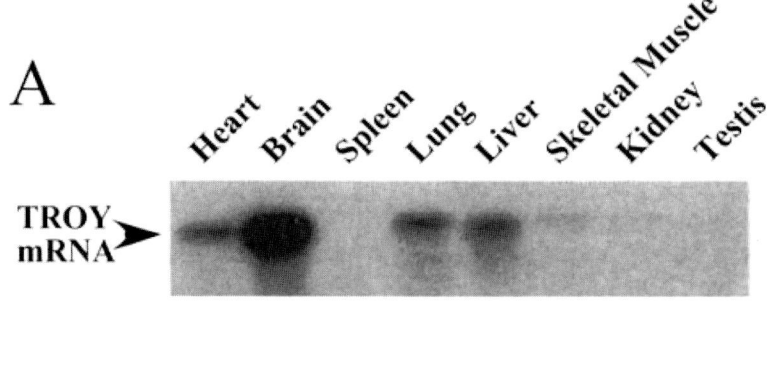
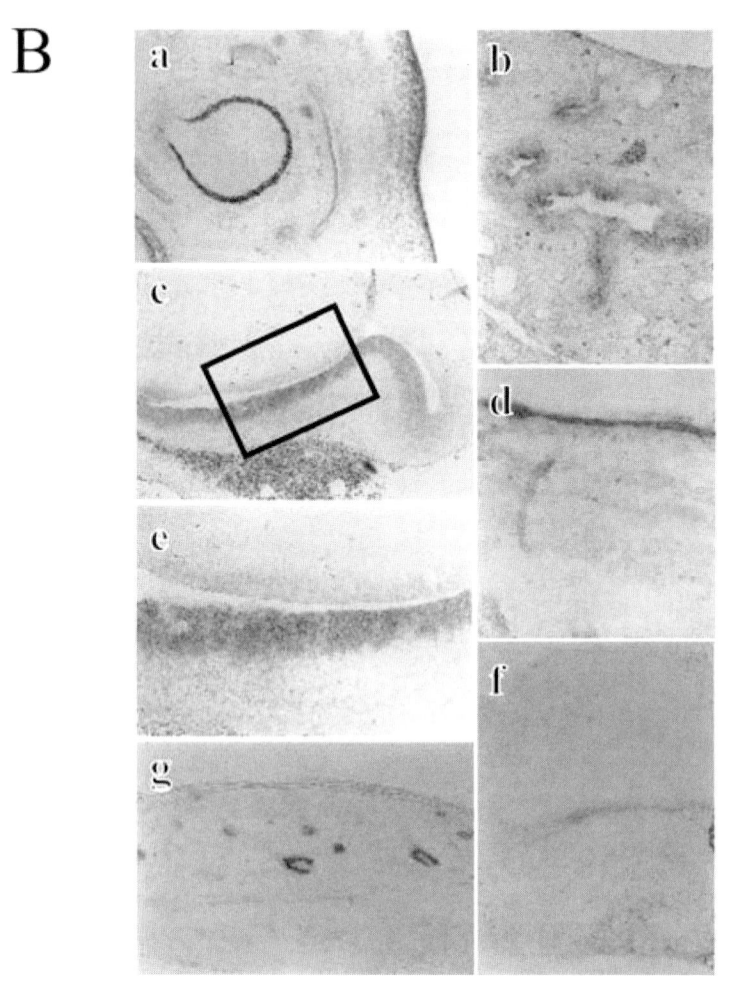

EXPRESSION OF TROY IN THE DEVELOPING CENTRAL NERVOUS SYSTEM (CNS)

To examine the expression pattern of TROY in the developing CNS, we performed *in situ* hybridization histochemistry.[7] At E11.5, TROY mRNA was highly expressed in neuroepithelia, from which the neocortex, the thalamus, the hypothalamus, and the amygdala develop (FIG. 3A, B). However, no expression of TROY mRNA was detected in the striatal neuroepithelium at E11.5 and in later stages. At E13.5, strong expression of TROY was detected in the anterior thalamic neuroepithelium, the mammillary body neuroepithelium, the inferior colliculus neuroepithelium, and the precerebellar neuroepithelium. The level of its expression in the ventricular zone (VZ) of the neocortical neuroepithelium and the epithalamic neuroepithelium was greatly decreased at E13.5.

In the hippocampal neuroepithelium, TROY mRNA was first detected at E13.5. At E15.5, its strong expression was seen in the VZ and the dentate subventricular zone (SVZ) of the hippocampus (FIG. 3C). At E17.5, its expression in the dentate SVZ, the migratory stream, and the dentate gyrus was more distinct than that in the VZ. After birth, although expression level declined, TROY mRNA was detected in some cells in the CA1 and the CA3 as well as in the dentate gyrus (FIG. 3D). In the SVZ, expression of TROY mRNA was weak at E17.5 and became more intense in the dorsolateral corner of the SVZ at the neonatal stage (FIG. 3E).

Thus, TROY mRNA was highly expressed in the VZ of various regions, including the neocortex, the thalamus, and the hippocampus at early stages of neurogenesis and was downregulated by the mid to late stages of neurogenesis. From this expression pattern, we speculate that TROY may be involved in the regulation of precursor proliferation or the maintenance of their undifferentiated state.

FIGURE 2. (**A**) Northern blot analysis for detection of TROY transcripts. Blotted membranes were probed with a [^{32}P]dCTP-labeled TROY cDNA. (**B**) *In situ* hybridization of TROY. Sections of an embryo at E13.5 (*a*, oculus; *b*, lung; *c* & *e*, frontal lobe; *d*, skin) and neonate (*f*, cerebrum; *g*, skin) were incubated with digoxigenin-labeled cRNA probe and then stained with an alkaline phosphatase-conjugated anti-digoxigenin Fab fragment, nitro blue tetrazolium, 5-bromo-4-chloro-3-indolyl phosphate, and levamisole. Panel *e* shows a higher magnification of the *boxed region* in *c*. TROY mRNA is seen as blue signals in the epithelium (i.e., conjunctiva (*a*), bronchiolar epithelium (*b*), neuroepithelium (*c*, *e*), and epidermis of the skin (*d*)); whereas in neonatal mice, TROY mRNA is detected in the brain (*f*) and in hair follicles (*g*) but not in the epidermis of the skin (*g*). (In color in Annals online.)

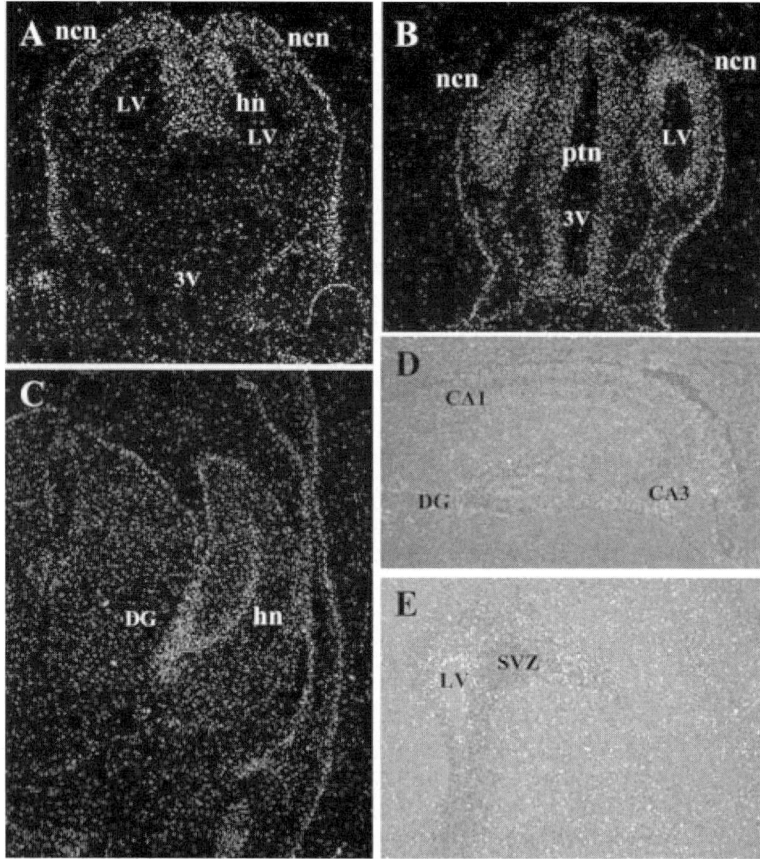

FIGURE 3. Expression of TROY mRNA in the brain of embryonic and neonatal mice. (A–C) Coronal sections of the embryonic brain at E11.5 (A, B) and E15.5 (C) are shown in dark-field view. The hybridization signals are seen as *white dots*. TROY mRNA was expressed in the neuroepithelium of the various regions. (D, E) Coronal sections of the neonatal forebrain are shown in half-bright-field view. Positive signals were observed in the hippocampus (D) and the SVZ (E). 3V = 3rd ventricle; CA1, 3, hippocampal area CA1, 3 (Ammon's Horn); DG = dentate gyrus; hn = hippocampal neuroepithelium; LV = lateral ventricle; ncn = neocortical neuroepithelium; ptn = posterior thalamic neuroepithelium; SVZ = subventricular zone.

CHARACTERIZATION OF TROY-EXPRESSING CELLS IN THE DEVELOPING AND ADULT OLFACTORY SYSTEM

In addition to the expression of TROY in the VZ and SVZ, TROY mRNA was expressed in the developing olfactory bulb from E13.5 to neonate. Therefore, we focused on the characterization of TROY-expressing cells in the olfactory system.[8]

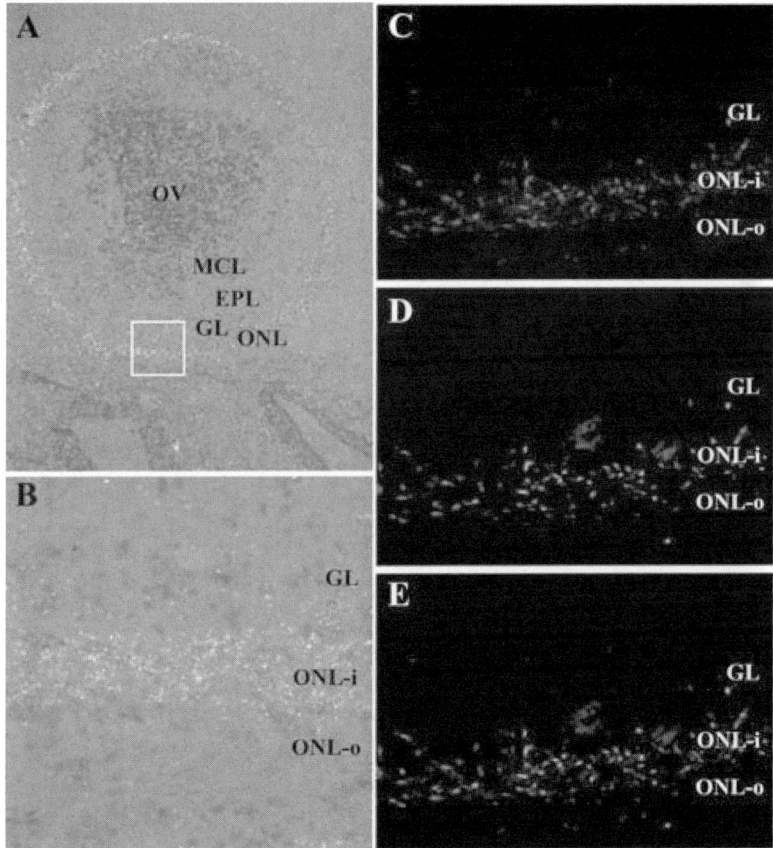

FIGURE 4. Expression of TROY in the ONL of the neonatal olfactory bulb. **(A, B)** *In situ* hybridization of TROY revealed that TROY mRNA was expressed in the ONL. At high magnification of the *boxed region* in A, intense hybridization signals were observed in the ONL-i, but not in the ONL-o. **(C, D)** Double-immunofluorescent staining for TROY (C, red) and NPY (D, green) in the ONL-i. The cells of colocalization appear yellow in the merged images **(E)**. EPL = external plexiform layer; MCL = mitral cell layer; GL = glomerular layer; ONL = olfactory nerve layer; ONL-i = inner olfactory nerve layer; ONL-o = outer olfactory nerve layer; OV = olfactory ventricle. (In color in Annals online.)

At E13.5, TROY was detected in the olfactory nerve layer (ONL) and in tissues surrounding the olfactory ventricle, whereas no signals were found in the olfactory epithelium. At E15.5, the expression of TROY became more distinct in the ONL. From E17.5 to adulthood, ONL forms two distinct layers, the inner ONL (ONL-i) and the outer ONL (ONL-o).[9–11] At E17.5 and P0, TROY was clearly detected in the ONL-i, but not in the ONL-o. From P7 to adulthood, TROY was localized in the glomerular layer (GL) in addition to the ONL-i (FIG. 4A, B).

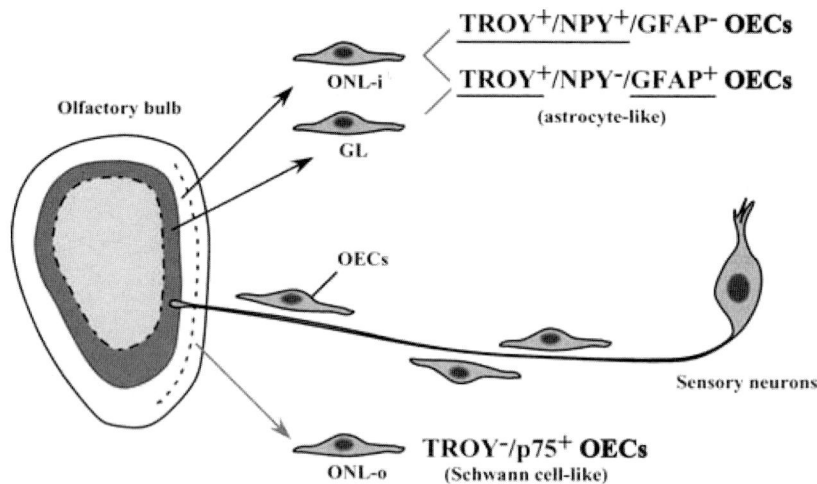

FIGURE 5. Summary of antigenic profiles of TROY-expressing cells in the olfactory bulb. GL = glomerular layer; OECs = olfactory ensheathing cells; ONL = olfactory nerve layer; ONL-i = inner olfactory nerve layer; ONL-o = outer olfactory nerve layer.

For further characterization of TROY-expressing cells in the ONL, we performed double-immunofluorescent staining using some glial markers, including S100, p75 low-affinity nerve growth factor receptor (p75), and glial fibrillary acidic protein (GFAP). S100 is useful as a marker for the olfactory ensheathing cells (OECs) of the ONL-o,[9,10,12] and p75 is positive in the OECs of the ONL-o, the glomeruli, and glial cells of the GL from late embryonic stages to adult stages.[10,13] Although GFAP is expressed in the astrocytes of the ONL-i and the GL during postnatal stages,[11,14] it has been reported that GFAP is also contained in OECs of the ONL.[15,16] TROY was highly observed in the ONL-i and was colocalized with GFAP in the GL and the ONL-i and with p75 in some glial cells of the GL of adult mice. However, no colocalization of TROY with S100 was observed in the ONL of adult mice nor in E17.5 embryo. These findings indicated that TROY is not expressed in the OECs of the ONL-o, but is expressed in some glial cells of the GL.

Recently, it has been reported that neuropeptide Y (NPY) is localized in the OECs of the ONL-i from embryonic stages to adulthood.[11,12,17] As shown in FIGURE 4C–E, colocalization of TROY and NPY was observed in the ONL-i of embryonic and adult mice, indicating that TROY-expressing cells in the ONL-i were NPY-positive OECs.

Furthermore, we used the triple-immunofluorescence method to examine whether TROY in the ONL-i colocalizes with both GFAP and NPY, and it revealed that TROY colocalized with GFAP or NPY, but no triple-positive cells were observed in the ONL-i. Thus, two populations of TROY-expressing cells exist in the ONL-i; $NPY^-/GFAP^+$ and $NPY^+/GFAP^-$ cells (FIG. 5). In addition,

TROY-expressing cells in the GL were negative for tyrosine hydroxylase, γ-aminobutyric acid, and calbindin, indicating that TROY was not expressed in the periglomerular neurons.

CONCLUSIONS AND PERSPECTIVES

We identified TROY, a novel member of the TNFRSF, which was strongly expressed in the neuroepithelia of various regions of the developing CNS. In the ONL-i of the olfactory bulb, TROY was expressed not in neurons, but in two types of supporting cells, $NPY^+/GFAP^-$ and $NPY^-/GFAP^+$ (astrocyte-like) OECs. These results suggest that TROY plays some role in the formation and maintenance of the olfactory system. In addition, TROY was expressed in the SVZ of adult mouse brain. Further characterization of TROY-expressing cells in the adult SVZ revealed that subpopulations of uncommitted precursor cells expressed TROY.[18] Thus, TROY signaling may be involved in maintaining an undifferentiated state of precursor cells. More extensive analysis is required to reveal the roles of TROY signaling in the SVZ.

ACKNOWLEDGMENTS

This work was supported by a Grant-in-Aid for Scientific Research (B) from The Ministry of Education, Culture, Sports, Science and Technology (16300113) and a Research Grant on Priority Areas from Wakayama Medical University.

REFERENCES

1. LOCKSLEY, R.M., N. KILLEEN & M.J. LENARDO. 2001. The TNF and TNF receptor superfamilies: integrating mammalian biology. Cell **104:** 487–501.
2. KOJIMA, T. & T. KITAMURA. 1999. A signal sequence trap based on a constitutively active cytokine receptor. Nat. Biotechnol. **17:** 487–490.
3. KOJIMA, T., Y. MORIKAWA, N.G. COPELAND, *et al.* 2000. TROY, a newly identified member of the tumor necrosis factor receptor superfamily, exhibits a homology with Edar and is expressed in embryonic skin and hair follicles. J. Biol. Chem. **275:** 20742–20747.
4. HEADON, D.J. & P.A. OVERBEEK. 1999. Involvement of a novel Tnf receptor homologue in hair follicle induction. Nat. Genet. **22:** 370–374.
5. MONREAL, A.W., B.M. FERGUSON, D.J. HEADON, *et al.* 1999. Mutations in the human homologue of mouse dl cause autosomal recessive and dominant hypohidrotic ectodermal dysplasia. Nat. Genet. **22:** 366–369.
6. EBY, M.T., A. JASMIN, A. KUMAR, *et al.* 2000. TAJ, a novel member of the tumor necrosis factor receptor family, activates the c-Jun N-terminal kinase pathway and mediates caspase-independent cell death. J. Biol. Chem. **275:** 15336–15342.

7. HISAOKA, T., Y. MORIKAWA, T. KITAMURA, et al. 2003. Expression of a member of tumor necrosis factor receptor superfamily, TROY, in the developing mouse brain. Dev. Brain Res. **143:** 105–109.
8. HISAOKA, T., Y. MORIKAWA, T. KITAMURA, et al. 2004. Expression of a member of tumor necrosis factor receptor superfamily, TROY, in the developing olfactory system. Glia **45:** 313–324.
9. ASTIC, L., V. PELLIER-MONNIN & F. GODINOT. 1998. Spatio-temporal patterns of ensheathing cell differentiation in the rat olfactory system during development. Neuroscience **84:** 295–307.
10. BAILEY, M.S., A.C. PUCHE & M.T. SHIPLEY. 1999. Development of the olfactory bulb: evidence for glia-neuron interactions in glomerular formation. J. Comp. Neurol. **415:** 423–448.
11. AU, W.W., H.B. TRELOAR & C.A. GREER. 2002. Sublaminar organization of the mouse olfactory bulb nerve layer. J. Comp. Neurol. **446:** 68–80.
12. UBINK, R. & T. HÖKFELT. 2000. Expression of neuropeptide Y in olfactory ensheathing cells during prenatal development. J. Comp. Neurol. **423:** 13–25.
13. GONG, Q., M.S. BAILEY, S.K. PIXLEY, et al. 1994. Localization and regulation of low affinity nerve growth factor receptor expression in the rat olfactory system during development and regeneration. J. Comp. Neurol. **344:** 336–348.
14. FRANCESCHINI, I.A. & S.C. BARNETT. 1996. Low-affinity NGF-receptor and E-N-CAM expression define two types of olfactory nerve ensheathing cells that share a common lineage. Dev. Biol. **173:** 327–343.
15. BARBER, P.C. & R.M. LINDSAY. 1982. Schwann cells of the olfactory nerves contain glial fibrillary acidic protein and resemble astrocytes. Neuroscience **7:** 3077–3090.
16. BAILEY, M.S. & M.T. SHIPLEY. 1993. Astrocyte subtypes in the rat olfactory bulb: morphological heterogeneity and differential laminar distribution. J. Comp. Neurol. **328:** 501–526.
17. UBINK, R., N. HALASZ, X. ZHANG, et al. 1994. Neuropeptide tyrosine is expressed in ensheathing cells around the olfactory nerves in the rat olfactory bulb. Neuroscience **60:** 709–726.
18. HISAOKA, T., Y. MORIKAWA & E. SENBA. 2006. Characterization of TROY/TNFRSF19/TAJ-expressing cells in the adult mouse forebrain. Brain Res. **1110:** 81–94.

Erratum for Ann. N.Y. Acad. Sci. 1108: 505–514
Autoimmunity, Part D: Autoimmune Disease, Annus Mirabilis. Yehuda Shoenfeld and M. Eric Gershwin, Eds.

In the chapter cited in Reference 1, the authors' last names (except for Nicoletta Di Simone) on page 505 were inadvertently placed first. They should have read as follows:

NICOLETTA DI SIMONE,[a] PIER LUIGI MERONI,[b] MARCO D'ASTA,[a] FIORELLA DI NICUOLO,[a] SILVIA D'IPPOLITO,[a] MARIA CLARA D'ALESSIO,[a] AND ALESSANDRO CARUSO[a]

We apologize for this error.

Reference

1. DI SIMONE, N. *et al.* 2007. Pregnancies complicated with antiphospholipid syndrome: the pathogenic mechanism of antiphospholipid antibodies. A review of the literature. Ann. N.Y. Acad. Sci. **1108:** 505–514.

Errata for Ann. N.Y. Acad. Sci. 1111: 442–454

Coccidioidomycosis: Sixth International Symposium. Karl V. Clemons, Rafael Laniado-Laborin, and David A. Stevens, Eds.

In the chapter cited in Reference 1, the following corrections need to be considered when reading the final version.

Table 3 lacks the footnote, "From Stevens.[2] Reprinted by permission."

Table 5 lacks a horizontal line or space before the last line (the line beginning with the $\mathcal{N}$); the last line does not show a percent.

Legends for Figures 4, 5, 6, and 7 lack a superscript "8" after "Tucker *et al.*"

The footnote to Table 7 lacks a superscript "14" after "Catanzaro *et al.*"

In Table 9, the "Fluconazole group" column lacks "(%)" after "n/n."

In Table 9, the wrong reference number is given in the footnote. It should be "18."

In the legend to Figure 11, the wrong reference number is given after "Kamberi *et al.*" It should be "20."

We apologize for these errors.

Reference

1. STEVENS, D. A. & K. C. CLEMONS. 2007. Azole therapy of clinical and experimental coccidioidomycosis. Ann. N.Y. Acad. Sci. **1111**: 442–454.

Index of Contributors

Adrover, M., 235–240
Ahmed, U., 262–264
Ait-Ameur, L., 173–176
Almy, J., 216–219
Ames, J.M., 20–24
Asami, J., 320–324
Ayala, V., 315–319

Bachmeier, B.E., 283–287
Banfi, C., 295–299
Barto, R., 231–234
Baumann, M., 201–204
Baumeyer, A., 113–117
Baynes, J.W., 272–275
Beattie, J.R., 59–65
Bellmunt, M.J., 315–319
Bhat, M., 107–112
Bierhaus, A., 42–45, 76–80
Birlouez-Aragon, I., 173–176, 177–180
Biswas, A., 107–112
Blank, I., 241–243
Blatnik, M., 272–275
Blatt, T., 328–332
Bonfante, L., 295–299
Boudier, H.S., 201–204
Breer, H., 1–6
Brioschi, M., 295–299
Brouwers, O., 231–234
Buchner, J., 257–261
Buetler, T.M., 113–117
Bütler, T., 300–306

Cacabelos, D., 315–319
Cai, W., 46–52
Capuano, E., 89–100
Cayzeele, A., 173–176
Cerny, C., 66–71
Chen, X.-M., 220–224
Chu, F.L., 30–37
Clynes, R., 7–13
Collard, F., 81–88
Cooper, M.E., 101–106

Cosma, C., 295–299
Coughlan, M.T., 190–193
Cristoni, S., 295–299

Dai, Z., 81–88
Dalfo, E., 315–319
Davidek, T., 241–243
de Revel, G., 216–219
Delatour, T., 113–117, 300–306
Baumeyer, A., 113–117
Devaud, S., 241–243
Dobler, D., 262–264
Donoso, J., 235–240

Fan, X., 194–200
Federici, A., 166–172
Feng, Y., 42–45
Ferrer, I., 315–319
Feskens, E.J., 162–165
Fiedler, T., 210–215
Fleming, T., 280–282
Fogliano, V., 89–100
Forbes, J.M., 101–106, 190–193
Förster, A., 300–306
Friess, U., 283–287
Frolov, A., 253–256
Fujiwara, Y., 38–41, 152–154, 155–157, 325–327

Gaens, K.H.J., 162–165
Gans, R., 42–45
Garbe, L.A., 244–247
Gawlowski, T., 276–279
Germanová, A., 268–271
Glenn, J.V., 59–65
Gouázec, E., 241–243
Götting, C., 276–279
Grune, T., 328–332

Hammes, H.-P., 42–45
Hartog, J.W.L., 225–230
Haslbeck, M., 257–261
Hayase, F., 53–58

Heemann, U., 201–204
Hegele, J., 300–306
Henle, T., 118–123, 248–252, 300–306
Herold, K., 7–13
Hidalgo, F.J., 25–29
Ho, C.-T., 72–75
Hoffmann, R., 253–256
Hofmann, T., 310–314
Humpert, P.M., 76–80

Ibusuki, D., 288–290, 291–294
Ikeda, T., 38–41, 152–154
Ilieva, E., 315–319
Inagi, R., 265–267
Ishii, A., 320–324
Ito, T., 53–58
Ivanov, I., 181–184
Iwao, Y., 38–41
Izuhara, Y., 141–146

Jáchymová, M., 268–271

Kalousová, M., 268–271
Kitts, D.D., 220–224
Kiyota, N., 152–154
Kleesiek, K., 276–279
Klenovicsová, K., 177–180
Koschinsky, T., 276–279
Krause, R., 300–306
Kroh, L.W., 210–215
Kueper, T., 328–332
Kumagai, T., 265–267

Lang, R., 310–314
Lapolla, A., 295–299
Larkin, S.J., 262–264
Latado, H., 113–117
Lecerf, J.-M., 173–176
Lenz, H., 328–332
Li, C., 118–123
Lo, C.-Y., 72–75
Lukic, I.K., 76–80

Machida, T., 53–58
Maczurek, A., 147–151

Maeda, K., 320–324
Mathiron, D., 158–161
Mauprivez, H., 173–176
Maurer, S., 300–306
McGarvey, J.J., 59–65
Meltretter, J., 134–140
Mende, S., 248–252
Mera, K., 38–41, 152–154, 155–157, 325–327
Mestek, O., 268–271
Mibus, A.L., 190–193
Miller, A., 107–112
Mironova, R., 181–184
Miyata, N., 320–324
Miyata, T., 141–146
Miyazawa, T., 288–290, 291–294
Monacelli, F., 166–172
Monnier, V.M., 81–88, 194–200, 205–209
Morales, F., 89–100
Motomura, K., 38–41, 152–154
Muñoz, F., 235–240
Muhr, G.-M., 328–332
Münch, G., 147–151
Murata, M., 307–309

Nagai, M., 38–41, 325–327
Nagai, R., 38–41, 152–154, 155–157, 325–327
Nagaraj, R.H., 107–112
Nakagawa, J., 320–324
Nakagawa, K., 288–290, 291–294
Nakamura, A., 320–324
Nangaku, M., 265–267
Nawroth, P.P., 76–80
Negrean, M., 276–279
Nemet, I., 81–88
Nerlich, A.G., 283–287
Niquet, C., 158–161
Nishitani, N., 53–58
Niwa, T., 181–184
Nogueiras, R., 14–19

Odani, H., 320–324
Odetti, P., 166–172

Oide, K., 320–324
Ono, H., 307–309
Ono, Y., 53–58
Otagiri, M., 38–41, 155–157, 325–327
Otsuka, N., 320–324
Oya-Ito, T., 107–112

Pamplona, R., 315–319
Parisod, V., 300–306
Park, Y.S., 185–189
Pawlak, A.M., 59–65
Penndorf, I., 118–123
Pfister, F., 42–45
Piechotta, C.T., 244–247
Pilard, S., 158–161
Pischetsrieder, M., 134–140
Portero-Otin, M., 315–319
Pouillart, P., 173–176

Rabbani, N., 124–127, 262–264, 280–282
Ramasamy, R., 7–13
Richoz, J., 300–306
Rohrbach, H., 283–287
Rubach, M., 310–314

Saavedra, G., 177–180
Sakata, N., 155–157
Sang, S., 72–75
Sanguineti, R., 166–172
Schalkwijk, C.G., 162–165, 201–204, 231–234
Scheijen, J., 201–204
Schleicher, E.D., xiii, 283–287
Schieberle, P., xiii
Schmid, K., 257–261
Schmidt, A.M., 7–13
Schwarzenbolz, U., 118–123, 248–252
Šebeková, K., 177–180
Sell, D.R., 81–88, 205–209
Seraglia, R., 295–299
Shanmugam, K., 147–151
Shen, W., 205–209

Shimohira, K., 53–58
Shirahashi, Y., 53–58
Smit, A.J., 225–230
Somoza, V., xiii, 177–180, 257–261, 310–314
Sourris, K.C., 101–106
Spiteller, G., 128–133
Sredovska, A., 181–184
Stäb, F., 328–332
Stehouwer, C.D.A., 162–165, 201–204, 231–234
Stirban, A., 276–279
Stitt, A.W., 59–65
Storace, D., 166–172
Stratmann, B., 276–279
Strauch, C.M., 205–209
Striker, G., 46–52
Sudo, T., 320–324

Takeo, K., 38–41
Takeya, M., 152–154
Tan, D., 72–75
Taniguchi, N., 185–189
Teerlink, T., 231–234
Tesař, V., 268–271
Tessier, F.J., 158–161, 173–176
Thornalley, P.J., 124–127, 262–264, 280–282
Thorpe, S.R., 272–275
Totsuka, H., 307–309
Traldi, P., 295–299
Tremoli, E., 295–299
Tressl, R., 244–247
Tschöp, M.H., 14–19
Tschoepe, D., 276–279
Tsurushima, K., 38–41

Uribarri, J., 46–52, 276–279
Usui, T., 53–58

Van Bezu, J., 231–234
van Greevenbroek, M.M.J., 162–165
van Veldhuisen, D.J., 225–230
van der Kallen, C.J.H., 162–165
Vilanova, B., 235–240

Vlassara, H., 46–52, 276–279
vom Hagen, F., 42–45
Voors, A.A., 225–230

Wagner, J., 210–215
Wang, Y., 42–45, 72–75
Watanabe, H., 53–58
Wenck, H., 328–332
Wittern, K.-P., 328–332
Würtz, A., 244–247

Yamashita, S., 288–290, 291–294
Yan, S.F., 7–13
Yaylayan, V.A., 30–37
Yoshitomi, M., 38–41

Zamora, R., 25–29
Zhang, J., 81–88
Zigman, J.M., 14–19
Zima, T., 268–271
Zumpe, C., 177–180